International Conference on Frontiers of Materials and Smart System Technologies (FMSST 2019)

IOP Conference Series: Materials Science and Engineering
Volume 590

Tamil Nadu, India
10 April 2019

ISBN: 978-1-5108-9814-1
ISSN: 1757-8981

Printed from e-media with permission by:

Curran Associates, Inc.
57 Morehouse Lane
Red Hook, NY 12571

Some format issues inherent in the e-media version may also appear in this print version.

Institute of Physics
Dirac House, Temple Back
Bristol BS1 6BE UK

Phone: 44 1 17 929 7481
Fax: 44 1 17 920 0979

techtracking@iop.org

Additional copies of this publication are available from:

Curran Associates, Inc.
57 Morehouse Lane
Red Hook, NY 12571 USA
Phone: 845-758-0400
Fax: 845-758-2633
Email: curran@proceedings.com
Web: www.proceedings.com

Table of contents

Volume 590

International Conference on Frontiers in Materials and Smart System Technologies

10 April 2019, Tamil Nadu, India

Accepted papers received: 12 July 2019
Published online: 15 October 2019

Preface

Papers

IOP Conf. Series: Materials Science and Engineering **590** (2019) 011001 doi:10.1088/1757-899X/590/1/011001

PREFACE

FMSST 2019, the first edition of International Conference on Frontiers in Materials and Smart System Technologies held on 10th April 2019 in Tamilnadu India. This conference provides a lively platform for delivering the innovative ideas in the research and application of Smart Systems. We are pleased to present the proceedings of the conference as its published record.

This conference that provides a unique platform for scientists, researchers, and professionals across multiple disciplines to share their research advancements and critical ideas to address the issues in smart systems and its material processing. For more than a decade, by recognising that problems require an interdisciplinary approach to reach a holistic solution, the conference has facilitated the communication between scientists across disciplines in material and smsart systems to discuss the contemporary issues and exchanging ideas to give solutions that can contribute to this planet.

We hope that this conference will further stimulate research in smart systems, and provide better techniques, algorithms, and hardware for deployment. We feel honored and privileged to provide the best recent developments this field to you through this great event.

Best regards,

Dr. Suresh P
Dr. Saravanakumar U

IOP Conf. Series: Materials Science and Engineering **590** (2019) 011002 doi:10.1088/1757-899X/590/1/011002

ABOUT THE CONFERENCE

International Conference on Frontiers in Material and Smart Systems Technologies allows the designers to build a specific system out of components which is able to obtain the information from the surroundings, to process it electronically, to communicate signals and data and to give enabled feedback signal to the ambience.

"A system or material which has built-in intrinsic sensor(s), actuator(s) and control mechanism(s) whereby it is capable of sensing a stimulus, responding to it in a predetermined manner and extent, in a short/appropriate time, and reverting to its original state as soon as the stimulus is removed".

Smart systems are a network of physical objects or things that are embedded with electronics, software, sensors, and network connectivity - which enable the object to collect and exchange data. Rapid proliferation of IoT is driving the intelligence in things used daily in homes, workplaces and industry. The IoT devices typically communicate via radio frequency (RF), such as WiFi, Bluetooth and other communication networks.

Smart systems and automation are ambiguous terms used in reference to a wide range of solutions for controlling, monitoring and automating functions in the home and industries. Smart devices that only can be controlled with switches, timers, sensors and remote controls are thus not included in the scope of this study in various automations covered in various applications throughout this book. Smart systems can be grouped into different categories such as energy management and climate control systems; security and access control systems; lighting, window and appliance control systems; home appliances; audio-visual and entertainment systems; and healthcare and assisted living systems.

The whole system is subject to the trend of today's world: miniaturization, networking capability, energy-autonomy and reliability. Due to increasing complexity and multidisciplinary developing teams in fields of microsystems- and nanotechnology, electronics, wireless communication technologies, optics, biology, medical science, and fluidics face highest demands. There are several different communication protocols and media used for lighting control available. They have different architecture and cannot communicate between each other. Here lies the biggest challenge for us as a world.

IOP Conf. Series: Materials Science and Engineering **590** (2019) 011002 doi:10.1088/1757-899X/590/1/011002

The conference will provide a platform for researchers and practitioners interested in the theory and practice of smart systems and technologies related to electrical engineering, communications, computer science and engineering, control systems, robotics, as well as interdisciplinary research and applications.

Best regards,

Dr. Suresh P

Dr. Saravanakumar U

IOP Conf. Series: Materials Science and Engineering **590** (2019) 011003 doi:10.1088/1757-899X/590/1/011003

Peer review statement

All papers published in this volume of *IOP Conference Series: Materials Science and Engineering* have been peer reviewed through processes administered by the proceedings Editors. Reviews were conducted by expert referees to the professional and scientific standards expected of a proceedings journal published by IOP Publishing.

CART- A Statistical Model for Predicting QoE using Machine Learning in Smartphones

Ruth Christina[1],Greeshma Liz Shajan[2],B.Ankayarkanni[3]
[1,2]UG Student, [3]Assistant Professor, School of Computing, Sathyabama Institute of Science and Technology, Chennai.

[1]christinathomas.2697@gmail.com,[2]greeshma.lizshajan@gmail.com,
[3]ankayarkanni@yahoo.com

Abstract— The proliferation of fast, cheap, and ubiquitous network access, in particular on mobile devices, has significantly changed the way user's access online benefits on an everyday premise. Alongside the expanded accessibility of a dependably on Internet, new advancements, for example, LTE have empowered the utilization of a wide range of administrations, with stringent necessities regarding system execution and limit requests. We consider different QoE appraisal models in view of administered machine learning strategies, which are proficient to foresee the QoE experienced by the end client of well-known cell phone applications (e.g., YouTube and Face book), utilizing as info the inactive in-gadget estimations. In this extraordinary issue, we investigate the connection between the Quality of Service (QoS) that administrators screen and oversee and the Quality of Experience (QoE) that the clients really get. Utilizing a rich QoE dataset got from field trials in operational cell systems, we benchmark the execution of various machine learning based indicators, and build a choice tree based model which is able to foresee the per-client general understanding and administration worthiness with a win rate of 91% and 98% separately. To the best of our knowledge, this is the initial proposal using end-user, in-device passive measurements and machine learning models to predict the QoE of smartphone users in operational cellular networks.

Keywords—QoE, Machine Learning, End-device Measurements, Mobile Apps.

1.Introduction

More than 75% of all mobile network traffic will be consumed and generated by smartphones, by 2020. In this paper, to get a distributed Quality we present a set of monitoring tools and machine learning models to enable distributed Quality of Experience (QoE) monitoring and prediction, relying exclusively on in-device passive measurements. In-device network and traffic monitoring provides highly valuable information to network operators, as measurements are taken as close as possible to the end user, shedding light into particular phenomena not observable from traditional, in-network monitoring. Using end-devices as vantage points becomes especially attractive in current traffic encryption context, as end-to-end encryption is rapidly growing– especially due to the massive adoption of HTTPS, obfuscating network traffic analysis.

1.1 QoE Challenges in Cellular Networks

- A constrained ghostly data transmission to be shared, causes impedance.
- Communication joins are time shifting, recurrence specific channels.
- Packet deferrals and Application nerves causes' client an irritating background.
- Error rate of wireless channel is higher due to mobility, interference from other media, multi-path fading. So mobile hosts may experience different channel rates in the same or different cells.
- Different applications have different requirements for bandwidth, delay, jitter, delay.

The most common approach to evaluate the performance of networks and services from a QoE perspective is to conduct experiments in the lab [2],[4]. Their key benefits rely on the full controllability on the evaluation process. Nevertheless, experiments in the lab normally introduce differences as com- pared to field studies [5], as they might overlook important QoE-relevant features such as contextual information, device usability, or user preferences. Field trials would therefore result in better, more representative evaluations, to the cost of higher complexity in acquiring and processing the results.

We study the QoE of popular apps in smartphones (YouTube, Face book, and Google Maps), combining passive traffic QoS measurements collected directly in smartphones with crowd sourced QoE feedbacks provided by the end user in a field trial. . While we are completely mindful that exact QoE estimation requires estimations gathered at numerous levels of the correspondences stack – including system, application and end-client layers, we adopt a handy strategy and give models to delineate activity QoS measurements to QoE straightforwardly.

We do as such to boost the use of bland system activity estimations which can be gathered in cell phones, independently of the particular application. Application level checking is for the most part more bulky, as few out of every odd application gives APIs to get to significant measurements, and much of the time, gadget root get to must be allowed to perform estimations profoundly into the application, obstructing huge scale detached observing.

Utilizing the individual cell phones and information contracts of the field trial members, the field trail keeps running in genuine operational cell systems. We built up a portion of the instruments, for example, aloof checking apparatus (to quantify the movement of the field trial members at their end gadgets) and a QOE-input application (which accumulates client encounter information in a group sourced design .Machine learning(ML) innovation is utilized as a part of this paper.

1.2 Machine Learning

It is a subfield of most interesting topic on artificial intellectual competence The goal of machine adjusting generally is to understand the structure of data and fit that data into models that can be fathomed and utilized by people. Notwithstanding the way that machine learning is a field inside programming building, it differs from customary computational procedures. In customary enlisting, figuring are sets of unequivocally changed bearings used by PCs to figure or issue appreciate. Machine learning computations rather mull over PCs to get ready on data information sources and use quantifiable examination remembering the ultimate objective to yield regards that fall inside a specific range. Thusly, machine learning supports PCs in building models from test data with a particular ultimate objective to motorize fundamental initiative structures in light of data inputs.

Optical character acknowledgment (OCR) innovation changes over pictures of content into versatile kind. Proposal motors, fueled by machine learning, recommend what films or network shows to watch next in view of client inclinations. Self-driving autos that depend on machine figuring out how to explore may soon be accessible to shopper's .Machine learning is a ceaselessly creating field. Along these lines, there are a few contemplations to remember as you work with machine learning approaches', or break down the effect of machine learning forms.

1.3 Supervised Learning

In managed taking in, the PC is furnished with case inputs that are named with their coveted yields. The reason for this strategy is for the calculation to have the capacity to "learn" by contrasting its genuine yield and the "instructed" yields to discover blunders, and alter the model likewise. Directed adapting hence utilizes examples to anticipate mark esteems on extra unlabeled information.

IOP Conf. Series: Materials Science and Engineering **590** (2019) 012001 doi:10.1088/1757-899X/590/1/012001

For instance, with regulated taking in, a calculation might be nourished information with pictures of sharks named as fish and pictures of seas marked as water. By being prepared on this information, the managed learning calculation ought to have the capacity to later recognize unlabeled shark pictures as fish and unlabeled sea pictures as water. A typical utilize instance of managed learning is to utilize recorded information to foresee factually likely future occasions. It might utilize recorded securities exchange data to envision up and coming changes, or be utilized to sift through spam messages. In directed learning, labeled photographs of pooches can be utilized as info information to arrange untagged photographs of puppies.

1.4 Unsupervised Learning

In unsupervised learning, data is unlabeled, so the learning count is left to find shared qualities among its data. As unlabeled data are richer than stamped data, machine learning methodologies that energize unsupervised learning are particularly gainful. The goal of unsupervised learning may be as immediate as finding covered plans inside a dataset, in any case it may in like manner have a goal of feature acknowledging, which empowers the computational machine to normally discover the depictions that are required to arrange unrefined data.

Unsupervised learning is conventionally used for esteem based data. You may have an extensive dataset of clients and their buys, yet as a human you will probably not have the capacity to understand what comparative qualities can be drawn from client profiles and their kinds of buys. With this information sustained into an unsupervised learning calculation, it might be resolved that ladies of a particular age extend who purchase unscented cleansers are probably going to be pregnant, and along these lines an advertising effort identified with pregnancy and child items can be centered to this gathering of spectators in order to fabricate their number of purchases.

Without being told a "right" answer, unsupervised learning strategies can look at complex data that is more expansive and obviously unimportant remembering the ultimate objective to form it in possibly huge ways. Unsupervised learning is consistently used for quirk acknowledgment including for false MasterCard purchases, and recommender structures that endorse what things to buy straightaway. In unsupervised learning, untagged photographs of canines can be utilized as information for the calculation to discover similarities and arrange pooch photographs together.

Utilizing the gathered estimations, we prepare a few directed ML models to anticipate the QOE experienced by the end client of these apps.ML gives a promising other option to QOE expectation and appraisal in light of the consolidated investigation of different movement descriptors or highlights.

The rest of the paper is as per the following:

1 .Overviews the related work, concentrating on the particular instance of cell phones.
2. Portrays the field trial setup, the instruments created to quantify QoS-and QoE-related measurements toward the end-gadgets, and the machine-learning models utilized as indicators. Sec.
3. Gives a concise outline on the field trial estimations, and spotlights on the assessment of the forecast execution acquired by various ML-based models, including an investigation of the pertinence of the distinctive info includes on quality expectation.
4. Finishes up this work.

IOP Conf. Series: Materials Science and Engineering **590** (2019) 012001 doi:10.1088/1757-899X/590/1/012001

2. Related Work

The investigation of the QoE for present day online applications as the ones we focus in this paper has a considerable rundown of new and late references [9]. Because of its fame, the particular instance of YouTube QoE merits specific consideration. Past papers [10]– [13] have demonstrated that stalling's, starting playback postponements and quality switches are the most essential KPIs for versatile gushing YouTube QoE. A complete review on versatile video gushing QoE is given in [14].

Concerning observing investigations in cell systems and cell phones, there is a varying rundown of instruments to gauge organize execution: In [18], [19] creators presented YoMoApp, an application to latently screen YouTube QoE-related highlights in cell phones. In [20] creators depict an on-line checking framework for YouTube QoE in cell systems utilizing as a part of system estimations as it were. QoE Doctor [21] measures portable application QoE, utilizing dynamic estimations at both application and system layers.

Other comparative papers contemplate Web QoE in cell systems [22], and video [23]. Near our work, creators in [24] propose a way to deal with assess versatile applications QoE utilizing inactive in-organize estimations and machine-learning models mapping QoE to QoE. The principle distinction to our approach is the use of in-organize estimations, in opposition to in-gadget ones.

While the lion's share of these papers just spotlight on QoE-applicable target measurements, for example, re-buffering, page stack times and collaboration latencies, they do not have an immediate reference to client criticism, e.g., Mean Opinion Scores. Here we take input from genuine clients and gather organize estimations to give an all-encompassing point of view to the issue of QoE observing in cell phones. This paper expands over past work on QoE for cell systems [7], [8]. Specifically, we consider a machine-learning based point of view to the issue of QoE expectation and appraisal, which isn't done in [7], [8].

2.1. Mobile QoE in The Field

The field trial occurred in Vienna in 2015; 30 clients furnished with their own particular gadgets associated with their favored cell administrators assessed the three applications as a major aspect of their typical day by day Internet movement. Members were told to perform free errands for each application. For YouTube, they were asked for to observe short (around 2 minutes in length) HD YouTube recordings. For Face book, they were asked for to get to the application, and peruse both the course of events and some photograph collections of a phony client. For Gmaps, they needed to peruse maps of various urban areas utilizing the satellite-perspective of the Google maps application.

Table 1. Metrics Collected For Each Data Flow

Metric ID	Metric Name	Units	Example
1	device id (IMEI)	-	352668049725157
2	flow start time	s	1430825689
3	flow direction (up/down)	-	downlink
4	flow duration	s	10,24
5	flow size	KB	4041,00
6	avg. flow throughput	kbps	3157,03
7	max. flow throughput	kbps	4320,15
8	app (Android API package)	-	com.android.browser
9	signal strength	dBm	-71
10	operator (MCC.MNC)	-	295.4
11	cell id	-	16815
12	cell location (lat-lon)	deg (°)	{40,198-12,347}
13	RAT	-	LTE

QoE criticism was accounted for every session through a modified QoE crowd sourcing application, for an aggregate traverse of two weeks. In this paper we just spotlight on the general experience announced by members and the administration worthiness, however the QoE criticism gave incorporates more data that we intend to assess later on. The general experience is appraised by a discrete, 5-levels ACR Mean Opinion Score (MOS) scale [2], extending from "terrible" (i.e., MOS = 1) to "phenomenal" (i.e., MOS = 5). Worthiness is a twofold input, expressing whether the client would keep utilizing the application under the comparing conditions or not. Applications arrange activity was latently observed and broke down at their gadgets with the apparatuses portrayed straightaway. Members additionally showed their area at the season of the test (e.g., underground, auto, home, road, and so on.).

3. Proposed System

In this proposed framework utilizes a ML based on QoE indicator. This QoE Model uses a choice tree calculation known as **C**lassification **A**nd **R**elapse **T**rees (**CART**) calculation used to distinguish the factual model that has greatest exactness for foreseeing the estimation of a straight out ward variable from a dataset comprising of clear cut and persistent factors.

CART classification algorithm characterizes the occurrences (Data) by more than once apportioning the info space, to fabricate a tree whose hubs are as unadulterated as could reasonably be expected. Truck calculation utilizes choice tree based approach which are straightforward yet quick and powerful. Considering the order speed is a foremost resource when thinking in expansive scale checking situations, and choice trees are notable for their speed.

Classification Trees

Where the objective variable is all out and the tree is utilized to distinguish the "class" inside which an objective variable would likely fall into that and the same is shown in Figure 1.

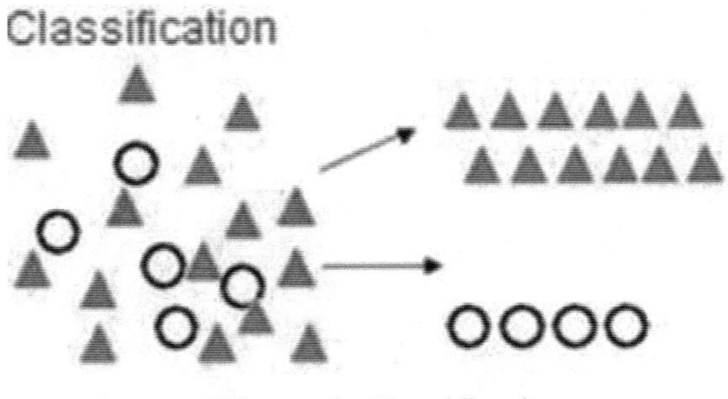

Figure 1. Classification tree

Regression Trees

Where the objective variable is nonstop and tree is utilized to foresee it's esteem and shown in Figure 2.

IOP Conf. Series: Materials Science and Engineering **590** (2019) 012001 doi:10.1088/1757-899X/590/1/012001

.

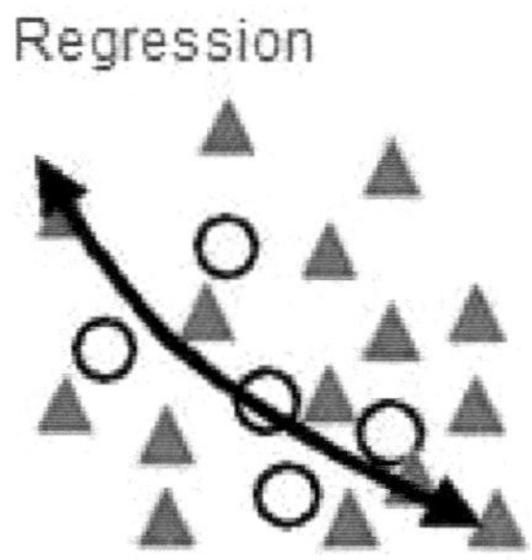

Figure 2. Regression Tree

The CART calculation is organized as a succession of inquiries, the responses to which figure out what the following inquiry, if any ought to be. The consequence of these inquiries is a tree like structure where the closures are terminal hubs and soon thereafter there are no more inquiries. The principle components of CART (and any choice tree calculation as in Figure 3) are:

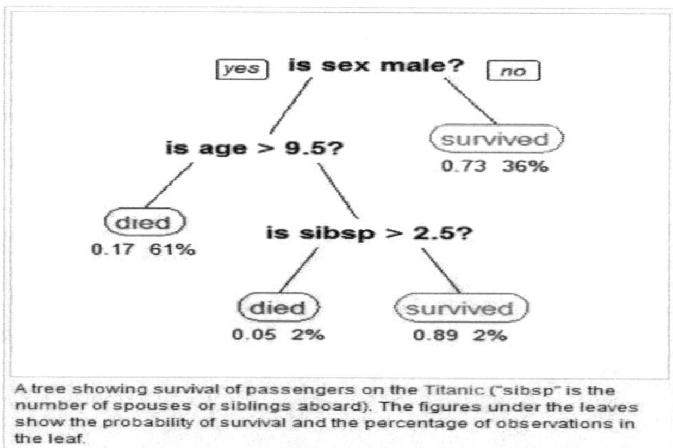

Figure 3. Tree calculation

The main elements of CART (and any decision tree algorithm) are:

1. Rules for part information at a hub in view of the estimation of one variable;
2. Stopping standards for choosing when a branch is terminal and can be part no more; and
3. Finally, an expectation for the objective variable in every terminal hub.

Dataset Metrics Collection

Keeping in mind the end goal to quantify Quality of Experience of Face book, Google Maps and YouTube, the detached QoS estimations for every individual Application should be gathered, sorted out, and detailed.

IOP Conf. Series: Materials Science and Engineering **590** (2019) 012001 doi:10.1088/1757-899X/590/1/012001

Passive Metrics

The detached approach watch the activity is the application as it cruises by. These Metrics are then handled by an instrument known as Sniffer, they are incorporated with the client's gadgets (Mobile Phones) and the measurements are embedded into a dataset. A dataset from Data. World – A non-benefit Organization is utilized as a part of the proposed demonstrate.

Table 2. Sample Metrics

Metric Name	Units	Sample Data
Device id (IMEI)	-	332647846534
Flow Start Time	-	11:02:55 (HH:MM:SS)
Flow End Time	-	11:03:55(HH:MM:SS)
Flow Duration	Seconds	60
Flow Size	KB	1255
Avg Flow Throughput	KBPS	404100
Max Flow Throughput	KBPS	4563214
App (API)	-	Com.android.browser
Signal Strength	dBm	-71
Operator	-	MC123
Cell id	-	32165
Cell Location	Deg	{20.5937° N, 78.9629° E}

4. Result and Discussions

This session gives a clear on how we assess the QOE and QOS with the basic errors we have collected. So here we have collected data from World-A organization and classified into 10 codes on the basis of the different classifications. We can randomly chose any data on the basis of the error codes and send the problem to the developer. Once if there is a solution there itself we can rectify it directly or otherwise the main admin would upload the fix and then the error gets rectified. The error is chosen from the data and the social application used, the id and the error code is noted were shown in Figure 4 and Figure 5.

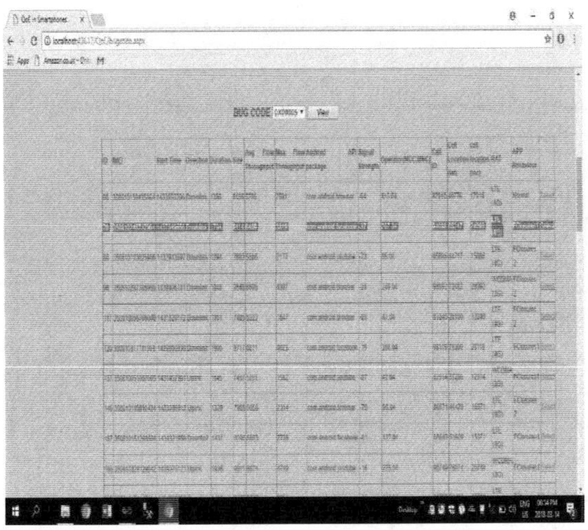

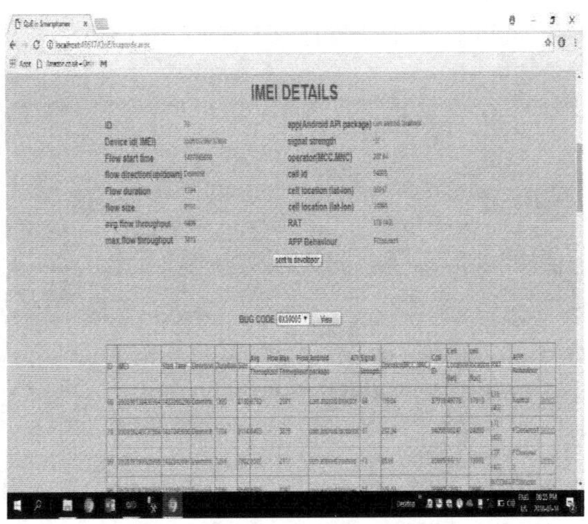

Figure.4 Choosing the error from the data collected and Report sent to the developer

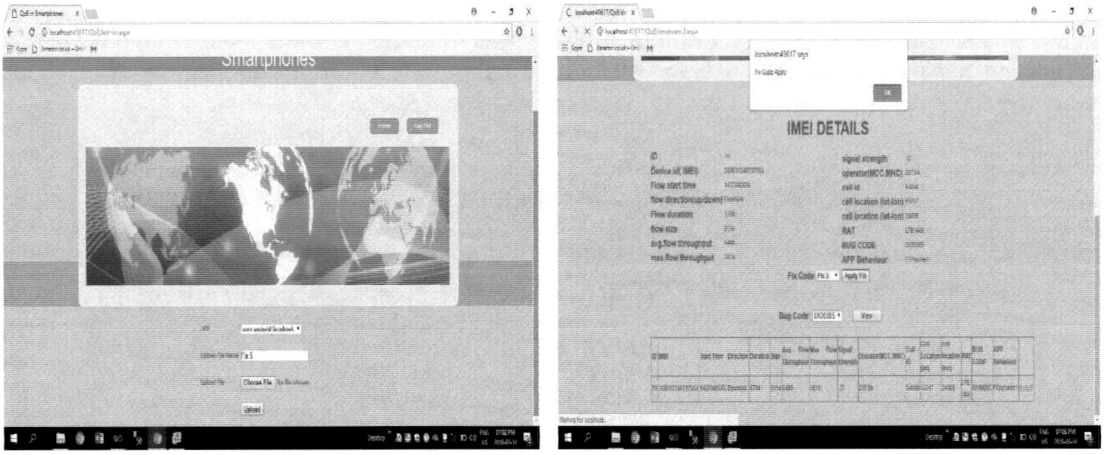

Figure.5 The developer uploads the fix for the problem and the error gets rectified.

5. Conclusion

In this paper, the problem of QoE monitoring, assessment and prediction in cellular networks, relying on in-smartphone QoS Passive traffic measurements are addressed. Using a rich dataset derived from a field study by a non-profit organization 'Data. World' is applied as part of the demonstration of this proposed methodology. We have trained different QoE prediction models based on supervised machine learning technique CART. The final selected model is based on decision trees, which are very attractive for large scale QoE/QoS assessment scenarios, not only because of their excellent performance but also due to their prediction speed.

References

[1] Thorsten Joachims "Text categorization with Support Vector Machines: Learning with many relevant features",June 2016.
[2] Hiroshi Sakai,Chenxi Liu, Michinori Nakata ,"Information Dilution: Granule-Based Information Hiding in Table Data - A Case of Lenses Data Set in UCI Machine Learning Repository", August 2017.
[3] Shan Huang, Zhicheng Xie,Jun Han, Xiaoyang Zeng,"A flexible low-power machine learning accelerator for healthcare applications" October 2016.
[4] Poonam Pandey,Radhika Prabhakar ," An analysis of machine learning techniques (J48 & AdaBoost) for classification",August 2016.
[5] Anku Jaiswal, A. S. Manjunatha, B. R. Madhu,P. Chidananda Murthy, "Predicting unlabeled traffic fointrusion detection using semi-supervised machine learning", December 2016.
[6] Kai Zhang, Liang Lan , James T. Kwok , Slobodan Vucetic , Bahram Parvi," Scaling Up Graph-Based Semi supervised Learning via Vector Machines", April 2014.

[7] Asha P, Albert Mayan J, Canessane A, "Efficient Mining of Positive and Negative Item sets Using K-Means Clustering to Access the Risk of Cancer Patients", Communications in Computer and Information Science ,ICSCS 2018, Kollam, India, 2018,pp.373-382, 2018.
[8] P.Casas, B.Grdlo, M.Seufert, F.Wamser, R.Schatz,"Taming QOE in cellular networks",2015.
[9] P.Casas, R.Schatz,"Quality of Experiences in cloud services", 2014.
[10] T.HobfeldM.Seufrt,M.Hirth,T.Zinner,P.Tran-gia,R.Schatz,"Quantification of You-tube QoE via crowd sourcing", 2011.
[11] Mayan J.A, Arifa S, Pavithra R,"Semantic based multi lexical ranking technique for an effective search in protected cloud,2016 International Conference on Control, Instrumentation, Communication and Computational Technologies (ICCICCT), Kumaracoil, 2016, pp. 570-576.
[12] M.Seufert, S.Egger, m.Slanina, T.Zinner, T.Hobfeld, P.Tran-gia," A survey on quality of experience of http adaptive streaming "2015.
[13] A.Nikravesh, H.Yao, S.Xu, D.Choffnes, Z.M.Mao,"Mobilyser:An open platform for controllable mobile network measuremrnts",2015.
[14] C.Kreibich,N.Weaver,B.Nechaev,V.Paxson,"Netalyzr:Illuminating the edge network",2010.
[15] Venkatakrishna D., Ankayarkanni B,"A description of content based image retrieval using from block truncation coding and image content description",Indian Journal of Science and Technology, Vol 9, No.21, pp.1-5, June 2016.

IOP Conf. Series: Materials Science and Engineering **590** (2019) 012002 doi:10.1088/1757-899X/590/1/012002

TECHNICAL ERA: An Online Web Application

Ajay M.D.V[1], Adithya N[2], Mary Posonia A[3]

[1,2] U.G Student, Department of Computer Science and Engineering, Sathyabama Institute of Science and Technology, Chennai
[3]Assistant Professor, School of Computing, Sathyabama Institute of Science and Technology, Chennai.

adithyachowdary02@gmail.com[1] , ajaymuttamsetty007@gmail.com[2], soniadelicate@gmail.com[3]

Abstract: The web has significantly affected all parts of every one of our students. As our students depend increasingly on the web, the constancy of web applications has turned out to be progressively critical. To make these applications more tried and true, we will make a web application with respect to placement training. In general we have Indiabix and placement season are providing a wide range of both technical and aptitude questions in order to prepare a student for the placement. But somehow those questions are repeatedly asked are the standard questions are given. To improve this situation, in this paper, we propose a model of a web application. This structure utilizes the offline application capacity with all the questions are put away in the database and the questions will be refreshed progressively on a period premise with the assistance of the placement coordinator in the college. Every student can sign up and practice all the questions that are provided with respect to the subjects.

Keywords: JQuery, java script, database, dynamic webpage, mcq questions

1. Introduction

As of late, learning framework over the Internet winds up well known and beginning to be imperative, especially in an educational environment, for example, school or educational institutes. ELearning framework can be utilized as a full separation learning style in which, students can learn remotely and freely by taking practice and assets provided. Our main objective is to provide training to the students in order to meet the requirements of the industries. As 'IT' industries are expecting a lot from the students we are providing the platform to prepare for the upcoming placements.

2. Existing System

HTML5 started in 2004 and is presently being done by W3C HTML WG and the WHATWG is a current improvement of HTML standard, an increase dialect that assembled a large portion of the web content. This standard has been advanced as a request of developing Internet clients and its applications and has been bolstered by most significant merchants of web programs. HTML5 is acquainted with giving a superior client encounter and better help for application programming interface and mixed media capacity. Right now, the detail is still in advance while some significant merchants had actualized a piece of its particular and it's not been utilized by the significant number of the Technical sites.

The existing system consists of either offline access to the website or online quiz oriented access and the level of questions which we see in online websites is pretty high and it cannot be used for a fresher. In the present existing websites, there is a problem of redundancy where questions have been repeated for years. For example, if we consider Indiabix it consists of the same questions from the years without any change. When it comes to colleges Most of the students are given access to the placement portals only in the final year.

3. Proposed System

Here we proposed a system that gives access to all the students so that they will aware of the questions that are asking the companies during campus recruitment process. This system is purely on technical side questions along with explanation and answers. In our website, we are also providing the discussion forum which is useful to all the students in which they can discuss their queries with their friends and even also with the respective subject staff. Dynamically the technical MCQ questions are also updated on time basis by the system administrator.

In these, we are also providing an explanation block beside the questions which provides the explanation about the questions. Some of the tutorial video links about the topic are given in the page itself so that students can make use of it during their preparations. These questions are updated by the university staff every time. Even the staff can create an assignment to the students if necessary so that the staff can identify the students on their performance basis.

3.1 Description of proposed system:

The complete website will be accessed only after login into the page. Hence login page should be created to make it more effective and good user interface. In the website, the entire signup and the login implementation is done by using the language JQuery in order to use the toggle function which makes the presentation good and simple. The entire user details will be stored in the database as shown in Figure 1.

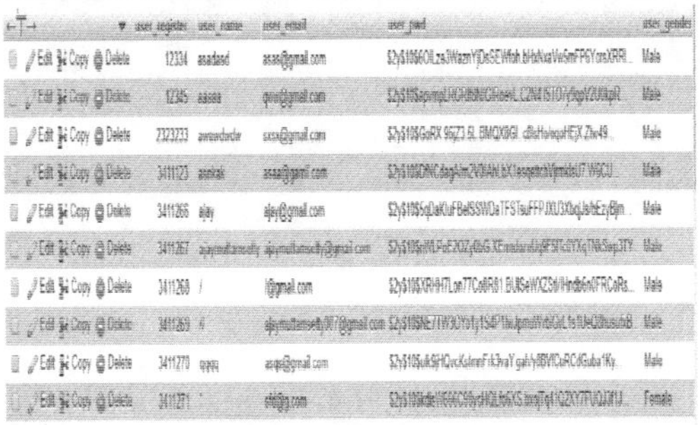

	user_register	user_name	user_email	user_pwd	user_gender
Edit Copy Delete	12334	asadasd	asas@gmail.com	[illegible]	Male
Edit Copy Delete	12345	aaaaa	qwe@gmail.com	[illegible]	Male
Edit Copy Delete	7323733	awawdwdw	sxsx@gmail.com	[illegible]	Male
Edit Copy Delete	3411123	aaakak	asaa@gmail.com	[illegible]	Male
Edit Copy Delete	3411266	ajay	ajay@gmail.com	[illegible]	Male
Edit Copy Delete	3411267	[illegible]	[illegible]	[illegible]	Male
Edit Copy Delete	3411268	/	l@gmail.com	[illegible]	Male
Edit Copy Delete	3411269	'	[illegible]	[illegible]	Male
Edit Copy Delete	3411270	qqqq	asqs@gmail.com	[illegible]	Male
Edit Copy Delete	3411271	'	[illegible]	[illegible]	Female

Fig 1 database storage

The password of the individual user will be stored in encrypted format for the security purpose. The login page is designed such a way that its user name should be the register number of the college such that there will be no possibility for a user to create multiple logins.

The Questions in the respective technical subjects will be uploaded by the Administrator. The questions will not be posted on the static page as it leads to the redundancy and there will be no chance to upload new questions. Hence we are introducing the new format of entering the questions which will be in the input format and these will be stored in the database as shown in Figure 2.

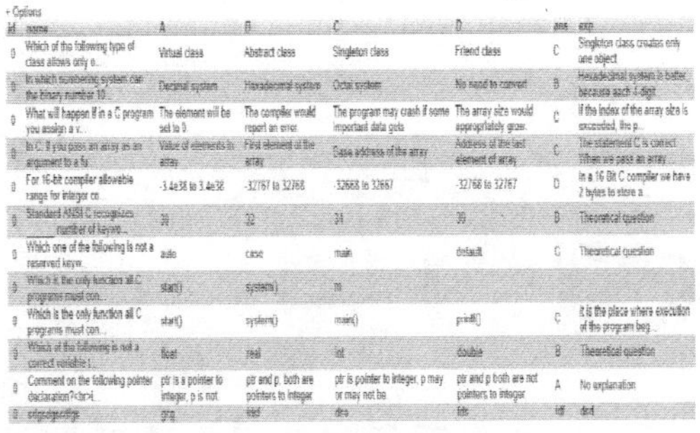

id	name	A	B	C	D	ans	exp
0	Which of the following type of class allows only o...	Virtual class	Abstract class	Singleton class	Friend class	C	Singleton class creates only one object
0	In which numbering system can the binary number 10...	Decimal system	Hexadecimal system	Octal system	No need to convert	B	Hexadecimal system is better because each 4-digit ...
0	What will happen if in a C program you assign a v...	The element will be set to 0	The compiler would report an error	The program may crash if some important data gets	The array size would appropriately grow	C	If the index of the array size is exceeded, the p...
0	In C, if you pass an array as an argument to a fu...	Value of elements in array	First element of the array	Base address of the array	Address of the last element of array	C	The statement C is correct. When we pass an array ...
0	For 16-bit compiler allowable range for integer co...	-3.4e38 to 3.4e38	-32767 to 32768	-32668 to 32667	-32768 to 32767	D	In a 16 Bit C compiler we have 2 bytes to store a ...
0	Standard ANSI C recognizes ____ number of keywo...	30	32	34	39	D	Theoretical question
0	Which one of the following is not a reserved keyw...	auto	case	main	default	C	Theoretical question
0	Which is the only function all C programs must con...	start()	system()	m			
0	Which is the only function all C programs must con...	start()	system()	main()	printf()	C	It is the place where execution of the program beg...
0	Which of the following is not a correct variable t...	float	real	int	double	B	Theoretical question
0	Comment on the following pointer declaration? i...	ptr is a pointer to integer, p is not	ptr and p, both are pointers to integer	ptr is pointer to integer, p may or may not be	ptr and p both are not pointers to integer	A	No explanation
0	[illegible]	[illegible]	[illegible]	[illegible]	[illegible]	[illegible]	[illegible]

Fig 2 Enrolling questions into database

The questions in the database will be retrieved by using PHP code and will be displayed continuously. When the administrator wants to delete a question in the database can simply enter the question number and submit. Which leads to deleting a question in the respective database for the entire database storage we are using the local host server XAMPP. The questions in the quiz are displayed one by one so that the user may not skip the question and the questions in the quiz are stored in the JAVASCRIPT array which will be useful for displaying one by one in a simple manner and final score will be calculated by writing the code in the JAVASCRIPT.

The video tutorials which will be useful for the placement purpose and understanding the concept of the respected Technical subjects will be posted along the questions on the website. These video tutorials will have the videos that are taught by the respective staff members in the college so that the students can clearly understand the concepts.

In this module, there is a unique way of uploading questions. When it comes to the other Technical websites all the questions are uploaded in the static page of HTML but in this website, the questions are uploaded in the database such that all the questions are retrieved at a time by using loop condition. Due to this, there is a very less chance of having repeating questions and is shown in Figure 3.

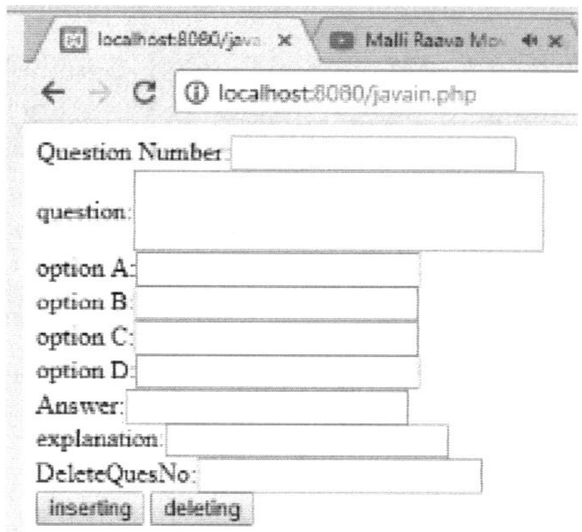

Fig 3 Uploading questions into database

4. Module Description

The languages that we used to develop the website are mentioned below

FRONT END:HTML, JAVASCRIPT, JQUERY & CSS
BACK END:PHP& MYSQL for database connections
The entire website is divided into 3 modules mainly login page, homepage, about page, contact page, department page.

4.1 User Interface:

The basic workflow of the website is shown below.

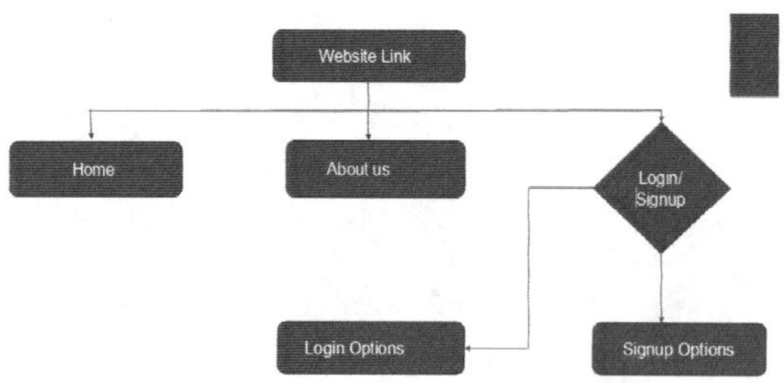

Fig.4 Architecture of user interface

The fig.4 illustrates the architecture for the login page and signup page. Signup page is to registering into the website before taking the questions. All the data is stored in a database for future use.

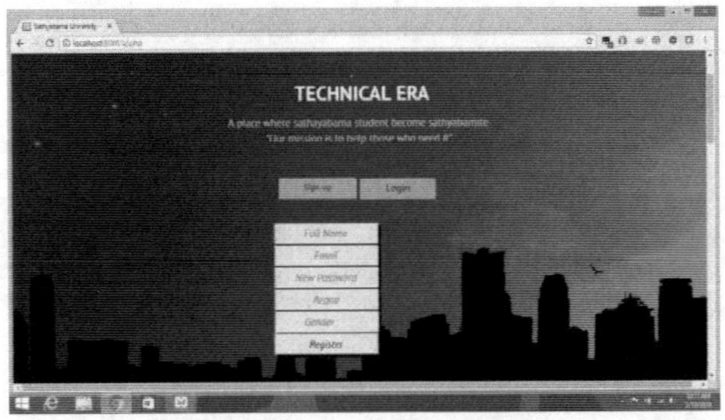

Fig 5 overview of sign up page

In the sign up page and login page we have used the jquery toggle method as shown in Figure 5

```
$("button").click(function(){
   $("p").toggle();});
```

in order make the user interface attractive and easy for the users and the same is given in Figure 6.

Fig.6 Overview of user interface page

4.2 Overview:

Overview of the website and the workflow is shown below

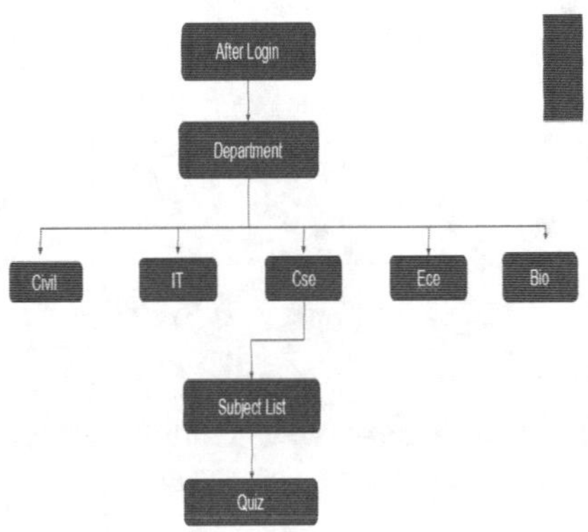

Fig.7 Architecture of department page

The Figure 7 illustrates the architecture for the department page where all the department modules exist. In each department we have a list of all the technical subjects where all the questions and the video tutorial links are present in it Like Figure 8.

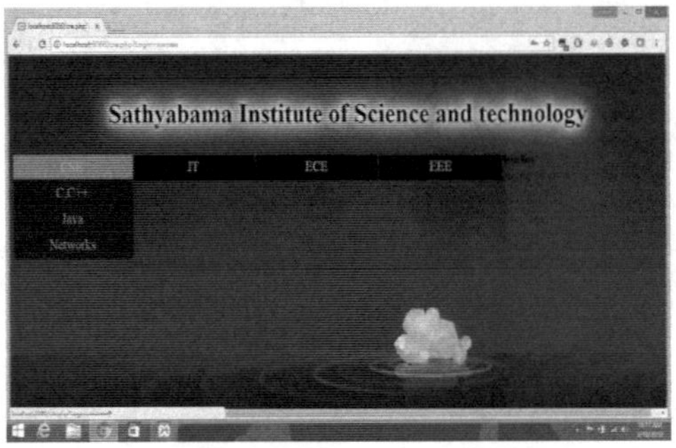

Fig.8 overview of department page

4.3 Department Module:

It consists of several departments for example after selecting CSE Module. The following work flow will be done and the same will be done in other modules also

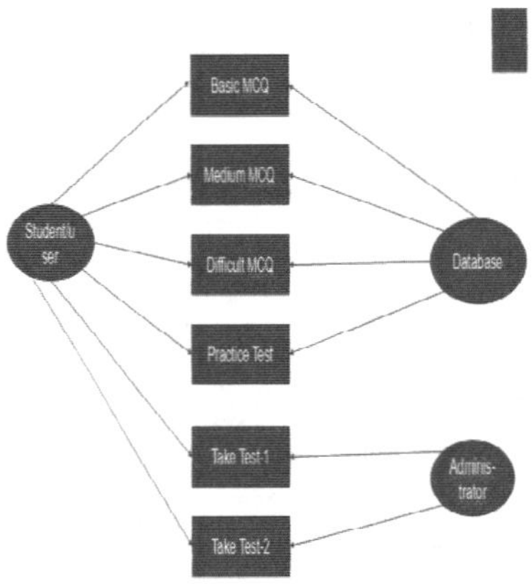

Fig.9 Architecture of Subject list

The above Figure 9 illustrates the architecture for the MCQ questions which are asked repeatedly during the recruitment process by the companies.

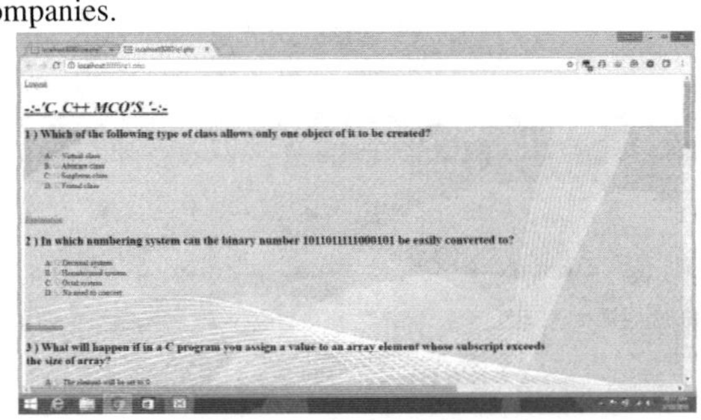

Fig.10 overview of MCQ questions

4.4 Quiz Application:

The quiz application consists of several questions on various concepts and levels and after completing the quiz the respective score will be displayed. The quiz questions are prepared based on the current placement requirements.

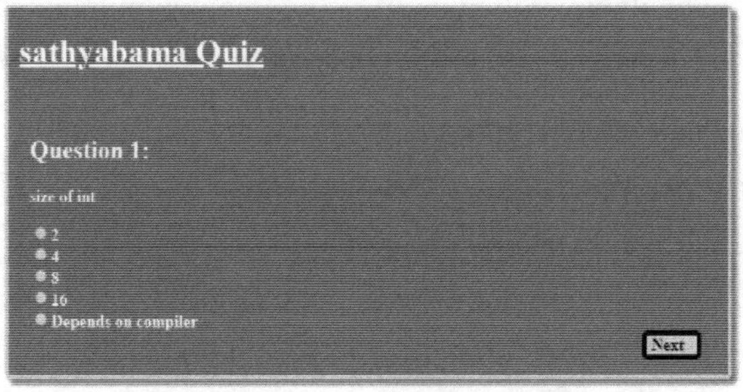

Fig 11. Overview of the Quiz Applicatio

5. V. Conclusion:

In this paper, we outlined a site for all the engineering students. It is exceptionally helpful for every one of the students for the most part amid the placements. The MCQ'S and quiz applications are exceptionally valuable to the students.

References:

[1] Japan Offline web application and quiz synchronization for e-learning activity for mobile browser. "Royyana M. Ijtihadie, Yoshifumi Chisaki, Tsuyoshi Usagawa" Graduate School Of Science And Technology Kumamoto University.

[2] Asha P, Albert Mayan J, Canessane A (2018), "Efficient Mining of Positive and Negative Itemsets Using K-Means Clustering to Access the Risk of Cancer Patients",Communications in Computer and Information Science ,ICSCS 2018, Kollam, 2018,pp.373-382.

[3] Surendar E, Thomas V.M, Posonia A.M ,"Animal tracking using background subtraction on multi threshold segmentation ", Proceedings of IEEE International Conference on Circuit, Power and Computing Technologies, ICCPCT, 2016 .

[4] Mayan J.A, Arifa S, Pavithra R, "Semantic based multi lexical ranking technique for an effective search in protected cloud,2016 International Conference on Control, Instrumentation, Communication and Computational Technologies, Kumaracoil, 2016, pp. 570-576.

[5] M.D. AntoPraveena and Dr.R. JosephManoj, "A webServices Authentication System Based On Web Server Log Analysis",International journal of Pharmacy and Technology,Vol. 8,No.3, pp.4545-4557, 2016

[6] Alier, M., Casany, M. Moodbile: Extending Moodle to the Mobile on/offline Scenario. Proceedings of IADIS International Conference Mobile Learning. Algarve, pp.11-18, 2008.

[7] Sandipkumar M Vaniya, B. Bharathi,"Exploring object segmentation methods in visual surveillance for human activity recognition", International Conference on Global Trends in Signal Processing, Information Computing and Communication, pp.520-525, 2016.

[8] Albert Mayan J, Kuldeep Anand D.S, Neha Sadhvi(2017),"Efficient and secure server migration on cloud storage with VSM and dropbox services", International Conference on Information Communication and Embedded Systems (ICICES),Chennai pp. 1-5.

[9] B.Ankayarkanni, Dr.A.Ezil Sam Leni,"A Technique For Classification Of High Resolution Satellite Images Using Object-Based Segmentation", Journal Of Theoretical And Applied Information Technology , Volume 68 ,No.2, October 2014,pp-275-286.

[10] Mary Posonia .A, V.L.Jyothi,"Efficient XML Keyword search using H-Reduction factor and Interactive Algorithm" A. Mary Posonia, V.L.Jyothi (2016)," Extraction of perfect protein sequences with minimal processing cost using enhanced B+ tree algorithm", Biomedical Research, special issue on S12345-S6789

[11] A. Mary Posonia, Dr. V.L.Jyothi(2015), "Improving Data Access Performance by Reverse Indexing", International Journal of engineering and Technology(IJET),Vol 7 No 3,pp-1057- 1061

[12] A. Mary Posonia, Dr. V. L. Jyothi, "XML Document Retrieval by Developing an Effective Indexing Technique", in IEEE International Conference on IcoAC, MIT, Chennai, 2014, IEEE , **DOI:** 10.1109/ICoAC.2014.7229758, ISSN - 2377-6927

[13] Vimal Kumar S., Vasudevan S. and Mary Posonia A, "Urban Mode of Dispatching Students from Hostel", ARPN Journal of Engineering and Applied Science ,2017 ,Vol.12, No. 13.

IOP Conf. Series: Materials Science and Engineering **590** (2019) 012003 doi:10.1088/1757-899X/590/1/012003

A SURVEY ON OPINION MINING AND SENTIMENT ANALYSIS

R.Nimesh[1], P. Veera Raghava [2], Dr. S. Prince Mary[3], Dr. B. Bharathi[4]

[1] [2] Student, Department of Computer Science and Engineeering, Sathyabama University
[3] Assistant Professor, Department of Computer Science and Engineeering, Sathyabama University, Chennai,
[4] Professor, Department of Computer Science and Engineeering, Sathyabama University, Chennai,
princemary26@gmail.com

Abstract. Utilizing on the web shopper surveys as electronic verbal exchange to help buy basic leadership has turned out to be progressively prevalent. The Web gives a broad wellspring of shopper audits; however one can scarcely read all surveys to acquire a reasonable assessment of an item or administration. A content preparing system that can condense surveys would hence be attractive. A subtask to be performed by such a structure is locate the general angle classes tended to in survey sentences, for which this paper presents two techniques. As opposed to most existing methodologies, the principal strategy exhibited is an unsupervised technique that applies affiliation control mining on co-event recurrence information acquired from a corpus to discover these angle classifications. While not comparable to best in class directed strategies, the proposed unsupervised strategy performs superior to a few straightforward baselines, a comparative however administered technique, and a managed standard, with a F1-score of 67%. The second technique is a regulated variation that beats existing strategies with a F1-score of 84%.

1. INTRODUCTION

For purchasing a product mouth talk has been influence a lot on customer decision making. Family and friends are asked usually for product purchase. This can make the customer influence on the product purchase. In last few years mouth talk has been increasing more popular. The important forms of mouth talk are product and service review posted on internet by consumers. Retail Company such as Amazon Flip kart and Myntra has numerous products and review on internet. The information that can be obtained from products and services reviews is not only for consumers but also helpful for companies for developing new products. Knowing what is posting on the web will helpful for the company.

2. SURVEY OF EXISTING RESEARCH

Kim Schouten said that on this paper we've presented two strategies for detecting aspect categories, that is useful for online evaluation summarization The first, unsupervised, method, uses spreading activation over a graph built from word co-prevalence information, allowing the use of both direct members of the family among words. This consequences in every word having an and oblique activation fee for every class that represents how possibly it is to imply that lass. Even as other technique want categorized want categorized training records to operate, this technique works unsupervised. The essential disadvantage of this method is that some parameters want to be set ahead, and mainly the class firing thresholds need to be carefully set to gain a good overall performance. We have given heuristics on how these parameters can be set.

Onne van der Weijde, Flavius Frasincar said that the second, supervised, method makes use of a as a substitute sincere co- incidence incidence approach where the co- occurrence frequency between occurrence frequency between annotated aspect classes and both lemmas and dependencies is used to calculate conditional probabilities. If the maximum conditional possibility is higher than the associated, educated, threshold, the category is assigned to that sentence. Comparing this technique at the official SemEval-2014 test set indicates a excessive F1-ratingof 83%.

Rommert Dekker said that In phrases of future paintings, we would really like to analyze how injecting outside knowledge would enhance the outcomes whilst lexicons are a terrific manner of doing that, as proven by means of Kiritchenko et al, we are mainly inquisitive about exploiting extra semantic alternatives, like ontology's or other semantic networks. Additionally, as we are managing unbalanced statistics, we plan to explore system learning strategies that cope with this trouble.

3. PROPOSED WORK

In our future work we proposed to have genuine reviews for purchasing the products. In this process we will generate a OTP for giving comments without generating of OTP you can't give reviews for the products. The modules that are to present are 1.User login\Registration 2.Buying products 3.Generation of OTP 4.Comments and 5.Graphs

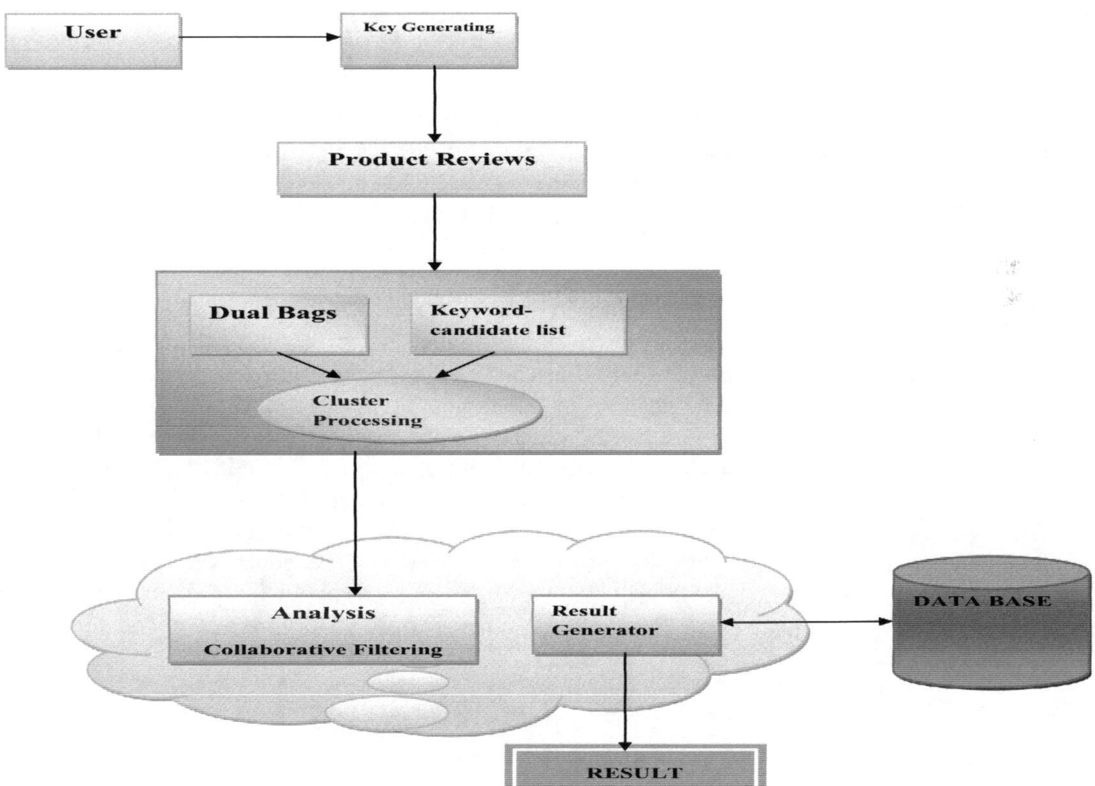

Figure1. System Architecture

In Figure 1shows the architecture, in which the user in order to give rating for the product after buying we will generate a key then the user can review about the product after commenting based

IOP Conf. Series: Materials Science and Engineering **590** (2019) 012003 doi:10.1088/1757-899X/590/1/012003

on the reviews the clustering algorithm will analyze the reviews into positive and negative comments and graph will be generated based on review by user then it will be stored in data base after that final result will be declared. The first module contains login for admin and user. This page is used by admin for adding the products, for user to view the products and admin only has right to add remove and update the products in database.

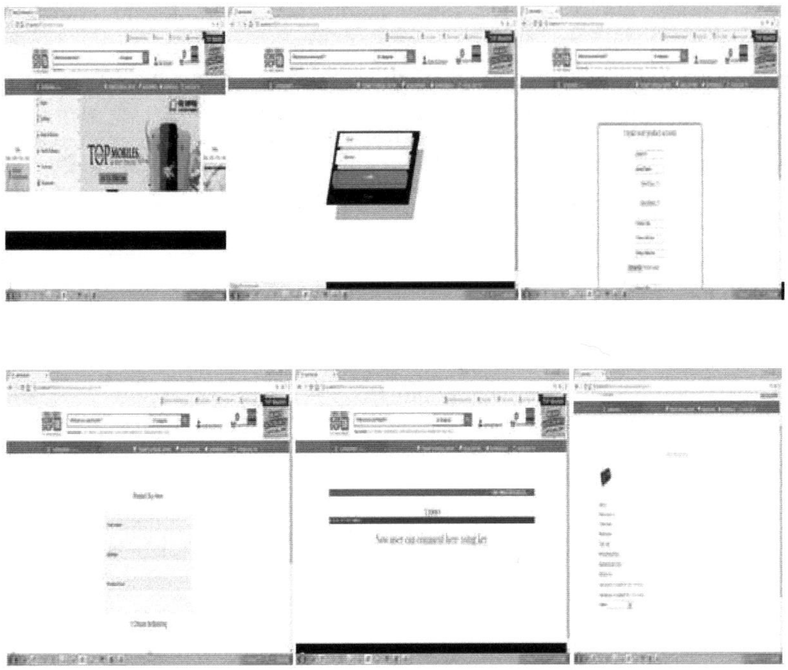

Figure 2. Implementation

Figure 2 shows the implementation of the proposed system. Initially the login page created for both user and admin for accessing the products, Admin after login he/she will add products and user will create account so that he/she can buy products from the login page. Admin will add the products user can’t add the products and user can buy the products from web page and in the product creation admin will give clear explanation about the products and details of the product. After selecting the product, he will go to the product payment, in order to buy the product we need to have a bank account for payment, after creating bank account then you should add delivery address, payment should be done using net banking. After the payment in order to give the rating we will generate an OTP with that only you can give command about the products, without the OTP you can’t command about the product. After buying the products and OTP will be generated, with the OTP we can comment about the product. Without OTP we can’t comment about the product. After commenting graph will be generating. Figure 3. is about the final review of the product in this review the graph shows that product has more positive rating.

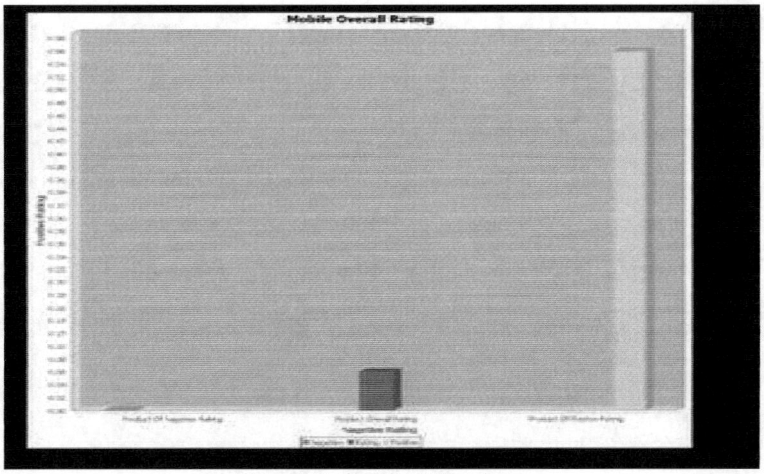

Figure 3. Product Reviews

4. CONCLUSION

In this project we added one time password for guanine reviews about the product by the customer and we developed Buying products based on genuine reviews given by customers on internet. For giving reviews we generated OTP only after purchasing the product and after commenting graph will be generated for the product.

REFERENCES

[1]. P. F. Bone, "Word-of-mouth effects on short-term and long-term product judgments," J. Bus. Res., vol. 32, no. 3, pp. 213–223, 1995.

[2]. R. Feldman, "Techniques and applications for sentiment analysis," Commun. ACM, vol. 56, no. 4, pp. 82–89, 2013.

[3]. S. Sen and D. Lerman, "Why are you telling me this? An examination into negative consumer reviews on the Web," J. Interact. Marketing, vol. 21, no. 4, pp. 76–94, 2007.

[4]. Avinaash Ram S.P., Albert Mayan .J, "Mobile Attendance Management and Employee Registration", ARPN Journal of Engineering and Applied Sciences, Vol:10, Issue 8, pp: 3727-3730,May 2015 , ISSN : 1819-6608.

[5]. Venkatakrishna D., Ankayarkanni B,"A description of content based image retrieval using from block truncation coding and image content description",Indian Journal of Science and Technology, Vol 9, No.21, pp.1-5, June 2016

[6]. M. T. Adjei, S. M. Noble, and C. H. Noble, "The influence of C2C communications in online brand communities on customer purchase behavior," J. Acad. Marketing Sci., vol. 38, no. 5, pp. 634–653, 2010.

[7]. Albert Mayan .J , Sharmila Latha T , Kislay Sinha ," Security Analysis of Three Factor Authentication Schemes for Banking", ARPN Journal of Engineering and Applied Sciences, Vol:10, Issue 8, pp: 3504-3509,May 2015 , ISSN : 1819-6608.

[8]. D.Usha Nandini, Dr.Ezil Sam Leni, "Shadow Identification using Ant Colony Optimization", Journal of Theoretical and Applied Information Technology, vol. 78,no.2, pp.195-200, 2015.

IOP Conf. Series: Materials Science and Engineering **590** (2019) 012004 doi:10.1088/1757-899X/590/1/012004

Track Me & Unlock: Secured Mutual Authentication System of Integrating Phone Unlock & Women's Safety Application using MEMS

S.Vengaten[1], M.Wilferd Roshan[2], Dr.S.Prince Mary[3], and D. Usha Nandini[4]

[1] [2] Student, Department of Computer Science and Engineeering, Sathyabama University
[3] Assistant Professor, Department of Computer Science and Engineeering, Sathyabama University, Chennai,
[4] Assistant Professor, Department of Computer Science and Engineeering, Sathyabama University, Chennai,
princemary26@gmail.com

Abstract. In the Existing System, Mobile must be unlocked before using. The widely usage of unlocking types are PIN, password and graph. In the Proposed System, we propose a novel token-based mobile device unlocking approach: transferring the authentication state between two devices by briefly shaking them conjointly. In this system; we propose 4 attributes for phone unlocking. 1. Hand waving pattern in mobile, 2. Creating information from embedded hardware through Bluetooth, 3. MEMS waving pattern by the hardware device, 4.Synchronization of both the device. The unlocking pattern of the mobile will be changing every time for safety reasons, It will be changing randomly. The major advantage of this system, even if hacker knows one of the above specified unlocking password, he have to provide answer based on the required pattern so this is more secured application. We also implement Women's Safety application along with this implementation. We include two patterns for this application. Pressing the switches in the hardware or Phone's waving pattern is matched immediately system will understand that Women safety application is initiated. GPS is initiated to save the women by sending location info both to the Guardian and Police.

Keywords: shaking them conjointly, hand waving, synchronization, changing randomly, answer based, GPS initiated.

1. INTRODUCTION

The unlocking mechanism in mobiles is PIN, graphical pattern and password, swipe. Biometrics approaches most widely used in mobile devices which includes fingerprint scanner faces detection, or voice detector. The security level of biometric is very low, and easy to access. Biometrics has to be changed, which decreases the use of leakage theft. In addition to biometric features, hardware helps to find and process biometrics on mobile phones is often proprietary. Unlocking techniques are rarely used on mobile phones, in contrast to knowledge-based authentication and biometrics. Many proposed ideas so far are based on token proximity and device to perform the unlock.

There is more possibility to hack the device and can unlock when the user is near to the attacker. For example, with Wi-Fi- or Bluetooth-based techniques it might be sufficient to be in the same room with the user to successfully unlock the device. When the hacker is near to the user he will be to access over the mobile phone, there is a chance unlocking the mobile immediately will be possible before leaving the scene. While using token-based authentication user is required with a token, when they want to use their mobile phones. Depending on the token location, it could be possible to obtain control over token and device at same time, and later apply the token to unlock the device. In case of theft, if token itself is locked to prevent illegal usage, the entire issue is moved from the mobile device to the token a sun locking the token could be done using knowledge-, biometrics or token based authentication. We propose a token-based mobile phone unlocking technique to address the above problems:

- By shaking the device conjointly there is a transfer in action station between two devices. This idea is that personal mobile can be unlocked in different ways each time; one could act as a token, allowing transferring authentication state between devices.
- We are going to provide four typeof unlocking by creating we can make sure that the device is protected properly. The security can be advanced by adding some security questions so that if people access our device with permission he cannot pass through the security question. The four types of unlocking system differ from each other hand waving, knock, using MEMS, Bluetooth device.

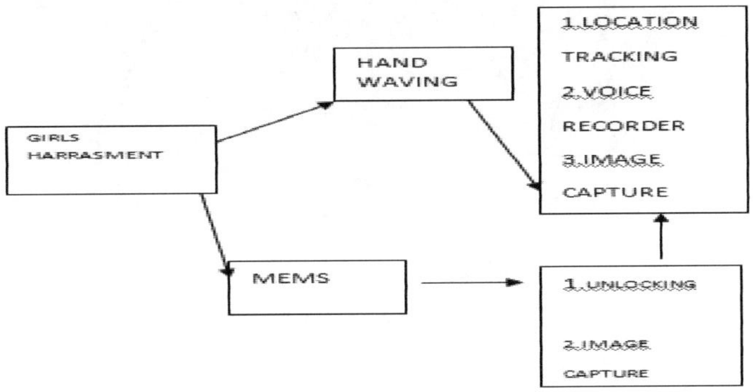

Fig.1: System Diagram

Fig.1 shows the flow diagram of women safety and unlocking with the use of MEMS sensor and hand waving principle. Here there is location tracking, Voice recorder, and Image capture during emergency.

2. PROPOSED SYSTEM

We propose a idea based on mobile device unlocking types, sharing the action state between two devices by shaking them conjointly. The key plan is that personal mobile devices can remain unlocked for different time of period; one could act as a token, allowing sharing action state between two devices. For example, a mobile device should lock by its own as soon as it put aside while a smart watch will remain unlocked as far it is strapped to the wrist and automatically lock by itself when removed. Example: unlocked once in the morning when attached to the wrist and automatically lock by it when removed.

In this system; we propose 4 attributes for phone unlocking. We need to construct Embedded Hardware consisting of MEMS Accelerometer Sensor for identifying Phone's shaking pattern, Bluetooth hardware for Communication, 2 Switch for Signalling the Phone devise. Using this above system user can set 4 types of password for phone unlocking.

- Hand waving pattern,
- Creating information from embedded hardware through Bluetooth
- MEMS waving pattern by the hardware device
- Knock-knock. There will be a change of unlocking pattern every time during unlocking.

The major advantage of this system, even if hacker knows one of the above specified unlocking password, he have to provide answer based on the required pattern so this is more secured application. We also implement Women's Safety application along with this implementation. We

IOP Conf. Series: Materials Science and Engineering **590** (2019) 012004 doi:10.1088/1757-899X/590/1/012004

include two patterns for this application. By pressing the switches in hardware or by waving the mobile is matched immediately system will understand that Women safety application is initiated. GPS is initiated to save the women by sending location info both to the Guardian and Police shown.

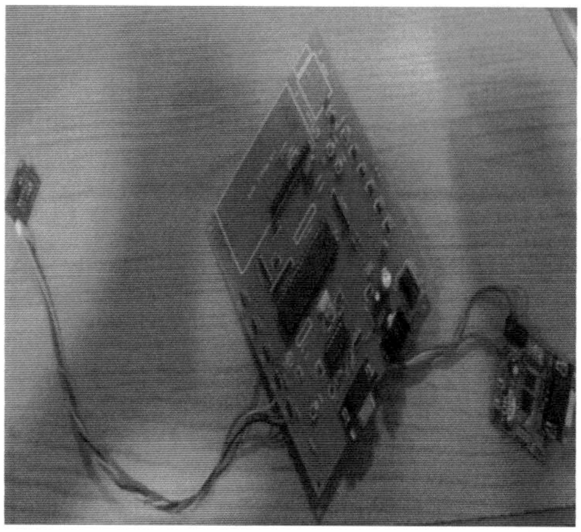

Fig.2. Hardware Device

Fig.2 is the hardware device with the Bluetooth and MEMS sensor connected to the board. This is used to unlock device.

3. IMPLEMENTATION:

In the module we create an application for secured authentication. The android application is created and then installed in User's Android Mobile Phone. The first page in this application consists of user registration process. We create the User login page by Button and Text Field class in the Android studio. User has three types of registration first they have to register their Name , Guardian Number, Police Number, Ambulance Number, Alternate number to send the message. Secondly have to register four types of unlocking system. Finally some static questions will be asked by application for security reasons. We use MEMS, Bluetooth and switch to unlock the device.

MEMS-Micro Electro Mechanical Systems .MEMS is define as combination of a number of micro components on single chip. This allows micro system to sense and function the environment. MEMS will analyze the motion to the hardware and image of MEMS is shown in Fig.3. Bluetooth will transfer the value to the server to verify the values that we already registered. Switch is a button to unlock the device when we press that button from hardware kit it will generate the value and that value will send via Bluetooth

Fig.3: MEMS SENSOR

Switch based authentication is a smart hand hold device is connected with user. A switch is attached with that device with Bluetooth. When the user press that switch it will generate a value to unlock. But mobile will not unlock after the switch is pressed. The generated value will send through Bluetooth and that value will be verified by the system after that only the device will unlock.

Synchronized authentication is both the MEMS and the hand waving should synchronize each other. That is we already register the mobile tilde and the motion of the Mims's if the user when unlock the mobile it will read both mobile tilde and MEMS motion. After the synchronization of the both device mobile will unlock automatically.

WOMEN'S SAFETY-In this, unlocking system will make only in mobile. In home page of application two buttons will be displayed on the screen. One for emergency and another for forgot password. If the user forgot the password a four set of questions will be asked by the application. After the correct answers givenby the user the mobile will unlock. If the answer is not correct mobile will consider the system is hacked by someone so a message is sent to alternate number of the user. Another button is for emergency if user press the button a GPS location is sent to the register mobile number. If she had physical harassment message will be sent to local guardian and police. Implementation result screens are shown in fig.4.

Fig.4. Implementation

4. CONCLUSION

We provide four types of unlocking system to the user. This system will give more security than the existing system. As we mentioned above we provided better security system for android devices. By including the four types of unlocking systems, now it is made sure that android device can be

IOP Conf. Series: Materials Science and Engineering **590** (2019) 012004 doi:10.1088/1757-899X/590/1/012004

unlocked and secured. Since security questions have to be inserted in this unlocking systems this application is made sure that the user's mobile cannot be breached easily. And also the above proposed system provide a security to the girls when their in emergency situation. GPS is initiated to save the women by sending location info both to the Guardian and Police so that the harassment can be controlled.

REFERENCES

[1]. R. D. Findling, M. Muaaz, D. Hintze, and R. Mayrhofer,"ShakeUnlock: Securely unlock mobile devices by shaking them together," in Proc. 12th Int. Conf. Adv. Mob. Comput. Multimedia, Dec. 2014, pp. 165–174.
[2]. M. Swan, "Sensor mania! the internet of things, wearable computing, objective metrics, and the quantified self 2.0," J. Sensor Actuator Netw., vol. 1, no. 3, pp. 217–253, Nov. 2012.
[3]. Z. Ni, Y. Shi, N. Ansari, and S. Wei, "Reversible data hiding," IEEE Trans. Circuits Syst. Video Technol., vol. 16, no. 3, pp. 354–362, Mar. 2006.
[4]. W.Hong, T.Chen and H.Wu, "An improved reversible data hiding in encrypted images using side match," IEEE. Signal process. Lett ,vall.19,No.4,pp.199-202,apr.2012.
[5]. Christian M. Toader,"User authentication for Pico"IEEE,No.4,pp,199-202,Apr 2011.
[6]. Bandaru R, Albert Mayan J (2016) ,"Novel approach for whole test suite generation using metamorphic relations",Indian Journal of Science and Technology, Vol 9,No.10, pp.1-7.
[7]. AntoPraveena M.D., Joseph Manoj R, Shanthi V (2014) , "An authentication method for secure web services access with preventing tautology type SQL injection ",International Journal of Applied Engineering Research,Vol.9, No.20,pp. 7531-7543.
[8]. Mary S.Prince, Dr.Baburaj.E, "Constraint informative Rules For Genetic Algorithm-Based Web Page Recommendation System", Journal Of Computer Science 9 (11), pp:1589-1601,2013.
[9]. D.Usha Nandini, Dr.Ezil Sam Leni, A Survey of the Existing Shadow Detection Techniques, International Conference on Control, Instrumentation, Communication and Computational Technologies, pp.175-177, 2014.

INTELLIGENT MEDICINE MANAGEMENT SYSTEM AND SURVEILLANCE IN IOT ENVIRONMENT

A. Jesudoss, M. Jacob Daniel, and J. Jerom Richard

Dept. of CSE, School of Computing, Sathyabama Institute of Science and Technology, Chennai, Tamil Nadu, India

jesudossas@gmail.com, dannyjacob066@gmail.com, jeromrichard001@gmail.com

Abstract. Medicine management based on Internet of Things (IoT) has great potential in medical field. We propose a surveillance security and maintenance for medicine in IoT environment. The proposed platform has an intelligent medicine security box that gives update of the medicine room and alerts when trespassers are prohibited. The medicine box is wirelessly connected to VNC viewer application. It is an android Application which runs through raspberry pi and it gives timely updates about medicines by the sensor connected to it. The surveillance camera capture the image of the person enters the room. If any person enters the room it automatically gives an SMS alert.

Keywords-Medicine monitoring, Surveillance security, SMS alert system.

1. Introduction

In existing system there is no proper maintenance for medicine, only LED indication is possible is existing system. There is no sensor in medicine monitoring in existing system. In heath monitoring there is availability of sensor, so I took the concept and applied in medicine monitoring. We propose a new medicine monitoring, surveillance security and SMS alert for medicine management system in the Internet of Things environment which is used for checking the authenticity of pharmaceutical products and intruders. The security scheme is efficient with respect to computation and communication cost..

The raspberry pi is connected with BMP180 to monitor pressure and temperature of the medicines, it also check the availability of the medicine's in medicine box by measuring pressure. PIR sensor will check the human are entered are not. The Temperature sensor, PH level sensor, Gas sensor, Liquid level sensor are used to monitor the status of the medicines. By this concept we get more advantages like live streaming is possible, we can receive the details through email, we can check Medicine stock availability, etc. The existing system is health monitoring shows the health conditions of a person continuously with some sensor. It can be used as home healthcare services and hospital management. We use sensor for medicine management and security purpose of the medicine.

According to World Health Organization (WHO), the commonly counterfeited drugs are antibiotics, anti-materials, hormones and steroids, and now anticancer and antiviral drugs are also included in the list. The counterfeiting of medicines causes the serious threat to the society like fake medicine, expiry date of medicine, manufacture defect in medicine etc. Due to this human beings get affected and the future generation will also get affected.

IOP Conf. Series: Materials Science and Engineering **590** (2019) 012005 doi:10.1088/1757-899X/590/1/012005

BMP180 is connected with Arduino to monitor weight of the medicines, it will check the availability of the medicine's in medicine box. PIR motion sensor, PH level sensor, Gas sensor, Liquid level sensor, Vibration sensor, Light sensor are used to monitor the status of the medicines and connected in raspberry pi. It can be used in Hospital Management Systems, material maintenance based applications, stock room maintenance..

2. Related Works

According to the author G. Subhramanya Sarma (et.al) [1] proposed health care and in addition of sensor for a person safety to this monitoring person use for travelling and elderly peoples. They had many drawbacks like healthcare is done but the medicine cannot be maintained. Healthcare Applications are used by the general public so it must be developed with utmost care and security features [1a]. Enhanced Authentication Scheme provides protection against replay and password-guessing attacks [1b]. The author Moeen Hassanalieragh (et.al) [2] implies IOT application enabled for smart phone and connected health care is a particularly important one but the cost is too high for this process. The healthcare applications can also be provided as web service so that it will be platform-independent [2a]. The IoT has played a crucial role in agriculture also. Raspberry Pi has been used for monitoring Agricultural activities [2b]. Author K. Natarajan (et.al) [3] proves IoT application connecting various objects together through to form a complete system but the system is not well defined. The author Jezna G Jose (et.al) [4] a modern healthcare IoT platform with an intelligent medicine box along with sensors for medicine monitoring is proposed here. Health care services based on Internet of Things have great potential in medical field. More delay occurred while transmissions of data. The author A. J. Jara (et.al) [5] proves for getting sensor values Zigbee can transfer sensor values effectively but when there is a need of continuous data transmission zigbee cannot be used. According to The author Suryadevara (et.al) [6] wireless sensor can be carried anywhere by the person but some medicine cannot be carried everywhere with sensor. Wireless sensor for home based safety of elderly person. This H. Cheung (et.al) [7] proposed anti counterfeiting systems for find the fake medicine like drugs and impure medicine. They 1st processed by placing Computers in Industry, to check the medicines. Author S. H. Choi (et.al) [8] proposed the security system for the anti counterfeiting medicine by maintaining in the stock room under the surveillance. This system is used hospital management. The author B. Yang (et.al) [9] proposed tracking system for medicine like "RFID tag data processing by tracing anti-counterfeiting" by this process medicine is secured in outdoors. Then author S. Coyle (et.al) [10] proposed sensor in health care IOT platform, like Biological sensor to shows the variation in human beings by sensors. Author Kim (et.al) [11] proposed an application-level technique for anti-counterfeiting medicine which employs a RFID reader available to a consumer's device. According to author Gipsa Alex (et.al) [12] proposed a medicine box in IOT with LED and buzzer. Through this concept we propose medicine box with different types of sensors..

3. Proposed System

Raspberry pi is used as the controller for the system and it accesses the direct power supply. Gas sensor detect there is any gas leakage in room. PIR motion sensor every person enter the room, PH sensor detect the impurities and level in it. Water Level Sensor detects the level of the liquid. Vibration sensor detect the vibration occur in the room. All the sensors are connected in raspberry pi. Arduino is also connected through raspberry pi for power supply. BMP180 is used measure the temperature and pressure of room. A DC motor is fixed for the security purpose by detecting a person by PIR motion sensor the motor work, if a intruder enter the room motion sensor detect the person and automatic door shutting system work.

Python program is stored in a SD card and placed in raspberry pi. The python program has been created for the senor level that had to maintain for a medicine. In our process we can measure only one medicine at a time, medicine level, state has been entered in the program.

Fig. 1 Raspberry Pi

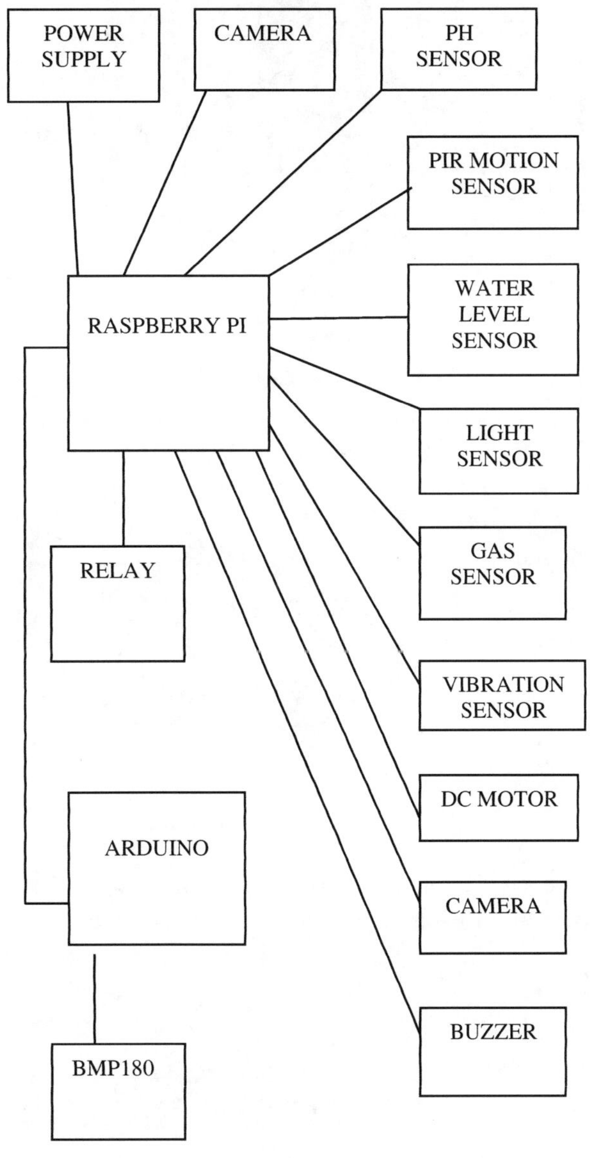

Fig. 2 Block Diagram of the Proposed Work

4. Pin Connection

Water level sensor is conneted in 3.3volt pin. DC motor is connected in 5volt pin and GPIO2 pin. Buzzer is coonected in GPIO3 pin and GPIO4 pin. PH sensor is connected in GPIO26 pin, GPIO13 pin and GPIO6 pin. Vibration sensor is connected in GPIO5 pin. PIR motion sensor is connected in GPIO7 pin. Light sensor is connected inID_SC pin. Gas sensor is connected in GPIO12.

5. Pin Configuration

For sensors power supply is given according to the Pin configuration.

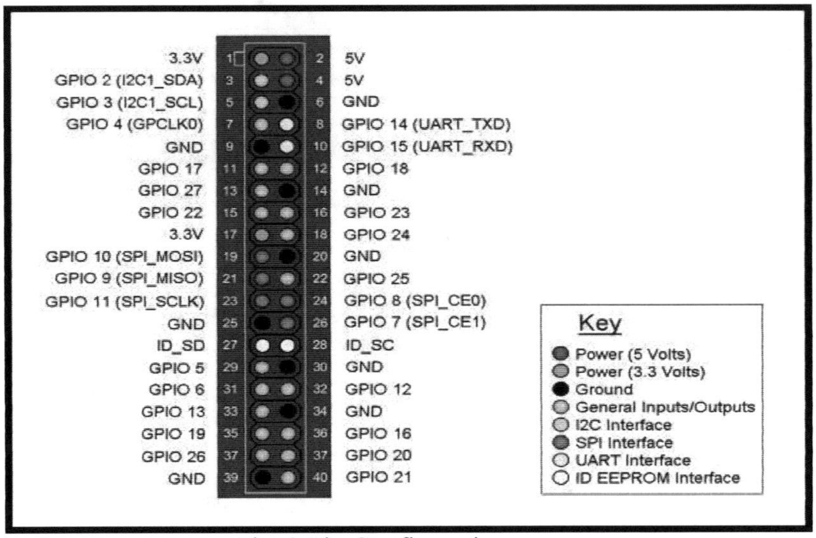

Fig. 3 Pin Configuration

Fig. 4 Arduino

Arduino is connected in raspberry pi for power supply. Arduino is getting 5volt from raspberry. It is only used for BMP180 connectivity. It is used for ADC (Analog to Digital convertor).

Fig. 5 BMP180

IOP Conf. Series: Materials Science and Engineering **590** (2019) 012005 doi:10.1088/1757-899X/590/1/012005

It is connected to the Arduino. BMP180 it will calculate the pressure and temperature at a time of the medicine box.

BMP180 has default weight when we give pressure on BMP180 weight value will automatically increase. If it is default value then the box is empty. The program code is stored in raspberry pi so it will be fully automatic.

Fig. 6 PIR Motion Sensor

A PIR detector is a motion detector very effective in enhancing home security systems. The sensor detect the person passing by through infrared race, it is simply sensitive to the infrared energy emitted by every living things. When a person walks inside the room, the sensor detects the person and captures the picture of the person. If the person is an intruder it gives an alert.

PH Sensor is used to detect the liquid state of the medicine in correct PH level or not. If the PH level is not is correct range, it will automatically show in system screen.

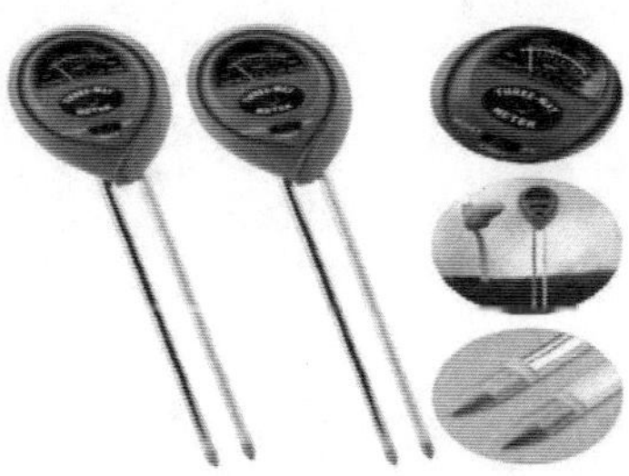

Fig 7. PH Sensor

Gas sensor is used to detect any leakage in gas medicine around the box. It also detect there is leakage is oxygen cylinder or smoke. It is connected to raspberry pi with 5volt power supply.

Fig 8. Gas Sensor

Level of liquids and other fluids are detected by level sensor. Such substances include liquids like water, oil, slurries, etc. It also measures the settled solid medicine inside the liquid. Level sensors measure the level of the medicine which is already entered.

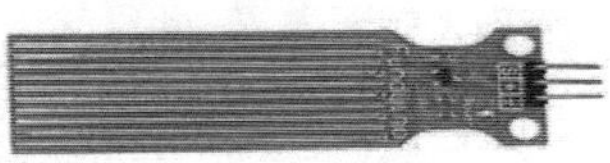

Fig. 9. Water Level Sensor

IOP Conf. Series: Materials Science and Engineering **590** (2019) 012005 doi:10.1088/1757-899X/590/1/012005

We use LDR module as light sensor. It is used has some medicine should be maintained in dark room without light. If there is light presented it gives us automatic notification.

Fig. 10. Light Sensor

It is for security when the box is kept in room. So anybody like child touches the box it gives us alert. It will alert with a sound.

Fig. 11. Buzzer

We use DC motor for security purpose. It any intruder enter the motion sensor detect the person gives us notification and automatically closes the door. It is a 12 volt DC motor. DC motor is connected to relay and from relay connected to raspberry pi.

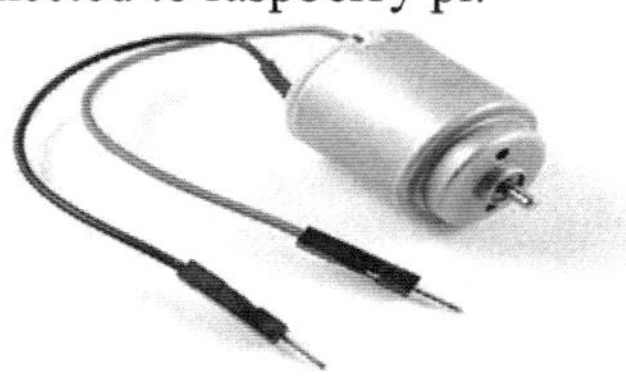

Fig. 12 DC Motor

Relay is used for switching purpose. It is mainly used for motor. DC motor needs more than 5volt. Hence, it is connected to relay.

Fig. 13 Relay

We use Logitech webcam C100 camera for security purpose as surveillance of the medicine box and the stock room.

IOP Conf. Series: Materials Science and Engineering **590** (2019) 012005 doi:10.1088/1757-899X/590/1/012005

Fig. 14 Camera

6. Implementation

VNC viewer is an android application which support Raspberry pi, in raspberry pi we have constructed a python program and saved in a SD card. The steps of implementation are given below:

STEP 1: sudo python ma.py is the command for executing the temperature and pressure sensor.

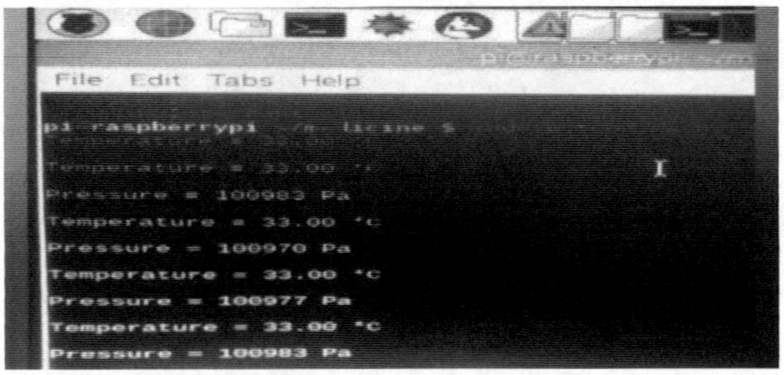

STEP 2: sudo python main1.py is the command for executing the light sensor, smoke sensor, liquid sensor, vibration sensor, PH sensor, PIR motion sensor.

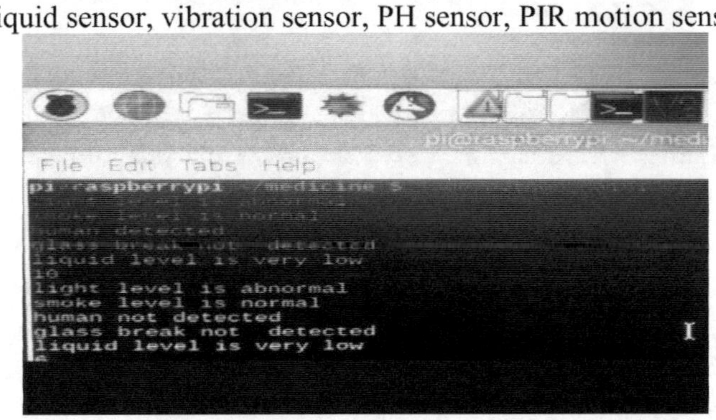

STEP 3: sudo python objectsms1.py is the command for message alert system.

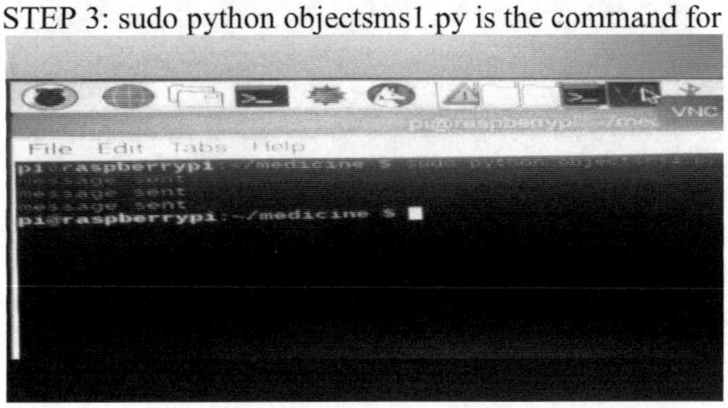

STEP 4: 192.168.43.88;8081 is the web cam Ip address used for online monitoring .

Nowadays there are so many anti counterfeiting of medicines causes the serious threat to the society like expiry date of medicine, manufacture defects of medicine etc.

IOP Conf. Series: Materials Science and Engineering **590** (2019) 012005 doi:10.1088/1757-899X/590/1/012005

Through this human get affected and the future generation affect through them. There are so many hospitals with no proper maintenance of medicine and in home power medicine (like act rapid, paracetamol, etc) are kept in front of children. For the safety of a person we propose advance security system in medicine maintenance.

7. Conclusion & Future Work

The proposed scheme utilizes the Internet of things (IOT) and is suitable for mobile environment, which also provides efficient IOT update phase. The scheme is efficient with respect to computation and communication costs. By this process we get so many results than existing project. Advantages we can receive the details through email, we can check Medicine stock availability, live streaming is possible. Through this we process advance.

In future work, for one medicine monitoring we can do multiple medicine monitoring at a time by connecting more raspberry pi and to connect all raspberry pi some system has to be taken. It gives more utilization in medicine management. It reduces the human work in hospital management and security.

REFERENCES

[1] C. Taskforce, "WHO launches taskforce to fight counterfeit drugs," Bulletin of the World Health Organization, vol. 84, no. 9, pp. 689–694, 2006.

[2] Jesudoss A. and Subramaniam N.P., "EPBAS: Securing Cloud-Based Healthcare Information Systems using Enhanced Password-Based Authentication Scheme", Asian Journal of Information Technology, Vol. 15, Issue 14, 2016, pp. 2457-2463.

[3] Jesudoss A. and Subramaniam N.P., "Enhanced Kerberos Authentication for Distributed Environment", Journal of Theoretical and Applied Information Technology, 2014 Vol. 69, No. 2, pp. 368-374.

[4] Albert Mayan .J , Sharmila Latha T , Kislay Sinha ," Security Analysis of Three Factor Authentication Schemes for Banking", ARPN Journal of Engineering and Applied Sciences, Vol:10, Issue 8, pp: 3504-3509,May 2015 , ISSN : 1819-6608.

[5] Lakshmanan, L. and K.S. Dinesh Kumar, A,(2016), "Expanding web service suggestions using web service usage history", International Journal of Pharmacy and Technology, Vol. 8, No.4,pp. 22839-22847,India, 2017, ISSN: 0975766X8.

[6] Lakshmanan, L. and Balaji,C(2015), "Agricultural Robocop using Raspberry Pi", International journal of Applied Engineering Research, RIP India, Vol. 10, No.5,pp.12177-12186,India, 2015, ISSN: 1087--1090.

[7] "Substandard, spurious, falsely labelled, falsified and counterfeit (SSFFC) medical products," available http://www.un.org/africarenewal/magazine/may- 2013/counterfeitdrugs- raise-africa's-temperature. Accessed on March 2016.

[8] Ullah Kaleem, Munam Ali Shah, Sijing Zhang, "Effective ways to use Internet of Things in the field of medical and smart health care", 2016 International Conference on Intelligent systems Engineering, 2016.

[9] Gipsa Alex1 , Benitta Varghese2 , Jezna G Jose3 , AlbyMol Abraham developed mobile applicayion for health monitoring ISSN : 0975-3397 Vol. 8 No.4 Apr 2016.

[10] H. Cheung and S. Choi, "Implementation of a Telemonitoring ConceptBased on NFC-Enabled Mobile Phonesand Sensor Devices" vol. 16, no. 1, January 2012

[11] D. He, S. Zeadally, "Authentication protocol for an ambient assisted living system", IEEE Common. Mag., vol. 53, no. 1, pp. 71-77, Jan. 2015.

[12] Yovan Felix A , Jesudoss A ,Albert Mayan J (2017) ,"Entry and exit monitoring using license plate recognition ", IEEE International Conference on Smart Technologies and Management for Computing, Communication, Controls, Energy and Materials (ICSTM), pp.227-231.

[13] S. H. Choi, B. Yang, H. H. Cheung, Y. X. Yang, "RFID tag data processing in manufacturing for track-and-trace anti-counterfeiting", Computer. Ind., vol. 68, pp. 148-161, Apr. 2015
[14] S. H. Choi and C. H. Poon, “An RFID-based anti-counterfeiting system,” IAENG International Journal of Computer Science, vol. 35, no. 1, pp. 1–12, 2008. Maintenance in medical security.
[15] A. J. Jara, M. A. Zamora-Izquierdo, and A. F. Skarmeta, “Interconnection framework for health and remote monitoring based on the Internet of Things,” IEEE J. Sel. Areas Common., vol. 31, no. 9, pp. 47–65, Sep. 2013
[16] S. Coyle et al., “BIOTEX—Biosensing textiles for personalized healthcare management,” IEEE Trans. Inf. Technol. Biomed., vol. 14, No. 2, pp. 364–370, Mar. 2010
[17] Suryadevara N.K, Gad dam A, Rayudu R.K, Mukhopadhyay S.C, “Wireless Sensors Network Based Safe Home to Care Elderly People: Behavior Detection”, Elsevier-Sensors and Actuators: A: Physical, Vol.186, Pages 277-283, 2012.

Study of the Management of Tickets in IT Administration

Aditya Bhowmik[1], Samrat Paul[3]
U.G Student , Department of Computer Science & Engineering
Sathyabama Institute of Science and Technology ,Chennai, India
aditya.bhowmik96@gmail.com, sampaul9085@outlook.com

D.Usha Nandini[2], S.Prince Mary[4]
Assistant Professor , Department of Computer Science & Engineering
Sathyabama Institute of Science and Technology
Chennai, India
ushaduraisamy@yahoo.co.in, princemary26@gmail.com

Abstract:IT Services are very much crucial for the smooth functioning of the system. And so is their administration. IT Services should be monitored at periodically for maintaining an efficient and productive system at an IT Workplace. Errors or Incident are bound to occur, which are then converted into 'Tickets'. Every ticket has a general format in which they are created. Tickets are then sent to the Admin side for the further resolution procedures. Classification of tickets is an instrumental step in which the emerged tickets from a monitoring system are classified using context basis. After this, all the tickets have a correct hierarchy, they are then sent to separate ticket resolution teams for the incident resolution.

Keywords: System Monitoring,Domain Knowledge,Resolution,Ticket,IT Management, Classification.

1. Introduction

Automation of IT Management is the paramount goal of the management and administration.The complexity of IT environments dictates usage of analytical approaches combined with automation to enable fast and efficient delivery of IT services. Incident management, one of the most critical processes in IT Service Management [1], aims at resolving the incident and quickly restoring the provision of services while relying on monitoring or human intervention to detect the malfunction of a component. Thus, it is essential to provide an efficient architecture for the IT routine maintenance.

Automation of IT companies' maintenance procedures is an ultimate goal of IT service management. System as well as Bug Checking is a reliable way for IT problem detection which paves the way for the creation of a ticket. In light of the ticket description, the possible categories of the IT problem are determined, and the ticket is assigned to the corresponding processing teams for problem resolution. Automatic problem category determination acts as a very effective during performing maintenance. In practice, IT problem categories are naturally organized in a hierarchy by specialization. Utilizing the category hierarchy, our project comes up with a hierarchical multi-label classification method to classify the tickets.

Also the knowledge from the domain experts, which partially indicates some categories the given ticket may or may not belong to, can also be used to guide the hierarchical multi-label classification. SDN: The improved searching algorithm proposed in this paper can help address the accurate security margin problem of block ciphers.

This algorithm has shown that the security margins of the studied ciphers are slightly lower than previously published findings. Further improvements to the results can be accomplished with the aid of parallel computing.

Controller placement technique is additionally used in our proposed system. Controller Arrangement Procedure is essentially for the WAN - wide area network whose objective is to limit the average delay. The central idea is to partition the WAN into multiple smaller areas by using SCA - Spectral Clustering Algorithm and allocate a controller to each group.

2. Related Work

There are a lot of related works regarding the proposed application. Some of them are listed below.
SDN - Software Defined Networking marks anideal shift towards a centralized network plane [2]. A specifically significant task in Software Defined Networks architectures is that of placement of the controller which is the positioning of a limited number of resources within a specific network to meet various requirements. All the aforesaid requirements range from delayconstraints to error tolerance to load balancing. In most situations or circumstances, some of the aforesaid objectives are challenging. Therefore, any best placement strategy is not available, thus a need for balanced settlement becomes mandatory. POCO, which is a framework for Pareto based Optimal Controller Placement is presented from the above process. It provides operators with efficient placements w.r.t. various performance factors. In thebasic default system configuration, POCO does an exhaustive assessment of all the possible placements. Above approaches are useful when utilized in small to medium networks, whereas it is not as effective when large sized networks comes into play, thus an alternative approach should be there keeping in mind all the resource and time constraints. A heuristic approach, though having less accuracy, is used in the POCO toolset for the faster calculation times for the above circumstances. An assessment of the heuristic approach is carried out on a set of various real world Network Topologies. The error generated due to the heuristic approach can be quantified by using a measure which later paves the way for a perusal of the balanced settlement between accuracy and time. Further, the suggested approaches can be utilized in developing solutions to alike Virtual Functions Placement Problems in the context of NFVs which are Network Functions Virtualization.

Large scale networks such as Wide Area Networks are always partitioned into multiple small networks for deployment of Software Defined Networking (SDN) [3]. The major problemswhich come when the network portioning is done with SDN control planes are: reliability, latency and load balancing. The two major processes which are highlighted are: (a) Partitioning a network into sub domains and (b) Placement of the controller.In order to describe the above processes, a novel approach is proposed to assess Controller Placement Problem for networks efficiently, which uses Spectral Clustering Algorithm for the This approach uses the Spectral Clustering placement algorithm to segregate a large network into multiple sub SDN domains. Various Internet Topologies are used to assess our approach. A new framework is developed in order to verify the SDN controllers in Wide Area Networks, for testing purposes. Controller Placement Problem showed its effectiveness in the results.

The controller is at the central part of all the Software Defined – Networks also known as (SDN) [4]. While the Software Defined Networks escalates to Wide Area Networks (WAN), the factors such as Scalability and Performance which distinguish one controller from another play an instrumental role for the success of SDN for end user service management. Control flows and data flows are differentiated, and novel dynamic control plane architecture is proposed in order to distribute various control flows among various controller objects depending on the specific controller load and controller processor usage or on the data flow service type. The 'Control flow tables' are proposed which are embedded in the flow tables to distribute the control flows among various controller objects. Results from various experimentseveral that the advances in the performance of the data plane services are due to the proposed control flow management approaches wherethe main CPU or output of links between the switches and controller is the constraint.

The significant problem in SDNs – Software Defined Networks is Controller Placement [7].Earlier the solution to the aforesaid problem only highlighted on the delay in the propagation but also avoided the load in the controllers in the networks, an important factor in big networks. Currently a Capacitated Controller Placement problem (CCPP) is defined, which not only considers the load of controllers but also introduces an effective approach to solve the problem. The detailed assessment result shows that the new approach can (a) significantly reduce the amount of controllers, (b) reduce the load of the highest load controller, and finally (c) deducts the radius stretches when compared with the K-center approach with dynamic controller or scheduling.

IOP Conf. Series: Materials Science and Engineering **590** (2019) 012006 doi:10.1088/1757-899X/590/1/012006

DSDN (Distributed SDNs) controllers are developed to highlightresilience and performance issues [8]. The procedures used in Data Centers are developed on highly consistent state sharing among controllers whereas for Wide Area Networks and highly constrained networks are based on lenient consistent state. The problem of the failure for distributed SDN controllers is solved by development of two strategies for active neighbor controllers to have the control of open OpenFlow switches: (a) Development Greedy Algorithm and (2) Pre partitioning among controllers in the given network. Amodel prototype was developed with distributed 'Floodlight' controllers for the various strategy evaluations.

The outcomes exhibit that the failover duration with the greedy approach is corresponding to the amount of OpenFlowswitches whereas the second approach that is pre-partitioning approach, introduces little amount of additional traffic, having a reaction time less than 0.2s.

3. Proposed Work

It is formulated as a (MILP) - Mixed Integer Linear Program. The goal is to limit the maximum, for all switches, of the aggregate of the delay. The switch to the closest controller with enough limitsand the delay from the first reference controller to its closest controller with enough limits.

We additionally proposed a summed up model which can be utilized to limit the normal latency and expanded it for numerous controller faults. Moreover, a SAH - simulated annealing heuristic that viably takes care of the issues on large sized networks.

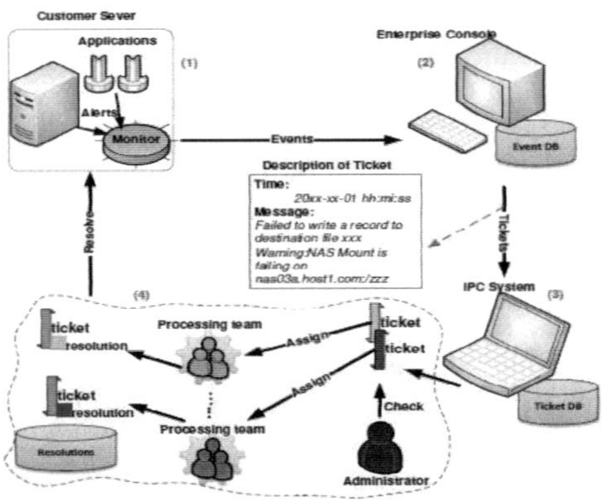

Fig. 1 IT Routine Maintenance Architecture

The various procedures used in the proposed system are:

- User Interface (UI)
- Events Monitoring & Generation of Ticket
- Label Classification
- Work Scheduling & Problem Resolutions

3.1. Events Monitoring and Ticket Generation

Here we are having doing two things:

- Monitoring of events
- Generation of ticket in the event of error

Events are monitored constantly with the help of event monitor. Events like database connectivity, network connectivity, basic and core system checks, etc., are monitored. Tickets are to be generated in the event of error detection. They are sent to the administration for problem resolution. Tickets have a basic format which include message body, IP Address and timestamp.

3.2 Label Classification

In this step, we are classifying the ticket data into a label hierarchy. Based on the ticket data, the possible labels are organized into a hierarchy. For this each and every text in the ticket data is parsed and the possible are identified. Therefore, the correct incident is identified and sent to the corresponding team for the problem resolution.

3.3 Work Scheduling & Problem Resolution

Work Scheduling is the process of giving priority to the tickets sent by the users. This is done by the system administrators. This is set by the timestamp of the ticket which is in the (DATE: HH: MM: SS) format and also the urgency/criticality of the error described in the ticket. An efficient factor is taken into account for the scheduling process.

Problem resolution is the final step in which the ticket is resolved by the teams and seeking for the user's approval.In the proposed system, the goal is to remove the system administrators to a minimum or automate the process as much as possible.

Eliminating system administrators from the process partially or completely where it is possible would reduce human error and accelerates the resolution process. An algorithm is used for the classification of domains based on the data provided in the ticket raised.The main advantage of the system is its presentation at a Graphical level which helps us to understand the process efficiently Also it reduces the involvement of the system administrators to a minimum.The proposed application would have some basic system requirements to run smoothly. The requirements are mentioned below.

Hardware Requirements

The following are preliminary hardware requirements:

- The processor should be Intel Core2Duo, 1.80 GHz
- RAM should be 4GB.
- Hard Disk space should be at least 2 GB.

Software Requirements

- Thefrontend of the proposed application will be made in JAVA (J2EE & SERVELETS).
- Thebackend of the proposed application will be using MySQL 5.5
- Operating System – Windows 7 / 10.
- Integrated Development Environment (IDE) used is NetBeans and Eclipse.

4. Results and Discussion

IOP Conf. Series: Materials Science and Engineering **590** (2019) 012006 doi:10.1088/1757-899X/590/1/012006

As a whole, in the existing system, it is observed that System Administrators have the sole work of problem resolution. That may lead to inefficiency in most of the cases. In our proposed work, we are mostly relying on the knowledge from domain experts which are more reliable and has much less response time for the purpose of problem resolution.

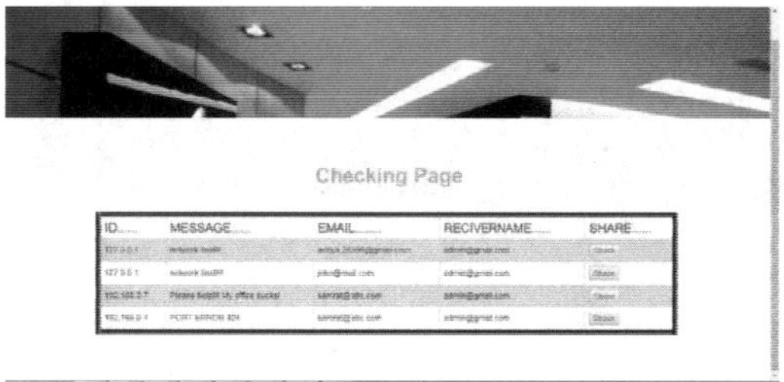

Fig. 2 Admin checking the User Tickets

Future Enrichments can be done to the proposed system eventually. Ranking mechanisms are one of the add on that can be implemented in order to rank the parses of the ticket description.

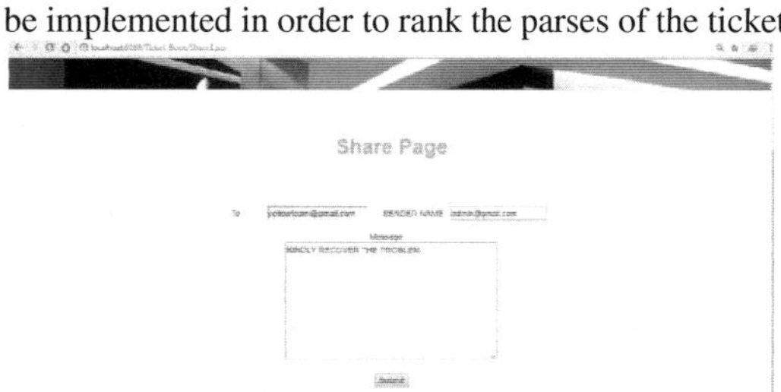

Fig. 3 Admin sharing to the respective team

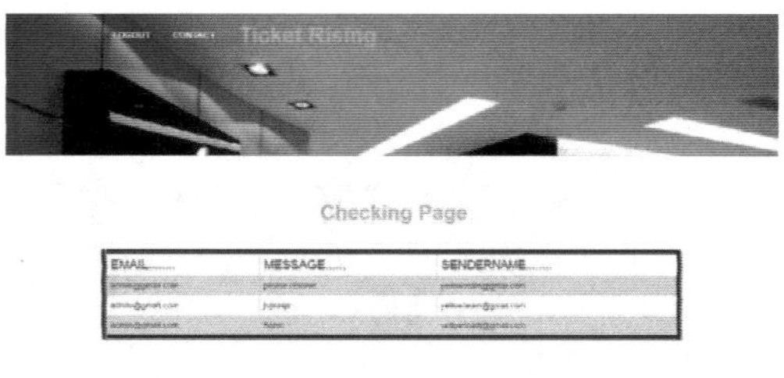

Fig. 4 Domain Team checking the Admin Messages

Also the proposed applications may provide some in built suggestions to the end users that may aid to the recovery of the problem. If not resolved, then the standard procedure is implemented.
Also in the classification process, DAG based hierarchy may be used in place of tree based. Also, more field based expert knowledge may be utilized for the IT problem categorization.In our proposed application, we are having a detailed classification process over the ticket data set fetched from a monitoring system in the end user node. The tickets are then classified into correct hierarchy for their further resolution.

References

[1] Chunqiu Zeng, Wubai Zhou, Tao Li, Larisa Shwartz, GenadyYa. Grabarnik, “Knowledge Guided Hierarchical Multi Label Classification over Ticket Data”, 2017
[2] M R AnnishBrislin, J Albert Mayan, R AroulCanessane, M R Anish Hamlin,"Blood donation and life saver app",2nd International Conference on Communication and Electronics Systems (ICCES), pp. 446 - 451,2017

[3] Stanislav Lange, "Heuristic Approaches to the Controller Placement Problem in Large Scale SDN Networks", 2015
[4] Kamalesh M. D., Albert Mayan J., Felix Y., Sumanth B. S., & Sai Tej B. (2018), " Magrisys: A Smart And Ubiquitous Controlled – Environment Agriculture System " , 2nd International Conference on I-SMAC (IoT in Social, Mobile, Analytics and Cloud) , pp: 1-5 , 2018
[5] AntoPraveena M.D., Joseph Manoj R, Shanthi V (2014) , "An authentication method for secure web services access with preventing tautology type SQL injection ",International Journal of Applied Engineering Research,Vol.9, No.20,pp. 7531-7543.
[6] BurakGorkemli, "Dynamic Management of Control Plane Performance in SDN", 2016
[7] Albert Mayan J, Kuldeep Anand D.S, Neha Sadhvi(2017),"Efficient and secure server migration on cloud storage with VSM and dropbox services", International Conference on Information Communication and Embedded Systems (ICICES),Chennai pp. 1-5.
[8] IBMTivoli, "IBM Tivoli Monitoring", 2016
[9] HP OpenView, "HP OpenView" and "Logic Monitor"
[10] Usha Nandini , Saravanan M , Albert Mayan J , MurariDevakannanKamalesh , Mohana Prasad K (2018) , " Automatic traffic control system using PCA based approach", International Conference on Energy, Communication, Data Analytics and Soft Computing (ICECDS-2017),pp.2387-2392.
[11] Guan Yao, "On the Capacited Placement Controller Problem" in Software Defined Networks", 2014
[12] D. Usha Nandini, Dr. A. Ezil Sam Leni, "Shadow identification using ant colony optimization", journal of theoretical and applied information technology, August 2015. Vol.78. No.2, pp.195-200
[13] Sandipkumar M Vaniya , B. Bharathi, "Exploring object segmentation methods in visual surveillance for human activity recognition" ,International Conference on Global Trends in Signal Processing, Information Computing and Communication , pp.520-525,2016.
[14] L. Tang, T. Li, L. ShwartzPinel, and G. Grabarnik, 'An Integrated Framework for Optimizing Automatic Monitoring Systems in Large IT Infrastructures," in 19th International Conference on knowledge discovery & Data Mining (2013).
[15] Y. Jiang, C.-S. Perng, T. Li, 'Intelligent Cloud Capacity Management', for (NOMS) - Network Operations and Management Symposium, IEEE, 2012, pp. 502–505.
[16] Jabez J, Gowri S, Vigneshwari S, Albert Mayan J and Srinivasulu S (2018)," Anomaly Detection by Using CFS Subset and Neural Network with WEKA Tools", Smart Innovation, Systems and Technologies,Vol.107, pp. 675–682.
[17] J. Knijff, F. Frasincar, and F. Hogenboom, 'Domain Taxonomy Learning from Text', Data & Knowledge Engineering, (2012).
[18] X. Liu, Y. Song, &S. Liu, 'Automatic Taxonomy Construction from Keywords',in 18th International Conference on knowledge discovery &Data Mining (2012).
[19] Kamalesh M. D., Albert Mayan J., Felix Y., Sumanth B. S., & Sai Tej B. (2018), " Magrisys: A Smart And Ubiquitous Controlled – Environment Agriculture System " , 2nd International Conference on I-SMAC (IoT in Social, Mobile, Analytics and Cloud) , pp: 1-5 , 2018
[20] N. Cesa-Bianchi, C. Gentile, and L. Zaniboni, 'Incremental Algorithms for Hierarchical Classification', The Journal of Machine Learning Research, (2006).
[21] N. Cesa-Bianchi and G. Valentini, 'Hierarchical Cost-Sensitive Algorithms for genome-wide gene function prediction',(2010).
[22] Ankayarkanni, B., and A. Ezil Sam Leni. "A new statistical rule model for image retrieval system of remote sensing images." In Circuit, Power and Computing Technologies (ICCPCT), 2016 International Conference on, pp. 1-5. IEEE, 2016.

IOP Conf. Series: Materials Science and Engineering **590** (2019) 012007 doi:10.1088/1757-899X/590/1/012007

A MULTI-SERVER DATA SECURITY WITH PUBLIC AUDITING IN CLOUD COMPUTING

Dinesh R[1] , Ezra Vivin A[2]
U.G Student , Department of Computer Science & Engineering
Sathyabama Institute of Science and Technology ,Chennai, India
[1]rjdinesh4@gmail.com,[2]vivinezra@gmail.com

Dr. Srinivasan N[3]
Professor , Department of Computer Science & Engineering
Sathyabama Institute of Science and Technology , Chennai, India
[3]professorsrini@gmail.com

Albert Mayan J[4]
Assistant Professor , Department of Computer Science & Engineering
Sathyabama Institute of Science and Technology, Chennai, India
[4]albertmayan@gmail.com

Abstract: In cloud computing, user share data while sharing data providing security and protecting the data from a unauthorized users is still a demanding problem, so we implement a multilevel data security popular a cloud computing in order to preserve the information using multi-security levels, and provide sufficient multilevel access and validate authorized identity. The recommended scheme is dynamic in multilevel data security, adaptable to certified resource share in, and protected against different malignant strike. Experimentation result checks the feasibleness. The processed data have multiple security level and access permission.

Keywords: Third Party Auditor, Authentication, Access control, Transparent Computing, Data Security.

1. INTRODUCTION

Over few years, computing has a great development in emerging technologies. Transparent computing is a development technology which helps the user to access user-controlled services. It carries a different variety of heterogeneous OS's and application dynamically in different devices. It allows the user to focus on the available services without depends upon the substantial devices that will be used and depends on which disk should be operated.

The new process comes with the advantages in multiple security aspect. This security system make the user data secure, lowers the theft and reduces the risk of data leaking using centralized management in server side. But still, challenging problem in providing security in different OS's and integrate between data application in servers are shared in cloud computing. The data has a different level of the security system and for example: Consider enterprises which introduce transparent computing as its system. In this module (e.g. data files and information) will be used as their daily activities for updating the information of the employee in the enterprises According to data security, common information can be shared with all public user whereas the sensitive information is been encrypted form only the Data Owners has permission to give the user to access the information .The user can access the protected files which have been encrypted by the owner, the owners give the access permission using a special key to some specific users. But third users cannot access the protected file which is encrypted by the owner. The authorized owner only has the access update, append and delete the files/information which he wants to perform.

Without the knowledge of the user, the data might be misused or embezzled by unauthorized access users. So this process is engaged to encrypt the private source of all users before stored the data in transparent servers. A multilevel security is implemented to protect the user information and provide exact access control to them. The precise scheme which was used to multilevel confidently and fine-grained access was Attribute-Based Encryption (ABE) but it, consumes more expensive due to bilinear map operation. Additionally, if the user revocation process is an unmanageable problem, so it should be processed to get data re-encrypted. We introduce an authentication server to perform identity authentication and multilevel access control with the user level data access, storage transversal in transparent computing.

2 RELATED WORKS

We haJ, Liu et al proposed a financially savvy unwavering quality administration instrument PRCR in light of a summed up information dependability demonstrate. It instate the information dependability while information requires least number of copy duplicate.

C. Wang et al (2013) proposed a protection safeguarding in broad daylight reviewing framework for information stockpiling security in distributed computing. It utilize direct reviewer and arbitrary concealing to acknowledge that TPA will recognize the server amid the procedure. This procedure is done for numerous clients for different data. So protection assurance is gotten to from numerous client security in distributed computing.

J. Kar2012 proposed "a provably secure on the web/disconnected total mark conspire in irregular prophet model" and this plan can be actualized on low processor gadgets, for example, savvy card, PDA and so on and WSNs applications. This scheme is based on identity-based signature using scour. It is used to detect the unauthorized user with no identity and in order to protect their content from the third party.

H. Jin et al in this paper, the data auditing of the web content based on their paper content and the title of the paper .we need to evaluate both content page and title to be related on the time of data evaluation to find the mismatched content. If any mismatched will find the server needs to give some alert messages to client system while the time of data uploaded on to the server. For these evaluations, we are using some arbitration protocol to finds the mismatched content on the web pages

S. g. work et alhe proposed to give a cryptanalysis in Work et al's. Conspire and demonstrate their plan has a security blemish. Misusing the security imperfection, a foe can discretionarily alter the cloud client's information while keeping away from the discovery. The formal security verification and execution investigation show that our plan is secure and as productive as work et al's. Plotted Security and performance analysis demonstrate that our scheme is secure and as efficient as work et al.'s schemed Security

J. Yu, K. Ren, C. Wang, and V.Varadharajan2015 proposed in that existed framework they are utilizing some multi-catchphrase pursuit to discover the client sought information. For that, they are taking the papers from the historical backdrop of hunt to effectively get the client intriguing data. The issue we have to process with a bigger measure of information. That is the reason we are executing our advantage models for each and every client with the assistance of semantic cosmology WorldNet to understand the client's need. To take care of the issue to take quite a while information handling.

3 PROPOSED SYSTEM

We utilize different security courses with a specific end goal to shield the information from third client different noxious assault. This plan is managed multilevel control in straightforward registering and it comprises of three gatherings: the client/translucent customer (TC), the verification server (AS) and the translucent Server (TS). In these, we start the client and straightforward customer as one section after the joined check between Authentication servers (AS), the Third Party Auditor (TPA), which is arranged amongst Tc and Ts.

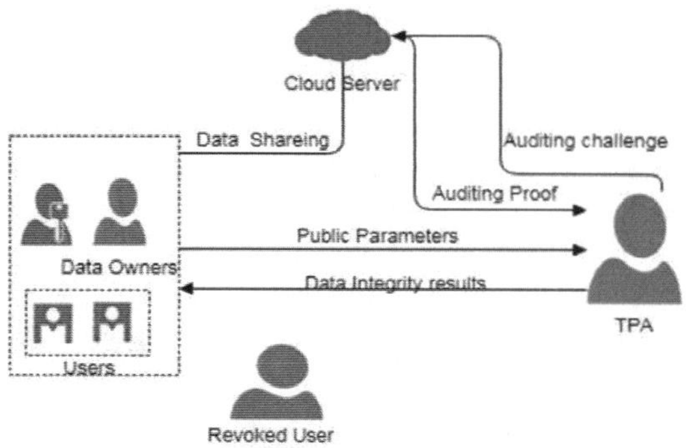

Fig.1 **Architecture Diagram for Multi Data Security.**

The main purpose of the authentication server is to verify the authorized individual and to check the read and write access permission to protect the data. Public information can be shared to all user, whereas the private/protected data are encrypted by the owner to protect the information from the revoked user. In order to view encrypted information, the third user request token to the owner and the owners respond to the specific user with the valid key. We distribute files/information into segments in order to provide fast and secure access transmission. The client can part the data either out in the open or ensured shape. Blunder location strategy to check whether the data gave by the client is applicable or not. The strategy utilized hub side level line mistake, edge side level line, hub side information lost blunder, edge information lost blunder, hub side too far out mistake, edge side beyond the field of play mistake, hub side spike mistake and edge side spike mistake.

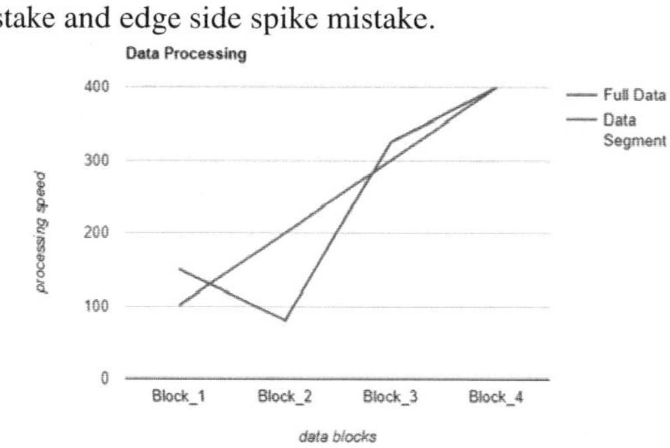

Fig2: Data Processing

4 TECHNIQUES PRELIMINARIES

4.1 MODULES

4.1 .1 Administrator

The administrator can view the new user details and give access to the user to upload the file in cloud storage. It can view all user details and the uploaded files.

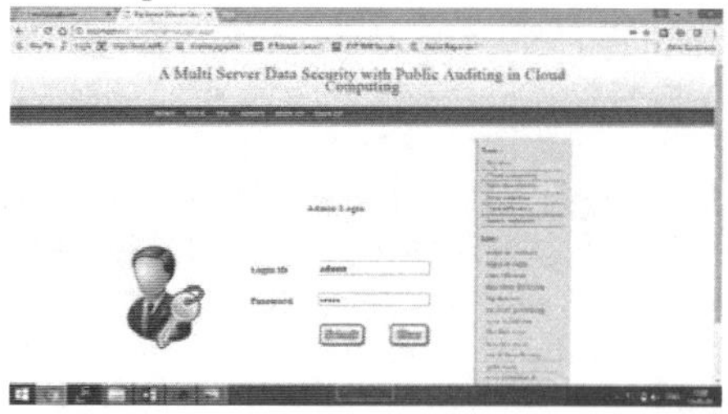

IOP Conf. Series: Materials Science and Engineering **590** (2019) 012007 doi:10.1088/1757-899X/590/1/012007

Fig 3 Administrator

User details and file details: The administrator view all the user and file detail. It can see the number of individual uploaded files. It creates a report based on uploaded file.

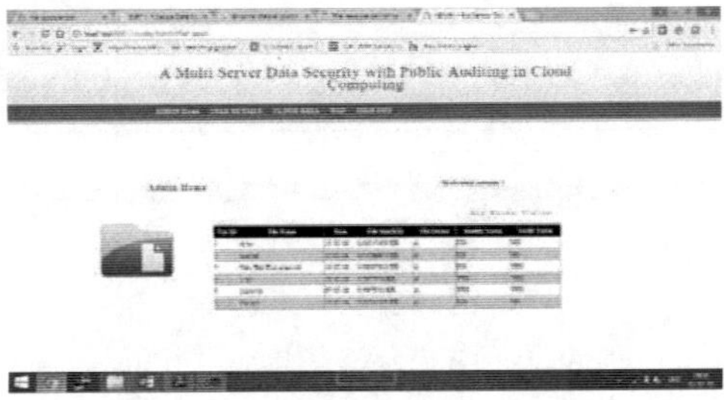

Fig 4 User details and file details.

4.1.2 Data Owner

New user: In this module, the administrator can view all new user details. It authorize the new user to upload the files. If the authorization is not processed by the admin, the process cannot be processed.

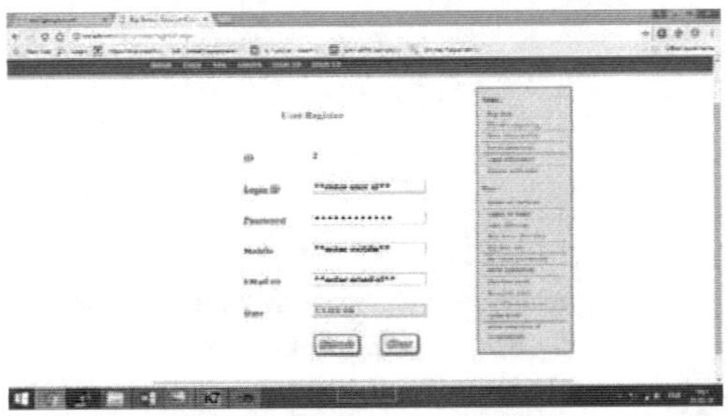

Fig 5 Data Owner(New User)

User phase:

In client stage, the new client movements to the login page where the client ought to enter the client name and secret word. The client gets a security enter with a specific end goal to process to the document transfer or download page.

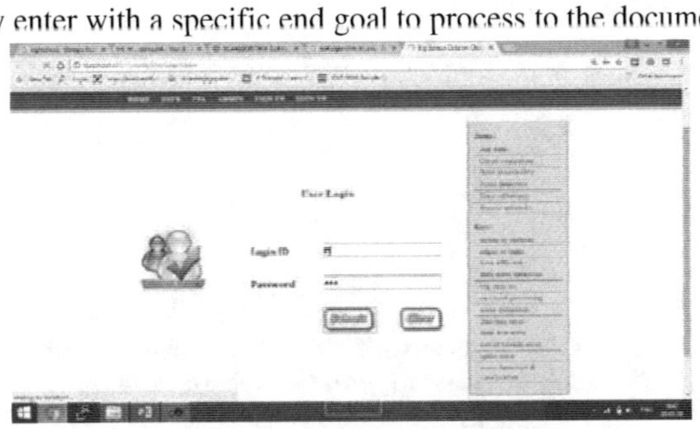

Fig 6 Data Owner (User phase)

4.1.3 File upload

The user module consists of file upload. In this module, the user can upload the file. While uploading, the user should share the secret key. The secret key and the file are positioned in encrypted form.

Fig 7 File Upload

4.1.4 File upload and append

If user want to download the file, the user must give the secret key. On the off chance that the client gives the right mystery key while downloading, the first record will be downloaded else the encoded document will be downloaded. On the off chance that the client needs to erase his transferred document, the client must give the mystery key else the client can't erase the client record.

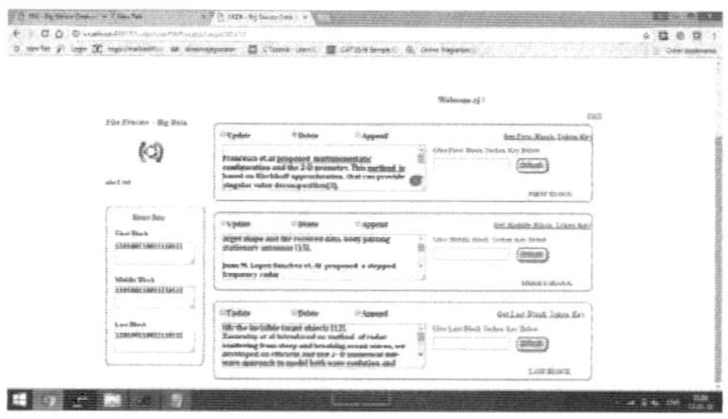

Fig 8 File upload and append

4.1.5 Authenticator Server Module

The main aim of the server is to be part of both user and cloud server. The server checks the valid user using a private key and if the key matches then he/she can access the check process to store the relevant information in the cloud storage. The server checks the information sent by the user is a valid information.

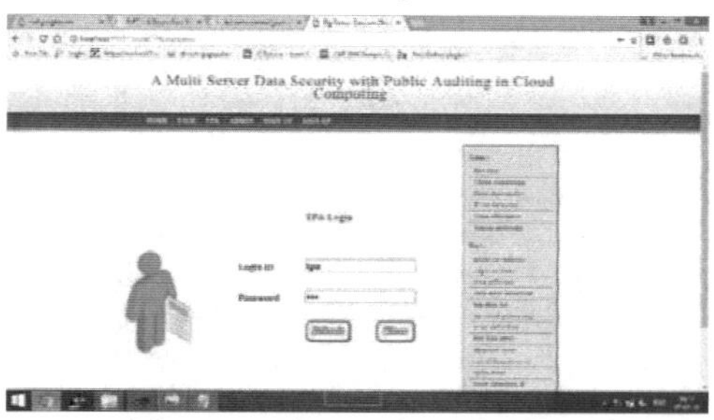

Fig 9 Authenticator Server Module

4.1.6 Error Detection

Cloud computing is an emerging technology which has vast improvement because it provides open, flexible, scalable, reconfigurable platform. The proposed error detection will be based on the nine types of numerical data abnormality/error are listed in our cloud error detection approach. The error detection process is trigger by defined error model. The contribution of our proposed detection is to achieve performance improvement in error detection without compromising.

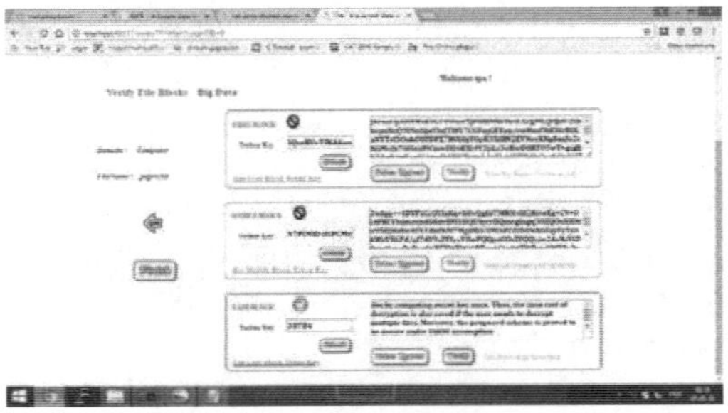

Fig 10 Authenticator Server Module

4.2 Algorithm

4.2.1 K-segmentation algorithm:

In segmentation algorithm, an example for describing this process we use dynamic programming. Firstly, let us consider a set of 'n' data points where data cannot be executed by least squares regression technique which is not possible by a single line data. Rather, we use the model for default line segments.To be more specific, we implement piece-wise regression for segmentation where ($1\leq k$) and ($k\leq n$) of n points. The least square techniques were repudiated because if (k=1), two degree of freedom is the best choice than one degree of freedom, and (k=n) does not fit the point as single line segments are retrieved for every single point.

Actually, the problem can be defined as:

Consider a set of n data points(x1,x2,.....xn) where a piecewise sequence is assigned to generate a line segment lsj where (j=1...k)with lsj=(sj,ej, pj).the line segment lsj case the expression initialize with xsj and end with xej. The overall standard of the fit is analyzed by reducing the total fit error.

$Es=\sum i=1\ nwi(x_i\text{-}x-i)2$

Where $x-i=b_j$ for sj $\leq i\leq e_j$ and wi is a measuring factor for each points.

4.2.2 Indexing and Merging Algorithm:

Indexing, an additional structure that accepts inverted list to access it randomly. In the part of merging techniques, there is a decrease in the requirement of disk space. Adding to it, the inverted list specify alternation of the structure and which is more significant for the collections of a vital document. We enhance merging techniques which are required to evaluate the data structure and the performance improvement. The inverted list that contains an entry for every works that which are in the stored document which consists of the document number and position information. It enables run length encoding and one pass compression techniques for better search and storage requirement. We implement an index structure as same as B-tree for the long list.

Run-length encoding and compression with a tree do not compress more space than a sequential list. The number of document starting from leaf node run-length. Dead spaceman leaf node terms document number. In every node, dead space should be represented. Unnecessary traversing of the tree should be eliminated. This technique implies that the area of the death of child node where the information stored in every non-leaf node,

{ Min; Max; ds; ref }

Where min and max is the minimum and a maximum number of subtree sourced by the ref, and ds is the dead space of the children node. The dead space is that the sum of the length extends from the node is not of the child node.

```
intsizeRemaining = (int)fiSource.Length;
for (inti = 0; i<OutputFiles; i++)
{
currPartPath = FolderOutputPath + "\\" + fiSource.Name + "."
+ String.Format(@"{0:D4}", i) + ".txt";
if (!File.Exists(currPartPath))
{
fsPart = newFileStream(currPartPath, FileMode.CreateNew);
sizeRemaining = (int)fiSource.Length-(i * partSize);
if (sizeRemaining<partSize)
{
partSize = sizeRemaining;
```

RC4 Algorithm:

The private key is sent to the client by the gathering supervisor in a scrambled arrangement utilizing RC4 calculation. The gathering supervisor plays out the underneath errands when a new client joins the gathering or a client has left the specific gathering,

- Update the entire client name list.
- Generate a safe key and scramble the key without activation and send to the updated user list.
- Update the rights in the cloud server.

```
Randomval = newRandom();
intrno = val.Next(12345, 54321);
cryptographykey = Convert.ToString(rno);
fileenccryptographykey=cr.Encrypt(cryptographykey);

splitfilesize = Convert.ToString(filesiz);
GenTocken();
Label16.Text = tocken;
Image6.ImageUrl = "Images/allowicon.png";
 TextBox3.Text=cr.Decrypt(TextBox3.Text);

TextBox3.ForeColor=System.Drawing.ColorTranslator.Fr
omHtml("#000000");
```

5 CONCLUSION

We enhance a method called multilevel access control in the transparent computing scheme to preserve user data with multiple security level. The main objectives maximum security for multi-data access, storage and processing in transparent computing also time efficient theme for locating and detecting error. In future works, we will enhance by initializing multiple authenticator server (AS) to resist traffic between the users (TS) and the server and to detect issues like error correction and recovery in cloud computing and to ensure the high accessibility of the system.

REFERENCES

1. C. Wang, S. S. M. Chow, Q. Wang, K. Ren, and W. Lou, “Privacy preserving public auditing for secure cloud storage,” IEEE Transactions on Computers 2013.
2. Albert Mayan.J, Surya.B, Pranoy Prabhakar, Prince kumar,” Department–Student Library Using Twig Pattern Query Processing Over Admin-User Login Privilege”, Pak. J. Biotechnol. Vol. 13 , Pp. 489 - 493,(2016).
3. D. He, S. Zeadally, and L. Wu, “Certificateless public auditing scheme for cloud-assisted wireless body area networks,” IEEESystems Journal, pp. 1–10, 2015.
4. Surya.V, J. Albert Mayan," A Secure Data Sharing Mechanism In Dynamic Cloud By Using KP-ABE", Research Journal of Pharmacy and Technology , Vol 10 , Issue 1 , pp:83-86,2017

5. H. Jin, H. Jiang, and K. Zhou, "Dynamic and public auditing with fair arbitration for cloud data," IEEE Transactions on Cloud Computing 2016.
6. Lakshmanan, L. and Shankar, (2015), "Dynamic cognitive system for recovering from vulnerable attacks in social network", International journal of Applied Engineering Research, RIP India, Vol. 10, No.4, pp.10365-10374, India, 2015, ISSN: 1087--1090.
7. C. Wang, S. S. M. Chow, Q. Wang, K. Ren, and W. Lou, "Privacy preserving public auditing for secure cloud storage," IEEE Transactions on Computers, vol. 2009.
8. Y. Zhang and Y. Zhou, "Transparent Computing: Spatio-temporal Extension onVon Neumann Architecture for Cloud Services," *Tsinghua Science andTechnology*, vol. 18, no. 1, 2013, pp. 10–21
9. Y. Zhang and Y. Zhou, "Transparent Computing: a New Paradigm for Pervasive Computing," *Ubiquitous Intelligence and Computing:2006 International Conf.*(UIC 06), 2006, pp. 1–11
10. G. Wang, Q. Liu, J.Wu, and M. Guo, "Hierarchical Attribute-based Encryption and Scalable User Revocation for Sharing Data in Cloud Servers," *Computers & Security*, vol. 30, no. 5, 2011, pp. 320–331.
11. Yovan Felix A , Jesudoss A ,Albert Mayan J (2017) ,"Entry and exit monitoring using license plate recognition ", IEEE International Conference on Smart Technologies and Management for Computing, Communication, Controls, Energy and Materials (ICSTM), pp.227-231.
12. Q. Liu, G. Wang, and J. Wu, "Time-based Proxy Re-encryption Scheme for Secure Data Sharing in a Cloud Environment," *Information Sciences*, vol. 258,2014, pp. 355–370.
13. Jabez J, Gowri S, Vigneshwari S, Albert Mayan J and Srinivasulu S (2018)," Anomaly Detection by Using CFS Subset and Neural Network with WEKA Tools", Smart Innovation, Systems and Technologies,Vol.107, pp. 675–682.
14. Kamalesh M. D., Albert Mayan J., Felix Y., Sumanth B. S., & Sai Tej B. (2018), " Magrisys: A Smart And Ubiquitous Controlled – Environment Agriculture System " , 2nd International Conference on I-SMAC (IoT in Social, Mobile, Analytics and Cloud) , pp: 1-5 , 2018
15. S. Muruganandam , N. Srinivasan ,"Appraisal of Felder - Silverman Learning Style Model with Discrete Data Sets",Indian Journal of Science and Technology, Vol 9(10), DOI: 10.17485/ijst/2016/v9i10/88992, 2016

IOP Conf. Series: Materials Science and Engineering **590** (2019) 012008 doi:10.1088/1757-899X/590/1/012008

A Survey Paper on Smart Parking System

B.Baron Sam [1], A.Lokesh Reddy[2], A.Prabhakar[3]

Assistant Professor[1],Student[2],Student[3]

baronsam1988@gmail.com [1],lokeshreddyjay@gmail.com[2], prabhakar.ankem@gmail.com[3]

Abstract. Start your abstract here… As there is a development in surge hour gridlock especially in the midst of the apex hours of the day, it is difficult to stop the vehicle in the halting opening. To handle this issue we are proposing sagacious halting system. So we are proposing a Shrewd ceasing system which enables the customer to find the nearest halting domain and gives openness of halting spaces in that individual ceasing zone. It prevalently revolve around diminishing the time in finding the halting openings and moreover it keeps up a vital separation from the unnecessary voyaging. Along these lines it decreases the fuel use. This system uses Cloud Computing and Internet of things

Keywords : Internet of Things; Cloud Computing; Ubidots ; Raspberry ; Cloud of Things

1. Introdution

The possibility of Internet of Things (IOT) started with things with identity particular contraptions. The contraptions could be taken after, controlled or checked using remote PCs related through Internet. IOT grows the use of Internet giving the correspondence, and thusly between arrangement of the contraptions and physical s, or Things. The two unmistakable words in IOT are web and things. Web infers a huge overall arrangement of related servers, PCs, tablets and mobiles using the universally used traditions and interfacing structures. Web engages sending, getting, or passing on of information. Thing in English has number of occupations and suggestions. Dictionary significance of Thing is a term used to reference to a physical inquiry, an action or thought, condition or activity.

IOT, all things considered includes between arrangement of the contraptions and physical , number of things can collect the data at remote regions and pass on to units directing, getting, orchestrating and analyzing the data in the methodology and organizations. It gives a fantasy where things end up sharp and carry on alive through recognizing, preparing and granting by embedded little contraptions which coordinate with remotes or individuals through system. The adaptable and solid nature of Cloud preparing is empowering architects to make and host their applications on it. Cloud goes about as a perfect assistant for IOT as it goes about as a phase where all the sensor data can be secured and got to from remote zones. These segments offered climb to the amalgamation of the two developments in along these lines provoking the game plan of another development called Cloud of Things (COT). In COT the things could be gotten to, checked and controlled from any remote region through the cloud. Due to high adaptability in cloud any number of center point could be

IOP Conf. Series: Materials Science and Engineering **590** (2019) 012008 doi:10.1088/1757-899X/590/1/012008

Incorporated or removed from the IOT structure reliably. In essential terms IOT can be illuminated in kind of a condition communicating: Physical Object + Controller, Sensor and Actuators + Internet = Internet of Things The ideal of making a Smart City is by and by getting the opportunity to be possible with the ascent of the Internet of Things. One of the key issues that sharp urban zones relate to are auto ceasing workplaces and development organization systems. In present day urban territories finding an open parking spot is continually troublesome for drivers, and it tends to wind up harder with reliably growing number of private auto customers. This situation can be seen as an open entryway for splendid urban groups to endeavour exercises all together update the profitability their halting resources thusly provoking diminishing in looking conditions, development obstruct and road setbacks.

Issues identifying with ceasing and movement stop up can be lit up if the drivers can be taught early about the availability of parking spaces at and around their normal objective. Late advances in trying, low-control introduced systems are helping originators to manufacture new applications for Internet of Things. Taken after by the headways in sensor advancement, various forefront urban groups have settled on passing on various IOT based systems in and around the urban territories to screen. A present report performed by the International Parking Institute reflects a development in number of innovative considerations related to ceasing structures. At appear there are certain ceasing structures that claim to occupants of passing on steady information about open parking spaces. Such structures require beneficial sensors to be sent in the ceasing zones for checking the inhabitance and also smart data planning units remembering the true objective to increment conventional bits of learning from data assembled over various sources. The shrewd ceasing structure that we propose is executed using a versatile application that is related with the cloud. The system empowers a customer to know the availability of parking spaces on a continuous start

2. Related works

Rico J et.al[1] involves generally on area sending of IOT respond in due order regarding screen and signalize the state of availability of each single parking space. This framework enhances stopping assets by open specialists. It utilizes information gathering for accessibility of stopping openings. It decreases time, contamination.

Zheng Y et.al[2] created utilizing sensor circuit, RFID and IOT. This framework encourages client to discover parking spot accessibility with help of IOT. RFID is utilized for burglary administration. Less human association, adaptability, security. Employable in air terminals, multiplex stopping.

Zhou et.al[3] here they have developed a unique course of action by using ZigBee Technology .They fabricated a system that use remote advancement of ZigBee

IPI et.al[6] built up a strategy for productive stopping is proficient lighting, direction in discovering stopping opening simpler, prebooking process, sustainable power source innovation and obliging electric vehicles.

Fastpark et.al[8] relies upon solitary halting sensors presented in each ceasing straight. The sensors control drivers by methods for electronic dynamic message signs (DMS). Once a vehicle stops over a sensor it is recognized and sensor remotely exchanges data to passage .Occupied spaces are noticeable by methods for an application through LED (Dynamic message signs).

Viswanath Y et.al[9] presents an estimation that extends capability of current cloud based sharp ceasing structure and develops a framework designing in perspective of IOT and finds out customer halting expense by considering the partition and total number of free places in each auto stop

IOP Conf. Series: Materials Science and Engineering **590** (2019) 012008 doi:10.1088/1757-899X/590/1/012008

Aishwarya et.al[10] built up an appropriate most brief way calculation is utilized to locate the base separation amongst client and every auto stop in the framework. Uses RFID innovation which is utilized as a part of robbery administration purposes. Also built up an android application which helps for the collaboration amongst client and parking spot

3. Proposed System

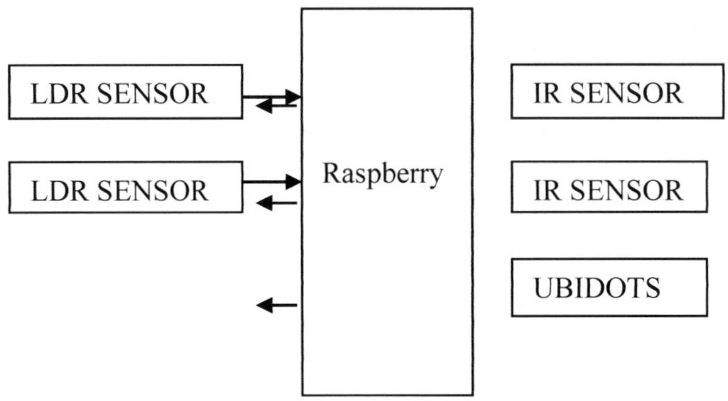

Fig 1: Block Diagram

The proposed framework utilizes cloud and IOT. Hardware prerequisites are IR sensor, LDR sensor Raspberry PI, Programming prerequisites are python.

3.1 Identification Of Vehicles Using IR Sensor

An infrared sensor is an electronic gadget, that emanates so as to detect a few parts of the environment. The auto stopping framework having the IR sensor to recognize the vehicles check and send it to the mail in the passage of the auto stopping opening. At whatever point the vehicle is identified, the unfilled space will be shown via the post office and The Car stopping accessibility will be demonstrated utilizing LED's insinuation.

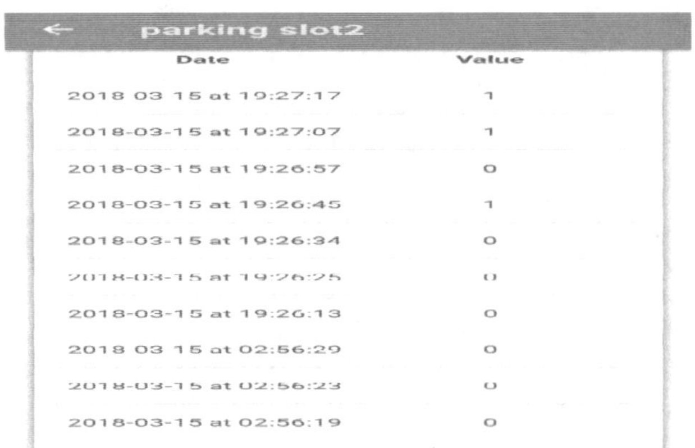

Fig2: Vehicles count in parking slot2

IOP Conf. Series: Materials Science and Engineering **590** (2019) 012008 doi:10.1088/1757-899X/590/1/012008

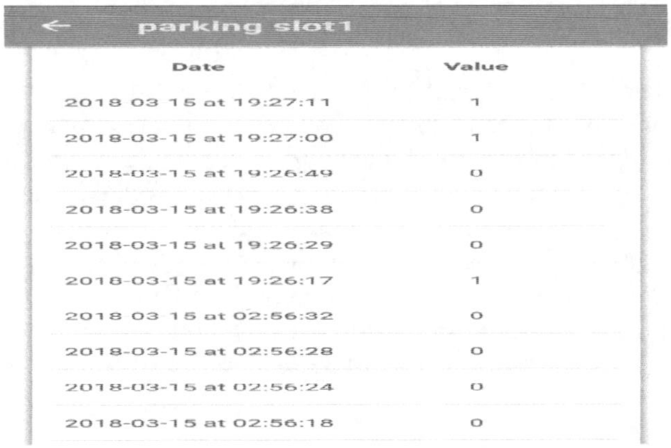

Date	Value
2018 03 15 at 19:27:11	1
2018-03-15 at 19:27:00	1
2018-03-15 at 19:26:49	0
2018-03-15 at 19:26:38	0
2018-03-15 at 19:26:29	0
2018-03-15 at 19:26:17	1
2018 03 15 at 02:56:32	0
2018-03-15 at 02:56:28	0
2018-03-15 at 02:56:24	0
2018-03-15 at 02:56:18	0

Fig 3 Vehicles count in parking slot 1

3.2 Detection of Light Using LDR Sensor

A Light Dependent Resistor (LDR) is similarly called a photo resistor or a cadmium sulfide (CdS) cell. When the vehicle is moving, we can prepared to switch on the light normally then ensuing to leaving the vehicle we can switch of the vehicle. At the day time, LDR sensor recognize thusly, it will switch of the light, by then at the night time, we can prepared to switch on the light. At the point when light are having any bother, we can sent the message to the organization of the different spots and besides we can prepared to screen the division between back of the vehicle and halting spot

3.3 Ubidots

In Ubidots IOT arrange we will make a GUI Interface to work the gateway which moreover contains sensors working. The whole sensors data is secured by time to time in our login in the ubidots and it can be found in it.

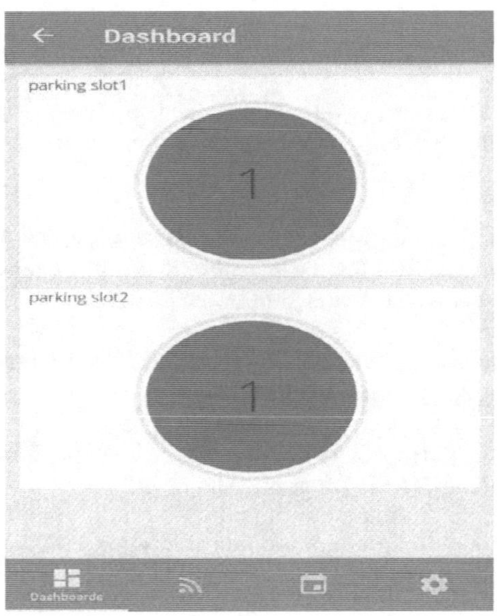

Fig 3: Ubidots Platform

CONCLUSION

Smart Cities have constantly been a dream for humankind. Since the current years broad movements have been made in making adroit urban groups a reality. The improvement of Internet of Things and Cloud progresses have offer climb to new potential results similarly as keen urban groups. Sharp ceasing workplaces and movement organization systems have reliably been at the focal point of creating shrewd urban groups. In this paper, we address the issue of ceasing and present an IOT based Cloud facilitated sharp halting structure. The system that we propose gives continuous information as for availability of halting openings in a ceasing an area. Customers from remote regions could book a halting opening for them by the usage of our flexible application. The undertakings made in this paper are indented to upgrade the ceasing workplaces of a city and thusly wanting to enhance the individual fulfilment of its family.

References

[1] Rico, J., Sancho, J., Cendon, B., and Camus, M. (2013, March). Stopping less demanding by utilizing setting data of a keen city: Enabling quick pursuit and administration of stopping assets. In Advanced Information Networking and Applications Workshops (WAINA), 2013 27th International Conference on (pp. 1380-1385). IEEE.

[2] Zheng, Y., Rajasegarar, S., and Leckie, C. (2015, April). Stopping accessibility forecast for sensor-empowered auto stops in keen urban areas. In Intelligent Sensors, Sensor Networks and Information Processing (ISSNIP), 2015 IEEE Tenth International Conference on (pp. 1-6). IEEE.

[3] Yovan Felix A , Jesudoss A ,Albert Mayan J (2017) ,"Entry and exit monitoring using license plate recognition ", IEEE International Conference on Smart Technologies and Management for Computing, Communication, Controls, Energy and Materials (ICSTM), pp.227-231.

[4] Zhou, F., and Li, Q. ,“ Stopping Guidance System Based on ZigBee and Geomagnetic Sensor Technology. In Distributed Computing and Applications to Business, Engineering and Science (DCABES), 2014 thirteenth International Symposium on (pp. 268-271). IEEE.

[5] Botta, A., de Donato, W., Persico, V., and Pescapé, A. ,” On the Integration of Cloud Computing and Internet of Things. In Future Internet of Things and Cloud (FiCloud), pp. 23-30, IEEE,2014.

[6] Asha P, Albert Mayan J, Canessane A (2018),"Efficient Mining of Positive and Negative Itemsets Using K-Means Clustering to Access the Risk of Cancer Patients",Communications in Computer and Information Science ,ICSCS 2018, Kollam, India, 2018,pp.373-382.

[7] Ballon, P., Glidden, J., Kranas, P., Menychtas, A., Ruston, S., and Van Der Graaf, S. (2011, October). Is there a Need for a Cloud Platform for European Smart Cities?. In eChallenges e-2011 Conference Proceedings, IIMC International Information Management Corporation.

[8] Fox, G. C., Kamburugamuve, S., and Hartman, R. D. (2012, May). Engineering and estimated attributes of an International Conference on (pp. 329-333). IEEE.

[9] Han, D. M., and Lim, J. H. (2010). Savvy home vitality administration framework utilizing IEEE 802.15. 4 and ZigBee. Purchaser Electronics, IEEE Transactions on, 56(3),

[10] Parwekar, P. (2011, September). From Internet of Things towards billow of things. In Computer and Communication Technology (ICCCT), 2011 second International Conference on (pp. 329-333). IEEE.

[11] Anish Hamlin M R, Albert Mayan J,"Blood donation and life saver-blood donation app",International Conference on Control, Instrumentation, Communication and Computational Technologies ,pp. 625 - 628, 2016.

[12] Suciu, Vulpe, Halunga, Fratu, O., Todoran, G., and Suciu, V. (2013, May). Keen urban areas based on versatile distributed computing and secure web of things. In Control Systems and Computer Science (CSCS), 2013 nineteenth International Conference on (pp. 513-518). IEEE

[13] B.Baron Sam, A.V.K.Shanthi, "Query Oriented User Profile Management In Web Search Engine", International Journal of Applied Engineering Research, ISSN 0973-4562 Volume 10, Number 76(2015) Special Issues, pp-295-300, IF 0.015, Snip 0.034, SJR 0.13

[14] B. Baron Sam , S. Balaji , A. Anthuvan Jerald Majella , K. Ashokkumar, "Using Mobile Robots To Act As Surveillance In The Crop Field", International Journal of Applied Engineering Research ISSN 0973-4562 Volume 10, Number 6 (2015) pp. 15825-15832

[15] Maria Anu V, Aroul Canessane R, Mary Gladence L, "Optimization of query processing time using Taguchi method for RFID data management", Indian Journal of Science and Technology,Vol 9(25), DOI: 10.17485/ijst/2016/v9i25/95141, July 2016.

[16] S.Dhamodaran, Dr.M.Lakshmi (2017), "Design and analysis of spatial-temporal model using hydrological techniques", International Conference on Computation of Power, Energy Information and Commuincation (ICCPEIC),pp. 023 - 028.

[17] S.Karthikeyan, S.Jayashri, "Proficient Data Gathering by Mobile Agent using Jade",i-manager's Journal on wireless communication networks,Vol.1, No. 2, 2012, pp. 32-40.

A PROFICIENT DDoS OVERFLOW ATTACK DISCOVERY AND AVOIDANCE SCHEME

A. Jesudoss, H. Manoj Kumar, and Manchala L. Sai Charan Reddy

Dept. of CSE, School of Computing, Sathyabama Institute of Science and Technology, Chennai, Tamil Nadu, India

jesudossas@gmail.com, manojkumarxii345@gmail.com, manchalasaicharanreddy@gmail.com

Abstract. The DDoS(distributed denial-of-service) attack happens when several clients seek the web server to a particular system which may lead to overflowing with abnormal traffic. The proxy server which measures the traffic on the server and classifies the attacks with floods. The current problem faced by web servers is that the anonymous user enters and done some malicious tasks, causes the server crashing. We provide a solution to blacklist and restrict the access. If the anonymous user tries to access the server, the mechanism will check with the backlist data and if it exists means it will block and provide a message to us by sending an alert to the server.

Keywords- Intrusion detection system (IDS), proxy server, DDoS, Network Anomaly Detection.

1. Introduction

The differences between dos attack and DDoS attack stands for a denial of service attack which means a single assailant assaulting a single target. This implies that one computer and one web the association is utilized to surge the server with bundles, the aggressor sends an expansive number of legitimate-looking demands to the server in a way that the server cannot truly recognize between substantial and non-valid demands. It will overpower the framework to a point that the server cannot handle the capacity anymore. This will make the server inaccessible to others. In another aspect, the DDoS attack requires may systems to done the attack behind the server. To begin with, the step of propelling a DDoS assault is to select an armed force of bots in arrange to turn a computer into a bot. the assailant creates specialized malware which spreads to ants numerous helpless computers as conceivable malware can spread by means of compromised websites email connection or through an organization's organize clients who were deceived to running such malware intentioned turn their computer into a bot and give get to focuses to the assailants to their computers.

Once a computer turns into a bot it interfacing to the aggressors command and control server and begins to recognize orders from the command and control server joins the course for moving a trap from the bots malware to specific target and the chosen assaulting strategy an outfitted constrain of ways is named as botnet and as a run the show comprises of thousands of bots any time the botnet proprietors require to alacrity an attack they send messages to the server with the instruction to perform the attack on particular target any sullied machine in the botnet will comply by moving all encouraged will time passed on ambush known as a DDoS attack. The primary contrast between dos and DDoS assault is that there's as it were one assailant in dos assault but in a DDoS assault, the target server will be over-burden by hundreds or indeed thousands of demands, hence, it is much harder for the server. In our mechanism, we introducing an alternative server called proxy. A proxy server acts as a mediator between the client and the network. A DDoS attack which tries to send more network traffic to the internet, which the system can't process it. Proxy detects abnormal traffic or any load balancing problems. Here the attack classifier classifies the attacks and made a blacklist for the user to restrict and secure the server safe.

2. Related Works

According to Qiao Yan (et.al) [1], The concept of SDN has been introduced to defeat the DDoS assaults in the cloud environment. SDN can significantly enhance cloud sensibility, adaptability, controllability and dynamism. But cloud computing would not be possible without the underneath support of networking.

The author Wei Wei (et.al) [2] says there are many controlled hosts that flood the victim's site with massive packages. Distributed Reflection DDoS (DRDoS) will trick innocent servers (reflectors) into hasty packages for the victim.The Rank Correlation based Detection (RCD) calculation is proposed to separate reflection streams from authentic ones productively and viable. In RCD,compute cost will be affected. EAM is a simple and single solution scheme for providing security to banking and finanacial applications where monetary transactions are done [3]. The digital certificates enable the server and client to protect against various types of security attacks. The countermeasures for these security attacks are discussed in Certificate-based Authentication work [4]. The work proposed by lakshmanan shows how the applications can be provided more security and also to improve the performance by resolving issues [5,6].

The author Paulo E. Ayres et al [7] says packet score has been implemented to detect the DDoS assaults which differentiate from packet scoring. Alpi, a new system that extends the concept of scoring packages with time reduction and improved performance. It is a proactive defense system by nature, able to confirm and block previous unseen attacks, and filters DDoS attacks based on the packet dialing approach.

The author Angelos D. Keromytis et al [8] says use simple analytical models, evaluate the probability a host may launch a DDoS attack against a protected SOS network.Run-time estimated by using a usage model show an end-to-end inertial expansion of a factor of two for the general case. A case is a database that maintains good or secret data, for example, construction structure reports, information, task updates or vital data.

The author Xiao Feng Wang et al [9] conveys a proposed way to use the website's graphics structure to alleviate flood attacks on a site, the WRAPS concept (preferred service web reference architecture) was introduced. It won't need any modification of the web client software and is extremely light for the reference sites, which facilitates its implementation. This mechanism gives you at a minimum cost and helps clients retrieve the internet reference relationships.

The Author Ying Xuan et al [10] says about the group technique to analyse the attack at backend. For this new group test (GT) approach implemented in back-end servers, it offers a practical method to receive a cut detection delay and a low false positive / negative rate. Implementation issues and performance improvements are also available to see their potential.

The author Khaled Salah et al [11] says about the predicting the overview of the firewall, which is crucial for network security engineers. It's effectiveness and resistance is explained. If network firewalls are poorly designed to support DDoS attacks, the overall security of the protected network will be compromised.

The author Wanchun Dou et al [12] says the study about the period of attack and the period without attack. Collect legitimate packages in the period without attack, retrieving pairs of attributes to recreate the data file. It has a high speed of marking, but it has a small storage size and this is the main disadvantage for an acceptable filtering accuracy.

According to Sanjeev Khanna et al [13] ASV method has emerged to use data transmission productively and won't need any server status or presumption of system lock. This can require a critical state of the server and is powerless against enemies that can cause real blue clients to back off while attackers ignore rejections.

According to Rodrigo Braga et al [14] The major network security problems during the last decade, especially threatening public web servers. This paper presents easy method to detect DDoS attacks based on the characteristics of the traffic flow, decrypted information is carried out with a very low overload compared to traditional approaches.

3. Proposed System

The Proposed Framework has "Welcome the Assaults" system. Usage of this strategy gives a fruitful response for increment the security and faithful nature of the framework. The method for sending the proposition to the host recognizes the development as an assault on the server and it's at that point directions to a cache server. Utilizing a system irregularity recognition, calculation (Nothing) to identify the unusual movement in the stream. This investigation will anticipate the blunder and distinguish the different assaults. Our calculation can identify an oddity caused either by real movement or by flooding assaults. We can secure our framework by recognizing the assault introduced in that data which was sent by an accessing user and the sort of assault will be boycotted for the future utilization will personality simultaneously uninformed that they are not utilizing real server.

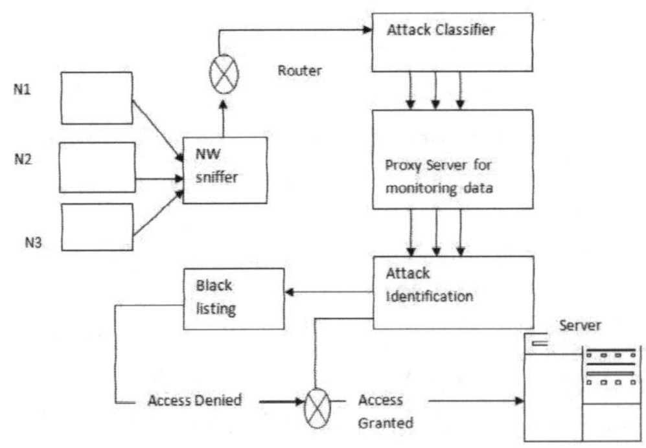

Fig. 1 System Architecture

In the proposed system, the user basis can be a host based or a network based system, as the interaction is basically performed in a network. The solution to the specific network intrusion is fully automated and the problem is addressed by the Secure Direct method as this can eliminate the process of decision making which is done by humans and the response times by humans is thus reduced. Thus a digital record of time for each connection is defined. The time stamp for each connection on receiving the packet is updated. Then, the powerful detection tool namely the Network Sniffer which can be used to intercept a network, log onto a network and can also be used to dissolve the traffic in a network is processed. The valid data can be picked using the Network Sniffer. The hackers and the attackers mainly use the effective method known as the Sniffing, as it is a passive attack which is difficult to detect and the process is done as if it is not seen nor heard.Next the router is defined as a device that forwards data packets along the networks.In arrange for the working of a switch, it has to be associated to the slightest of two systems, to be specific two LAN's or WAN's or a LAN and its ISP arrange. It works as a door where there is an association between two or more systems. To decide the finest way for sending bundles, the switches utilize the headers and the sending tables.

The communication of the packets and the setup of the most excellent course between any two has is done by the utilize of the conventions named as ICMP. The filtering of data done through the routers is very little.In Attack Classifier, the main intention is to redirect the consideration of the assailant from the genuine organize, in a way that the fundamental data assets are not compromised. To construct aggressor profiles in arrange to distinguish their favored assault strategies comparative to criminal profiles utilized by law authorization offices in arrange to recognize a criminal's modus operandi. In get to remittance, stipend pattern which is fueled by insights along with the plan of assault classifier. The yield created by the classifier creates an energetic list of assaults, which are at that point lined in the proposed intermediary server built with a neural arrange to get it the different approach to conduct and designs of the assailant. The organize chairman collects all such significant data over the arrange itself permitting the inbound organize association from the assailant to do so. The framework makes a crossover system to anticipate the likelihood of helpless and unfriendly circumstance over the organize indeed some time recently the assault occasion is performed by the aggressor. In run the show creation, the energetic run the show creation instrument exceptionally effortlessly a suspicious interloper and interruptions can be recognized based on the conduct and setting boycotting of the resource/host/IP/network can be done without numerous overheads.The final process is blacklisting which blacklists to access control. There is a mechanism which allows everyone to access the server and denies for the blacklist users. The whitelist which allows the whitelist members only. The grey list which notices the temporary accepted and blocked the incoming connection. These grey list can be reviewed for future white and blacklists.

4. Detection and Prevention

Host-Based: HIDS put on a single have and screens the framework records and OS. It looks for any evil-minded or doubtful exercises in its discrete have and employments the mailing time to conclude the nearness of non-essential exercises. We present a caution for the movement to hint each time. Signature-based and event-based methods are taken after through the handle. It is put in numerous places inside the organize to screen the activity. For case working environments, individual computers, mobiles. Prevention ways: Arrange conduct is dissected utilizing the irregularity discovery strategy. By making an elective server called an intermediary, we can ensure our unique server.

Network-Based: It evaluates the data captured from the internet and analyze the packets which travel across the network. It is placed in many places within a network to monitor the traffic. For example workplaces, personal computers, mobiles and many more. It will perform examination on the activity as it is flown around the arrange and matches the activity to a list of known assaults. At that point the director can explore it advance too they can be arranged by the chairman to characterize an ordinary organize such as typical activity stack, typical bundle estimate.

IOP Conf. Series: Materials Science and Engineering **590** (2019) 012009 doi:10.1088/1757-899X/590/1/012009

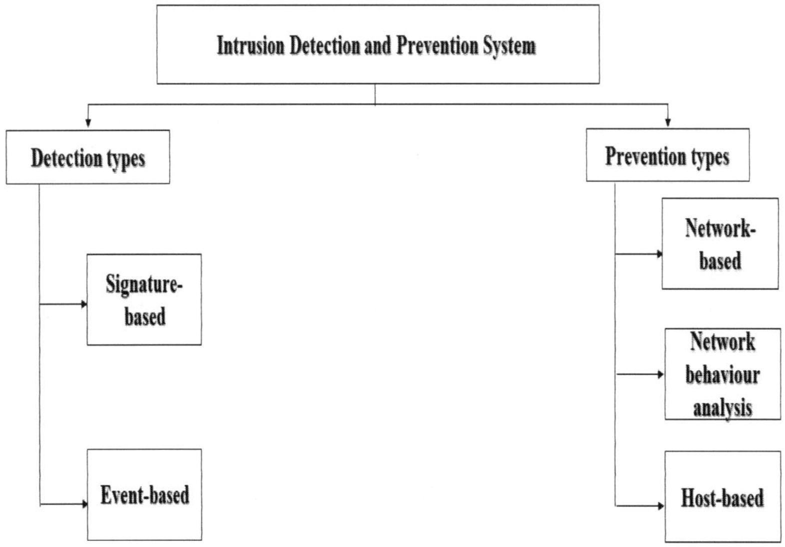

Fig. 2 Classification of Intrusion Detection and Prevention Systems

5. Chaos Theory

Chaos hypothesis is a hypothesis whose primary preface was that little changes in introductory conditions could result in tremendous contrasts in the last yield. In this venture, utilizing chaos we made a little alter in the arrange could allow distant way better comes about than before.

5.1 Pseudo Code for Chaos Theory

```
procedure Protocol Analyzer
if diff = k then
rate= packetcount/k
break
if number of packets > Ω then
while End of File not reached do
if packet= ARP then
arp++
else if packet=ICMP then
icmp++
else if packet=TCP then
tcp++
else if packet=UDP then
udp++
else if packet=ACK then
ack++
end if
else
packetcount++
Continue
end if
end procedure
```

IOP Conf. Series: Materials Science and Engineering **590** (2019) 012009 doi:10.1088/1757-899X/590/1/012009

5.1.1 Network Anomaly Detection Algorithm

Network anomaly detection algorithm is proposed to monitor the behaviour and traffic of the network. It is a multi-order, multi-point of comparison and multi-level of IP aggregation approach. It has two points. One is to detect and typecast the traffic inconsistency, the second phase targets the inconsistency flow by fully typecasting them. In this paper, According to the Claffyetal flow definition, traffic inconsistency is done at the network layers called TCP/IP stack using packet flow data.

$$N = \{a_1, a_2, ..., a_n\}, a_i = \{\# packets | \# byts | \# flows\} / \Delta$$

$$Q = \{b_1, b_2, ..., b_{n-1}\}, Q_i = x_{i+1} - x_i$$

$$\{ \; qi >= E(q) + k\sigma, select$$

$$\{ \; qi < E(q) + k\sigma, reject$$

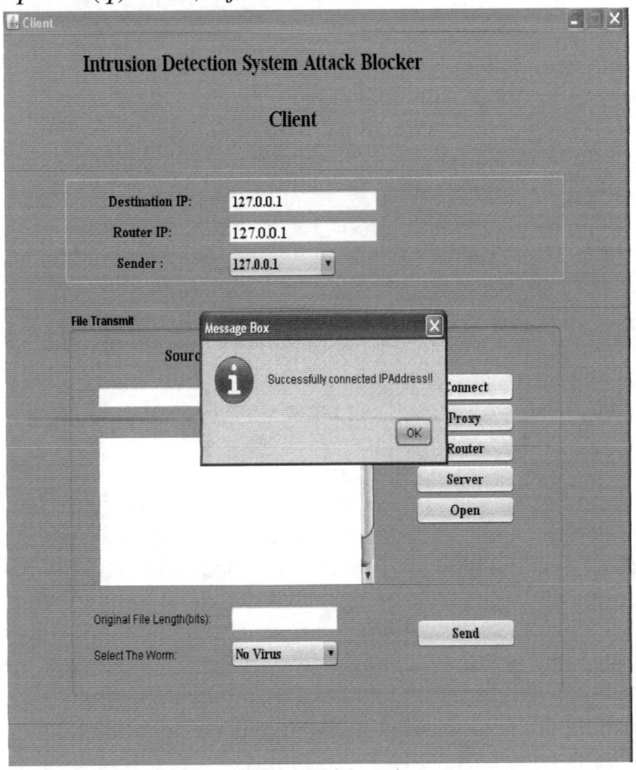

Fig. 3 Client Side Diagram

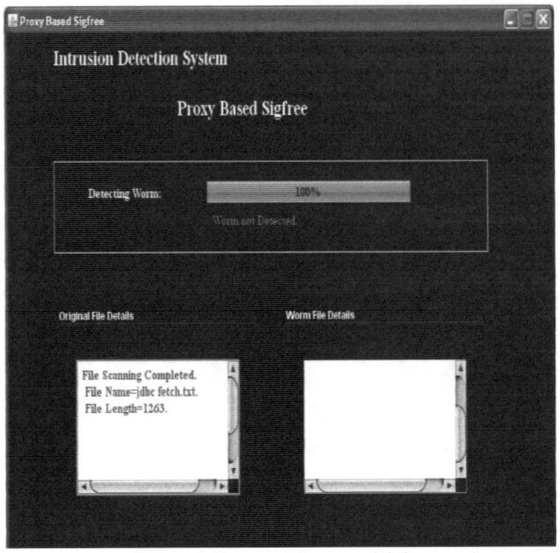

Fig. 4 Proxy Server Side

6. Conclusion & Future Enhancement

In this exploration work, we have proposed a framework where the system manager will watch and examine the different manners of assaulting inclinations starting from a various source in arranging. The procedure fundamentally comprehend the example and conduct of the antagonistic conditions over the system and afterward it makes the profiles of the assailants in view of this example investigation, which will ensure the system arrangement of the association by boycotting the beginning of the asset profiling over the system itself in the manner of proxy and guaranteeing the system to be the most secure one in any future like hood of system dangers from those aggressors.

In this work, we have depicted a portion of the past endeavours to quantify IDS, and we have laid out a portion of the troubles that have been experienced. We trust that an intermittent, far-reaching assessment of IDSs could be important for arrange directors, data security officers and information supervisors. Be that as it may, in light of the fact that both ordinary and assault activity are so factor from site to site, and that typical and assault movement develop after some time Solving the issue of high accessibility and security all the while offers the open door for more unwavering quality than frameworks which tackle the issues independently, notwithstanding being simpler to execute and offering expanded open door for recording and information investigation. The coordination of these two advancements, in any case, is a non-insignificant undertaking. A fine adjust must be accomplished amongst speed and strength of IDS highlights – it is similarly terrible to have the web server crash in fact that an assault was missed, or drop a lot of activity because of an excessively through IDS turning into a bottleneck.

REFERENCES:

[1] Q. Yan, F. R. Yu, Q. Gong, and J. Li, "Software-defined networking (SDN) and distributed denial of service (DDoS) attacks in cloud computing environments: A survey, some research issues, and challenges," IEEE Communications Surveys & Tutorials, vol. 18, pp. 602-622, 2016.
[2] Yovan Felix A , Jesudoss A ,Albert Mayan J , "Entry and exit monitoring using license plate recognition ", IEEE International Conference on Smart Technologies and Management for Computing, Communication, Controls, Energy and Materials (ICSTM), pp.227-231,2017.

IOP Conf. Series: Materials Science and Engineering **590** (2019) 012009 doi:10.1088/1757-899X/590/1/012009

[3] Jesudoss A., and Subramaniam, N.P., "EAM: Architecting Efficient Authentication Model for Internet Security using Image-Based One Time Password Technique", Indian Journal of Science and Technology, Vol. 9 Issue 7, Feb. 2016, pp. 1-6.

[4] Jesudoss A. and Subramaniam N.P., "Enhanced Certificate-Based Authentication for Distributed Environment", Springer Advances in Intelligent Systems and Computing, Vol. 325, 2015, pp. 675-685.

[5]Lakshmanan, L. and Tomar, D.C. (2015), "Analysis of Shortest Route for heterogeneous node in Wireless Sensor Network" in the proceeding of the Artificial Intelligence and Evolutionary Algorithms in Engineering Systems, Advances in Intelligent Systems and Computing Series,Vol.325, pp.815-821,Springer.

[6]Lakshmanan, L. and Tomar, D.C. (2014), "Optimizing Localization Route Using Particle Swarm-A Genetic Approach", American Journal of Applied Science(AJAS), Science Publication, Vol. 11, No. 3,pp.520-527,USA, 2014, ISSN: 1546-9239.

[7] P. E. Ayres, H. Sun, H. J. Chao, and W. C. Lau, "Alpi: A DdoS defence system for high-speed networks," IEEE Journal on Selected Areas in Communications, vol. 24, pp. 1864-1876, 2006.

[8] D. Keromytis, V. Misra, and D. Rubenstein, "SOS: An architecture for mitigating DDoS attacks," IEEE Journal on selected areas in communications, vol. 22, pp. 176-188, 2004.

[9]X. Wang and M. K. Reiter, "Using web-referral architectures to mitigate denial-of-service threats," IEEE Transactions on Dependable and secure computing, vol. 7, pp. 203-216, 2010.

[10] W. Wei, F. Chen, Y. Xia, and G. Jin, "A rank correlation based detection against distributed reflection DoS attacks," IEEE Communications Letters, vol. 17, pp. 173-175, 2013.

[11] Albert Mayan J, Kuldeep Anand D.S, Neha Sadhvi,"Efficient and secure server migration on cloud storage with VSM and dropbox services", International Conference on Information Communication and Embedded Systems (ICICES),Chennai pp. 1-,2017.

[12] Y. Xuan, I. Shin, M. T. Thai, and T. Znati, "Detecting application denial-of-service attacks: A group-testing-based approach," IEEE Transactions on parallel and distributed systems, vol. 21, pp. 1203-1216, 2010.

[13] K. Salah, K. Elbadawi, and R. Boutaba, "Performance modelling and analysis of network firewalls," IEEE Transactions on network and service management, vol. 9, pp. 12-21, 2012.

[14] W. Dou, Q. Chen, and J. Chen, "A confidence-based filtering method for DDoS attack defence ina cloud environment," Future Generation Computer Systems, vol. 29, pp. 1838-1850, 2013.

[15] S. Khanna, S. S. Venkatesh, O. Fatemieh, F. Khan, and C. A. Gunter, "Adaptive selective verification: An efficient adaptive countermeasure to thwart dos attacks," IEEE/ACM Transactions On Networking, vol. 20, pp. 715-728, 2012.

[16] R. Braga, E. Mota, and A. Passito, "Lightweight DDoS flooding attack detection using NOX/OpenFlow," in Local Computer Networks (LCN), 2010 IEEE 35th Conference on, 2010, pp. 408-415

QR CODE AND BIOMETRIC BASED AUTHENTICATION SYSTEM FOR TRAINS

Muthukumar B[1]
Professor , Department of Computer Science & Engineering
Chennai Institute of Technology,Chennai, India
muthukumarb@citchennai.net

Albert Mayan J[2]
Assistant Professor , Department of Computer Science & Engineering
Sathyabama Institute of Science and Technology , Chennai, India
albertmayan@gmail.com

Gokul Nambiar[3]**, Denil Nair**[4]
U.G Students, Department of Computer Science & Engineering
Sathyabama Institute of Science and Technology, Chennai, India
gokulcn1996@gmail.com,denil007nair@gmail.com

ABSTRACT :The utility of a QR Code, is plenty in terms of usage to identify a particular entity. They are barcode of sorts that can hold a certain data set. Biometric security measures are one of the safest security measures, with fingerprint based biometric security being a rather easy and efficient way. In this paper, we discuss a way to efficiently check-in to a seat, during train travel by using a QR Code and fingerprint-based authentication system. A Quick Response Code (QR Code) can be used to mark one's attendance at their seat. Fingerprint authentication is the means to authenticate a person's identity without having to carry any form of physical identity proof along with them, during travel. Given the present system to check tickets is rather redundant and time consuming, having to manually check the ticket and identity of each passenger. By using this system, we can create a more efficient and less time-consuming system, that has much lesser probabilities of human error to occur.

Keyword : QR Code , Biometric, Ticketing System,Smart Phone Application

1. INTRODUCTION

The current train ticket checking system comprises of many challenges that is the tickets must be checked manually by the conductor of the train, the passenger has to carry a physical identification proof of theirs, in their person, without which they cannot travel. The conductor of the train takes care of the entire process of checking tickets, and managing any vacancies of a seat that may take place in the course of the journey. In case of any contingencies during the travel, there is no alert system present, that can notify the respective passenger about the same. If any confirmed passenger does not board the train for the journey, their seat has to be manually reallocated to any passenger who may request for the seat, without either of the party getting any notification of the same.

With this paper, we intend on creating a system that allows for the passenger to check themselves into their seat, without having to check with the conductor. This checking in is achieved using a QR Code.Thisis more of a square barcode, that can be easily generated to and fed a certain data, which remains stored within the code. A QR Code can be easily scanned using any smartphone camera. On doing so, the values that are already stored in the code are retrieved and is received on the device used to scan the code. Instead of carrying an identity proof, the passenger can simply scan their finger on a fingerprint reader, in their coach. Fingerprint is one of many Biometric security measures that are available today. Others include retinal scans, DNA based scanning etc. However, given that fingerprint based biometrics is a lot easier and effective, in comparison to the other security measures, we employ fingerprint based biometrics here. The system will send an alert to the user, prior to the train's departure, and will also notify the passenger to acknowledge their presence in the train. However, if the passenger does not board the train, the system will automatically allot their seat to the very next person on the waiting list.

All the related reference work done with respect to this paper, is detailed further in Section II of this paper, followed by a detailed description of the overall system and each of the functional units, in Section III. Section IV illustrates the various pages displayed during execution, in the final android application. Section V concludes the paper with appropriate mention of the overall achievement made possible by using this system, with Section VI finally summing up the references and the authors, referred in this paper.

2. RELATED WORKS

SmitaPatil et al [1], has proposed that the biggest challenge faced today in the ticketing system is having to wait in queue for purchasing tickets. Their work is primarily intended toward booking suburban train tickets which is more of a challenge, in comparison to booking tickets for long journey, using the existing system. In their system, tickets are booked through a smartphone application and the ticket information is stored within a QR code. A time-based technique is used to automatically delete the ticket after a specific period of time, once the user reaches the respective destination. They employed a Cloud-based database to store all the information of every user, for security purpose which is not available in the current suburban railway system database for ticket checking. Also, the ticket checkers are given a QR code scanner, using which they were able to retrieve all the details of the passenger's ticket. For generating the QR code, transition id was used. When this transition id gets scanned by the ticket checker, using the reader, a request is sent to the server, to retrieve the user's data into the checker's phone. In this way the checker can easily check the user's ticket.

Ravi Subbanet al [2], has further done research in their paper, about the viability of fingerprint as a biometric security technique. Fingerprint of any individual does very well serve as a unique way to identify the person's identity, in order to authenticate who they are. Fingerprint identification is also a very popular biometric security technique, as it is relatively much easier to acquire, and also the sheer availability of plentiful resources in comparison to other biometric security techniques. The plentiful resources mentioned earlier is nothing but a reference to ten fingers for an average human being, and to their established usage. Their paper further summarizes all the research that was done in the fingerprint matching techniques, their recognition methods and also the detailed performance analysis.

Xiong Wei et al [3], has described in their paper about a system that handled the problem of recording the attendance of a given set of audience, using smartphones. The system they proposed is a combination of two applications, one of it is for generating a QR Code by entering the student details and the second application is for taking the attendance for the QR Code and generating the attendance sheet in CSV or XLS format. The teacher had to scan the student's QR code in order to confirm their attendance. Their paper details about how the system verifies each student identity in order to eliminate false attendance registrations. Their system deals with managing and evaluating the attendance of all students. A student's QR Code was presented to the professor for marking their attendance. The respective professor, handling the subject was in charge of marking the attendance for all students of their group or class. The attendance that was marked, were in the form of binary 0 or 1, where 0 is for absent, and 1 is for present. This attendance value gets stored in a database of the particular student row in the respective table. The final attendance sheet was generated in CSV and XLS sheet for any further reference.

3. PROJECT DESCRIPTION

An android mobile application is developed through which ticket for trains can be bought online. The system consists of the travel details and a Quick Response code (QR code). If a user buys a ticket through this application, the details of the seat that is booked, like the seat number, coach number, etc. are stored against this user's credentials in the database. The user will have to register their name, mobile number, unique identification number, etc., while signing up for the application. The train seat will also contain QR code printed for each seat. The user has to scan the QR code on seat using a camera on their phone, through the application, and punch their fingerprint in a fingerprint reader, which is installed in each coach. The details of the seat fetched while scanning the code, is cross-checked with the stored seat details within the application, for the purpose of verifying the seat. The seat will be allotted if both the values, from the scan and those that are stored in the application match. This allocation is marked as attendance at that seat and is reflected on the database. Through this admin will know about the user's presence on the seat.Fingerprint of the passenger can be scanned by them through the fingerprint reader. We use fingerprint as it is one of the best and most efficient means of authentication, used all over the world. Their use for human identification is based mainly on two premises: (a)permanence, and (b)uniqueness or distinctiveness. Thus the entire system is broken down into the following functional units:

3.1 ANDROID APPLICATION:

Client-side interaction is implemented by means of an Android application which is created and installed on the User's Android Mobile Phone. We develop this application using Android Studio. Once we design the page we have to write the code for each functionality, within the application. Once we have created the full mobile application, it will be generated as an Android Platform Kit (APK) file. This APK file is installed in the User's Mobile Phone as an Application. User has to register their details within the mobile application along with their Aadhar number.

3.2 SERVER:

The Server will maintain the User's information in the database and verify them when required. It will also monitor the User's information, for any changes made, within the database. Also it has to establish the connection to communicate with the User. Furthermore, it will update each User's activities in its database in real time like the transactions being done by the user, etc. The Server will also authenticate each user before they access the Application, by means of a login password that the user had entered while registering. Thus the Server can prevent any User without authorization, from accessing the user's credentials.

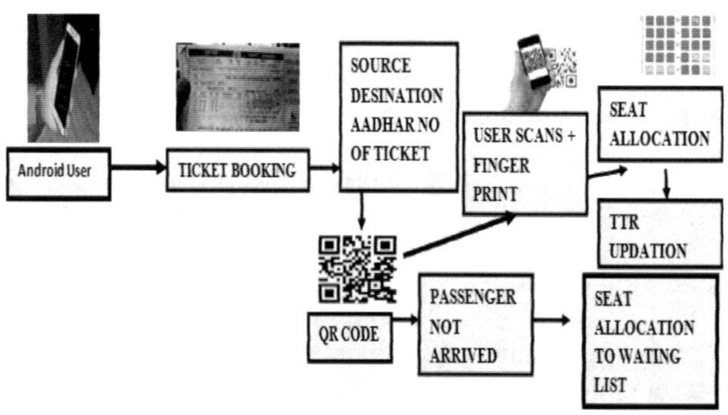

Fig 1: Architecture Diagram

3.3 TICKET BOOKING AND MOBILE WALLET:

The user will have to book their ticket to travel on train through the application. While booking a ticket, system will show how many seats are available at that time in train. The passenger will book their ticket and the amount for ticket will be debited to their mobile wallet on their mobile phone. We use a mobile wallet to carry our credit or debit card information, in digital form and use this to pay for the various transactions done. This is a minimal, secure, and a more efficient way to process money transactions on the go.

3.4 QR CODE:

QR code is a type of matrix barcode, or two-dimensional barcode. This code is a machine-readable label that contains related information about whichever item it is attached to. A QR code uses numeric, alphanumeric, binary, and kanji format, to efficiently store data. Here we are using a QR code to store seat number, coach number and train number. This is a permanent QR barcode which is attached to a seat. While people book ticket to travel, the same three values from the ticket are also stored in the application. When the user scans QR code, it will retrieve the stored values within the QR code and compare it with the values in the booked ticket. If they match, it will allot that seat to the passenger.

3.5 FINGERPRINT AUTHENTICATION:

User has to scan their fingerprint before they sit by means of a fingerprint reader in each coach. System will have a set of stored fingerprints from prior scans pulled from the Aadhar database. When the user scans their fingerprint, the system will cross-check this print with those stored prints and authenticates the user if they match.

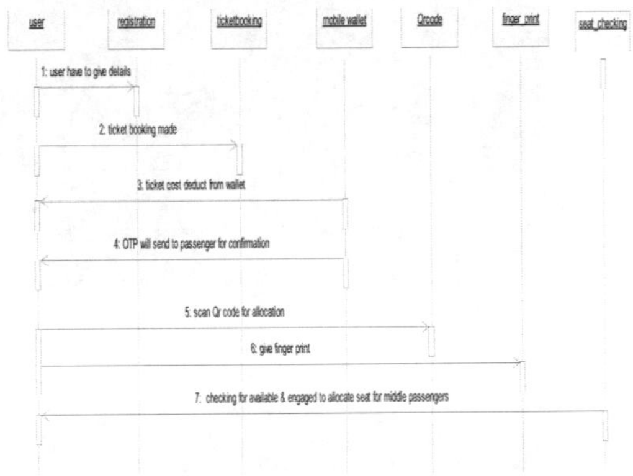

Fig 2: Sequence Diagram

3.6 CHECKING AVAILABILITY OF SEATS:

Admin will have an application to check the availability of seats. In the application there are two views on the status of a seat - one is available, the other one is occupied. While checking for availability of seats, if any seat is unoccupied, admin can allot this vacant seat to any passenger requesting for it.

4. RESULT

We develop the application through Android Developer Kit, available in Eclipse IDE. The registration page is followed by the main menu as shown below in Fig 4. The user can then choose from a variety of operations. The operations include booking a new ticket, e-wallet transactions details, user's personal information display page, view ticket page, and logout. Fig 3, illustrates the "Apply Ticket" page that is opened from the main menu.

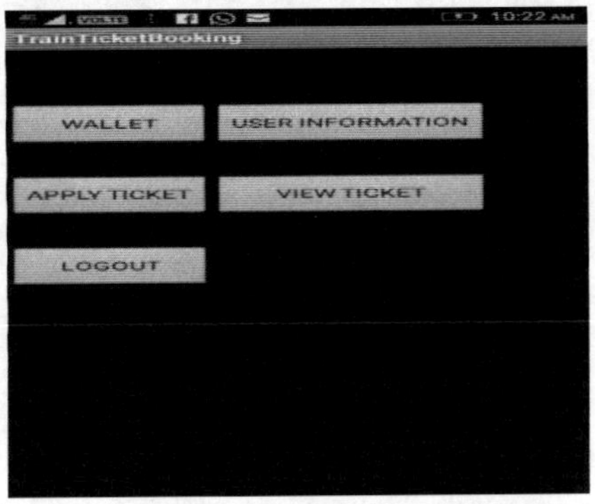

Fig 3:Main Menu After Login

Also from the Main Menu, user has an option to view their personal account details, through "User Information". This User details page is illustrated in Fig 4. User has an e-Wallet available within the application, through which they can make payments to book their ticket.

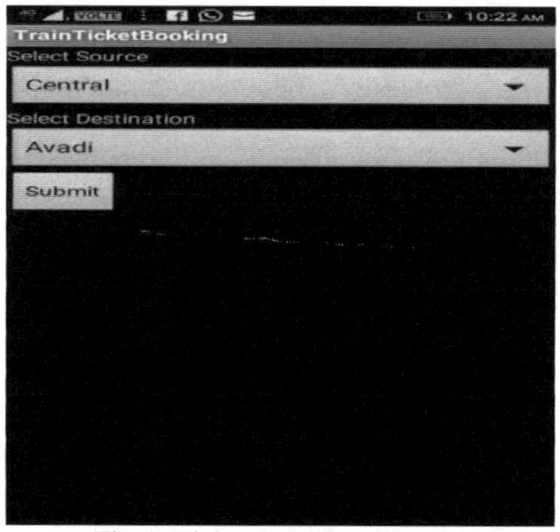

Fig 4: Ticket Booking Page

The user can also recharge their mobile wallet account, view the available balance, and check transaction history of their wallet, from the "Wallet" page of the main menu. This E-Wallet transaction page is as shown in Fig 5.

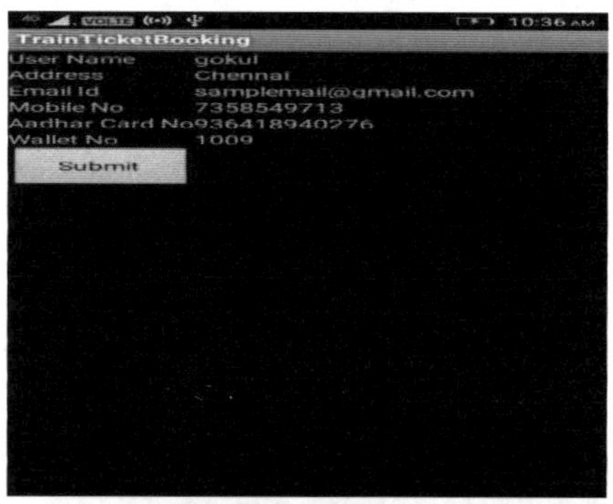

Fig 5: User Details

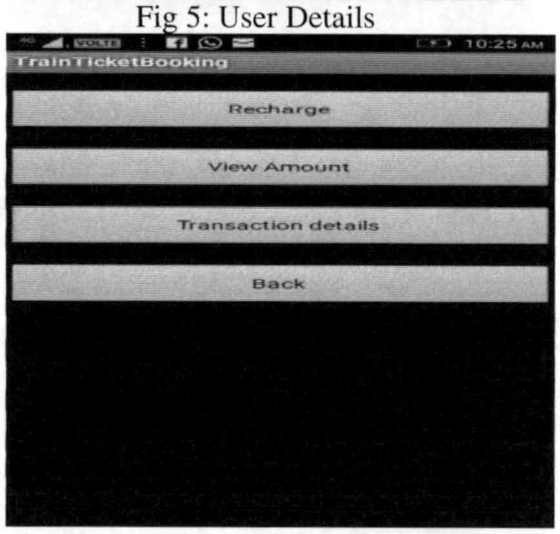

Fig 6: E-Wallet Transactions

After having boarded the train with a confirmed ticket, the user has to scan the QR code at their seat, followed by scanning their fingerprint using the fingerprint reader, on the coach. If both the fingerprint as well as QR code match to their corresponding identity and seat, respectively, the ticket is validated. The valid message displayed to the user's screen is detailed in Fig 7.

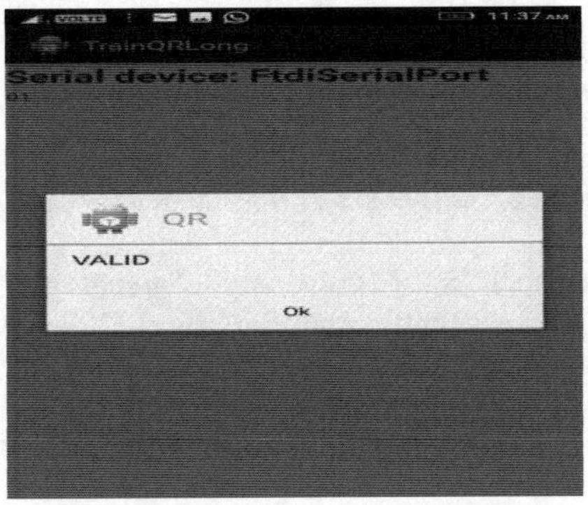

Fig 7: Fingerprint validation And QR Confirmation

5. CONCLUSION

Thus we implement a secure travel system by enabling each passenger to check themselves in, to their allotted train seats, by means of QR code. Also, a hassle-free means of verifying their identity is implemented by means of a biometric based authentication system. Furthermore, we provide an efficient means to book train tickets for the purpose of train travel. By this system, we are providing an easier way for verifying the occupants of the seat in the train, at a given instant. Finally we are also able to implement an automated system that allocates any vacant or absentee's seat to the next set of passengers on the waiting list, from the start of the list.

REFERENCES

1. SmitaPatil, ShrutiDesurkar,DipaliSanas; "An Intelligent Ticket Checker Application for Train using QR Code"; International Journal of Computer Applications, (2016), pp.15-20

2. Ravi Subban and Dattatreya P. Mankame; "A Study of Biometric Approach Using Fingerprint Recognition"; Lecture Notes on Software Engineering, Vol. 1, No. 2, May 2013, pp.209-213

3. Xiong Wei, AnupamManori, NandgopalDevnath, NitinPasi, and Vivek Kumar; "QR Code Based Smart Attendance System"; International Journal of Smart Business and Technology Vol. 5, No. 1, (2017), pp.1-10

4 Jabez J, Gowri S, Vigneshwari S, Albert Mayan J and Srinivasulu S (2018)," Anomaly Detection by Using CFS Subset and Neural Network with WEKA Tools", Smart Innovation, Systems and Technologies,Vol.107, pp. 675–682..

5. S. Istyaq and M. S. Umar, "Encoding passwords using QR image for authentication," 2016 2nd International Conference on Next Generation Computing Technologies (NGCT), Dehradun, 2016, pp. 818-823.

6. Y. G. Kim and M. S. Jun, "A design of user authentication system using QR code identifying method," 2011 6th International Conference on Computer Sciences and Convergence Information Technology (ICCIT), Seogwipo, pp. 31-35.

7. Albert Mayan .J , SharmilaLatha T , Kislay Sinha ," Security Analysis of Three Factor Authentication Schemes for Banking", ARPN Journal of Engineering and Applied Sciences, Vol:10, Issue 8, pp: 3504-3509,2015.

8. A. S. Shinde and V. Bendre, "An Embedded Fingerprint Authentication System," 2015 International Conference on Computing Communication Control and Automation, Pune, 2015, pp. 205-208.

9. A. L. H. Jin, A. Chekima, J. A. Dargham and Liau Chung Fan, "Fingerprint identification and recognition using backpropagation neural network," Student Conference on Research and Development, 2002, pp. 98-101.

10. S. Karthick. and A. Velmurugan., "Android suburban railway ticketing with GPS as ticket checker," 2012 IEEE International Conference on Advanced Communication Control and Computing Technologies (ICACCCT), Ramanathapuram, 2012, pp. 63-66.

11. Sankari.A,AlbertMayan.J,"Retrieving call logs and SMS by messaging services",International Journal of Pharmacy &Technology,Vol. 8 , No.4 , pp.22951-22958,Dec 2016.

IOP Conf. Series: Materials Science and Engineering **590** (2019) 012010 doi:10.1088/1757-899X/590/1/012010

12. G. M. D'silva, A. K. Scariah, L. R. Pannapara and J. J. Joseph, "Smart ticketing system for railways in smart cities using software as a service architecture," 2017 International Conference on I-SMAC (IoT in Social, Mobile, Analytics and Cloud) (I-SMAC), Palladam, 2017, pp. 828-833.
13. Le Hoang Thai and Ha Nhat Tam; "Fingerprint recognition using standardized fingerprint model"; IJCSI International Journal of Computer Science Issues, Vol. 7, Issue 3, No 7, May 2010; ISSN (Online): 1694-0784; ISSN (Print): 1694-0814.
14. Kamalesh M. D., Albert Mayan J., Felix Y., Sumanth B. S., & Sai Tej B. (2018), " Magrisys: A Smart And Ubiquitous Controlled – Environment Agriculture System " , 2nd International Conference on I-SMAC (IoT in Social, Mobile, Analytics and Cloud) , pp: 1-5 , 2018
15. RishabhDudheria; "Evaluating Features and Effectiveness of Secure QR Code Scanners"; 2017 International Conference on Cyber-Enabled Distributed Computing and Knowledge Discovery.
16. M.Ramya, Mrs.M.JayaSheela; "VLSI Implementation of Hybrid QR Code Generation System"; 2014 International Conference on Electronics and Communication System (ICECS -2014).
17. Asha P, Albert Mayan J, Canessane A (2018),"Efficient Mining of Positive and Negative Itemsets Using K-Means Clustering to Access the Risk of Cancer Patients", Communications in Computer and Information Science ,ICSCS 2018, Kollam, India, 2018,pp.373-382.
18. Lei Fu; "Design of QR CodebasedMall Shopping Guide System"; International Conference on Information Science and Technology; March 26-28, 2011 Nanjing, China.
19. Muthukumar B, Praveen Kumar R, and Nagarajan G.(2015),"Hybrid Intrusion Detection System for Private Cloud: A Systematic Approach",Procedia in Computer Science, vol 48, pp. 325-329
20.Yovan Felix A ,Jesudoss A ,Albert Mayan J (2017) , "Entry and exit monitoring using license plate recognition ", IEEE International Conference on Smart Technologies and Management for Computing, Communication, Controls, Energy and Materials (ICSTM), pp.227-231.
21. B.Bharathi and Mahesh kumar (2016), 'Non invasive BG scrutinizer system', Global Journal of Pure and Applied Mathematics, vol.12, No: 8, pp. 5123 – 5125
22. Albert Mayan J, AntoPraveena M D ,Telkar Bharath Rao , Uravakonda Uday Sagar (2018), "Optimized test data generation over suspicious implementation of oracle problem", International Conference on Power, Control, Signals & Instrumentation Engineering , pp.2559-2563.
23. .Muthukumar B and Jabezs J. (2015)," Anomaly Detection System in Determination of Heavy Metal Pollution using K-medoid Algorithm", Journal of Pure and Applied Microbiology, vol. 9, no.4, pp. 1-4

IOP Conf. Series: Materials Science and Engineering **590** (2019) 012011 doi:10.1088/1757-899X/590/1/012011

Discovering Clusters of nearby Routes in Spatial Networks

Vellampalli Vikas[1],Debasish Das[2], B.Bharathi[3]
[1,2]student, [3]professor,
[1,2,3]Department of Computer Science and Engineering,
[1,2,3]Sathyabama Institute of Science and Technology

[1]bharathivaradhu@Gmail.Com,[2]vikas.Vellampalli@Gmail.Com,[3]dasdebasish1995@Gmail.Com,

Abstract:Now-a-days finding which is the best path to travel from one place to another place is difficult and also confusing. So TSR(Trajectory search by regions of interest) query is proposed.TSR query takes set of inputs called places and returns best possible path to travel between this two places. To find this best path, decision tree(best first search) and naïve bayes classification along with sentiment analysis are being used.

KEYWORDS—**TSR query, decision tree, naïve bayes classification.**

1. INTRODUCTION

At present, to travel from one particular place to another place, there are a lot of parameters to look after such as distance, safety, comfort etc,. and current existing systems shows us a lot of paths based on the parameter of distance(i.e which is the shortest path)but whether it is good to travel or not nobody says. So TSR query is proposed to meet that goals. In TSR query, two areas of interest are taken as input to it and it will return a single best Path between these two areas.

Here in this proposal user reviews are considered as another parameter along with distance as parameter. So here best path is obtained based on the reviews and as well as distance factors. In this model, optimal path may not be the shortest path but user will get good and better path based on people experiences and in terms of distance as well. This model will helps in tour planning to unknown places and regions and also location based services etc,. In this model, decision tree(greedy best first search) is applied for getting distance as well as path and naïve bayes classification along with sentiment analysis for classification of user reviews. So optimal path is obtained based on this procedure.

2. RELATED WORK

There are a lot of related works regarding the proposed application. Some of them are listed below:

2.1 On map-matching vehicle tracking data:

Vehicle following information is a basic main material for an expansive scope of utilisations, for example, movement administration and control, directing, and route. An essential issue with this information is its exactness. The technique for examining vehicular development utilizing GPS is influenced by two mistake sources and subsequently creates incorrect direction information. To wind up valuable, the information must be identified with the basic street organize by methods for outline calculations.Show three such calculations that consider particularly the direction idea of the information instead of just the present position as in the commonplace guide coordinating case. An incremental calculation is recommended that matches back to back parts of the direction to the street organize, adequately exchanging exactness for speed of calculation. Interestingly, the two worldwide calculations contrast the whole direction with hopeful ways in the street organize. The calculations are assessed as far as (I) their running time and (ii) the nature of their coordinating outcome. Two novel quality measures using the Fréchet separate are presented and in this manner utilized as a part of an exploratory assessment to survey the nature of coordinating genuine following information to a street organize.

2.2 Robust and fast similarity search for moving object trajectories:

An imperative thought in closeness based recovery of moving item directions is the meaning of a separation work. The current separation capacities are normally delicate to clamor, moves and scaling of information that generally happen because of sensor disappointments, blunders in recognition procedures, aggravation signals, and diverse examining rates. Cleaning information to dispose of these isn't generally conceivable. In this paper, Let us present a novel separation work, Edit Distance on Real succession (EDR) which is hearty against these information blemishes. Examination and correlation of EDR with other well known separation capacities, for example, Euclidean separation, Dynamic Time Warping (DTW), Edit remove with Real Penalty (ERP), and Longest Common Subsequences (LCSS), demonstrate that EDR is more powerful than Euclidean separation, DTW and ERP, and it is overall half more precise than LCSS. Additionally create three pruning systems to enhance the recovery proficiency of EDR and demonstrate that these methods can be joined viably in a pursuit, expanding the pruning power altogether. The trial comes about affirm the prevalent productivity of the consolidated strategies.

2.3 Searching trajectories by locations: an efficiency study:

Direction look has for some time been an alluring and testing subject which sprouts different intriguing applications in spatial-transient databases. In this work, consider another issue of seeking directions by areas, in which setting the question is just a little arrangement of areas with or without a request determined, while the objective is to discover the k Best Connected Trajectories (k-BCT) from a database to such an extent that the k-BCT best interface the assigned areas topographically. Unique in relation to the traditional direction scan that searches for comparable directions with respect to shape or other criteria by utilizing an example question direction, concentrate around the integrity of association gave by a direction to the predetermined inquiry areas. This new inquiry can profit clients in numerous novel applications, for example, trip arranging. In this work, initially characterize another likeness work for estimating how well a direction associates the inquiry areas, with both spatial separation and request requirement being considered. Upon the perception that the quantity of question areas is typically little (e.g. 10 or less) since it is unreasonable for a client to enter excessively numerous areas, dissect the plausibility of utilizing a broadly useful spatial list to accomplish productive k-BCT look, in view of a straightforward Incremental kNN based Algorithm (IKNN). The IKNN successfully prunes and refines directions by utilizing the contrived lower bound and upper bound of similitude. L.Chen commitments for the most part lie in adjusting the best-first and profundity first k-NN calculations to the fundamental IKNN appropriately, and all the more imperatively guaranteeing the productivity in both pursuit exertion and memory use. A top to bottom examination on the adaption and its effectiveness is given. Encourage streamlining is likewise exhibited to quicken the IKNN calculation. At long last, check the proficiency of the calculation by broad trials

2.4 Algorithms for nearest neighbour search on moving object trajectories:

Nearest Neighbor (NN) seek has been in the center of spatial and spatio temporal database explore amid the most recent decade. The writing on NN inquiry handling calculations so far manages either stationary or moving question focuses over static datasets or future (anticipated) areas over an arrangement of ceaselessly moving

IOP Conf. Series: Materials Science and Engineering **590** (2019) 012011 doi:10.1088/1757-899X/590/1/012011

focuses. With the expanding number of Mobile Location Services (MLS), the requirement for viable k-NN question handling over verifiable direction information has turned into the vehicle for information investigation, along these lines enhancing existing or notwithstanding proposing new administrations. In this paper, explore systems to perform NN seek on R-tree-like structures putting away recorded data about moving article directions. The proposed (profundity first and best-first) calculations fluctuate regarding the kind of the question protest (stationary or moving point) and in addition the sort of the inquiry result (verifiable persistent or not), in this way bringing about four sorts of NN inquiries. N.Pelekis likewise propose novel measurements to help our hunt requesting and pruning systems. Utilizing the execution of the proposed calculations on two individuals from the R-tree family for direction information (to be specific, the TB-tree and the 3D-R-tree), to show their adaptability and effectiveness through a broad exploratory investigation utilizing substantial manufactured and genuine datasets.

3. PROPOSED SYSTEM

TSR query is proposed in this method. It is using decision tree (greedy best first search)algorithm by taking location data from Google map API. Data can be accessed through API key. Later apply naïve bayes classification along with sentiment analysis. In sentiment analysis consider user experiences and find out whether it is good path to travel or not. Finally a single best path to travel from initial place to final destination is obtained.

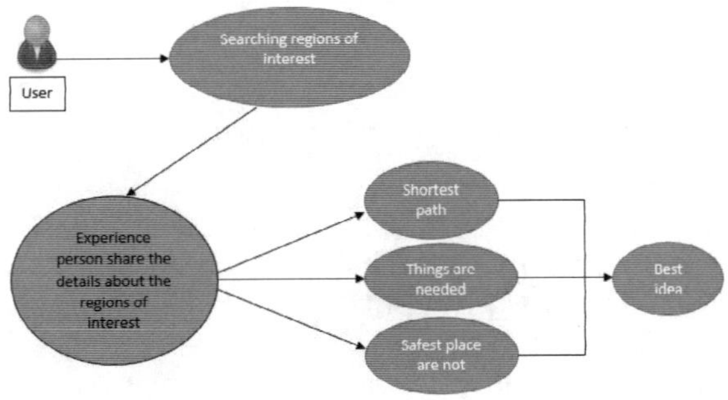

Fig.1. overview of proposed system

Various modules proposed in this system are:

1.user interface:

Here user can interact with the application, user can register and login the application.

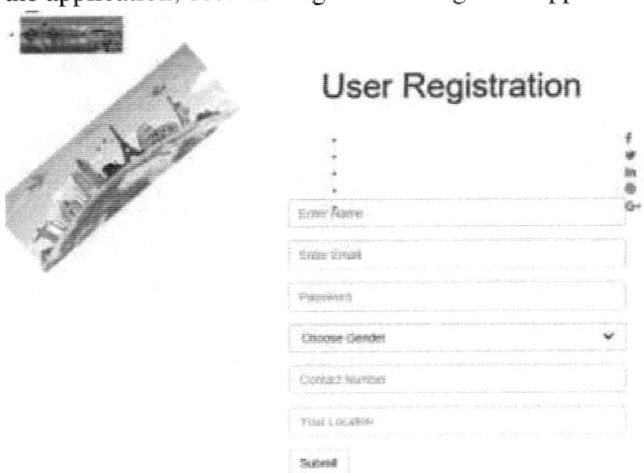

Fig.2. user registration form

In this user had to enter his name, email, password, gender, contact number, current location to access this model.

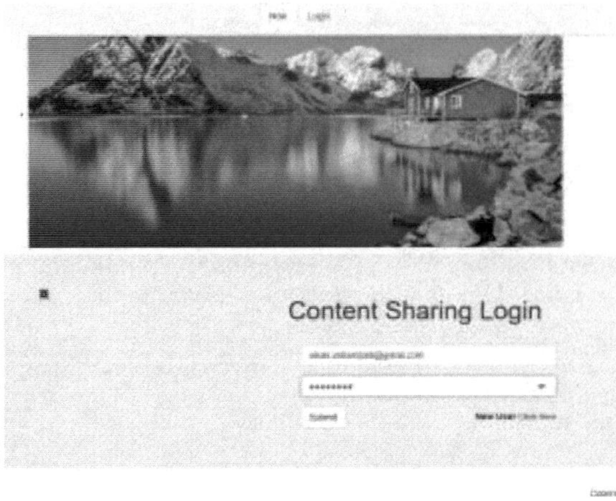

Fig3: Login details

Once any user registered in this model, they can access it directly using his email-id, password.

2.path finding module:

User can find the path between the regions of interest.

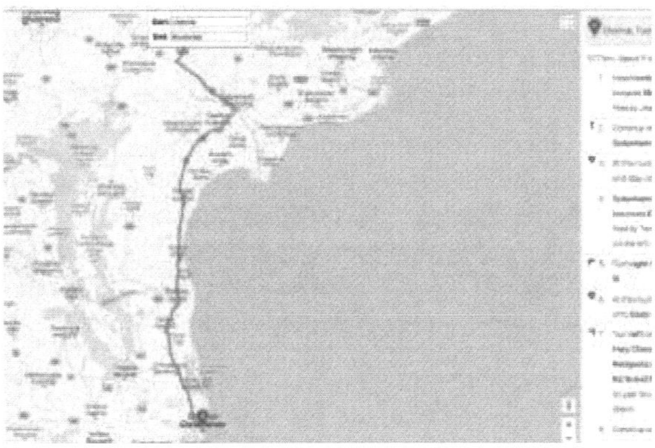

Fig4: obtaining the shortest path required

Here user gets his required shortest path between two locations. Data is taken from Google map API and decision tree(greedy best first search) is applied to get the required path.

3. user reviews:

User can get the experience person reviews, based on that reviews user can make a plan decision. Here Naïve bayes algorithm along with sentiment analysis is used to get status of all given reviews. Probability values are calculated and based on this values, it will generate the final status. If probability value>0.5 then it is called as positive status and if Probability value<0.5 then it is called as negative status and finally if probability value==0.5 or 0.0 then it is called as neutral status. Main objective of this system is to obtain the best possible optimal path based on user experiences and distance as parameters.
Here user can check the final status of the path to decide whether to go or not in this track.

4. ADVANTAGES & DISADVANTAGES

Advantages:

1. Complete visibility,
2.True knowledge capture
3.Maximize Efficiency.
Disadvantages:
1.Genuine user reviews should be given.

2.Google map API has so many limitations in free version.

5. CONCLUSION & FUTURE WORK

Contrasted with existing investigations of direction look by areas, The concept of query region and the density of spatial objects into account is new. This sort of inquiry is valuable in numerous well known applications, for example, trip arranging and suggestion, and area based administrations by and large. To figure the TSR inquiry efficiently, one can build up a best-first algorithm that endeavors upper and lower limits to prune the hunt space and receives a query source choice system, and a heuristic methodology. The execution of the TSR query was researched through broad analyses on both genuine and manufactured spatial information. It is important to consider transient data and further expand the TSR inquiry into a spatio temporal question. It is important to ponder how to successfully part and join directions with a specific end goal to return better outcomes.

REFERENCES

[1] S. Brakatsoulas, D. Pfoser, R. Salas, and C. Wenk. On map-matching vehicle tracking data. In VLDB, pages 853–864, 2005.
[2] L. Chen, M. T. Ozsu, and V. Oria. Robust and fast similarity search for moving object trajectories. In SIGMOD, pages 491–502, 2005.
[3] Z. Chen, H. T. Shen, X. Zhou, Y. Zheng, and X. Xie. Searching trajectories by locations: an efficiency study. In SIGMOD, pages 255– 266, 2010.
[4] Asha P, Albert Mayan J, Canessane A (2018),"Efficient Mining of Positive and Negative Itemsets Using K-Means Clustering to Access the Risk of Cancer Patients", Communications in Computer and Information Science ,ICSCS 2018, Kollam, India, 2018,pp.373-382.
[5] R. Agrawal, C. Faloutsos, and A. N. Swami. Efficient similarity search in sequence databases. In FODO, pages 69–84, 1993.
[6] E. W. Dijkstra. A note on two problems in connection with graphs. Numerische Math, 1:269–271, 1959.
[7] E. Frentzos, K. Gratsias, and Y. Theodoridis. Index-based most similar trajectory search. In ICDE, pages 816–825, 2007.
[8] H. Gonzalez, J. Han, X. Li, M. Myslinska, and J. Sondag. Adaptive fastest path computation on a road network: A traffic mining approach. In VLDB, pages 794–805, 2007.
[9] C. Guo, Y. Ma, B. Yang, C. S. Jensen, and M. Kaul. Ecomark: evaluating models of vehicular environmental impact. In SIGSPATIAL/GIS, pages 269–278, 2012.
[10] A. Guttman. R-trees: a dynamic index structure for spatial searching. In SIGMOD, pages 47–57, 1984.
[11] Julian Menezes .R, Albert Mayan .J, M. Breezely George," Development of a Functionality Testing Tool for Windows Phones", Indian Journal of Science and Technology,Vol:8,Issue:22,pp: 1-7, 2015, ISSN : 0974-6846
[11] Kamalesh M.D., Bharathi B,"Slicing an efficient transaction data publication and for data publishing", Indian Journal of Science and Technology, Vol 8(S8), pp.306-309, 2015.

IOP Conf. Series: Materials Science and Engineering **590** (2019) 012012 doi:10.1088/1757-899X/590/1/012012

INDUSTRIAL AUTOMATION WITH BIDIRECTIONAL VISITOR COUNTER

Dr. K Mohana Prasad
Department of Computer Science & Engineering
Sathyabama Institute of Science and Technology ,Chennai, India
Mohanaprasad1983@gmail.com

Pranjal Dhar
Department of Computer Science & Engineering
Sathyabama Institute of Science and Technology
Chennai, India
pranjaldhar234@gmail.com

Naseem
Department of Computer Science & Engineering
Sathyabama Institute of Science and Technology
Chennai, India
mr.naseem512@gmail.com

Abstract: This paper displays the plan of Industrial Automation utilizing Arduino. This circuit will do the task of tallying the number of persons or workers in an industry bi directionally. This count will be very accurate and it will beep a warning alarm if the number of people exceeds the limit. The aggregate number of people will appear on the LCD. In the circuit an Arduino UNO Board is utilized. This will help in the accurate measurement of the visitors and is less complex compared to a microcontroller. The arduino will get signals from the sensors and those signals work under the control of a programming code which is put away in the ROM of the arduino. The Infrared Receivers will continuously monitor any entity which passes both outside the building and inside the building. According to the number of guest inside the industry, electronic appliances like lights,fans and coolers will be trigerred on and off. This automation will save lots of electricity. There will also be a temperature and humidity sensor which will detect the temperature of an industry. If the temperature exceeds for any reason the circuit will beep, a warning alarm

Keywords: Arduino, Digital bidirectional visitor counter, IR Rays, LCD, DHT11, Alarm, electronic appliances.

1. INTRODUCTION

Industries require a huge number of work force and maintenance. Here in this given paper we have taken steps to ease the industry work by reducing the consumption of electricity and providing safety features which will not lead to dangerous situations. This will ease the work of electricity maintenance. Over many years guest, counters are used in many areas. Here microcontroller is replaced with an arduino board for better results and less complexity. The sensors used in this circuit are very efficient in terms of execution in comparison to the manual tally counter. This circuit can be used in many places like classroom halls, an auditorium, libraries, community halls, etc. It is a necessity to monitor the visitors for better human traffic management. The initial method for counting the visitors

IOP Conf. Series: Materials Science and Engineering **590** (2019) 012012 doi:10.1088/1757-899X/590/1/012012

involves hiring people to stand and manually calculate the number of guests or workers who enters or exits by a specific location.

The human counting was unreliable and came at a great cost. It may be confusing for a person to tally the quantity of individual entering and leaving at a same time. Our intension is to design and develop this system to manage human traffic in a large industry and also conserve the usage of electricity. Our main aim in this paper adds constructing a visitor counter which will make a controller circuit model to count and calculate the number of guest in a building or room at a particular time and all the electrical appliances will be turned on and off accordingly. It is also our aim that this controller base circuit model beeps a warning alarm for safety purposes when the capacity of the building and the temperature exceeds.

2. METHODOLOGY

This section involves the methodology of DBVC from the system overview of each and every component used to congregate the guest counter to allow effective crowd management as in monitoring and controlling. The Arduino based guest counter is created to react to the defects in the executions of already-occurred counters. The design in its sense has four (4) main sections and circuits as displayed in Figure1. These include detection section (IR sensor circuitry), Arduino section, alerting sections (LCD and Buzzer) and power supply circuit.

3. EXISTING SYSTEM:

Many already made systems for temperature measuring and analyzing, generally uses an microcontroller ATMEL 89C512. When the calculated signals are converted to digital form are called the examining(sampling) instants; the time in between continuous samplings is called the examining(sampling) period. The output recieved from the process is a continuous time signal. The output is then changed into digital form by the A-D converter. The alteration is done during the examining times.

DISADVANTAGES OF EXISTING SYSTEM:

a) Theoretical concepts may require few changes in practical implementation.
b) Circuits with low range cannot be used to work at large areas.
c) While continuous changes in the count value may look confusing.
d) Uses micro-Controller ATMEL 89C51.
e) Circuit gets bigger in size
f) Programming for microcontroller 8051 is difficult.
g) Circuit gets complex, difficult and confusing.

4. PROPOSED SYSTEM

In this model we are using DHT11 sensor into a correct voltage. This voltage is applied to arduino and according to program it processes the analog signal into digital signal and forms a particular voltage for a particular temperature. We used an arduino (microcontroller) to implement this system. IR sensors will count the numbers of guests and automatically run the lights and fans accordingly. An LCD display will show the results as output.

ADVANTAGES

a) This model can be implemented as an automated switch to increase energy efficiency.
b) The model can be applied at the entrance of a room to operate the lights and other appliances.
c) When the count of people is zero in a room, the circuit will automatically put off the power supply and thus conserving the use of electricity.
d) It will greatly help the physically challenged people.
e) It is easily operated and not complex like micro-controller.

4.1. APPLICATIONS

a) This model can be implemented in party or community halls where the number of people attending a party can be calculated easily.
b) It can be applied at offices.
c) It can be used at homes and classrooms to calculate the number of guests getting inside.
d) It can also be used for electricity conservation by automatically switching off the lights and fans when not in use.

4.2. COMPONENTS USED:

Here in this model we have used the following components

1. *Arduino UNO as the control unit*: The Arduino Uno is an open source platform of electronics. The micro-controller board runs on the ATmega328. It consists of 14 digital input/output pins, 6 analog inputs, 16 MHz ceramic resonator, a USB port, a power port, an ICSP header, a reset button. It consists of necessary items needed to run the microcontroller; just we have to connect the arduino to a computer with a USB cable or power it with an AC-to-DC adapter or a battery to initiate.

2. *Infrared sensor module*: An infrared sensor emits infra rays in order to detect any object of its surroundings. An IR sensor can both detect the heat and motion of an object. The rays are not being seen by our eyes and can be identified by an IR. The emitter is a LED (Light Emitting Diode) and the detector is a photodiode.

3. *Liquid Crystal Display*: Liquid Crystal Display screen is a digital display module. It is found in a colossal range of applications. It is 16*2 LCD display basic module. LCDs are easily programmable; have no limitation of displaying special & even custom characters (unlike in seven segments), animations and so on. An 16*2 LCD can display 16 characters per line and there are two lines.

4. *Digital Humidity and Temperature Sensor (DHT11)*: The DHT11 sensor is an ultralow-cost temperature and humidity sensor. It uses a capacitance dampness sensor and a thermostat to calculate nearby surrounding, and throws out an output signal on the data pin. It can be used very easily, but needs correct timing to achieve data.

5. *Relay Switch*: A relay is an electrically operated switch. Many relays are used to control various electrical appliances safely. Relays are used in a circuit controls by a separate less power signal and also many circuits can be controlled by only one signal.

6. *Buzzer*: A buzzer or alarm is an audio device, which may be of the following types, mechanical, electromechanical, or piezoelectric (*piezo* for short). Typical uses of buzzers include alarm devices and safety. It will notify the user if there is some kind of overload.

7. *Electronic Devices*: Electronic devices like bulb and fan are used to test the system. Many other electronic devices can be used in this module in the future.

4.3 HARDWARE PLATFORM

In this model our main intension is to propose a model that consists a guest counter which will automatically switch on the electrical devices and also, the temperature and dampness will be measured inside an industry. The Proposed system's architecture diagram is given in Figure. An IR beam is used as the source of a light beam. Bidirectional Guest Counter, which helps in the automatic electrical devices controller and Arduino as the main control unit has two sections. DHT11 sensor is used for detecting room temperature and a buzzer will notify if there is highly increase of both people and temperature.

IOP Conf. Series: Materials Science and Engineering **590** (2019) 012012 doi:10.1088/1757-899X/590/1/012012

A. Monitoring Arduino Reading:
Arduino helps clients to screen different sorts of sensors, for example, IR sensor and motion sensors. The GPIO pins inbuilt on Arduino board can help as universal use I/P and O/P pins (GPIO). The ATmega328 micro-controller on the Arduino board carries out the analog-to-digital conversion, which changes the I/P signal from 0 to 1023.

B. Controller:
The activating is at last done by the Arduino door. The Arduino makes a move continuously to operate the on/off of the LED and controlling the buzzer. In this model it has two parts, one the transmitter section where the light output is given. The other is the receiver end where light is taken as input and applied on entry sensor circuit and exit sensor circuit.

C. System Protection
Often random power supply may lead to damaging the components of the circuit. When the arduino cable is fed with the I/P power source then it sends the output of components. It provides a regular and static power supply and also protects from any type of power surges.

D. Infrared Sensing System
The IR module holds IR diodes and LEDs. Potentiometer is utilized for setting similar voltage at comparator's one terminal and IR sensor detects any individual and give an adjustment in voltage at comparator's second terminal. At that point comparator looks at other two voltages and a digital signal at output one is send for entry sensor and second for an exit sensor circuit.

E. Counter Display
In this segment we have utilized LCD to show the number of guests in the room. The framework is outlined utilizing the Arduino(IDE) stage. At the point when Arduino check for zero condition (Zero condition implies nobody in the room) and discovers it is genuine then the Arduino kills the globule by deactivating the transfer through a transistor. Here there are two capacities for in and out. This addition or decrement is executed in LCD.

4.4 BLOCK DIAGRAM

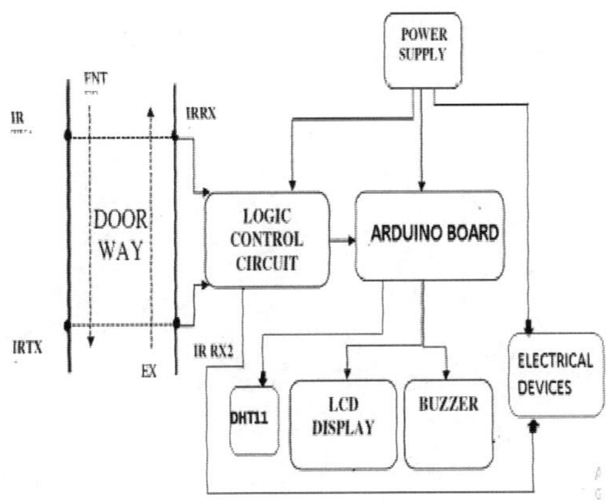

Fig1: Block Diagram

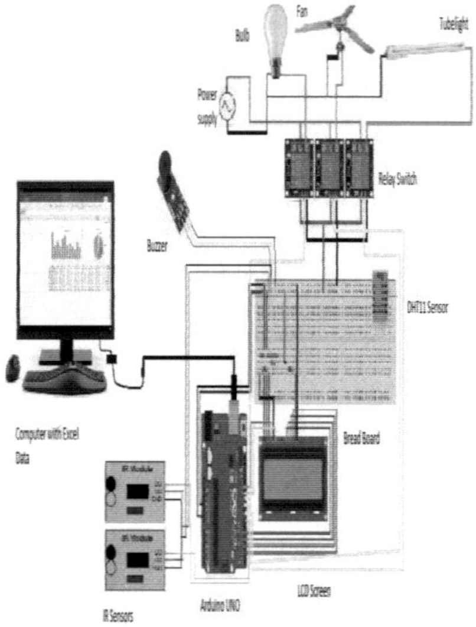

Fig2: Circuit Diagram

4.5 FLOWCHART OF WORKING MODEL

In the beginning the sensor 1 is interfered first then the arduino will search for the sensor 2. Furthermore, on the off chance that is intruded on the arduino will add the count. When the first individual goes inside the room then the counter goes to 1 and that time the FAN/LIGHT will turn on. Again the sensor 2 is interfered with first then the arduino will search for the sensor 1. And if it is interrupted then the count will be decreased.

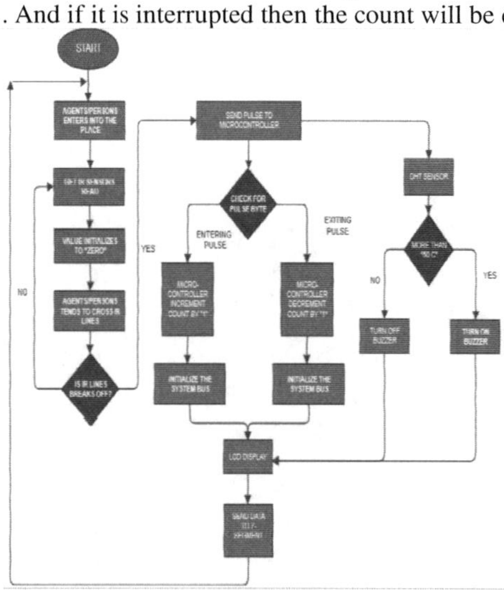

Fig3: Flowchart of the working model system

When all people in the room leaves then counter goes to 0 and then the FAN/LIGHT will turn off. If there are excess numbers of persons the buzzer will start beeping and will stop buzzing once the excess members leaves the room. Also if the temperature of the room exceeds than certain limit the buzzer will start beeping and this will add as a safety feature. The figure is shown in fig3

4.6 GRAPH OF DATA STORED:

We put away the information of the mugginess and temperature. The information is constantly received over excel sheet. The information is then appeared as a graph. This data can be utilized as checking and recording for security reason. In the future this data can be additionally applied for data analytics.

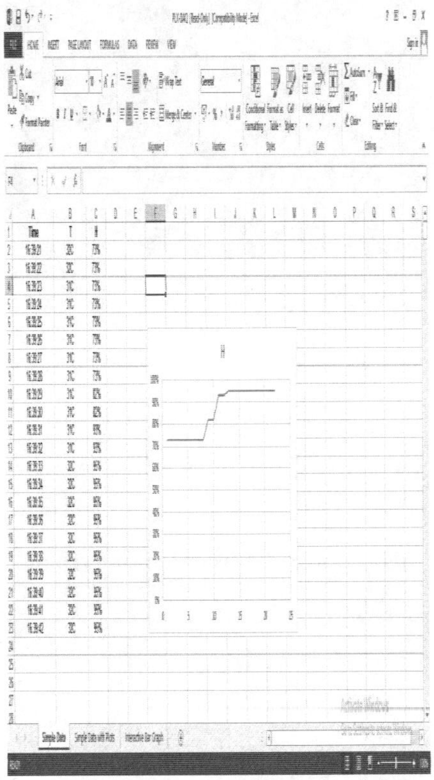

Fig4: Data stored over excel sheet and graph

5. EXPERIMENTAL EVALUATION

So as to actualize and exhibit the framework grew hypothetically, we made a model that speaks to the framework. Thus the whole system that is being developed is given below in the fig4.

Fig5: Complete Circuit

For future work, some suggestions like, expansion of cameras through with considering the picture can be put away accurately. Remote availability will be implemented to the framework, by controlling the Wi-Fi modules. The entire framework can be manufactured as commercial business equipment bundle.

6. DISCUSSIONS AND CONCLUSIONS

A novel engineering for a financial bidirectional Visitor Counter, automated electrical devices controller along with humidity and temperature detector with security alarm is proposed and executed in this paper. It gives an essential thought of how to run the bidirectional guest counter and room light controller utilizing Arduino circuit and Arduino IDE). The cost of this innovation is exceptionally prudent.

REFERENCES

[1]Avinaash Ram S.P, Albert Mayan.J, "Mobile Attendance Management and Employee Registration", ARPN Journal of Engineering and Applied Sciences, Vol:10, Issue 8, PP:3727-3730, May 2015.

[2]B.Bharathi, Mahesh kumar (2016), "Non invasive BG scrutinizer system", Global Journal of Pure and Applied Mathematics, Vol.12, No: 8, PP. 5123-5125.

[3] Asha P, Albert Mayan J, Canessane A (2018),"Efficient Mining of Positive and Negative Itemsets Using K-Means Clustering to Access the Risk of Cancer Patients", Communications in Computer and Information Science ,ICSCS 2018, Kollam, India, 2018,pp.373-382.

[4]Dr. K. Mohana Prasad, Dr. R. Sabitha "Meta Physical Algorithmic Representation for Flawless Clustering" Journal of Theoretical and Applied Information Technology (JATIT), ISSN:1992-8645.

[5]Dr .K. Mohana Prasad, Dr. R. Sabitha "Evolution of an Algorithm for Formulating Efficient Clusters To Eliminate Limitations" International Journal of Applied Engineering Research (IJAER), ISSN: 00973-4562.

[6]Dr. K. Mohana Prasad, Dr. R. Sabitha "Yoking of Algorithms for Effective Clustering" Indian Journal of Science and Technology, ISSN: 0974-6846, September 2015

[7] Yovan Felix A , Jesudoss A ,Albert Mayan J (2017) ,"Entry and exit monitoring using license plate recognition ", IEEE International Conference on Smart Technologies and Management for Computing, Communication, Controls, Energy and Materials (ICSTM), pp.227-231.

[8] S.Dhamodaran, Dr.M.Lakshmi (2017), "Design and analysis of spatial-temporal model using hydrological techniques", International Conference on Computation of Power, Energy Information and Commuincation (ICCPEIC),pp. 023 - 028.

[9] S.Karthikeyan, S.Jayashri, "Proficient Data Gathering by Mobile Agent using Jade",i-manager's Journal on wireless communication networks,Vol.1, No. 2, 2012, pp. 32-40.

[10] Sankari.A,Albert Mayan.J,"Retrieving call logs and SMS by messaging services",International Journal of Pharmacy & Technology,Vol. 8 , Issue No.4 , pp.22951-22958,Dec 2016.

IOP Conf. Series: Materials Science and Engineering **590** (2019) 012013 doi:10.1088/1757-899X/590/1/012013

VIRTUAL FITTING SPACE FOR DRESS TRIALS

Dr.K. Mohan Prasad,
Associate Professor
Dept of cse
School of Computing
Sathyabama Institute of Science and Technology
Chennai 600119, Tamil Nadu, India
mohanaprasad1983@gmail.com

R. Sri Kavya,
Dept of cse
School of Computing
Sathyabama Institute of Science and Technology
Chennai 600119, Tamil Nadu, India
rkavya125@gmail.com

S.Bhuvaneswari Devi
Dept of cse
School of Computing
Sathyabama Institute of Science and Technology
Chennai 600119, Tamil Nadu, India
bhuvanasayapaneni8888@gmail.com

Abstract-People buy clothes in stores which may take a lot of time for them to try their fitting and in online shopping there is no possibility where people can try-on the clothes which they want to buy. Our intention is that by creating a virtual fitting room environment we can improvise approachability of trying clothes on and need not have to waste time we need to have the correct measurement of the user and the dress model with correct rotation, measurement, location. Even people are interested more in online shopping and do buy a lot of things, but buying clothes online is still considered as ineffective way because people don't have enough knowledge about dress fittings in online shopping and it is not even possible to try- on the clothes. In order to improvise this drawback we can use the virtual fitting room which would provide the online customers with a wide range of options and a great delight to the vendors online. This concept of e-clothing helps you decide whether the dress fits you or suits you in an easy and efficient manner. Our analysis about customers provides a great fitting of dress to the person as per their requirement. To identify a human we use skeleton detection and to find the outline of human we use human contour detection method. Different positions of human body are detected using Haar cascade classification method and warping process is used to resize the outfit according to the size of the human body.,

KEYWORDS: Skeleton detection, Human contour detection, Haar cascade classification, warping.

1. Introduction

The computers show an image or video as a 2D representation of a 3D image. As in Fig 1 a Kinect sensor consist of an infrared light source, RGB camera and four microphones. Unlike computers Kinect sensor can use all the data produced by cameras and microphones to detect the objects infront of it, determine the direction of sound produced by using signal processing hardware.

IOP Conf. Series: Materials Science and Engineering **590** (2019) 012013 doi:10.1088/1757-899X/590/1/012013

Fig 1: Kinect Sensor

A part from the two cameras in the middle there is a light source in the left and the microphones are present along the bottom of the Kinect sensor. The Kinect sensor has the ability to see object in 3D view. It has a depth sensor that can measure create its own depth map of area present before it.RGB camera is used by most of multimedia applications because of the depth information provided by it. This camera helps us in creating a visual dressing model. Each pixel has a colour. The colour is a 32-bit integer. As in Fig 2 Red is determined by the first eight bits, green by the next eight bits, blue bythe next eight bits, and the remaining eight bits the transparency of the Pixels.

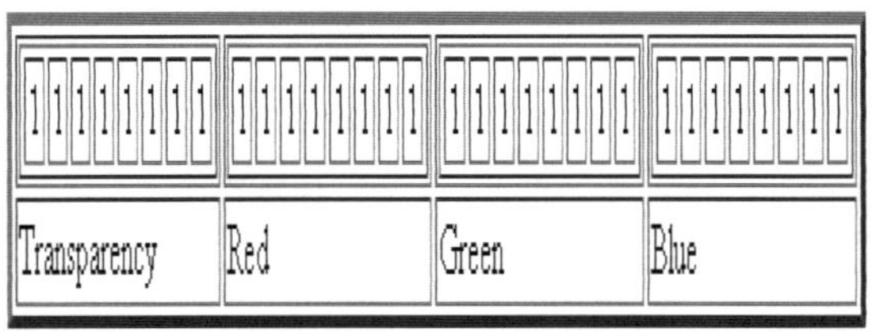

Fig 2: 32-bit integer colour pixels

Visual dressing is nothing but the user wearing the 3D designed outfit instead of manually trying it out. Visual dressing helps in me efficiency and majorly fills the gap in the existing online based shopping system by helping a customer to analyse if outfit suits them or not. There are a few challenges that we face during virtual dressing. In the existing model, we only have a 2D model. To develop a 3D model for virtual dressing we need to develop a skeleton detection method, and many methods have been proposed for skeleton detection, and they work very effectively. But most of them are inefficient when they are used in virtual dressing. This is due to the lack of synchronisation of the human body and outfit model. It is important to have an effective deformation model. In our proposed work, we would like to introduce a 3D model for virtual dressing. The formation of this model can be done by using a Kinect sensor. It is used for skeleton detection through the joints of the human body. We use Mat lab acquisition pre-processing and DAC algorithm for 3D model development. It is also an important task to check the synchronization of the human body and the developed 3D outfit model. This smart shopping method can be very useful to people which helps in me efficiency and gives effective way of shopping.

2. Existing Method

3D model and 2D image are the two major models in the existing virtual fitting model. Certain systems give results of 3D simulations of the human model and outfit using the size limitations of the outfit.In recent dressing simulations that can generate a perfect fitting on various different body shapes. In the above systems the generation of outfit are done with the help of a predefined human model .Depth camera is used to give a good 3Dimensional model of human body. Virtual fitting room concept works by imposing the 3Dimension model or picture with live video of the customer. In the video view the customer can feel the outfit or the accessory virtually according to the movement of the customer with the help of superimposed 3D model or picture. Real 3D fitting rooms has the properties of both 3D models and photo accurate fitting. A 3D figure is generated based on the photo and simple body measurements, in which the customer can visualize themselves in theirs chosen outfit. Human Contour detection it will detect the edges with respect to the coordinates or high peak value.

Warping serially randomly will capture the body and merge part by part .KNN Classifier (Need to set same body position with dress position)

3. Literature Survey

3.1 Dynamic context capture and distributed video arrays for intelligent spaces:

Human and machines combine to accomplish a task, for such kind of combinations can be possible by using intelligent environments.Human and computer interactions can be possible in multiple ways by maintaining a list of wide range of events. Some important functionalityincludes intruder detection, body pose, person identification, movement analysis. Useful intelligent spaces are identified by novel multi model sensor system. Cameras and microphones distributed all over will capture the audio and visual signals. Classification algorithm is used for taking high level semantic information from observed signals.Each signal have a specific task. Individual video analysis modules as well as to evaluate basic feasibility system are calculated by an experimental study in this paper
Advantage: Time delay.
Disadvantage: Uniform distribution.

3.2 Computer modelling, analysis and synthesis of dressed humans:

CAD/CAM systems are used in garment design and manufacturing based on customer's choice. They must be able to order dresses in the virtual way by trying over internet. In this model, Image formation scheme is used to represent the 3D structure of a human body. For this, two visual cues are used: occluding contours and stereo cue. They are used to monitor part positioning, orientation and deformation. 3D points provided by the stereo cue can filter the surface estimate and improve fitting of the dress. Arbitrary camera configuration, avoids overlapping of re-constructed surface. From the reconstructed human body, the body measurements are to be taken automatically and used in dress design on a virtual environment.
Advantages:Less distortion.
Disadvantage:Less speed.

3.3 Application of virtual reality technology in online fashion store development:

In recent times mannequins and dressing rooms have improved to a virtual environment. Virtual fitting environment provides a great help to online shopping. In few circumstances the clothes bought on online may disappoint the customer very badly. There may be a great difference of size, change of colour and may be dissatisfied with return policies. Virtual fitting clothes, consist of combination of virtual model creation, dress databases and methods of fitting the clothes. This research aims to develop a virtual fitting room toolkit.It consists of 3 main parts: mannequin modelling,clothes construction and clothes fitting.Finally,the whole process is integrated into an online fashion shop.
Advantage: Pixel calculation with high speed.
Disadvantage: x,y direction feature extraction.

3.4 A feasible face pose estimation by evaluating 3D facial feature vectors from 2D features:

Application for vision based human computer interaction is possible by having the head pose correctly from give input, Using features like skin colour and hair a 3D head pose is estimated The 3D head pose of the 2D input face images is estimated using the primitive facial features. Skin colour is adapted by YCbCr model. By using 2D eye point, nose are used to get 3D head pose of input face image. This works with 3majour steps:

1. Facial region is separated by skin tone model.
2. Hair like features are used fordetecting nose and eyes.
3. The head pose is modelled as mock face model with better result.

Advantage: Feature extraction of face in 3D
Disadvantage:No random distribution.

IOP Conf. Series: Materials Science and Engineering **590** (2019) 012013 doi:10.1088/1757-899X/590/1/012013

4. Proposed Method

In our proposed method to find the coordinates of a human body we use skeleton detection method and to find out the edges and outlines we use human contour detection. We use Kinect sensors for the detection of joints as in Fig 3.

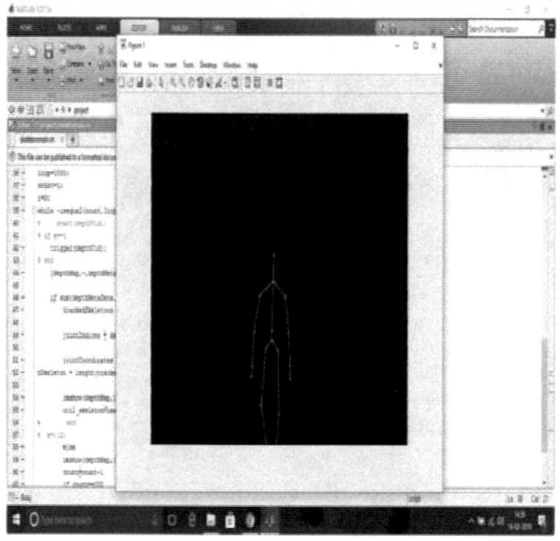

Fig 3: Skeleton Detection

For warping we use DRLBP and DRLTP (they extract pattern features of coordinates and will do warping parallel). Haar cascade classification is used to manage different positions of human body. Warping is a process in which we resize the outfit into the size of the person. The sample dress selected for fitting must be resized according to the size of person, this resizing is done by considering the feature points of the dress such that we can superimpose it over other similar texture. We perform interpolation technique on the dress that is divided to different regions of distinguished variations.

When we superimpose the warped dress on person as in Fig 4 there may be few flaws like hand size mismatch, hip size, overlay of dress etc… Firstly using the process of image fusion by blending. We impose the dress over the person.Skin matrix is used to correct the problems of overlaid sample dress and original image. Finally for the person who is virtually dressed we can use the same background or we can create a new background for the new and different look of the imposed outfit.

Fig 4: Super imposing of Dress

5. Conclusion

The skeleton of the customer is detected by the Kinect device using IR-camera and RGB-camera based on the skeleton joints of the body. After knowing the joints we will frame the human structure. The dresses are then viewed on the human structure in a 3Dimension manner to give a lively look of dress on the customer. The dress image moves according to the motion of the human.

References

[1]. B. Wu, T. Mei, W.-H. Cheng, and Y. Zhang, "Unfolding temporal dynamics: Predicting social media popularity using multi-scale temporal decomposition," in Pr oc. 13th AAAI Conf. Artif. Intell. , 2016, pp. 272–278.

[2]. Sankari.A, Albert Mayan.J, "Retrieving call logs and SMS by messaging services",International Journal of Pharmacy & Technology, Vol. 8 , Issue No.4 , pp.22951-22958,2016.

[3]. W. H. Hsu, L. S. Kennedy, and S.-F. Chang, "Reranking methods for visual search," IEEEMultimedia, vol. 14, no. 3, pp. 14–22, Jul./Sep. 2007.

[4]. W. H. Hsu, L. S. Kennedy, and S.-F. Chang, "Video search reranking through random walk over document-level context graph," in Proc. 15thACM Int. Conf. Multimedia, 2007, pp. 971–980.

[5]. R. Yan, A. Hauptmann, and R. Jin, "Multimedia search with pseudo relevance feedback," in Image and Video Retrieval. Berlin, Germany: Springer, 2003, pp. 238–247.

[6]. M. Wang, H. Li, D. Tao, K. Lu, and X. Wu, "Multimodal graph-based reranking for Web image search," IEEE Trans. Image Process. , vol. 21, no. 11, pp. 4649–4661, Nov. 2013.

[7]. J. Cui, F. Wen, and X. Tang, "Real time Google and live image search re-ranking," in Proc. 16th ACM Int. Conf. Multimedia , 2008, pp. 729–732.

[8]. Y. Rui, T. S. Huang, M. Ortega, and S. Mehrotra, "Relevance feedback: A power tool for interactive content-based image retrieval," IEEE Trans.Circuits Syst. Video Technol. , vol. 8, no. 5, pp. 644–655, Sep. 1998.

[9]. M R Annish Brislin, J Albert Mayan, R Aroul Canessane, M R Anish Hamlin,"Blood donation and life saver app",2017 2nd International Conference on Communication and Electronics Systems (ICCES), pp. 446 - 451,2017

[10]. Usha Nandini , Saravanan M , Albert Mayan J , Murari Devakannan Kamalesh , Mohana Prasad K (2018)," Automatic traffic control system using PCA based approach ", International Conference on Energy, Communication, Data Analytics and Soft Computing, pp.2387-2392.

[11]. X. S. Zhou and T. S. Huang, "Relevance feedback in image retrieval: A comprehensive review," Multimedia Syst., vol. 8, no. 6, pp. 536–544, Apr. 2003.

[12]. K. Mohana Prasad, Dr. R. Sabitha "Meta Physical Algorithmic Representation for Flawless Clustering" Journal of Theoretical and Applied Information Technology (JATIT), ISSN: 1992-8645, Volume 76, NO 1,PP 82-87.

[13]. K. Mohana Prasad, Dr. R. Sabitha "Evolution Of An Algorithm For Formulating Efficient Clusters To Eliminate Limitations" International Journal of Applied Engineering Research (IJAER), ISSN 0973 – 4562, Volume 9, Issue 23, pp. 20111-20118.

[14]. Albert Mayan J, Kuldeep Anand D.S, Neha Sadhvi(2017),"Efficient and secure server migration on cloud storage with VSM and dropbox services", International Conference on Information Communication and Embedded Systems (ICICES), pp. 1-5.

[15]. K. Mohana Prasad, Dr. R. Sabitha," Yoking of Algorithms for Effective Clustering", Indian Journal of Science and Technology, ISSN: 0974-6846 Vol 8(22), IPL0269, September 2015, pp 1-4.

[16]. S.Karthikeyan, S.Jayashri, "Proficient Data Gathering by Mobile Agent using Jade",i-manager's Journal on wireless communication networks,Vol.1, No. 2, 2012, pp. 32-40.

[17]. S.Dhamodaran, Dr.M.Lakshmi (2017), "Design and analysis of spatial-temporal model using hydrological techniques", International Conference on Computation of Power, Energy Information and Commuincation (ICCPEIC),pp. 023 - 028.

[18]. Mohana Prasad, K.,"Efficient dynamic clustering mechanism through unfathomed clustering techniques" Journal of Advanced Research in Dynamical and Control Systems, 06-Special Issue, 2018,pp 934-939.

Data Analysis And Prediction On Cloud Computing For Enhancing Productivity In Agriculture

K.Ashokkumar[1], D.DivyaChowdary[2], Ch.DivyaSree[3]
[1,2,3]Department of Computer Science and Engineering
Sathyabama Institute of Science & Technology
e-mail: ashokkumar.cse@sathyabama.ac.in

Abstract: Introducing a concept to increase productivity and predict the crops from the diseases. To find the leaf characteristics using image processing to detect the diseases and pests that are present in leaves. In agriculture the main important aspect is to take proper steps to increase the production, which means to know the state of leaves condition in the field. In this undertaking the programmed system recognizes the qualities of the leaves. The programmed leaf attributes recognition is basic one in observing extensive fields of yields, and to naturally distinguish indications of leaf attributes when they show up on takes off. The basic leadership framework uses the picture portrayal and managed classifier of Neural Network. Picture preparing procedures for this sort of choice examination includes preprocessing, highlight extraction and characterization organize. In this procedure, the picture is taken and that can be resized if required the locale of intrigue choice can be performed. Here, shading and surface highlights are extricated from a contribution for organize preparing and grouping. Shading highlights like mean, standard deviation of HSV shading space and surface highlights like vitality, differentiation, homogeneity and connection. The framework will be utilized to order the test pictures naturally to choose leaf attributes. For this approach, programmed classifier Neural Network (NN) is utilized for grouping in light of learning with some preparation tests of that same classification. This system utilizes digression sigmoid capacity as portion work. At long last, the mimicked result demonstrates that utilized system classifier gives least blunder amid preparing and better exactness in order. This helps to read the classifier (Characteristics difference of the input image).By using the Threshold values we can detect the pests/diseases and also helps to give the required data about the diseases, details of pesticides and other like required quantity of the pesticides and displays the present market prices for the selected crop.

Keywords:- ***Image processing, K-means clustering, Neural Network, GLCM, Preprocessing, Image classifications ,Gaussian image, Gray image, Bilateral filtered image, Extracted Regions, Segmented image, Regions detected image.***

1. Introduction

From past few generations we are aware that farmers are facing lots of problems in their agriculture field. The basic concept of this project is to find the diseases/pests which causes major damage to the fields. That makes them to get less profit or sometimes they even get the whole crop to be infected by the disease. That causes the farmer to earn less.

So, here by using the image processing and IOT concepts we can get the data of the plants and fields which are affected. In the image processing a picture can be taken through the camera. The

IOP Conf. Series: Materials Science and Engineering **590** (2019) 012014 doi:10.1088/1757-899X/590/1/012014

image captured was the back side of the leaves. Because, the veins can be showed clearly. At first the image is taken and the image classification takes places to perform the clustering methods which helps to get the required components in one place to get the characteristic values for the detection of diseases/pests. There consists of two datasets one is the input image and the other is the saved images .At first the input leaf .The image GLCM classification like gaussian image, segmented image, region detected, bilateral filtered image, gray image and extracted region which helps to form the perfect structure of the leaf without any noise. This filtering method helps to know the disease part in a very clear pattern based on color, texture and features can be applied for the effective implementation.

In this process it takes the feature extraction, NN training .After taking these steps the trained network classifies the input leaf image and saved image which helps to know the diseases and other damages that reduces the production rate .By using the above process we can increase their production. At last it displays the characteristic values that help to get the data about disease name and the pesticides that can help us to cure the diseases.

Not only image processing helps but IOT plays a vital role for not getting diseases. Because many diseases can be caused due to water content present in their soil, temperature and humidity. Using sensors we can get the data. By collecting the data from the fields we can reduce the disease rate. Soil moisture sensor, humidity sensor and temperature sensor takes place in the IOT process. And not only collecting data about the fields from sensors but also gives the market price for the selected crop.

2. Related Work

To increase the productivity in farming land we are combining both Image Processing and IOT. In Image Processing we will take the picture of a leaf and it will check whether the disease is present in that leaf or not. Before displaying the disease name, it undergoes a back ground process. The back ground process is it will perform the k-means clustering, Preprocessing, Bilateral filtering, segmentation and it will train the network. It will display images separately for every classification, in each and every step one classification technique will be done. It displays a gray scale image, bilateral filtered image, regions detected image, extracted regions image, segmented image and gaussian image. It will display the disease name and the pesticides which can be used to cure the disease which occurs for that particular crop. By giving the pesticides name the farmer will know more about the disease and how to cure it. After this process we will use IOT, using IOT we are going to find the humidity, moisture and temperature that are present in the soil by using humidity sensor, soil moisture sensor and temperature sensor. It will display all the values and the current market price for that particular selected crop.

The dataset that helps to store the images of the leaf that are effected and non-effected leaf. The dataset images undergo the same image characteristics process which helps to get the required details to detect the input image is effected or not effected. The data from both the input image and dataset by using comparison.

By comparing both the complete data and the input image with their characteristic values. If the values are compared and characteristic values are same then by that we can detect the disease is present in the leaf or not and by getting the disease name we give the details about the pesticides that helps to reduce the disease rate in the leaf.

By using these pictures and required algorithms we can detect the diseases/pests that cause damage to the crop. The algorithms are GLCM,K-means clustering, Segmentation and Neural Network and an understandable scripting language "PYTHON" is used.

Second part is about the IOT process. If the crop have excess water or less water that damages the farming land and effects the production rate. IOT helps farmers to know the basic knowledge about their own field .To collect details the IOT can be used in effective manner. Here the sensors plays an vital role for collecting data. The data is about temperature, humidity and soil moisture.

Soil moisture sensor will display whether the soil moisture is sufficient for the crop or not. If required how much amount of water should be added. By displaying the details about the humidity, temperature and soil moisture which are collected from the sensors. To run the sensors Raspberry pi can be used and other devices like bread board, connectors etc., can be used. The details of the required Soil moisture quantity can be given. It displays the current market price for the selected crop

3. Leaf Image Partitioning

Leaf images are in different types of colors. Collecting large amount of data set can be used in this method. By using gray scale and other models it makes easy to run the entire process in an effective manner.

3.1. K-Means Clustering:

Grouping similar data together. One of the best example is search engine. Basically, to collect the similar data the centroids are introduced (at any place of the data).After initializing the centroids, the centroids attracts the near data. To get the exact similar data, the boundary line is drawn between the centroids. It helps to collect the similar data. If the data is not in the correct format the iteration helps to get the data to which it belongs.

3.2. GLCM:

It is also known as Gray Level Dependence Matrix. It is used for reading the values of the features like colour, texture.

3.3. Threshold Algorithm:

It consists of settings exactly those pixels to white whose value is above the threshold value it can be displayed in black colour.

Probabilistic Neural Network

It is a bolster forward neural system. It was gotten from the Bayesian system. At the point when an information is available, the principal layer processes the separation from the information vector to the preparation input vectors. This delivers a vector where its components show how shut the info is to the preparation input. The second layer entireties the commitment for each class of sources of info and produces its net yield as a vector of probabilities. At long last, a contend exchange work on the yield of the second layer picks the most extreme of these probabilities, and produces a 1 (positive distinguishing proof) for that class and a 0 (negative recognizable proof) for non-focused on classes.

4. Results And Discussions

IOP Conf. Series: Materials Science and Engineering **590** (2019) 012014 doi:10.1088/1757-899X/590/1/012014

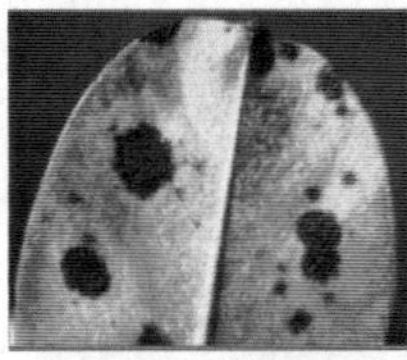
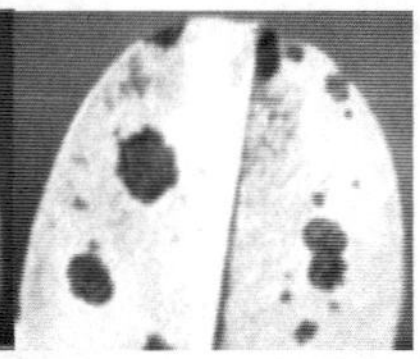

Fig 1 ***Fig 2***

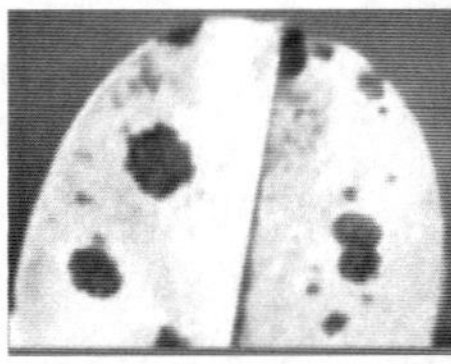
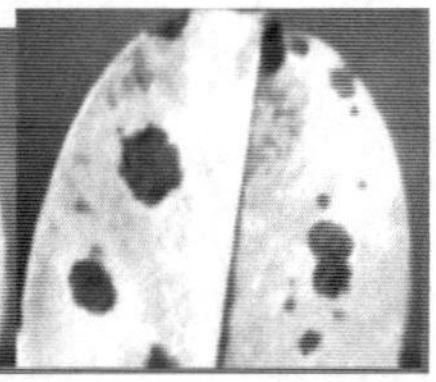

Fig 3 ***Fig 4***

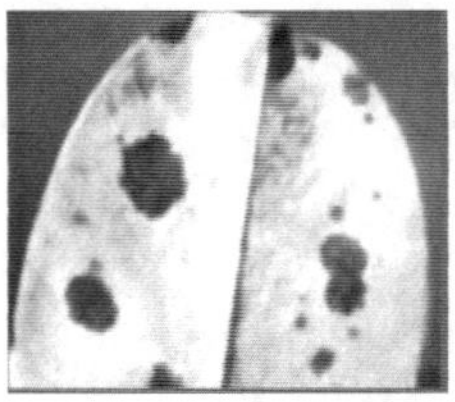

Fig 5 ***Fig 6***

Fig 7

Fig 1: Input Image:

The taken picture is a data picture. Isolating procedure finds the leaf is affected or not. Filtering is a method for changing and enhancing a photo.

Fig 2: Gaussian image:

The photo that channels the clatter and makes the photo clear, can be generally called Gaussian smoothing. (In picture taking care of, a Gaussian darken (generally called Gaussian smoothing) is the delayed consequence of clouding a photo by a Gaussian limit. It is a comprehensively used effect in outlines programming, typically to diminish picture confusion and abatement detail.)

Fig 3: Bilateral Filtered image:

The photo that channels the clatter and makes the photo clear. (A particular channel is a non-straight, edge-sparing, and disturbance decreasing smoothing channel for pictures. It replaces the power of

IOP Conf. Series: Materials Science and Engineering **590** (2019) 012014 doi:10.1088/1757-899X/590/1/012014

each pixel with a weighted ordinary of energy regards from near to pixels. This weight can be established on a Gaussian transport.

Fig 4: Gray Scale image:

This photo is of only two tints very differentiating. It recognizes the mischief easily by using parallel as 0 and 1.As addressed 0 as dull and 1 as white. As, showed up in picture underneath the disease parts will be in dull shading and strong part with white shading.

Fig 5: Segmented image:

In PC vision, picture division is the path toward isolating a mechanized picture into various areas. The goal of division is to revise and also change the depiction of a photo into something that is more critical and less requesting to look at.

Fig 6: Regions detected:

By recognizing the districts it makes easy to decrease or discard the establishment .It lessens the multifaceted idea of the photo taking care of to find the required material.

Fig 7: Extracted Region:

The evacuated picture urges or makes easy to enhance the proficiency or life of the item.

Table .1. Characteristic Values

SNO	Characteristics	Values
1	Gaussian Filtering	6.68093517e+01
2	Bilateral Filtering	9.55553209e-01
3	Gray Scale	7.51696284e+02
4	Segmentation	8.48471389e-01
5	Regions detected	3.05843487e+01
6	Extracted Regions	2.93997579e+03

```
[ 4.12256980e-01  2.82531189e+00  8.79025553e-01  1.16769166e+01
  9.31246897e-01  4.71034105e+00  4.38823544e+01  2.16143247e+00
  2.49454244e+00  8.95526997e-03  7.24667226e-01 -7.19294476e-01
  9.68931484e-01]
[[2.36752185]] [[3.41941342]]
5.116241538330417
```

Data set values:

1.############
[5.50062383e-01 6.68093517e+01 9.55553209e-01 7.51696284e+02
8.48471389e-01 3.05843487e+01 2.93997579e+03 2.61295081e+00
3.33728211e+00 3.18052731e-03 1.41783243e+00 -5.58955058e-01
9.61426703e-01]
[[15.36545562]] [[27.4565894]]
256.1343746606766
2.############
[4.80208581e-01 9.72510577e+01 9.31168461e-01 7.06569428e+02
8.10466606e-01 3.33920265e+01 2.72902665e+03 2.92388761e+00
3.91374357e+00 2.94800005e-03 1.78313247e+00 -4.83092086e-01
9.57298984e-01]
[[16.78591207]] [[26.61387879]]
241.3972478286338
3.############
[4.12256980e-01 2.82531189e+00 8.79025553e-01 1.16769166e+01
9.31246897e-01 4.71034105e+00 4.38823544e+01 2.16143247e+00
2.49454244e+00 8.95526997e-03 7.24667226e-01 -7.19294476e-01
9.68931484e-01]
[[2.36752185]] [[3.41941342]]
5.116241538330417

From the above data analysis, Fig 1 characteristic values are similar to the third set from the dataset. Combination of image processing and IOT makes an easy way to increase the production means reducing the diseases and pests by giving the information of the particular disease required pesticides and also the required quantity(approximately) which is very useful for farmers in their Agriculture field.

5. Conclusion

We are using many image classifications and IOT to make it effective and also using some of the algorithms to detect the diseases/pests. By taking these into considerations the clustering plays the major role to detect. Comparing the data plays a major role. The process has to be done automatically with more efficiency and flexibility. By using it we can increase the production rate and also pesticides can be used as per the required quantity. IOT helps us to get the details of the farmer's field. By combining both the data in the process we can easily grow the crops in an effective way. That helps the production rate.

References:

[1] P.Tamilselvi, K.Ashok kumar, 2017, Optimized Neural Network for Color-Based Infected Leaves Classification, international converence on computing, ieeexplore. 2017.
[2] Jayamala K. Patil, Raj Kumar, 2011, Advances In Image Processing For Detection of Plant Diseases‖, JABAR, , 2(2), 135-141.
[3] P.Revathi, M.Hemalatha,2012, Classification of Cotton Leaf Spot Diseases Using Image Processing Edge Detection Techniques‖, 169-173, IEEE.
[4] Albert Mayan .J , Sharmila Latha T , Kislay Sinha , 2015, Security Analysis of Three Factor

Authentication Schemes for Banking", ARPN Journal of Engineering and Applied Sciences, Vol:10, Issue 8, pp: 3504-3509.
[5] Piyush Chaudhary, Anand K. Chaudhari, 2012, Color Transform Based Approach for Disease Spot Detection on Plant Leafǁ, IJCST, Vol 3,No.6, pp.65-70.
[6] S. Arivazhagan, R. Newlin Shebiah, S. Ananthi, S. Vishnu Varthini, ,2013, Detection of unhealthy region of plant leaves and classification of plant leaf diseases using texture featuresǁ, CIGR, 15(1), 211-217.
[7] Ashok Kumar, K. Thamizharasi, K, 2015, Gesture controlled robot using MEMS accelerometer for eradication of weeds , Indian Journal of Science and Technology, Vol 8(5), 460–465, DOI:10.17485/ijst/2015/v8i5/60481.

Sign Board Monitoring and Vehicle Accident Detection System Using IoT

K.Ashokkumar[1], Ch. Venkata Deepak[2], D.VamsiRattaiah Chowdary[3]
[1,2,3]Department of Computer Science and Engineering
Sathyabama Institute of Science & Technology
e-mail: ashokkumar.cse@sathyabama.ac.in

Abstarct: The Rapid growth of technology and infrastructure has made our lives easier. The advent of technology has also increased the traffic hazards and the road accidents take place frequently which causes huge loss of life and property because of the poor emergency facilities. Our project will provide an optimum solution to this draw back. According to this project when a vehicle meets with an accident immediately Vibration sensor will detect the signal or if a car rolls over, and vibration sensor will detects the signal and sends it to RASPBERRY PI controller. Alcohol detection, eye blink is performed by the RASPBERRY PI Microcontroller sends, it alert mail through the IOT to the parents or a rescue team. So the person can immediately then after conforming the location necessary action will be taken. If the person meets with a small accident or if there is no serious threat to anyone`s life, then the alert message can be terminated by the driver by a switch provided in order to avoid wasting the valuable time of the medical rescue team. This paper is useful in detecting the accident precisely by means of both vibration sensor and Alcohol detection, eye blink sensor.. As there is a scope for improvement and as a future implementation we can add a wireless webcam for capturing the images which will help in providing driver`s assistance.

Key words: eye blink sensor, vibration sensor, alcohol sensor, web cam, buzzer, raspberry pi

1. Introduction

The Rapid growth of technology and infrastructure has made our lives easier. The advent of technology has also increased the traffic hazards and the road accidents take place frequently which causes huge loss of life and property because of the poor emergency facilities. Our project will provide an optimum solution to this draw back. According to this project when a vehicle meets with an accident immediately Vibration sensor will detect the signal or if a car rolls over, and vibration sensor will detects the signal and sends it to RASPBERRY PI controller. Alcohol detection, eye blink is performed by the RASPBERRY PI Microcontroller sends, it alert mail through the IOT to the parents or a rescue team. So the person can immediately then after conforming the location necessary action will be taken. If the person meets with a small accident or if there is no serious threat to anyone`s life, then the alert message can be terminated by the driver by a switch provided in order to avoid wasting the valuable time of the medical rescue team. This paper is useful in detecting the accident precisely by means of both vibration sensor and Alcohol detection, eye blink sensor.. As there is a scope for improvement and as a future implementation we can add a wireless webcam for capturing the images which will help in providing driver`s assistance.

2. Related Work

Existing System:

In the running world there is a growing demand for the users to convert the printed documents in to electronic documents for maintaining the security of their data. Hence the basic OCR system was invented to convert the data available on papers in to computer process able documents, So that the documents can be editable and reusable. The existing system/the previous system of OCR on a grid infrastructure is just OCR without grid functionality. That is the existing system deals with the homogeneous character recognition or character recognition of single languages.

Alerts To Driver:

Eye blink sensor used to detect the drowsiness of a driver at first the status of driver will be analyzed by eye blink sensor if eye is closed the output of eye blink sensor is high and so buzzer fixed nearby driver buzzer gives sound . Alcohol sensor used to detect the drunk position of driver at first the status of driver will be analyzed by alcohol sensor if the alcohol percentage in high taken by driver then also buzzer fixed nearby driver it will gives sound and car will automatic stops

Rescue Alert:

In this module, we are observing whether the vehicle is safe or not by monitoring the vehicle with vibration sensor. Whenever we are travelling, there is a chance for accidents to occur. If the vehicle is met with an accident, the vibration sensor will be activated. With the values received from the vibration sensor, the threshold value is calculated. The threshold value is the current value, which is displayed by the serial monitor. The value ranges from 0 % to 100%. The threshold value is split into three classes If the threshold value ranges between 0 and 30%, the accident occurred can be ignored. So, no alert will be given If the threshold value ranges between 30% and 70%, then alert message will send to parents and rescue team

3. Meterials & Methods

3.1 Architecture:

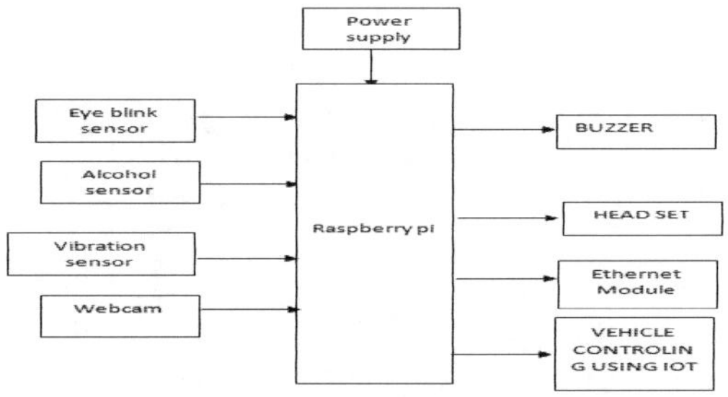

Figure 1. Vehicle Accident Detection System

3.2 Eye Blink Sensor

It is important in our attempting to locate the flickering of eye, since it is utilized to drive the gadget and to work occasions. So flicker discovery must be done, for which we can profit promptly accessible squint identifiers in showcase (Catalog No. 9008 of Enable gadgets) or we can join it with an exceptional direction written in picture handling that, if there is no student found for the specific time of pre-decided i.e. time more prominent than the human eye flickering time at that point consider an occasion called "squint", for which the arrangement of activities will be taken after. Here, for this situation we have to set time as 1 second or above it, as "squint occasion" is not the same as "should be expected eye flickering". We have to perform testing for just flicker occasion estimation, and not to discover typical eye squinting.

3.3 Alcohol Sensor

The MQ-3 sensor will detect the liquor content from human breath and gives its incentive to aurdino. The MQ-3 is helpful for recognizing liquor. SnO2 is delicate component which is utilized to detect the liquor.

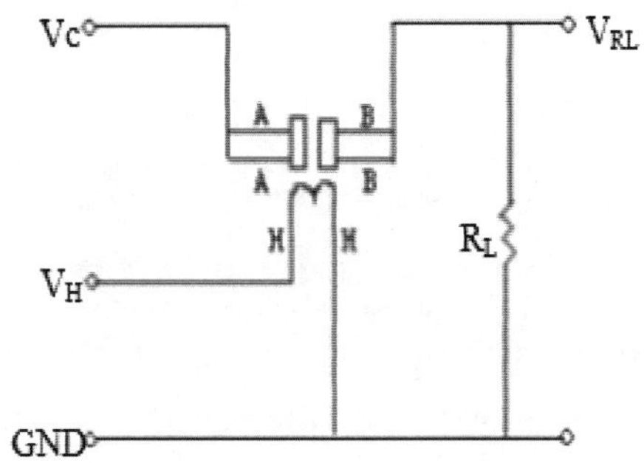

Fig. 2 Alcohol Sensor

IOP Conf. Series: Materials Science and Engineering **590** (2019) 012015 doi:10.1088/1757-899X/590/1/012015

3.4 Buzzer

Ringer is sound flagging gadget, it is utilized as a part of family unit apparatuses and car framework. It comprises of two transistor and ringer ON and OFF is controlled by the match of transistor.

3.5 Raspberry pi

The Raspberry Pi is manufactured in two board configurations through licensed manufacturing deals with Newark element14 (Premier Farnell), RS Components and Egoman. These companies sell the Raspberry Pi online. Egoman produces a version for distribution solely in China and Taiwan, which can be distinguished from other Pis by their red coloring and lack of FCC/CE marks. The hardware is the same across all manufacturers.

Fig.3. Raspberry pi kit for Vehicle Accident Detection System

The Raspberry Pi has a Broadcom BCM2835 system on a chip (SoC), which includes an ARM1176JZF-S 700 MHz processor, Video Core IV GPU, and was originally shipped with 256 megabytes of RAM, later upgraded to 512 MB. It does not include a built-in hard disk or solid-state drive, but uses an SD card for booting and persistent storage.

3.6 Webcam

A webcam is a video camera which feeds its images in real time to a computer or computer network, often via USB, Ethernet or Wi-Fi. Their most popular use is the establishment of video links, permitting computers to act as videophones or videoconference stations.

This common use as a video camera for the World Wide Web gave the webcam its name. Other popular uses include security surveillance and computer vision. Webcams are known for their low manufacturing cost and flexibility, making them the lowest cost form of video telephony.

They have also become a source of security and privacy issues, as some built-in webcams can be remotely activated via spyware.

IOP Conf. Series: Materials Science and Engineering **590** (2019) 012015 doi:10.1088/1757-899X/590/1/012015

4. Results and Discussion

This project we are going to explaining about when driver was sleep position drunk position he can't able see the sign board and directions then our project is helpful to reduce the accidents the processor is to capture sign board by web cam and it will sends to raspberry pi in the raspberry pi the image is decoded by using OCR algorithm and it will gives voice to driver about directions and sign boards

The driver was in drowsiness position the eye blink sensor will detected the eyes and gives signal to raspberry pi it will give signal to buzzer then automatically buzzer will gives sound to driver and also give estimation car owner by using Ethernet module

The driver was in drunk the alcohol sensor will detect and gives signal to raspberry pi it will gives signal to buzzer then it automatically buzzer will gives sound to driver and also give estimation car owner by using Ethernet module

The car was met with accident then vibration sensor will detect then it will gives signal to raspberry pi then it will gives estimation to driver and owner of car by using Ethernet module this was done in this module method

Fig.4. a) working of Vehicle Accident Detection System

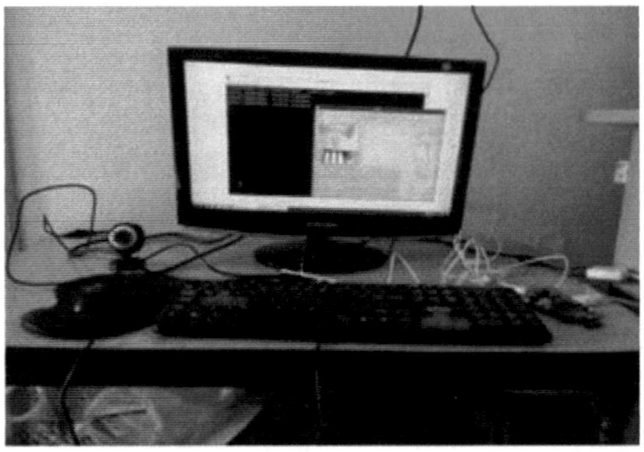

Fig.4. b) working of Vehicle Accident - Detection sensor

the aim of project is obtain using web cam and it is connected to raspberry pi then it should be connected to battery then web cam will capture image and give voice to driver the real time accident can be detect and message will sends to persons

The aim of this project is obtain using vibration ,eye blink, alcohol sensors are connected to raspberry pi then power supply should be given then buzzer sound can be hear by driver if it not a serious threat then driver will automatically stop the buzzer

5. Conclusion

In this project we have success fully designed sign board monitoring and accident detection system using IOT when accident occur it sends the intimation to owner of a car and also guardian of driver. It is the real fact that implementation of system will increase the some cost of the vehicle but it is better to have some present safety rather than having no percent of safety. the proposed system can be used for estimation of driver condition will intimated to certain guardian of driver

References

1] Abid Khan, Ravi Mishra —GPS – GSM Based Tracking System‖, International Journal of Engineering Trends and Technology, Volume3, Issue2, Pp: 161-169, 2012.
[2] S.P. Bhumkar, V.V. Deotare, R.V.Babar —Intelligent Car System for Accident Prevention Using RASPBERRY PI-7‖, International Journal of Emerging Technology and Advanced Engineering, Volume 2, Issue 4,Pp: 56-78, 2012.
[3] B.G .Nagaraja, Ravi Rayappa, M.Mahesh, Chandrasekhar M Patil, Dr TC Manjunath:"Design and Development of GSM based vehicle theft control system" Advanced Computer Control ICACC "09 International conference.pp148.
[4] Xu Li, Wei Shu, Minglu Li, Hong-Yu Huang, Pei-En Luo, and MinYou Wu, "Performance Evaluation of Vehicle-Based Mobile Sensor Networks for Traffic Monitoring" IEEE 2009.[5] Young, Kar-Keung D. "Automated navigation and mobile vehicle control using wireless sensor network technology," Proceedings of the IEEE International Conference on Industrial Technology, 2008.

IOP Conf. Series: Materials Science and Engineering **590** (2019) 012015 doi:10.1088/1757-899X/590/1/012015

[5] PoojaChandran, Aravind S, JishaGopinath And Saranya S S ," Design And Implementation Of Speech Generation System Using Matlab " Publish In International Journal Of Engineering And Innovative Technology (Ijeit) Volume 4, Issue 6, December 2014.
[6] Albert Mayan .J , Yovan Felix .A ,Chintalapudi M, Balaji R, Vigneswaran R,"Fault-Open Minded Resource Allocation Rescheduling Algorithm and Expense Minimization for Cloud Systems", Global Journal of Pure and Applied Mathematics,Vol 11,Issue:6,pp: 4111-4117,Dec 2015, ISSN: 0973-1768
[7 DevikaSharma ,RanjuKanwar ," Text To Speech Conversion With Language Translator Under Android Environment " Publish In International Journal Of Emerging Research In Management &Technology Issn: 2278-9359(Volume-4,Issue-6).
[8] Siddhant C. Joshi, Dr. A.N.Cheeran "Matlab Based Back-Propagation Neural Network For Automatic Speech Recognition", IJAREEIE(An Iso 3297: 2007 Certified Organization)Vol. 3, Issue 7, July 2014.
[9] Siddhant C. Joshi, Dr. A.N.Cheeran, "A Comparative Study of Feature Extraction Techniques for Speech Recognition System" ISSN: 2319- 8753, IJIRSET ,Vol. 3, Issue 12, December 2014.
[10] Dr.P. Tamilselvi, Dr. K. Ashok kumar , "Optimized Neural Network for Color-Based Infected Leaves Classification", , Sathyabama University.

IOP Conf. Series: Materials Science and Engineering **590** (2019) 012016 doi:10.1088/1757-899X/590/1/012016

Secured User Behaviour Based Access Framework for Web Service

R. Joseph Manoj[1], M.D.Anto Praveena[2], Manchikalapudi Anvesh[3], Murarisetty Pujith[4]

[1]Department of IT, St. Joseph's College of Engineering, Chennai, India.
[2 3 4]School of Computing, Sathyabama Institute of Science and Technology, Chennai, India
rjmanoj79@gmail.com[1],antopraveena@gmail.com[2],anveshm123@gmail.com[3],
puijth.murarisetty@gmail.com[4]
Abstract. There are many web applications around us providing their services using web services. In order to prevent malicious users from accessing the web services is an important challenge now. Hence this proposed access control model provides the user an effective way to prevent the malicious user by calculating the trust value of every user based on their behaviours such as success rate, failure rate, frequent of access, transaction timeout etc., in the web application. This model also notices and precludes IP address spoofing, SQL injection to allow only the authorized users and access appropriate information. Prototype implementation and simulation results show the proposed model is the effective one when it is compared with existing models.

Keywords: Web Services Access Control, User Behaviour, Trust Based Access

1. Introduction

An Access control framework of web service is to limit the web service user who tries to access the functions of the web services. Each model will have access control policies (rules) which are used to decide the credentials of specific user. Hence this is needed to know the movement of unauthorized users and grant the authorized users to access the web services. The following are the few traditional access control models [1]:

Role Based Access Control (RBAC) is the access control scheme which was widely used in web services. In RBAC, clients will be assigned roles which grants permission in order to provide an authorization to access a particular web service.

Attribute Based Access Control (ABAC) approach allows or denies the web services users based on their attributes such as location, role, attributes related to the environment.

Trust Based Access Control (TBAC) is varying from other access control models. In this model, web services access permission will be given based on the trust value of the user. Trust value will be calculated by the parameters such as user behaviours, success rate, failure rate etc., Hence, violation or misbehaving actions leads to decrease of trust value.

The following sections of the manuscript are organized as follows. In Section II, background concepts of trust based access control scheme is discussed. In Section III, the proposed work is presented. Section IV explains experimentation and performance analysis. Section V concludes the working of proposed access control model and future enhancement of system

2. Background of Literature Survey

There are many researchers have done their research in web services access control models and it approaches. Some of the researchers follows Naive Bayesian algorithm cantered trust calculation model [Hien et al 2010] for the selection of web service. There are three main sources for trust calculation such as QoS, direct experience and user recognition.

Vivekananth developed a user behaviour based trust calculation model that describes the reliability of user behavior [2].A trust threshold cost is used to grant permission or ignore the users from accessing the web services. If the value of trust is greater than threshold then the access will be allowed or denied. Trust ranges from 0 to 1.

Stefania Galizia proposed a trust based access scheme to access web service [3]. It has Third Party centered method for the grouping the web services based on the support of internet service tool.

Wang Meng presented a recommendation credibility based trust model [4] for web services access. This method varies honest and dishonest recommendation and dynamically corrects the value of trust.

Shangzhu Jin developed an access scheme in which trust value is computed using the factors time decay and feedback [5]. It is not dealing with assigning trust value for the newly registered users.

Surya Nepal proposed a trust management method based on trust to access web service [6]. But they failed in addressing issues like dishonest, propagation, trust bootstrapping.

Shashi Bhanwar proposed an entrance control demonstrate in light of trust by deciding notoriety and trust of the area on the premise history of past exchanges and appraised value from the feedback [7].

Srivaramangai proposed a model based on reputation of the users to improve consistency of transactions [8]. Consistency may be achieved by forming reciprocal trust between the service provider and requester.

Cesar Ali proposed another trust framework [9] to get to the web administrations in light of setting and part of the administrations requester. Here too they neglected to deal with new client trust esteem viably.

The above web services access models have not handled newly registered users trust value calculation effectively and study few parameters to evaluate the trust value for the user who request service. Also it is not considering the failed transaction due to server errors. Proposed work handles the issues by incorporating the trust value calculation using multivariate such as timeout, access frequency, network conditions, and success and failure rate. Also this proposed system has additional features such as IP address spoofing and SQL injection prevention to tighten the security features.

3. Proposed System

The proposed framework design is given in the Fig.1.It is partitioned as 1. Web Service requester is the individual or framework which gets benefits from web services 2.Web service provider will give the support to the requester and deals with the trust calculation. The different sub processes of service provider are given below

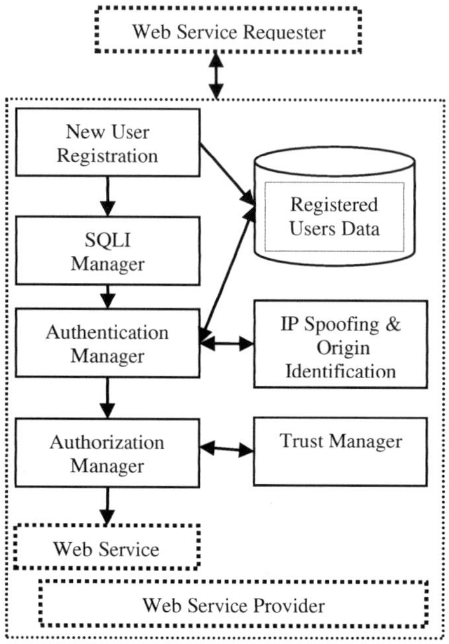

Fig 1: Proposed Framework

3.1 New User Registration

This component is used for registering new users and it has two sub processes.

1. User Registration: All new users have to register their environment and personal details with the web service provider to access the web service. While the new user registration, service requesters will be given unique id and passcode to use the web service. Once the registration is done, the new user details will be stored and request will be sent to SQL Injection Manager.

2. Registered Users Data: This is a repository where the new users data like id, passcode, IP address of last transaction, date and time of last transaction, requested pages, number of hits, transaction status code like '400', '408' and '200 and details of new users such as name, address, contact details etc., Later these data will be used in the authentication process.

3.2. SQL Injection (SQLI) Manager

This SQLI Manager Module detects and prevents SQL Injection occurs at web service provider side. After user registration process is completed the SQL request will be passed to the SQLI Manager. It has following sub processes to detect and prevent SQL Injection [13].

1.XML Generator: This process will be analysing the intercepted SQL query and generate XML file by converting the input values from SQL Query to detect SQL injections are present. This newly generated file will be used in validation process in order to detect the SQLI vulnerabilities.

For example, assume the SQL query select * from userdet where uname='Mathew' OR 1=1 - - 'password=''for login into the applications. The generated XML Code after the interruption of SQL query is shown in Fig 2 which has input values given in query. This file will be validated to check the presence of SQL Injection.

```
<? xml version="1.0" encoding= "utf-16"?>
<sqlexper>
<input>Mathew</input>
<input>'1' ='1</input>
</sqlexper>
```

IOP Conf. Series: Materials Science and Engineering **590** (2019) 012016 doi:10.1088/1757-899X/590/1/012016

```
</xml>
```

Fig 2: Sample XML file for the SQL Query

2. XML Validator: The Validator process is validating the XML file. Before validating XML file, XSchema will be created to verify the SQL injection depends upon the input parameters of XML file. Sample XSchema of given XML is given in following Fig.3:

```
<? xml version=”1.0” encoding= “utf-16”?>
<xs: element name=”input1”>
<xs: simpletype>
<xs: restriction base=”xs: string1”>
<xs: pattern value= “[a-zA-Z0-9]>
<xs: restriction>
<xs: simpletype>
<xs: element>
<xs: element name=”input2”>
<xs: simpletype>
<xs: restriction base=”xs: string1”>
<xs: pattern value=“ [a-zA-Z0-9]{8}>
<xs: restriction>
<xs: simpletype>
<xs: element>
<? xml>
```

Fig 3: XSchema File for XML Validation

The above XSchema file as in Fig 3 used to validate the XML file to detect the SQL injection. But this Schema can detect only tautology type SQL injection. If the XML validation is success then process concludes that there is no SQL injection and process is allowed to move the next authentication process. If failure occurs then this process will be terminated the user.

3.3. Authentication Manager

1. Data Categorization: The data such as uid, passcode, URL Requested, service requester IP Address are segrigated from server log file and call the authentication process by sending the filtered data

.

2. Origin & IP Spoofing Identification: The responsible of this process is to recognize Ingress Packet filtering based IP spoofing. Ingress packet filtering verifies the network range of IP address. If it violates the range of network address then user will not be allowed to access the web service.

3. Authentication: Final authentication will be taken place in this process by evaluating and sieving the details from of server. once it gathers the details of server log of requesters, it recovers the prior history of requester from the registered user data using user id. The authentication process has three cases as shown in the given algorithm [15]. Authorization processes will be initiated by passing user id, passcode, IP address and Requested URL Page once the user is authenticated.

```
UserAuthent (usrid, pcode, ipadd, requrlpage)
{
Cat 1: If usrid, pcode are legal and ipadd is
          registered and no spoofing
          Step1: Permit to authorization interface.
Cat 2: If usrid or pcode is  not valid and ipadd is
          registered and not spoofed
           Step 1:Verify the requrlpage  is visited
&   Total hits > threshold value &
Past incorrect req< threshold value  If yes
print "Enter name and password correctly"
Else
 print "Access Denied"
Cat 3: If usrid or pcode is not valid and ipadd is
not registered
              Step 3: Deny to access web service.
}
```

Fig 4: User Authentication Algorithm

3.4 Authorization Manager

It is playing a vital role in user authorization which decides whether to permit or deny the user from accessing of web service based on the trust value calculated by trust manager. Trust manager is the process which evaluates the trust value of requesters dynamically. This model defines the values trust from 0 to 10 and threshold level of trust as 0. If each user trust value is greater than the threshold level, then service user may access the web services else will be ignored. The trust manager is described follows [16].

1. Trust Assessment Point (TAP): This is the initial process of trust manager which verifies the trust value initially and avoids the user. TAP allows or denies user from accessing web services based on current trust value. If the user is new to the system, TAP will assign 0 as trust value and allows the requester to avail the web service.

2. Trust Managing Point (TMP): It is dynamically computes the value of trust based on below parameters that are accessed from the database of registered users and the types of users.

1. *Success Rate (St)* is the count of transactions of a user which are success during the specific period of time.Service provider will define the range of time.

2. *Failure Rate (Ft)* is the count of failed transactions of a user for a quantified period of time.

3. *Frequency of Access (Af)* is the frequency value of accessing the service by the requester. If the Af value of requester is lesser than threshold level which was assigned by service provider, then the service requester can be considered as lazy requester.

4. *Time-out (To)* bethe count of time-out occurs while requester is accessing the resource. It is used to identify the honest users.

5. *Average Time (At)* is the time spent while requester accessing the web service. It is also used to identify the honest requesters.

TMP classifies the requesters into four kinds in light of their conduct in the framework. They are 1.Honest and Active 2.Dishonest and Active 3.Dishonest and Active 4.Dishonest and idle. In the event that the time out (To) and Average time (At) are more noteworthy than its edge esteem (Vo), at that point requester is accepted as genuine else untrustworthy requester. In the event that recurrence of access (Af) is more noteworthy than its limit level, at that point the requester is considered as dynamic requester else idle requester. pf determines punishment factor. Framework expect punishment factor as 0.5. The proposed framework has following trust value computing techniques.

Type 1 If the requester is perceived as genuine and active, then requester's trust value (dt) will be increased as given below:

$$dt^{t+1} = (dt^{t} + (St+Af)*dt^{t}) / (St+Ft) \qquad (1)$$

Type 2: On the off chance that the requester is perceived as fair and idle or deceptive and dynamic, at that point requester's time decay value (dt) will be diminished as below:

$$dt^{t+1} = (dt^{t} - (St+Af)*dt^{t}) / (St+Ft) \qquad (2)$$

Type 3: In this type the user is realized as inactive and dishonest, then user's decayed trust value (dt) will be decreased by subtracting penalty factor as follows:

$$dt^{t+1} = (dt^{t} - (St+Af)*dt^{t}) / (St+Ft) - pf \qquad (3)$$

TMP checks the whether any server disappointments happens in past exchange. On the off chance that there is a server disappointment,TMP sends demand to TNP for trust esteem transaction with the goal that legit clients won't be rebuffed.

3. Trust Negotiation Point (TNP): This process is to negotiate the trust value of service requester who is affected with server failures. The following two factors will cause the trust value to negotiate.
1. *Server Error rate (Se)* is the value to identify the count of time-out or disconnect occurs in the transaction occurs due to Server failure. Since the server error occurs sometimes, the value of trust will be lessened to honest requester. To evade the problems, TNP checks whether the past access has correspondence failures because of server.
2. *User class (Uc)* identifies the class of user and reward them based on their consistency. Platinum user is a user who maintains trust value of 7 and above consistently for last 3 months.

Consequently if there are any server disappointments or customer continues with platinum customer class, TNP will counsel by including Negotiate Factor (Nf) of 0.5 with the trust esteem got from TMP. Game plan process will orchestrate honest to goodness and dynamic customers for 3 times to stimulate their honest to goodness and liveliness. At long last TNP sends last trust an incentive to TDP which sends the incentive to approval interface. Finally Authorization Interface picks whether to allow or deny the requester to get to the entrance of web benefit in light of the incentive from the Trust Manager.

4. Simulation and Result Analysis

To check the performance of the proposed system and obtain the experimental data, a framework was created. An e-Library application was developed on top level of the framework and implemented the functionality of the proposed system. A web service was created at low level of framework to search and download the necessary books. The e-library web application was tested in LAN environmental setup for 15 days with about 250 registered users. Simulation result was recorded and compared with existing models.

4.1. Experiment 1

In experiment 1, how the system is responding to handle malicious user based on ingress packet filtering technique for definite days. Hence the proposed work was purposely assigned few specific numbers of attackers to spoof IP address. The resultant data of experiment have been given in Table 1.

Table 1: Fake Fatalities Identification

Day s	Victims Count	Days	Victims Count
1	21	8	21
2	25	9	23
3	25	10	24
4	31	11	27
5	43	12	40
6	43	13	40
7	38	14	36

4.2. Experiment 2

In the analysis 2, consider the values [12] of dt=0.68, Ft=95, St = 75 and Frequent of access (Af) ranges from 15 to 100. Frequent of access (Af) edge an incentive for a particular period is accepted as 100. Here To, Af are defined that they are not as much as their limit esteem and Penalty factor (Pf) value is 0.5. Resultant trust esteems in view of frequent of access (Af), Time out (To), Access Time (At) and Penalty Factor (Pf) are recorded in Table 2.

Table 2: Resultof Including Multifactor in Trust Calculation

Access Frequenc y (Af)	Trust Value Based on Af	After including To, At with Af	After IncludingAf with To, At, Pf
15	0.43	0.40	0.39
25	0.47	0.46	0.41
35	0.43	0.39	0.37
45	0.48	0.41	0.39
55	0.50	0.59	047
60	0.54	0.51	0.47
70	0.57	0.54	0.50
80	0.61	0.59	0.54
90	0.63	0.60	0.57
100	0.65	0.76	0.62

IOP Conf. Series: Materials Science and Engineering **590** (2019) 012016 doi:10.1088/1757-899X/590/1/012016

From the data shown in table 2, it is concluded that value of trust is increased as value Af is increased. It implies that if the client is dishonest and inert, trust esteem will be influenced. The qualities from this table have been taken for the further investigation done beneath and demonstrated this proposed framework is limiting vindictive client effectively

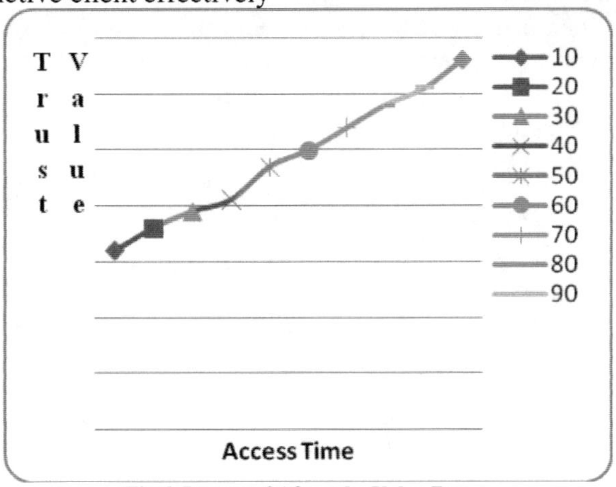

Fig 6. Impact of Af on the Value Dt

From the chart in Fig 6, we can see that the trust estimation of noxious requesters will be consistently diminished if the requester's frequent of access (Af) is lower than edge esteem generally trust esteem will be expanded. Henceforth we can reason that the lower the estimation of Af is the quicker decrease for the estimation of dt, which viably urges the requester to give legit get to and keep away from noxious access. Nonetheless, those pernicious practices can't be illegal essentially by Access recurrence.

4.3. Experiment 3

In this experiment, Trust values were ascertained utilizing the recipe and recorded in Table 3 in light of estimations of St, Ft of the client. Likewise estimations of Af, To, At and Pf are considered to figure the esteem.

TABLE 3: Impact of multivariate on trust value

St	Ft	Af	St,Ft,Af, To Based (Dt)	Af ,To, At Based on (Dt)	Af,To, At, Pf Based Dt
15	55	15	0.328	0.31	-0.233
25	45	25	0.556	0.521	0.05
35	35	35	0.834	0.81	0.42
45	25	45	1.16	0.92	0.52
55	15	55	1.46	1.19	0.79

It can be concluded that the rate of success will be increased as trust value also inreased or value will be decreased. In Table 3 above, dt is computed based on the factors St, Ft, Af, To and Pf and values are recorded.

The chart in Fig 7 demonstrates the adjustments in trust an incentive as indicated by multifaceted, for example, St, At, Af, To and At. This diagram has been drawn in light of 3 sets of information 1. In view of St, Ft, Af and To 2. In light of St, Ft, Af with To, At 3. In the view of St, Ft, Af with To, At, Pf. This diagram presumes that untrustworthy and idle requester's trust esteem is relentlessly diminished when trust esteem is computed utilizing the multifaceted, for example, St, Ft, Af with To, At, Pf. Likewise discovers that if service user is straightforward and dynamic esteem develops gets expanded. In the wake of including Penalty factor, untrustworthy and latent client trust esteem will be corrupted tremendously.

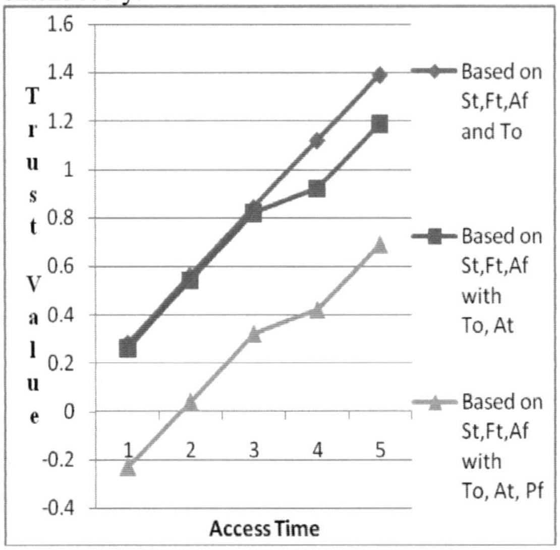

Fig 7. Multifactor Effect in trust computaion

In any case, Honest and dynamic users might be viewed as dishonest and idle user in view of server disappointments. To maintain a strategic distance from this issue rebuffed requester trust esteem will be included with Negotiate factor (Nf) so it will energize requesters. In our framework, 0.5 was set as Negotiate factor (Nf). Likewise if requesters achieve most extreme estimation of 10 and keep up a similar incentive for one month, those users may be identified as platinum client. Transaction process will likewise be improved the situation these clients for three times to support their genuine and liveliness.

4.4 Performance Analysis

This exhibits a correlation of proposed approach with prevailing web administrations get to control access models. Table 4 delineates the qualities and highlights upheld in the proposed access method and prevailing methods. The accompanying examination mirrors that our approach utilizes a system to appoint get to percentile to web administrations and utilizations. The chart in Fig 7 demonstrates the adjustments in trust an incentive as indicated by multifaceted, for example, St, At, Af, To and At. This diagram has been drawn in light of 3 sets of information 1. In view of St, Ft, Af and To 2. In light of St, Ft, Af with To, At 3. In the view of St, Ft, Af with To, At, Pf. This diagram presumes that untrustworthy and idle requester's trust esteem is relentlessly diminished when trust esteem is computed utilizing the multifaceted, for example, St, Ft, Af with To, At, Pf. Likewise discovers that if service user is straightforward and dynamic esteem develops gets expanded. In the wake of including Penalty factor, untrustworthy and latent client trust esteem will be corrupted tremendously.

IOP Conf. Series: Materials Science and Engineering **590** (2019) 012016 doi:10.1088/1757-899X/590/1/012016

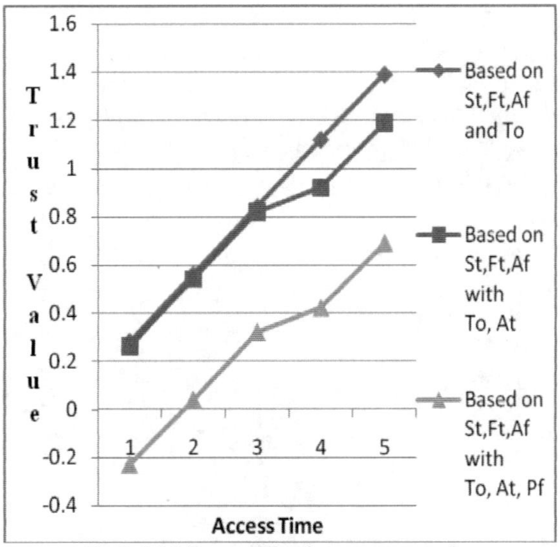

Fig 7. Multifactor Effect in trust computaion

In any case, Honest and dynamic users might be viewed as dishonest and idle user in view of server disappointments. To maintain a strategic distance from this issue rebuffed requester trust esteem will be included with Negotiate factor (Nf) so it will energize requesters. In our framework, 0.5 was set as Negotiate factor (Nf). Likewise if requesters achieve most extreme estimation of 10 and keep up a similar incentive for one month, those users may be identified as platinum client. Transaction process will likewise be improved the situation these clients for three times to support their genuine and liveliness.

Since it prevents SQL attack (Injection) and IP spoofing initially, the access to resources is lessened which brings about diminished approval work amid serving of access demands.

Table 4: Existing and Proposed Model Comparison

Characteristics	**RBAC**	**ABAC**	**Proposed system**
Distributed environment support	Yes	Yes	Yes
Trust level assignment	No	No	Yes
Reduced authorization work for access requests	No	No	Yes
Detect SQL Injection	No	No	Yes
Detect IP Address Spoofing	No	No	Yes
Policies specification & Maintenance	Simple	Complex	Simple

At long last the execution investigation of proposed framework was led with present conventional service control model, such as RBAC, ABAC and result [13] is appeared in Fig 8. Execution examination says that how effectively all the current models are confining malevolent clients or conduct. The investigation inferred that proposed show confines a larger number of clients than customary access control models.

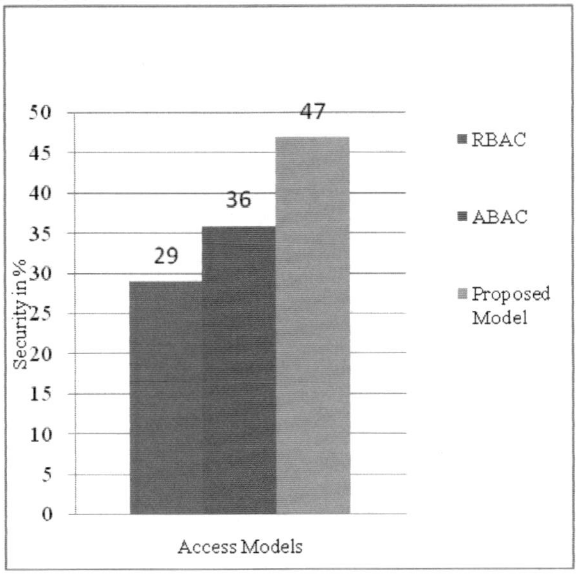

Fig 8: Comparing Security Parameter With Existing System

5. Conclusion

In this framework, a novel approach of web administrations get to control access of web services is proposed. Likewise talked about the energy of multifaceted, for example, Access recurrence, Time of access and Time out on trust an incentive to get to web benefit are examined. In the wake of presenting the SQL Injection and IP address parodying checking, security highlight is fix in the proposed framework. As an outcome, the security of web administrations got enhanced and this new model is proficient to rouse benefit requesters to participate in get to control framework effectively and sincerely. At last by investigating the execution of the proposed framework it is reasoned that the proposed demonstrate capacity to recognize and keep the malignant clients or practices is superior to any conventional access models.

This framework will find new ideas to implement the authentication and authorization process more efficiently. It will also find the new parameters to calculate the requesters trust value more effectively. Framework will locate the diverse method to isolate the clients as indicated by their trust esteems with the goal that legit and dynamic clients will get more advantages.

References

[1] R. Joseph Manoj, Dr.A.Chandrasekar, "An Literature Review on Trust Management in web services access control", International Journal on Web Services Computing, Vol 4, No 3, pp.1-19, Sep 2013.

[2] Vivekananth, "A Behavior Based Trust Model for Grid Security", International Journal of Computer Applications, Volume 5– No.6, 2010

[3] Stefania Galizia, Alessio Gugliotta and John Domingue,"A Trust Based Methodology for Web

Service Selection", International Conferences on Semantic Computing, 2007.

[4] Wang Meng, Hongxia Xia, Huazhu Song, "A Dynamic Trust Model Based on Recommendation Credibility in Grid Domain", International Conference CiSE, 2009.

[5] Usha Nandini , Saravanan M , Albert Mayan J , Murari Devakannan Kamalesh , Mohana Prasad K (2018) , " Automatic traffic control system using PCA based approach", International Conference on Energy, Communication, Data Analytics and Soft Computing , pp.2387-2392..

[6] Surya Nepal, Wanita Sherchan and Athman Bouguettaya,"A Behaviour-Based Trust Model for Service Web, IEEE international Conference on Service Oriented Computing and Applications 2010.

[7] Bhanwar, S, Bawa, S, "Establishing and Evaluating Trust in a Grid Environment", 10th International symposium ISPAN'09, 2009.

[8] Renagaramanujam, Srinivasan, Srivaramangai P,"A Comprehensive Trust Model for Improved Reliability in Grid.", International Journal of Computer Applications Vol 5, No.7, August 2010

[9] Cesar Ali "CATRAC: Context Aware Trust and Role based Access Control for composite web services" 10th IEEE international Conference on computer and information technology (CIT 2010).

[10] Hien Trang Nguyen, Weiiang Zhao, Jian Yang, "A Trust and Reputation Model Based on Bayesian Network for Web Services", 2010 IEEE International Conference on Web Services.

[11] Asha P, Albert Mayan J, Canessane A (2018),"Efficient Mining of Positive and Negative Itemsets Using K-Means Clustering to Access the Risk of Cancer Patients" ,Communications in Computer and Information Science ,ICSCS 2018, Kollam, 2018,pp.373-382.

[12] X. Wang, J. Luo, A. Song and T. Ma, "Semantic Access Control in Grid Computing". Proc. 11th International Conference on Parallel and Distributed Systems, 2005

[13] Shanshan Song, Kai Hwang, and MikinMcwan, "Fuzzy Trust Integration for Security Enforcement in Grid Computing" NPC '04, LNCS 3222, pp. 9-21.

[14] R. Joseph Manoj , Dr.A.Chandrasekar, M.D.Anto Praveena, "An Approach to detect and tautology type SQL injection in web services based on web services based on XSchema Validation", International Journal of Engineering and computer science, Vol 3, No 1, pp.3695-3699, Jan 2014.

[15] R. Joseph Manoj, Dr.A.Chandrasekar, "An Authentication System of web services based on Web Server Log Analysis" International Journal of Engineering and Technology,Vol 5, No 6, pp.4786-4793, Jan 2014.

[16] R. Joseph Manoj, Dr.A.Chandrasekar, "An Access control model of web services based on multifactor trust based management, International Review on Computers and software", Vol 8, No 10, pp.2460-2466, 2013.

[17] Albert Mayan J, Anto Praveena M D , Telkar Bharath Rao , Uravakonda Uday Sagar (2018) ,"Optimized test data generation over suspicious implementation of oracle problem",IEEE International Conference on Power, Control, Signals and Instrumentation Engineering ,pp.2559-2563.

LARYNGEAL TUMOR RECOGNITION AND CLASSIFICATION USING NEURAL NETWORKS

S. Karthikeyan
Associate Professor, DeptofECE,
Sathyabama institute of Science and technology , Chennai..
skarthi1879@gmail.com

P.Farhath
U.G Student, Dept. of ECE,
Sathyabama institute ofScience and technology , Chennai

M.Vineesha
U.G Student, Dept. of ECE,
Sathyabama institute ofScience and technology , Chennai

T. Ravi
Associate Prof., Dept. of ECE,
Sathyabama institute ofScience and technology , Chennai .

Abstract : One of the highly challenging fields is Biomedical Image Processing. Image Processing involves many techniques used to detect and classify the tumor more efficiently than doctors. Manual classification is a time consuming and gives inaccurate results compared to automatic classification. Imaging methods like CT (Computer Tomography) scans, MRI (Magnetic Resonance Image), X-rays plays a major role in tumor detection and classification. Out of these MRI is widely used because it does not involve any radiation and also identifies even the smallest abnormalities present in throat. The whole process of detecting the laryngeal tumor from these scanned images can be categorised into four steps: Pre-processing, Segmentation, Feature Extraction and Classification. In proposed method, classification involves hybrid classifiers to improve accuracy.

Keywords: MRI (Magnetic resonance imaging), laryngeal Tumor, Segmentation, Feature extraction, Classification.

1. INTRODUCTION

In today's world, cancer rate is increasing day by day. One of the most frequent types of cancer is laryngeal cancer. According to the estimation of American Cancer Society, about 13,150 new cases of laryngeal cancer may occur in 2018. Out of which 10,490 may occur in men and 2,660 in women. It is also estimated that 3,710 will die of laryngeal cancer [2]. Based on the location of larynx, the tumor is divided into many stages. If the tumor is not detected in earlier stage, it may spread to nearby lymph nodes or to distant parts [2]. Also, there will be many treatment options available if the tumor is detected sooner. Based on type of tissue and location of tumor, it can be categorised into two types: Benign (non- cancerous) and Malignant (cancerous). Cancerous tumors are again divided into primary and secondary tumors. Primary tumors occur due to uncontrolled growth of cells within in the laryngeal whereas secondary tumors occur due to uncontrolled growth of cells which spreads from

IOP Conf. Series: Materials Science and Engineering **590** (2019) 012017 doi:10.1088/1757-899X/590/1/012017

other parts of the body to laryngeal [9]. Manual detection and classification of laryngeal tumor is time consuming and challenging task because of its location, size and appearance of tumor in laryngeal [3]. Image Processing and artificial neural network techniques were used for automatic classification of tumor which helps to give right treatment. Tumor can be detected using various imaging tests like MRI (Magnetic Resonance Imaging), MRS (Magnetic Resonance Spectroscopy), CT (Computer Tomography) and Barium swallow [3]. Among all, MRI is the most frequently used imaging test to detect the laryngeal tumor. Because it provides a detailed location and shape of the tumor in throat without involving any radiation. Generally, MRI images contain noise. This noise has to be removed in order to enhance the image quality[4].

This process is classified into two types:

1. Spatial Domain
2. Frequency Domain

Spatial domain acts directly on the intensity values of the image. The techniques involved in spatial domain are Histogram equalisation, Threshold Method, Filtering techniques.(maximum, minimum, average, adaptive, gabor filter etc.,) It is also used for gray level transformations. Frequency domain is another process which directly acts on the transform coefficients of images. It is less complex and ease of viewing. Image smoothing can be done here. One of the main problem in tumor detection is segmentation. It is a process which splits an image into multiple set of pixels. Each pixel is assigned a label in which pixels having same labels are grouped together and also share certain characteristics in common. The segmented image contains the boundary of the objects and also gives features of tumor region like area, eccentricity, orientation, bounding box. The image segmentation mainly depends upon different techniques like region based methods, threshold, cluster, edge, pixels. After segmentation, features are extracted from that image. Feature Extraction is the process of extracting desired features of an image which can be used for further processing.

Normally, it can be classified into two types:

1. Texture feature extraction
2. shape feature extraction

Techniques like DWT (discrete wavelet transform), counter based methods, GLCM (Gray Level Co-occurrence matrix) and Region based are used to extract features. Classification of tumor depends upon the feature extraction[4]. Classification of tumor into normal and abnormal (Benign & Malignant) depends upon the classifiers. Some of the algorithms used are Artificial Neural Networks, Support vector machine, Adaboost Classifier, Convolutional Neural Network (CNN)[3].

The other part of the paper explains: Proposed system (discussed in section 2), its simulation results (discussed in section 3) and conclusion of the work in section 4.

2. EXISTING SYSTEM

In the existing system, the process of detection of tumor is explained as in the below Fig (1):

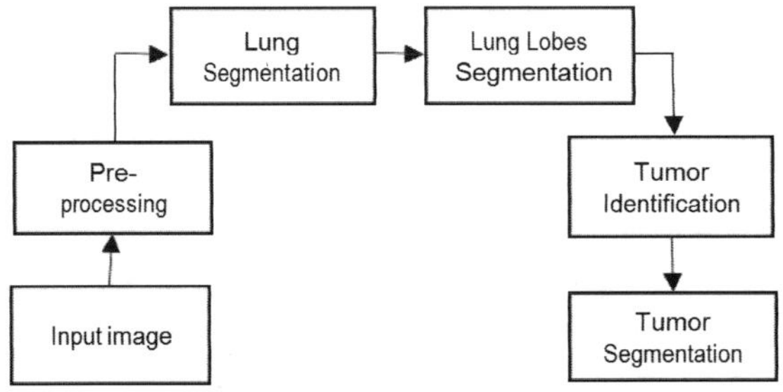

Fig. 1 Existing Method Block Diagram

Normally, in MRI images while converting optical signals into electrical signals and then into digital signals, noise is introduced. This noise has to removed using pre-processing since it changes the intensity values of the pixels. Pre-processing is a process to remove noise and enhance the image quality by using different filtering techniques. In pre-processing, MRI image is given as input and is converted into grayscale image. Image binarization is applied to convert gray scale image into black and white image. Later, image is segmented which splits the image into different parts. The features like contrast, correlation, clustering, energy, entropy ratio, kurtosis factor are extracted from MRI image. These texture features are used for training. While testing, the same features are extracted from the testing image and is compared with trained feature values. SVM (Support Vector Machine) is used for this purpose. The working principle of SVM can be easily understood by the following Figure (2). Training of SVM classifier is very easy. It is used to classify the tumor into normal or abnormal. Using these training features, the hyper plane is calculated which is used to differentiate the two classes. This classifier gives unique solution by using kernel functions which is used for non linear separation. But selecting the kernel function is tedious task. Also, SVM are not much effective if the boundaries are not known exactly. After classifying, the segmented tumor is displayed using EK means algorithm. This clustering permits each member to be part of many clusters. But it is difficult to predict k- value. Also, it is not effective when the clusters are of different size and density.

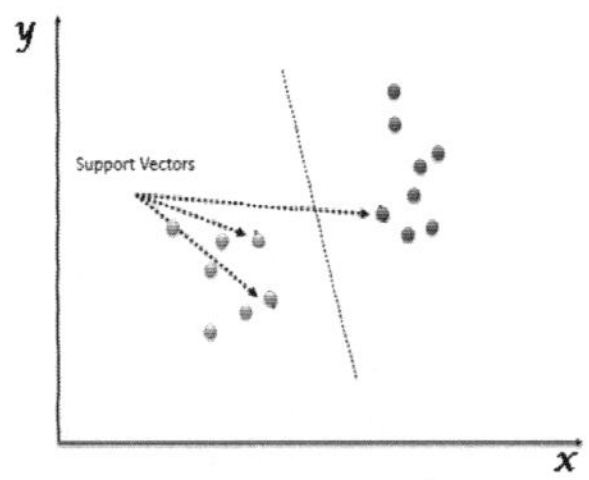

Fig. 2 SVM Classifier

3. PROPOSED APPROACH

In the proposed system, the process of detection of laryngeal tumor can be categorised into 4 steps:

- Pre-processing
- Feature Extraction
- Post-processing
- Classification

The block diagram of proposed system is shown in figure (1).

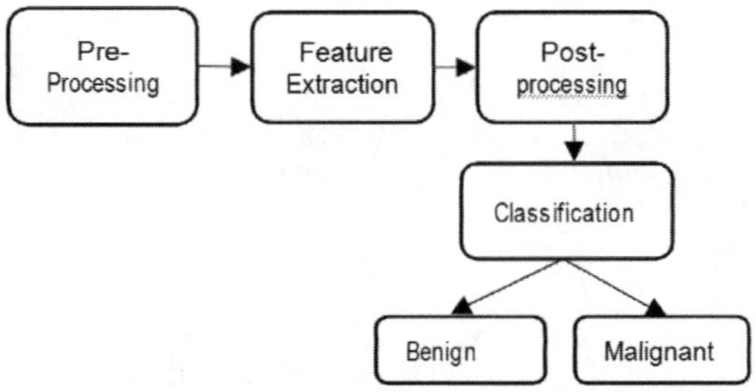

Fig.3. Block diagram of the proposed system

3.1.Pre-processing:

Firstly, MRI image is taken as input which is shown in Figure (2). But noise is introduced while converting optical signals into electrical signals and then into digital signals. This noise is removed in the process of pre-processing. Firstly, input image is resized and then converted into gray scale image as image may be colored.

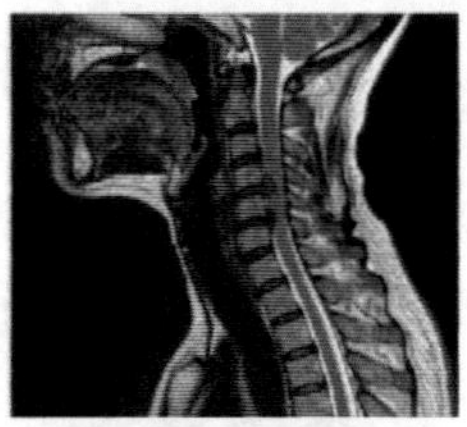

Fig.4. Input MRI image

This converts the 3D pixel value of an image into 1D pixel value thereby reducing the complexity. The mathematical parameters of the image are extracted and the image is segmented using canny edge detection.

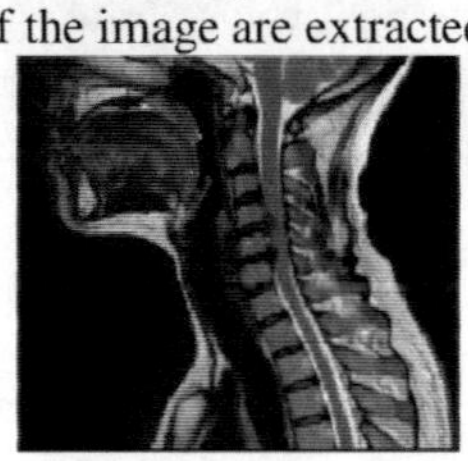

Fig.5. Filtered image

3.2.Feature Extraction:

Texture is one of the most important characteristics used to identify the region of interest. This texture features of the image are extracted using GLCM.(Gray Level Co-occurrence Matrix). It also calculates how often a pixel with gray scale intensity value i occurs either vertically, horizontally or diagonally with neighboring pixels of value j. The following features are extracted from GLCM

1. Energy: It detects disorders in textures i.e., it calculates the uniformity of the texture.

$$\text{Energy} = \sum_{i,j} p(i,j)^2 .$$

2. Contrast: Contrast is the difference between highest and lowest values of contiguous set of pixels.

$$\text{Contrast} = \sum_{i,j} |i-j|^2 \, p(i,j) .$$

3. Homogenity: It returns a value that calculates the closeness in distribution of elements in GLCM to that of GLCM to that of GLCM diagonal. It varies from 0 to 1.

$$\text{Homogeneity} = \sum_{i,j} \frac{p(i,j)}{1+|i-j|} .$$

IOP Conf. Series: Materials Science and Engineering **590** (2019) 012017 doi:10.1088/1757-899X/590/1/012017

4. Correlation: It returns the measure of how much a pixel is correlated to its neighbour over the complete image. It varies from -1 to +1

3.3. Post-processing:

The pre-processing of an image may be distorted with noise and texture. Morphological operations can be used to eliminate this distortions and enhance the result.). These operations process the image based on their shape. It depends upon the relative ordering of the pixels rather than the intensity value of the pixels. Here processing is done by using a small shape called structuring element which is positioned at different locations of the image and is compared with neighbourhood pixels. The two basic operations used are Dilation and Erosion. Pixel values are added to the boundary of objects in dilation and is removed in case of erosion. The image obtained after morphological operation is show in Fig(4).

Fig.6. Morphological image

Here, erosion operator is applied which shrinks the foreground by assigning minimum pixel values of neighbourhood pixel values in input image. Later, dilation operator is given which expands the foreground by assigning maximum pixel values of neighbourhood pixels values in input image. The following figure (5) shows the flow of proposed method:

3.4.Classification:

CNN (Convolutional Neural Network), KSONN (Kohonen's Neural Network), ANN (Artificial Neural Network) are used for this purpose. Using these hybrid classifiers, disadvantage of each classifier is overcomed with other classifiers. CNN is commonly used to identify the objects. These neural network learns directly from the image data without manual feature extraction. They have hundreds of layers which learn to detect the features of image by applying filters at different resolutions. The output of image is given as input to the next layer. It is convenient as it learns without much training dataset and complexity. Next to it, KSONN (Kohenen's self organising neural networks) is applied. It is ideal for obtaining low dimensional view of high dimensional data. It applies competitive learning to a set of input data opposing error correction learning like other neural networks. It is best for accomplishing functions on unknown data to describe hidden structures in it. At last, ANN is applied to classify the tumor into benign or malignant. In this way, it provides accurate results when compared to single neural networks which has accuracy more than 90%.

IOP Conf. Series: Materials Science and Engineering **590** (2019) 012017 doi:10.1088/1757-899X/590/1/012017

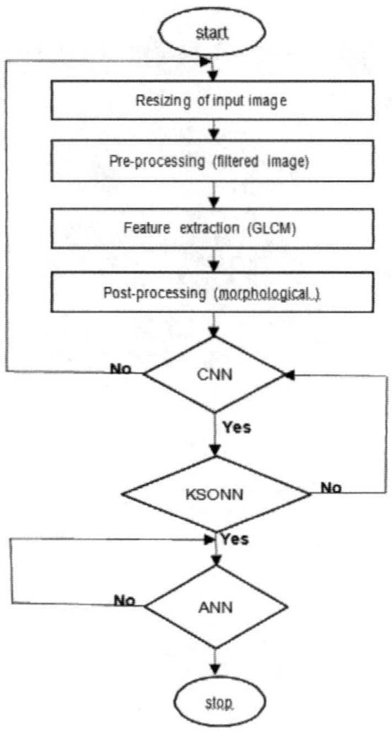

Fig.7. Flow of Proposed Method

4. SIMULATION RESULTS

The following table (1) shows the values obtained for different features extracted from different MRI images:

Parameters	MRI 1	MRI 2	MRI 3	MRI 4
Correlation	0.88575	0.8911	1.8103	0.9052
Energy	0.20295	0.21665	0.20435	0.20435
Contrast	0.84765	1.0863	0.96945	0.96925
Homogenity	0.81595	0.8278	0.81545	0.81545
PSNR	97.4983	97.7757	98.0473	97.7365
MSE	95.6417	95.9192	96.1907	95.8799
Accuracy	97.7483	98.0257	98.0473	97.9865

Table-1. Values of the features extracted from GLCM

The figure (8) displayed below shows the image obtained after performing pre-processing techniques:

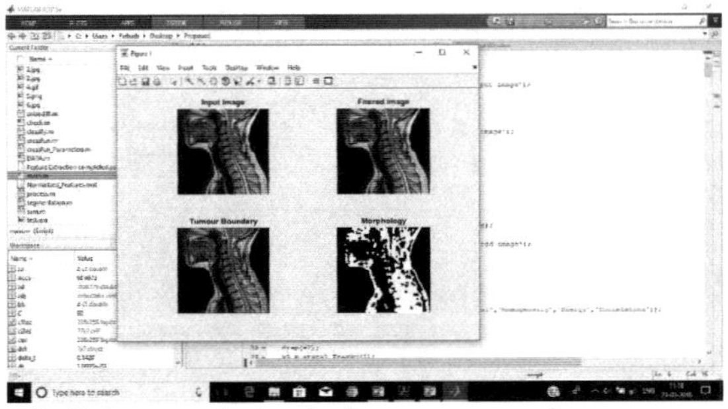

Fig.8. Images obtained after pre-processing

The figure (9) displayed below shows the image obtained after performing post-processing techniques:

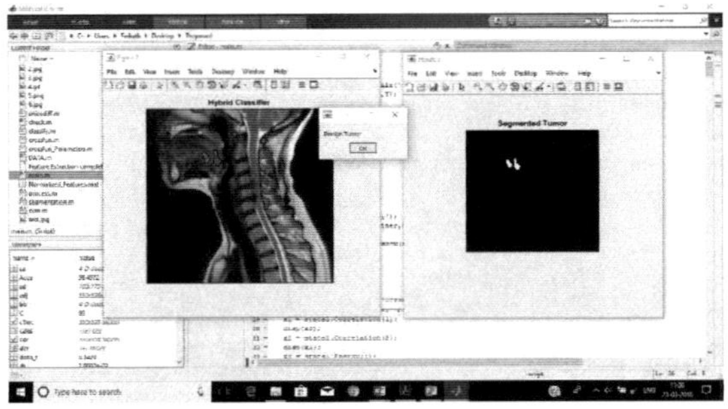

The figure (10) displayed below shows the values obtained from the extracted parameters:

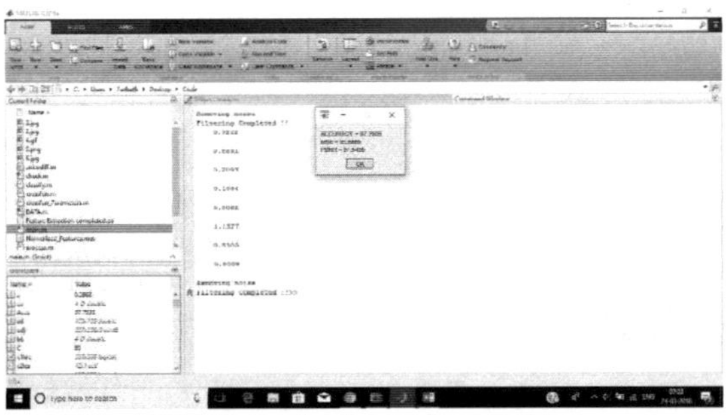

Fig.10. Values of extracted parameters

5. CONCLUSION

In proposed system, our goal is to automatically and statistically diagnose the type of tumor in the throat by using MR images. Region of interest, filtration and segmentation techniques were utilized to be the base of the information to get the statistical data for each case. It has been proven that a simple, harmless and accurate statistical technique can efficiently distinguish between the malignant and benign tumor (TCa). The huge advantage of this approach is that there will be no need to make any further tests or examinations on the patient after making the MRI. The diagnosis system achieves accuracy over 97.95% in differentiation between the different tumor types on various qualities of MR images. The results of the experiments and application of the proposed method may also be applied.

REFERENCES

[1] Gopi K, Selvakumar J, "Lung tumor area recognition and classification using EK-means clustering and SVM", IEEE-International Conference on Nextgen Electronic technologies, 2017.

[2] "Laryngeal Tumor Statistics"- Cancer.org. [Online]. Available: https://www.cancer.org./cancer/laryngeal-and-hypopharyngeal- cancer/about/key-statistics.html.

[3] Larisha Rayen, B., Monica, B., Karthikeyan, S., Sivakumar, V.G.," Multi-band circularly polarized antenna for 2.4/5.3/5.8 ghz WLAN and 3.5 ghz WI-MAX applications", Journal of Advanced Research in Dynamical and Control Systems, 2017,vol. No:16, PP: 1311-1322.

[4] Manikandan, R., Karthikeyan, S., Balaji, S.R., Vasudevan, V., Shalini, S. " An internal startup circuit for pacemakers using body temperature", International Journal of Pure and Applied Mathematics, Volume 117 No. 21, december 2017, PP:155-162.

[5] Karthikeyan, S., Venkata Nagi Reddy, M., Mazeed Ahamed, M., Sivakumar, V.G., "A novel approach to cost-aware energy based routing protocol for wireless sensor networks", International Journal of Pure and Applied Mathematics, Volume 117 No. 21 2017, PP: 107-112.

[5] Rajesh C. Patil, Bhalchandra A.S, "Brain Tumour Extraction from MRI Images Using MATLAB" International Journal of Electronics, Communication & Soft Computing Science and Engineering ISSN:2277- 9477, vol. 2, no. 1, April 2012.

[6] Daizy Deb, Bahnishikha Dutta and Sudipta Roy, "A noble approach for noise removal from brain image using Region Filling," 2014 IEEE International Conference on Advanced Communications Control and Computing Technologies, 2014.

[7] Dubey R, M. Hanmandlu, and S. Vasikarla, "Evaluation of Three Methods for MRI Brain Tumor Segmentation," 2011 Eighth International Conference on Information Technology: New Generations, 2011.

[8] Hadeel Abdullah N, Mustafa A. Habtr," Brain tumor extraction approach in MRI images based on soft computing techniques ",IEEE- International Conference on Intelligent networks and Intelligent systems,2015.

[9] Albert Mayan.J,Surya.B,Pranoy Prabhakar,Princekumar,"Department–Student Library Using Twig Pattern Query Processing Over Admin-User Login Privilege", Pak. J. Biotechnol. Vol. 13 , Pp. 489 - 493,(2016).

[10]Ravindra Sonavene, Poonam Sonar," Classification and segmentation of brain tumor using Adaboost classifier",IEEE- International Conference on Global Trends in Signal Processing, Information Computing and Communication,2016.

[11] Kanwarpreet Kaur, Gurjot Kaur, Jaspreet Kaur, "Detection of brain tumor using NNE approach",IEEE-International Conference On Recent Trends In Electronics Information Communication Technology,2016.

[12] Murthy T.S.D and Sadashivappa G, "Brain tumor segmentation using thresholding, morphological operations and extraction of features of tumor," 2014 International Conference on Advances in Electronics Computers and Communications, 2014.

[13] Sathies Kumar T, Rashmi,K etal," Brain tumor detection using SVM classifier",IEEE-International Conference on Sensing, Signal Processing and Security,2017.

[14] Selvakumar J, Lakshmi A and Arivoli T, "Brain Tumor Segmentation and Its AreaCalculation in Brain MR Images using K-Mean Clustering and Fuzzy C-Mean Algorithm" IEEE-International

Conference On Advances In Engineering, Science And Management (ICAESM-2012) March 30, 31, 2012.
[15] Nandhini J.; Shabatini K, .Karthikeyan S," Wireless colour sensing arm robot", International Conference on Robotics, Automation, Control and Embedded Systems (RACE) Year: 2015 Pages: 1 – 6.
[16]Gopika. S. Kumar, S. Karthikeyan and N. Ambily, "A Metaphorical Scrutiny in Banishing Salt and Pepper Noise",BIOSCIENCES BIOTECHNOLOGY RESEARCH ASIA,December 2014 Vol. 11(3).
[17]G. Karudaiyar, S.Karthikeyan and B. Sainath, "Encryption and Decryption Scheme by Using Finite State Machine", BIOSCIENCES BIOTECHNOLOGY RESEARCH ASIA, December 2014 Vol. 11(3), 1867-1872.
[18]S.Karthikeyan, S.Jayashri, "Energy Utilization Strategies using Novel Approach in Wireless Sensor Networks", International Conference on Software Engineering and Mobile Application Modelling and Development, Dec 2012, pp: 1-5.
[19]S.Karthikeyan, S.Jayashri, "Energy Efficient System for Heterogeneous Wireless sensor Networks", European Journal of Scientific Research, Volume 72,No 4 march 2012, PP:599-607. ISSN: 1450-216X.
[20]S.Karthikeyan, S.Shanmugapriya and S.Jayashri, "Sub-head Transmission of Heterogeneous Data by Cloned Agent to Android Mobile",Research Journal of Applied Sciences, Engineering and Technology, 8(1):pp.24-34, 2014
[21]S.Pavithra, S.Karthikeyan, V.J.K.KishorSonti, S.Jayashri, "Competent Realization of Co-operative spectrum sensing in cognitive radio systems", International journal of Engineering systems modeling and simulation, Inder science publishers, Vol. 7, No. 2,pp 103-110 2015.
[22] Mayan J.A, Arifa S, Pavithra R,"Semantic based multi lexical ranking technique for an effective search in protected cloud,2016 International Conference on Control, Instrumentation, Communication and Computational Technologies (ICCICCT), Kuma
[23]K. Bharathi,S.Karthikeyan, "A Novel Implementation of Image Segmentation for Extracting Abnormal Images in Medical Image Applications", Indian Journal of Science and Technology, Vol. 8(S8), 333–340, April 2015.
[24]S.karthikeyan, Merin Mary Koshy and V.G.Sivakumar, "Smart & automated robotic circular car parking using GSM",International Journal of Pure and Applied Mathematics Volume 117 No. 21 2017,PP: 81-86.
[25]Hari Krishna, K., Swamy, Y., Siva Kumar, V.G., Karthikeyan, S. "Integration of two way communication using gesture control hand movement and voice based text", Journal of Advanced Research in Dynamical and Control Systems, 2017, vol. No:17 , PP: 1124-1135
[26]Meenakshi. R, Pooja Suppaiah , Siva kumar .V.G, Karthikeyan.S,"COST EFFECTIVE REAL TIME HOME ENERGY MONITORING AND MANAGEMENT SYSTEM" Journal of Advanced Research in Dynamical and Control Systems, 2017, vol. No:17 , PP: 1103-1111
[27] Albert Mayan J, R. Julian Menezes, M. Breezely George, "Designing a Customized Testing Tool for Windows Phones Utilizing Background Agents", International Conference on Soft Computing Systems (ICSCS), Advances in Intelligent Systems and Computing , vol.398, Springer,pp.33-46.
[28]Aashik JPM , Ajith M ,Mr.V.G.Sivakumar,Mr.S. Karthickeyan, " AN INTERNET OF THINGS APPROACH FOR MOTION DETECTION AND CONTROLLING HOME APPLIANCES USING RASPBERRYPI WITHOUT CLOUD SERVER",Journal of Advanced Research in Dynamical and Control Systems, 2017, vol. No:17 , PP: 1112-1123

[29]T.V.N.Surendra Gopi, Illuru Phani Kumar, Dr. S.Karthikeyan , Dr. V.G.Sivakumar "A Novel Method Of Dynamic Mobile Data Collection Scheme By Fuzzy Logic", Journal of Advanced Research in Dynamical and Control Systems, 2017, vol. No:17, PP: 1136-1149

[30]Kumar, M.K., Surya, J., Kumar, V.G.S., Karthikeyan, S."Online prediction of driver distraction based on brain waves activity patterns", International Journal of Pure and Applied Mathematics, Volume 117 No. 21 2017, PP: 139-146.

DOORBELL SYSTEM IN HOME USING IOT

B.Baron Sam[1], K. Purna Chander[2], K. Vinay[3], A.Pio sajin[4]

Assistant Professor[1],Student[2],Student[3], Assistant Professor[4]

School of Computing, Sathyabama Institute of Science and Technology, Chennai.

baronsam1988@gmail.com[1],itsmepurna66@gmail.com[2],kompalli.vinay@gmail.com[3]

Abstract. The Internet of Things (IOT) is the communication between different objects—gadgets, structures, vehicles, and different things—implanted with hardware, programming, sensors, and system availability. Burglary appropriation in light of IOT gives a framework. Web of Things is relied upon to create high level of human to machine correspondence alongside machine to machine correspondence. The work characterizes common access focuses to a home as essential and optional access focuses relying upon their utilization. Rationale based detecting is executed by recognizing ordinary client conduct at these entrance focuses and asking for client confirmation when important .We will secure the home from any undesirable individual. In this we will utilize IR and PIR sensor which is utilized to recognize the individual. Vibration sensor is utilized to identify the undesirable individual who break the glass. Every one of the exercises are screen by camera and every one of the parameters are gathered by the raspberry pi. At the point when undesirable individual distinguished means it naturally send warning to the portable, for example, Gmail (various client) and drop box utilizing IOT. An IOT ubidots stage which contains sensors information and a gui surrounded in that stage to open the entryway when we get the notice. We will get to a record in that stage which speaks to the working of sensors in it. By utilizing the conventions we will send the data to the client. We are utilizing python for building up the interface utilized as a part of this. So if client requires he can work it. Each datum is stockpiled cloud avant-garde in our ubidots account. Along these lines we are creating entryway chime framework utilizing IOT for savvy homes.

Keywords: Raspberry pi, PIR Sensor, vibration sensor, camera, Raspberry OS, Python, Alarm.

1. Introduction

Scientists have been testing and enhancing the strategy for savvy house from 1970s. As innovation progressed with time, electronic gadgets and use of web has turned out to be more prominent and reasonable, so the idea of home mechanization and people's desire from a shrewd home has changed drastically. Presently a-days shrewd home turned into an advanced mix of numerous Smart Devices and Sensor/Actor Networks which are remote. All these new client desires, entangled gadgets and unusual client conduct conveyed new security difficulties to the home mechanization front. The module of house security has additionally created with time.

The Internet of Things (IOT) is the communication between different objects—gadgets, structures, vehicles, and different things—implanted with hardware, programming, sensors, and system availability. A hostile to robbery framework is any gadget or technique used to avoid or stop the unapproved apportionment of things thought about profitable. Burglary appropriation in view of IOT gives a framework. Web of Things is relied upon to deliver high level of human to machine correspondence alongside machine to machine correspondence.

IOP Conf. Series: Materials Science and Engineering **590** (2019) 012018 doi:10.1088/1757-899X/590/1/012018

In spite of savvy home security being basic there are a few vulnerabilities in the current frameworks . Throughout the years specialists showed different security issues related with the gadgets and innovation utilized as a part of present day brilliant homes. The remote sensor systems sent in current shrewd homes for gadget to gadget correspondence is helpless against different Routing and Wormhole assaults. Mainstream correspondence innovations like ZigBee and 802.15.4 utilized as a part of shrewd homes are defenseless to Replay assaults. Every one of these variables added to the fast ascent in home robberies over the previous decade and exhibits the significance of Home Security in the cutting edge world. Our past works in keen home security clarifies the changing part of present day home security frameworks and characterizes the part of an advanced home mechanization framework as, one fit for recognizing, cautioning and averting interruption endeavors in a home in the meantime safeguarding confirmation of the interruption or endeavored interruption so that the perpetuator or perpetuators can be distinguished and arraigned.

This venture proposes the security framework utilizing IOT, which anticipates robbery in home, bank and so on. The essential target of this undertaking is to lessen human work. Computerization has dependably been a prime factor for security framework. We pointed in the task is to outline and execute a security framework. Framework that offers controllability through a hand held cell phone by methods for IOT

2. Related Work

Crafted by B.N. Schilit et al. proposed the utilization of Infrared (IR) matrices and wearable recognizable proof (ID) identifications to distinguish the area of a client at home and foresee the setting of client activities. IR lattice ended up being hard to execute in a home domain while the wearable ID labels ended up being badly designed and given misdirecting data to unpracticed and thoughtless clients. The creators likewise proposed the utilization of static protest checking in which an occupants area is related to regard to a static question in a home. Static protest checking seriously constrains the adaptability of the home condition and when these static items are moved the proposed framework is effectively tricked and can't adjust. The exploration of J. Choi et al. used body temperature, beat, outward appearance, room temperature, time and area to foresee and learn client setting. Their work neglected to consider the way that, clients body temperature, heartbeat or outward appearance may differ contingent upon different components like perspective, sickness and so forth. Besides, their work utilizes cameras to peruse outward appearances which when traded off by a technically knowledgeable assailant conveys new security and protection issues to the home.

The idea of treat was brought into the setting of the web program in 1994, by Lou Montulli . Treat enabled webservers to store little measure of information on the meeting client's PC which is sent back to the server upon ask. The idea of treat was immediately grasped by program producers. Before long, assailants started to exploit treat's state-full nature. Outsider promoting destinations utilized treats to track clients over numerous sites which supported behavioral publicizing . This security abusing conduct got the consideration of the examination group , lawful group and was a reason for worry among the general population . Additionally, treats are defenseless against Cross Site Scripting and Cookie Stealing. The idea of treat was additionally extended to Flash treats and later to ever treat which is relatively difficult to expel; this further improved the protection concerns related with treats. A treat maintenance consider demonstrated that one of every three clients erased their first and outsider treats inside a month of going to a site. The above inquires about show the protection, security and inaccessibility issues related with utilizing treats to recognize a client. So using treats in Home Automation to recognize a client over the web doesn't appear like a sensible choice.

IOP Conf. Series: Materials Science and Engineering **590** (2019) 012018 doi:10.1088/1757-899X/590/1/012018

2.1 Existing System

In the current framework, the house is secured by utilizing sensors and is conveyed through message write as it were. In the current framework just gsm module is utilized. No pictures are caught in the current framework. Camera is utilized just to monitor reason. In existing, entryway is opened when the individual is inside the house. The given equipment framework isn't exceptionally dependable. It isn't reasonable for all circumstance.

3. Proposed System

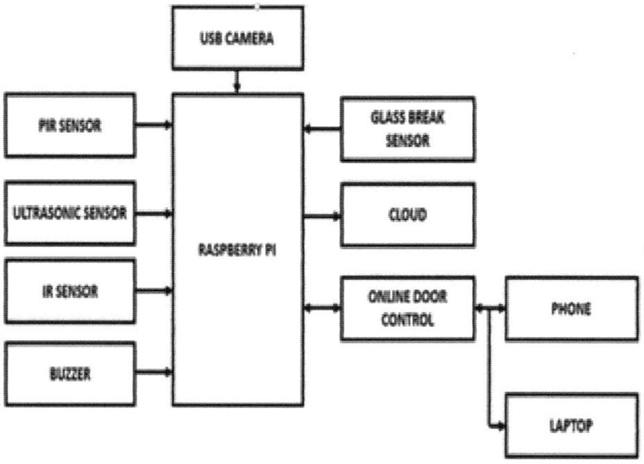

Fig1.Architecture

In the proposed venture we will utilize USB camera to catch the individual who are remain on the entryway side to catch that individual, and send them to the Gmail and drop box. To catch the individual utilizing two strategy. PIR sensor recognizes the human and catches that send to that validated individual. On the off chance that some of them wearing dark dress means the sensor can't recognize that on that time we actualize the picture preparing procedure called movement location it is utilized to identify the human and get catch the photograph. On the off chance that the individual is well-known means entryway open by confirmed individual and he is having the consent to open or close the entryway and furthermore naturally ringing the caution. The camcorder is additionally go through web. The IR and ultrasonic sensor used to distinguish the individual enter in the window side or in entryway side. The glass break sensor is utilized to distinguish whether any one break the glass of windows it is likewise gives the warning to the verification individual. These sensor are persistently checking and recording in cloud. Alongside PIR Sensor, we utilize vibration sensor, which can be useful amid an assault, it gives a ready when the glass separates in the house. IR sensor used to distinguish the individual who enter in the window side. These sensors persistently screen the house and the entire procedure records in cloud. By utilizing conventions we will send the message and it is executed in python. Fn Ubidots IOT stage we will make a GUI Interface to work the entryway which additionally contains sensors working. Subsequent to getting message ready we utilize the above interface to work the entryway.

Everything is put away in cloud and when the client gets the mail he will get a picture and the camera continues checking the environment and with the assistance of the engine utilized above opening and shutting of the entryway is controlled. The entire control of entryway that is opening and shutting is controlled by the motor. Buzzer is given when anything is distinguished as it is connected with the sensors and it gives some beep sound when it is detected. By utilizing conventions the data is sent to client and with the client's approval the entryway is opened from the encircled interface and the engine controls its movement. The entire sensors information is put away by time to time in our login in the ubidotS and it can be seen in it.

3.1. Ubidots Platform

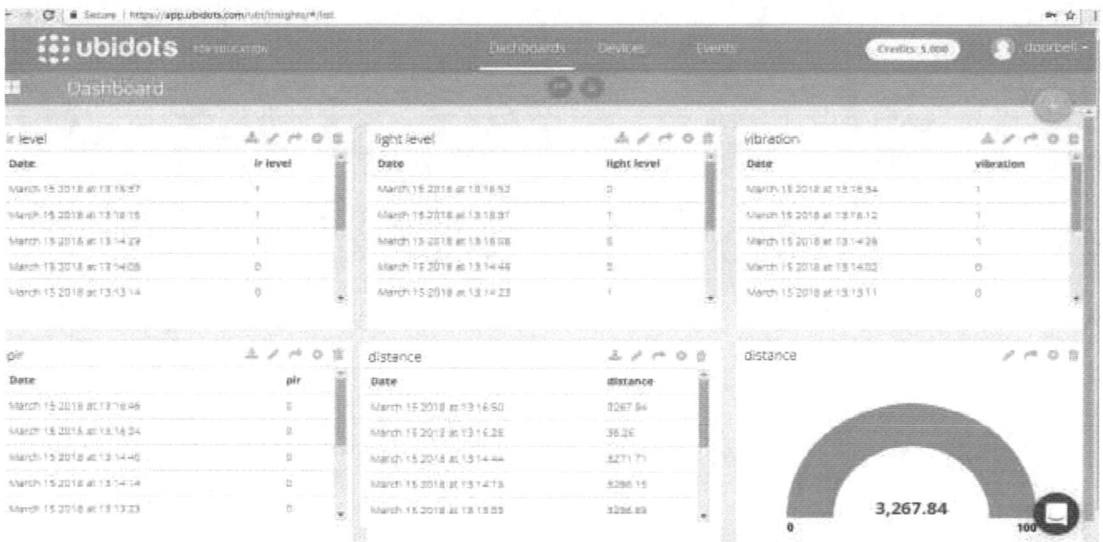

Fig2 ubidots dashboard

3.2. Methodology

3.2.1. Role of infrared sensor

An infrared sensor is an electronic gadget, that radiates so as to detect a few parts of the environment. An IR sensor can quantify the warmth of a question and identifies the movement. These sorts of sensors measures just infrared radiation, instead of emanating it that is called as a latent IR sensor. Normally in the infrared range, every one of the articles transmit some type of warm radiations. These kinds of radiations are undetectable to our eyes, that can be identified by an infrared sensor. The producer is essentially an IR LED (Light Emitting Diode) and the locator is just an IR photodiode which is delicate to IR light of an indistinguishable wavelength from that radiated by the IR LED. At the point when IR light falls on the photodiode, The protections and these yield voltages, change in extent to the size of the IR light got.

IOP Conf. Series: Materials Science and Engineering **590** (2019) 012018 doi:10.1088/1757-899X/590/1/012018

ir level

Date	ir level
March 15 2018 at 13:16:37	1
March 15 2018 at 13:16:15	1
March 15 2018 at 13:14:29	1
March 15 2018 at 13:14:05	0
March 15 2018 at 13:13:14	0

Fig3.IR sensor information

3.2.1Role of pir sensor

All articles with a temperature above outright zero transmit warm vitality as radiation. Typically this radiation isn't unmistakable to the human eye since it transmits at infrared wavelengths, however it can be distinguished by electronic gadgets intended for such a purpose.The expression inactive in this case alludes to the way that PIR gadgets don't produce or emanate vitality for discovery purposes. They work completely by identifying infrared radiation discharged by or reflected from objects. They don't recognize or measure "warm

pir

Date	pir
March 15 2018 at 13:16:46	0
March 15 2018 at 13:16:24	0
March 15 2018 at 13:14:40	0
March 15 2018 at 13:14:14	0
March 15 2018 at 13:13:23	0

Fig4.pir sensor information

It likewise measures the separation of the body from the entryway that is how much separation it is far from the door.It is additionally actualized in the ubidots stage.

IOP Conf. Series: Materials Science and Engineering **590** (2019) 012018 doi:10.1088/1757-899X/590/1/012018

distance

Date	distance
March 15 2018 at 13:16:50	3267.84
March 15 2018 at 13:16:28	36.26
March 15 2018 at 13:14:44	3271.71
March 15 2018 at 13:14:19	3286.15
March 15 2018 at 13:13:03	3236.89

Fig5.distance parameters

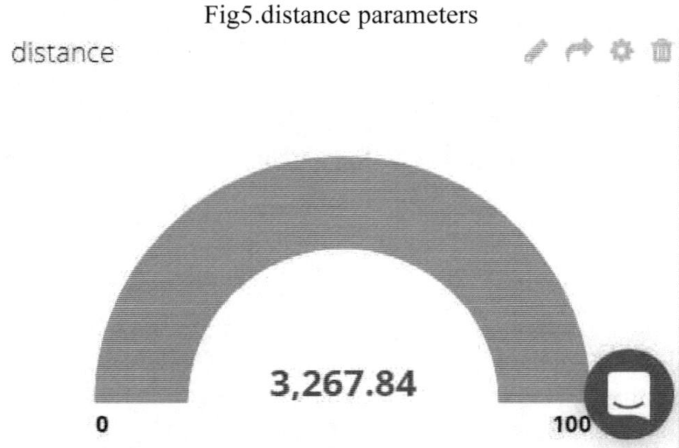

Fig 6. distance parameters

3.2.3 Role of vibration sensor

Voltage mode piezoelectric is the most famous accelerometer plan because of its abnormal state yield and wide data transfer capacity. We offer voltage mode accelerometers in the customary 3-wire or 2-wire (IEPE) designs. Charge-mode piezoelectric accelerometers measure stun and vibration in high temperature conditions. Notwithstanding its high temperature working ability when utilized with a superb charge enhancer, a charge mode accelerometer offers dynamic range versatility. To quantify movement (speed, relocation) precisely, an accelerometer with DC reaction is required. Joining MEMS innovations and the most recent simple and advanced ASICs, our DC accelerometers offer elite and excellent esteem. All items are EAR99 and RoHS consistent.

IOP Conf. Series: Materials Science and Engineering **590** (2019) 012018 doi:10.1088/1757-899X/590/1/012018

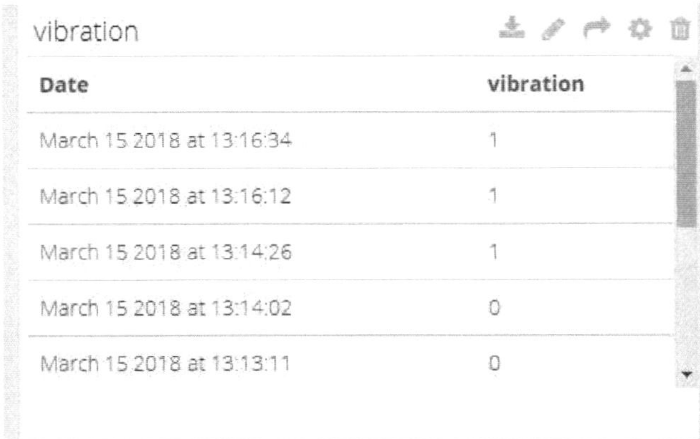
vibration

Date	vibration
March 15 2018 at 13:16:34	1
March 15 2018 at 13:16:12	1
March 15 2018 at 13:14:26	1
March 15 2018 at 13:14:02	0
March 15 2018 at 13:13:11	0

Fig7 vibration sensor information

3.2.4 Role of USB camera

A usb camera is a camcorder which is utilized to catch the pictures and sends an indistinguishable information to the enrolled email from the same information. It will be ceaselessly checking and at whatever point the sensor recognizes it sends the information to the client.

4. Conclusion

The paper additionally identifies all the client activities in the home and each sensor information is naturally put away in the cloud and it can saw when the client needs it. when the client needs the information straightforwardly in the ubidots stage there will get to login so it can be seen in it. For assist works, it discloses that to enhance client conduct forecast by breaking down different client activities inside the home to additionally enhance shrewd home security.

References

[1] M.A. Al.Khedher, "Half breed GPS-GSM confinement of vehicle following framework" International Journal of Computer Science and Information Technology, vol. 3, no. 6, pp. 75-85, Dec 2011.

[2] Usha Nandini , Saravanan M , Albert Mayan J , Murari Devakannan Kamalesh , Mohana Prasad K (2018) , " Automatic traffic control system using PCA based approach",International Conference on Energy, Communication, Data Analytics and Soft Computing (ICECDS-2017),pp.2387-2392.

[3] S.S.Pethakar, Srivastava, and S. D.Suryawanshi, "RFID, GPS and GSM based vehicle following and representative security framework", International Journal of Advanced Research in Computer Science and Electronics Engineering, vol. 1, no. 10, pp. 91-96, Dec. 2012.

[4] Pankaj Verma, J.S Bhatia "Outline and improvement of gps-gsm based following framework with Google delineate observing", International Journal of Computer Science, Engineering and Applications Vol.3, No.3, June 2013.

[5] Arun Sasi, Lakshmi R Nair "Vehicle against burglary framework in view of an installed stage", International Journal of Research in Engineering and Technology eISSN: 2319-1163 | pISSN: 2321-7308.

[6] C. Suh and Y.- B. Ko, "Plan and execution of keen home control frameworks in view of

dynamic sensor systems," IEEE Transactions on Consumer Electronics, vol. 54, no. 3, pp. 1177– 1184, 2008.

[7] B. Fouladi, S. Ghanoun, "Security Evaluation of the Z-Wave Wireless Protocol," Black cap USA, Aug. 2013.

[8] Wenye Wang, Zhuo Lu, "Digital security in the Smart Grid: Survey and difficulties," Computer Networks, Volume 57, Issue 5, Pages 1344-1371, April 2013.

[9] Asha P, Albert Mayan J, Canessane A (2018),"Efficient Mining of Positive and Negative Itemsets Using K-Means Clustering to Access the Risk of Cancer Patients",Communications in Computer and Information Science ,ICSCS 2018, Kollam, 2018,pp.373-382.

[10] A.C Jose, R. Malekian, N. Ye, "Enhancing Home Automation Security; Integrating Device Fingerprinting Into Smart Home", IEEE Access, vol. 4, October 2016.

[11] A.C Jose, R. Malekian, "Savvy Home Automation Security: A Literature Review", Smart Computing Review, Vol. 5, No. 4, pp. 269-285, August 31, 2015.

[12] B. Schilit, N. Adams, R. Need, "Setting Aware Computing Applications," WMCSA '94 Proceedings of the 1994 First Workshop on Mobile Computing Systems and Applications, pp. 85-90, 1994.

IOP Conf. Series: Materials Science and Engineering **590** (2019) 012019 doi:10.1088/1757-899X/590/1/012019

Identification of Selective Forwarding Attacks in Remote locator Network utilizing Adaptive Trust Framework

B.Baron sam[1], T.S.Sandeep[2], Sai Sreenath Reddy.V[3]

Assistant Professor[1], Student[2], Student[3]

School of Computing, Sathyabama Institute of Science and Technology
baronsam1988@gmail.com[1]

Abstract. Discovery of SFA in Wireless Sensing Networks utilizing Adaptive and Channel mindful Reputation Systems. Wireless detecting component systems (Wireless sensor networks) use unit defenseless against SFA. It will malignantly drop a subset of forward packets to corrupt system execution and also imperil the information trustworthiness. It represents a decent test to distinguish the malignant drop and conventional bundle misfortune. Amid this paper, we propose a Channel-mindful name System with versatile recognition edge (Adaptive - Channel - mindful notoriety framework) to discover SFA in Wireless sensor networks. The Adaptive - Channel - mindful notoriety framework eval Warning: Demo Version - reports are fragmented! Abnormal state of Plagiarism might be identified! Get your entire report: 1. Most point by point reports. 2. Moment permit initiation. 3. Lifetime bolster. uates the data forward behaviors of sensor hubs, in venture with the deviation of the observed bundle misfortune and furthermore the measurable conventional misfortune. To streamline the exact location of Adaptive - Channel - mindful notoriety framework, we tend to in principle determine the best limit for forward analysis Broad reproduction comes about show That Adaptive - Channel - mindful notoriety framework will precisely discover SFA and decide the bargained detecting component hubs, though the assault tolerant data forward theme will essentially enhance the data conveyance size connection of the system.

Keywords: CRS-A, bundle dropping, channel-mindful, particular forward assault (SFA) remote detecting component organize,.

1. Introduction

As a promising occasion perception and learning gathering system, remote gadget organize (Wireless sensor systems) has been wide connected to every military and regular citizen applications. Numerous Wireless sensor systems are sent in even antagonistic conditions and unattended situations to perform mission-basic undertakings, similar to autonomous organization perception and package observation mission. Notwithstanding, gadget hubs region unit basically traded off by foes, making Wireless sensor systems at risk to various security dangers because of the absence of physical assurance. one in everything about premier extreme dangers is particular forward attack, wherever the bargained hubs will malignantly drop an arrangement of forward packets to break down the information conveyance extent connection of the system.

It also has impressively negative effects to learning uprightness, especially for information touchy applications, e.g., social insurance and business perception. Then again, since Wireless sensor networks territory unit normally sent in open regions (e.g., native backwoods), the precarious remote channel and medium access impact will cause remarkable customary parcel misfortunes. The SFA square measure stowed away by the traditional bundle misfortunes, convoluting the assault recognition. In this way, it's hard to discover the SFA and enhance the system execution. The greater part of associated works target viewing the parcel misfortunes in each transmission interface and expository the hubs with high PLRs from the data forward path. These arrangements will enhance the data conveyance quantitative connection or system turnout however has almost no effect on police examination SFA.

Since the most test of assault Detection is to separate the malignant drop from customary parcel misfortune, the ordinary PLR of the transmission connect should be contemplated inside the forward analysis. For instance, a supply hub Ns sends ten bundles to the goal hub Dn through 2 forward nodes An and Bn, severally. An advances vi parcels to Dn, though Bn exclusively advances 5 bundles to Dn. Naturally, A carries on higher than Bn all through the information sending. Be that as it may, if the regular PLRs from Ns to An and Bn square measure 2 hundredth and five hundredth, severally, An ought to have the following opportunity to offense amid this data sending. Accordingly, we tend to consider the deviation between the ordinary misfortunes and genuine misfortunes on the grounds that the key issue to discover SFA.

In any case, for the Wireless sensor networks sent in unfriendly conditions where the remote channel is precarious, conventional PLR to a great degree relies upon the remote channel quality that changes spatially and transiently. In the event that we tend to utilize a deliberate or measurable ordinary PLR to discover SFA, some blameless hubs could likewise be inaccurately known as assailants due to the time varied channel condition. for instance, if a versatile snag short obstructs the information transmission of 2 gadget hubs, the shocking bundle misfortunes may deceive the assault identification.

Along these lines, an adaptable and fault tolerant investigation method is pivotal to precisely set up the assaults and traded off gadget hub. In the interim, because of the negative effects of SFA, data conveyance proportion of a system turns into the principal execution metric for opposing the assaults. Despite the fact that traded off gadget hubs will be precisely known, they're as yet possible contender to forward data for elective gadget hubs before physically restored or supplanted. In the event that a traded off hub dispatches assault with an espresso likelihood however has sensible channel condition, it ought to forward more data parcels than a standard hub with poor channel condition, despite the malignant drops.

During every investigation sum, gadget hubs assess the ordinary PLRs amongst themselves and their neighboring Nodes, and embrace the measurable PLRs to check the forward behaviors of its downstream neighbors on the information forward path. The gadget hubs getting into mischief in Data forward area unit berated with diminished name esteems by Adaptive - Channel – mindful notoriety framework. Once the name worth of a grown-up male hub is underneath an alert worth, it'd be known as a bargained hub by Adaptive - Channel - mindful notoriety framework. Contrasted with our past work, this paper has the consequent upgrades and new commitments. (I) we tend to propose Adaptive - Channel - mindful notoriety framework, that assesses the forward Behaviors of gadget hubs by using partner accommodating recognition edge. By on paper separating its execution, we derive relate perfect acknowledgment edge for surveying the Forward behaviors to enhance the exact recognizable proof of Adaptive - Channel - careful reputation structure.

The perfect area edge is decided for each transmission interface in an extraordinarily probabilistic approach, and may even be adaptable to the time-varied channel condition and besides the attack probability of the forward node. (ii) We tend to develop an appropriated and attack tolerant information forward scheme to cooperate with Adaptive - Channel - careful reputation structure for stimulating the forward cooperation of dealt centers and up the data conveyance quantitative association of the framework.

Rather than uninflected all the bargained hubs from data sending, it mutually considers the time varied channel condition and assault potential outcomes of neighboring hubs in choosing forward nodes. (iii) serious reproduction comes about exhibit that the arranged Adaptive - Channel - mindful notoriety framework with assault tolerant data forward theme can accomplish a precise location with every one of false and missed identification conceivable outcomes close to zero, and enhance extra than 100 percent data conveyance quantitative connection for the system.

2. Related Works

Expanding consideration has been paid to creating countermeasures against SFA, attributable to their negative effects on arrange execution and information trustworthiness. The key arrangement of existing works is to watch the forward behaviors of detecting component hubs, which may give proof and steerage for assault discovery and guard. In the accompanying writing audit, we tend to isolate the present works into 2 classifications: affirmation basically based and neighbor observation based plans, with regards to entirely unexpected watching methods for data sending.

2.1 Acknowledgment based Defense Techniques:

This sort of plans is to utilize affirmations from various hubs inside the directing way to see the PLR of each jump and watch the aggressors propose a topic that unpredictably picks assortment of transitional hubs on a forward path as checkpoints to come affirmations for each got bundle. In the event that suspicious conduct is recognized, it produces relate degree caution parcel and conveys it to the supply hub. Style and actualize relate degree intrusion detection framework, named expanded Adaptive Acknowledgments (EAACK), for portable unintended systems. Because of the high heap of bounce by-jump affirmations, EAACK consolidates a two-bounce affirmation topic relate degreed a conclusion to-end affirmation topic to watch the malignant practices and cut back the system overhead. Furthermore, EAACK embraces an advanced mark with affirmation to guarantee validation, uprightness, and non-renouncement. As an Elastic examination subject, name framework is moreover connected to assault discovery. Zhang et al. create relate degree review based bad behavior recognition framework to coordinate name administration, dependable course revelation, and distinguishing proof of getting rowdy hubs bolstered conduct reviews in unintended systems. In, the relationships between's connection blunders and pernicious drops region unit examined to watch SFA. So as to ensure honest count for the connections, they propose a homomorphism direct appraiser (HLA) based for the most part open examining outline that allows the indicator to check the honesties of affirmations reputed by hubs.

3. Framework Model and Design Goals:

3.1 Network Model

We mull over a Wireless sensor systems comprising of a gathering of aimlessly Distributed locator hubs, signified by N, and a sink hub to screen relate open space. Each finder hub sporadically faculties the intrigued data from nature, and transmits the recognized information to the sink by means of multi-bounce steering among indicator hubs. Identifier hubs speak with their neighboring hubs upheld the IEEE 802.11 DCF. The observed space has relate unsteady radio setting, making the PLRs all through the correspondences of identifier hubs extensively expanded and fluctuate every once in a while. Since locator hubs square measure conveyed in open space and need satisfactory physical assurance, they'll be bargained by foes through physical catch or bundle vulnerabilities to move in learning sending. We tend to utilize PM to mean the trading off probability of finder hub that is illustrated as the probability that an indicator hub is bargained by the enemy. In the interim, we tend to expect that identifier hubs will screen the information forward activity of their neighboring by hubs. Neighbor-reconnaissance based Defense Techniques

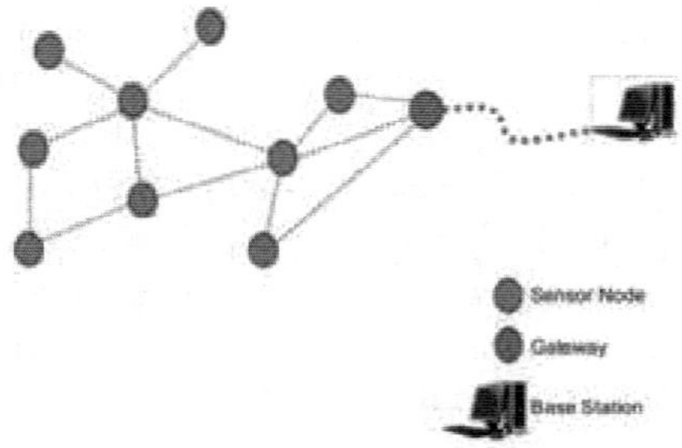

Fig 3.1 Network Model

With the Watchdog equipment, gadget hubs will screen the forward behaviors of their neighboring hubs and record the real parcel misfortune precisely. Researches a valuable name based generally dependable information accumulation method against SFA in grouped Wireless sensor networks. Every hub keeps up a name table to measure the practices of its neighbor hubs, upheld the forward observation of the neighboring hubs. The hubs with low name esteems are confined from the directing way. In any case, the name assessment is just bolstered the observed parcel misfortune all through the sending. Style a repetitive diversion based generally way to deal with inquires about the conspiracy on SFA in multi-bounce remote systems. Propose a feature Channel perception (SCM) topic to find SFA in remote coincidental systems. SCM utilize the hubs neighboring a data correspondence course, to constitute an aspect channel for perception the forward behaviors of the hubs in transit. When mischievous activities are distinguished, the perception hubs send alert parcels to the supply hub through each channel. A large portion of associated works said higher than will successfully relieve the negative effects of SFA on information trustworthiness and system execution. Be that as it may, they have limited ability to precisely see the assaults and build up the traded off detecting component hubs. Numerous current examinations consider the customary bundle misfortune into specific forward attack discovery for remote work systems. In any case, every one of the works utilize Associate in nursing measurable conventional PLR to judge the data forward behaviors over an extended sum. Such methodologies don't appear to be appropriate for the Wireless sensor networks in insecure radio air, wherever the high and time varied bundle misfortune could extensively reduce exact location. Additionally, in their plans, hubs are known as Associate in nursing guilty party once the measure of lost bundles all through its forward exceeds an exact cost. The one-time location can likewise make a larger than usual false discovery chance for the pure hubs. In our past work, a name

Framework is misused to see SFA by taking the traditional PLR into thought. Notwithstanding, it's bolstered a set investigation limit and effectively detaches all the bargained hubs from the information forward paths. Amid this paper, we tend to confirm Associate in nursing accommodating edge to assess the data forward behaviors, which may advance the precise location of the name framework. In addition, we create Associate in nursing assault tolerant directing subject working together with the name framework to animate the participation of bargained hubs for Associate in Nursing enhanced information conveyance quantitative connection.

3.2. Assault Model

Traded off detecting component hubs will dispatch SFA to debase the execution of the system. In particular, when a traded off detecting component hub gets a data bundle, it noxiously drops it with a possibility, said as assault likelihood. Since the resister will administration the assault conceivable outcomes of traded off hubs, it's hard to differentiate if the parcel misfortunes square measure caused by vacillated channel condition or malignant drops, especially for the hubs with low assault probabilities. Besides, numerous neighboring traded off detecting component hubs will work together with each other to dispatch advancement/downgrade assaults to acknowledge favorable circumstances.

IOP Conf. Series: Materials Science and Engineering **590** (2019) 012019 doi:10.1088/1757-899X/590/1/012019

For instance, if nuclear number 11 and Bn square measure 2 neighboring traded off detecting component hubs and information activity is from nuclear number 11 to Bn, nuclear number 11 could give a fractional assessment of how Bn carries on. Also, N a can report N b as a customary hub to its option neighboring hubs, despite Bn getting into mischief inside the information sending. We consider that crypto rationale strategies are used in the system to create adequate learning privacy and validation against the resister, and after that we can work in opposing SFA. Inside the accompanying, we watch out for choice the bargained detecting component hubs as malevolent hubs, and in this manner the elective sensor hubs as conventional hubs.

3.3 Design Goals

The target of this paper is to identify SFA in view of the observed forward movement data and enhance the information conveyance proportion for WSN's. In particular, the proposed conspire expects to accomplish the accompanying two objectives

3.3.1 Accurate location

An exact location should be accomplished for police examination SFA and unmistakable the vindictive hubs, which may be estimated by 2 measurements. The one is that the assaults should be precisely recognized once the pernicious hubs misdemeanant in information sending. The other is customary hubs can't be inaccurately identified as noxious hubs due to the varied conventional parcel misfortunes.

3.3.2 Data conveyance proportion change

Other than the identification of SFA, the data conveyance greatness connection of the system should be enhanced by the arranged subject to alleviate the negative effects caused by the assaults. Then, the arranged topic should have the capacity to part fortify the participation of malignant hubs in learning sending.

4. Versatile Detection Threshold for Adaptive - Channel - mindful notoriety framework: As we tend to say in Sec. IV-B, the precise location of Adaptive - Channel - mindful notoriety framework is impressively conservative by the acting up recognition edge for name examination. Amid this segment, we intend to decide the ideal examination edge for each attempt of neighboring hubs on the data forward path to enhance the exact discovery of Adaptive - Channel - mindful notoriety framework. In advance with the assault display, malevolent hubs will dispatch assaults with totally unique probabilities that show the recognition limit should be distinctive for each correspondence connect. Then, because of the idea of dynamic directing, the location limit should be versatile to the timed data activity and customary PLR of the connection. While not loss of simplification, we tend to represent considerable authority in definitive the optimal limit for the transmission from Ni to New Jersey all through the period Tt, inside the accompanying investigation.

5. Existing System

The fundamental arrangement of existing works is to watch the forward behaviors of detecting component hubs, which may offer verification and guiding for assault discovery and safeguard. Mainly there are the 2 existing strategies are utilized one is Acknowledgment based for the most part Defense Technique and another is Neighbor-observation based Defense Technique. Be that as it may, every one of the works utilizes a measurable customary PLR to check the information forward behaviors over an extended sum. Such methodologies don't appear to be pertinent for the Wireless sensor networks inside the temperamental radio setting, wherever the high and time-changed parcel misfortune may significantly reduce exact location.

6. PROPOSED SYSTEM:

To enhance precise location and parcel conveyance proportion we tend to propose the strategy named as Adaptive - Channel - mindful notoriety framework. In Adaptive - Channel - mindful notoriety framework, each finder hub keeps up a name table to measure the long forward behaviors of its neighboring hubs. The embodiment of Adaptive - Channel - mindful notoriety framework is to progressively refresh the name table bolstered the forward behavior investigation for the neighboring hubs, by taking the regular PLR into thought. Be that as it may, on the grounds that the insecure radio environment fabricate the standard of remote channel change with time; customary bundles could likewise be totally unique over an extended key amount. In this way, we tend to isolate the entire system life expectancy into a succession of investigation periods. Bolstered estimation each finder stores the bundle misfortune data thus indicator utilizes the hang on data to measure the forward behavior of neighbor.

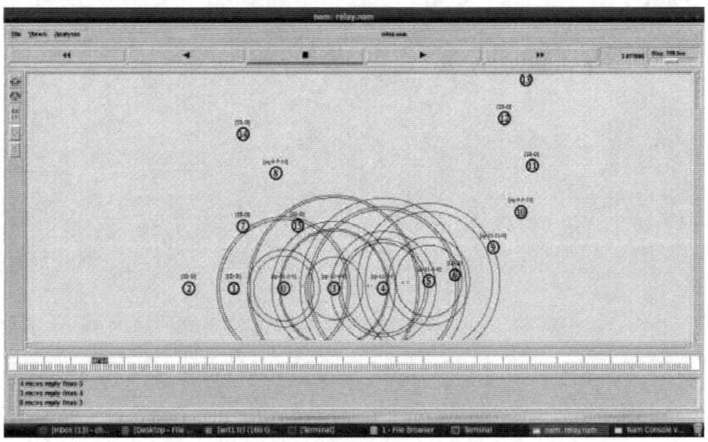

Fig 6.1 SFA assault distinguish at hub 5

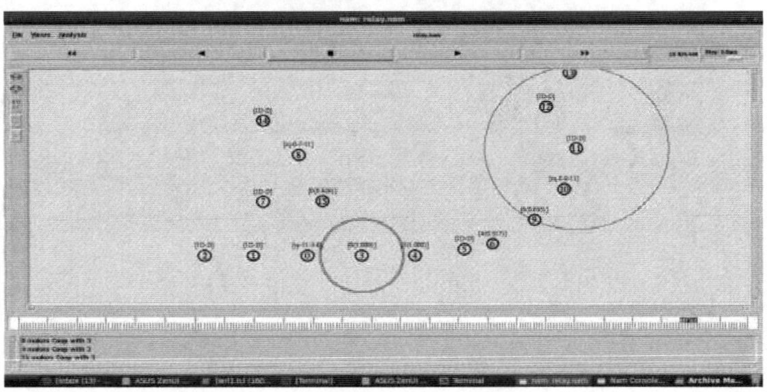

Fig 6.2 Packet estimation of SFA assault kept the procedure from hub 6 decrease the parcel

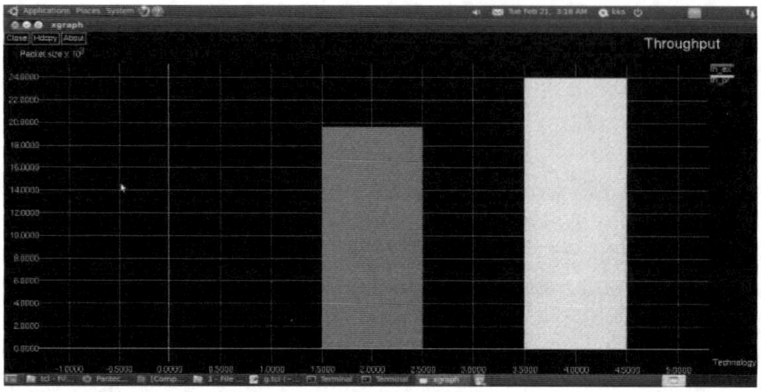

Fig 6.3 Throughput correlation (existing and proposed)

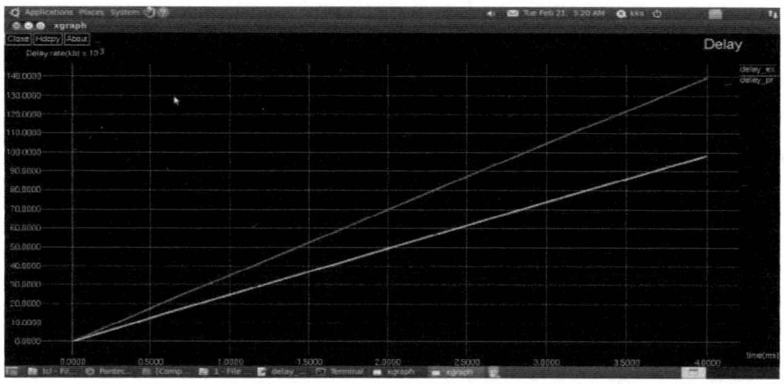

Fig 6.4 Delay correlation (existing and proposed)

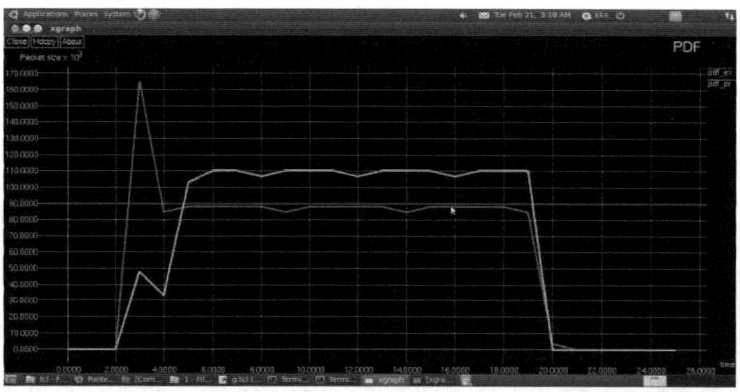

Fig 6.5 Pdf comparison(existing and proposed)

7. Conclusion:

A trust reputation structure with adaptable recognition constrain (trust component) to distinguish particular sending ambushes in WSNs. To accurately perceive particular sending strikes from the conventional package hardship, CRSA surveys the sending hones by the deviation between the quantifiable standard bundle incident and checked bundle adversity. To improve the disclosure accuracy of Trust System, we have any decided the best examination cutoff of Trust System out of an astoundingly probabilistic means, that is accommodative to the time-varied station condition and besides the attack potential results of exchanged off center points. Moreover, a passed on and attack tolerant getting the hang of sending design is made to cooperate with Trust System for invigorating the joint effort of exchanged off centers and rising the information movement significance association.

References

[1] I. Butun, S. Morgera, and R. Sankar, "A review of interruption identification frameworks in remote sensor systems," IEEE Commun. Surv. and Tutor., vol. 16, no. 1, pp. 266– 282, 2014.

[2] Y. Zou, X. Wang, and W. Shen, "Physical-layer security with multiuser planning for intellectual radio systems," IEEE Trans. Commun., vol. 61, no. 12, pp. 5103– 5113, 2013.

[3] Surya.V, J. Albert Mayan," A Secure Data Sharing Mechanism In Dynamic Cloud By Using KP-ABE", Research Journal of Pharmacy and Technology , Vol 10 , Issue 1 , pp:83-86,2017

[4] Y. Zhang, L. Lazos, and W. Kozma, "Amd: Audit-based bad conduct Discovery in remote impromptu systems," IEEE Trans. Horde. Comput., prePrints, distributed online in Sept. 2013.

[5] S. Ozdemir, "Useful notoriety based solid information collection and transm ission for remote sensor systems," Comput. Commun, vol. 31, no. 17, pp. 3941– 3953, 2008.

[6] D. Hao, X. Liao, A. Adhikari, K. Sakurai, and M. Yokoo, "A rehashed amusement approach for breaking down the conspiracy on specific sending in multihop remote systems," Comput. Commun., vol. 35, no. 17, pp. 2125– 2137, 2012.

[7] X. Liang, X. Lin, and X. Shen, "Empowering reliable administration assessment in benefit arranged versatile informal communities," IEEE Trans. Parallel Distr. Sys., vol. 25, no. 2, pp. 310– 320, 2014.

[8] J. Ren, Y. Zhang, K. Zhang, and X. Shen, "Sacrm: Social mindful crowdsourcing with notoriety administration in portable detecting," Computer Commun., vol. 65, no. 15, pp. 55– 65, 2015.

[9] Albert Mayan .J , Sharmila Latha T , Kislay Sinha ," Security Analysis of Three Factor Authentication Schemes for Banking", ARPN Journal of Engineering and Applied Sciences, Vol:10, Issue 8, pp: 3504-3509,May 2015 , ISSN : 1819-6608.

[10] J. Ren, Y. Zhang, K. Zhang, and X. Shen, "Misusing channel-mindful notoriety framework against particular sending assaults - in wsns," in Proc. IEEE GLOBECOM, 2014, pp. 330– 335.

[11] Usha Nandini , Saravanan M , Albert Mayan J , Murari Devakannan Kamalesh , Mohana Prasad K (2018) , " Automatic traffic control system using PCA based approach",International Conference on Energy, Communication, Data Analytics and Soft Computing (ICECDS-2017),pp.2387-2392.

[12] K. Liu, J. Deng, P. Varshney, and K. Balakrishnan, "An affirmation based approach for the recognition of directing trouble making in manets," IEEE Trans. Horde. Comput., vol. 6, no. 5, pp. 536– 550, 2007.

[13] E.Mahmoud and X. Shen, "An incorporated incitement and discipline instrument for obstructing bundle dropping assault in multihop remote systems," IEEE Trans. Vehic. Tech., vol. 60, no. 8, pp. 3947– 3962, 2011.

[14] E. Shakshuki, N. Kang, and T. Sheltami, "Eaacka secure intrusiondetection framework for manets," IEEE Trans. Ind. Electro., vol. 60, no. 3, pp. 1089– 1098, 2013.

[15] T. Shu and M. Krunz, "Identification of vindictive bundle dropping in remote impromptu systems in view of protection saving open evaluating," in Proc. ACM WiSec, 2012, pp. 87– 98.

[16] Asha P, Albert Mayan J, Canessane A (2018),"Efficient Mining of Positive and Negative Itemsets Using K-Means Clustering to Access the Risk of Cancer Patients",Communications in Computer and Information Science ,ICSCS 2018, Kollam, 2018,pp.373-382.

[17] S. Marti, T. J. Giuli, K. Lai, and M. Pastry specialist, "Relieving directing rowdiness in versatile impromptu systems," in Proc. ACM MobiCom, 2000, pp. 255– 265.

[18] B.Baron Sam, A.V.K.Shanthi, "Query Oriented User Profile Management In Web Search Engine", International Journal of Applied Engineering Research, ISSN 0973-4562 Volume 10, Number 76(2015) Special Issues, pp-295-300, IF 0.015, Snip 0.034, SJR 0.13

ADVANCED GARBAGE COLLECTION IN SMART CITIES USING IOT

A.Amarnadha Reddy[1], B. Gangadhar[2]
U.G Students, Department of Computer Science & Engineering
Sathyabama Institute of Science and Technology, Chennai, India
amar53.reddy@gmail.com[1], gangadhar219.bathula@gmail.com[2]

Muthukumar B[1]
Professor , Department of Computer Science & Engineering
Chennai Institute of Technology,Chennai, India
muthukumarb@citchennai.net

Albert Mayan J[2]
Assistant Professor , Department of Computer Science & Engineering
Sathyabama Institute of Science and Technology , Chennai, India
albertmayan@gmail.com

ABSTRACT: Now-a-days, we see that dustbins that are kept at open areas in the urban regions are flooding as a result of addition of the garbage persistently. It makes unhygienic condition and upsetting smell around the surroundings resulting in spreading some savage ailments and human diseases. To overcome such issues, Advanced Waste Collection is proposed for Smart Cities which depend on Internet of Things (IoT). It will take fewer efforts to clean all the bins over the city by checking whether the dustbins are full or not. The proposed work includes a smart dustbin that separates dry and wet waste so that the waste recycling can be done more efficiently. To find the efficient way to find garbage collecting routes, we use shortest path algorithms like Dijkstra's Algorithm.

Keywords*:* Internet of Things (IoT), Unhygienic, Smart cities

1. INTRODUCTION

The smart cities are improving living conditions. Governments and private companies also invest their budget, to research and development of smart cities. In our research it is to collect the garbage waste in smart cities using IOT technology. As the IOT technology is incrementing, efficient methods are used to make the process easily done by the systems. The project is based upon collecting the garbage using IOT technology. The process is to check whether the dustbins all over the city is full or not .In case if the bins are get filled that it detects over weight through moisture sensor and then workers come to collect the garbage.

In the greater part of the urban communities, the flooded garbage containers are making an unhygienic environment. This will additionally results in emerging of various sorts of infections. This will debase the way of life. To avoid these circumstances a proficient brilliant waste gathering framework must be created. As the extent of IoT is creating step by step, compelling strategies can be discovered effectively.

In the current framework, waste is gathered by metropolitan specialists by week by week once. Though the trash psychologists and floods from the container and spread over the streets and makes unhygienic condition.The scent will be overwhelming and creates air contamination and spreads infection. The animals eats the waste nourishment and spreads over the zone and makes grimy condition.

2. LITERATURE SURVEY

M. A. Mamun., et al. [1]haveproposed system forwaste management. A Waste-to-Energy (WTE) office particularly fourth era cremation office is been the best arrangement in Lebanon. The arrangement of a Waste-

Let go Power Plant in three destinations around Lebanon would process 2.6 million tons (MSW) every year and create 197.3 MW of power in addition to 470672 Btu/h of warmth to be utilized for abutting mechanical procedures or for circulation as locale warming. This procedure would furthermore limit squander volume to < 5% by transporting 7100 tons for every day of MSW specifically from family units and organizations to the WTE office as opposed to landfilling a large portion of these squanders. The establishment of best in class outflow control advances at the office would decrease vaporous discharges well underneath measures set up by the Lebanese Service of Condition. At long last the natural effects of WTE were measured.

P.P.Repoussis., et al. [2] have proposed system for Multiple Objective Decision Support SystemforSimulation ofGarbage Monitoring System, Managing the municipal waste overall is a strong test. When arranging a DSS framework, clashing goals, i.e., the framework working expenses and condition augmentation of asset use and at the same time considered. The ideal target is dependably not a decent decision for a presents a choice emotionally supportive network (DSS) administration arrangement in view of a multi co-ordinations show. The DSS comprises of a con and the relating scientific has been additionally delineated through an exhibited toward the paper ending.

L. Anthopoulos., et al. [3]have proposedsystem for developing Smart Cities using Internet of Things. For a creating nation, for example, India, which has very restricted innovation infiltration at the national level, an effective engineering for IoT should be founded on introduce innovation propels, capacities that give reasonable and economical arrangement, and entrepreneurial and social esteem. Shrewd city is an imperative idea for the improvement of any country. It is urgent for administration of India to offer distinctive administrations to its subjects and IoT encourages fundamentally accomplishing this reason. It will be conceivable to impart straightforwardly and consistently with huge number of homogeneous and heterogeneous frameworks, while having chosen access to information for outlining various advanced administrations. The basic role of this exploration paper is to contemplate the part of IoT being developed of Indian brilliant urban areas, comprehend the India IoT arrangement, discover the key drives and favourable circumstances of IoT based savvy city and distinguish the buyer inclinations and socioeconomics of Indian subjects who lean toward IoT based keen city arrangements.

Radek Fujdiak., et al. [4] have proposed system forAdvanced Municipal Waste Collection in Smart City using genetic algorithm. To enhance the strategic technique of waste accumulation, have utilized possess hereditary calculation execution. The exhibited arrangement gives computation of more efficient waste vehicle courses. As a yield, we give an arrangement of re-enactments concentrated on said zone. Every one of our calculations are executed inside the incorporated re-enactment structure which is produced as an open source arrangement concerning future modifications.

Theodoros Anagnostopoulos., et al. [5] have proposed system forChallenges and Opportunities of Waste Management in IoT enabled Smart Cities. In a city, this prompts Smart City structures. Wise administrations could be offered over such data identified with any part of people's exercises. A run of the mill case of administrations offered in the structure of Smart Cities is IoT-empowered waste administration. Squander administration includes not just the accumulation of the loss in the field yet in addition the vehicle and transfer to the proper areas. In this paper, we display a far reaching and exhaustive overview of ICT-empowered waste administration models.

Michal Ryes., et al. [6]haveproposed system for*Universal Smart Energy Communication Platform*,Online information accumulation is urgent for expanding effectiveness of vitality utilization and decreasing ecological effects. In addition, utility market progression and parallel EC prerequisites on shrewd estimation brought new difficulties for utility and media transmission suppliers. Further developed Metering Infrastructure (AMI) systems have been proposed with not clear broadly acknowledged arrangement. Our examination researches plausibility of cost proficient estimation engineering in light of as of now sent system and home access advancements. Created framework running on the brilliant home door goes about as a multipurpose empowering agent giving an on-going shrewd meter information accumulation and perception, yet can be utilized for any home mechanization benefits also.

3. PROPOSED WORK

Manage the waste in different type of box by using automation. In this system, we detect the dry and wet waste and separate them automatically using Moisture sensor. Every time the waste is thrown into the bin, ultrasonic sensor senses the distance of waste filled inside the bin dynamically. If the bin is filled up to threshold distance, red LED will ON indicating the filled bin. All the filled bins will be detected with the help of LED's using IoT.It indicates that the bin was filled and it should be cleaned.

When the garbage reaches the threshold distance, an SMS should be sent to garbage collecting authorities. The SMS includes the location of the filled bin and the type of bin filled (either Wet or Dry).After the bin reaches threshold distance, red LED will glow indicating the filled bin. After the timer gets started, for each interval of 1 hour, the bin is rechecked. If the waste is collected within the threshold time, a green LED will glow indicating the empty bin. For example, if the threshold time to collect the filled bins is set to 8 hours, then for every 1 hour, the status of bin is rechecked. If the filled waste is collected by the workers, green LED will glow indicating the empty bin. Otherwise, another SMS will be sent to higher officials regarding the negligence of the workers.

4. ARCHITECTURE

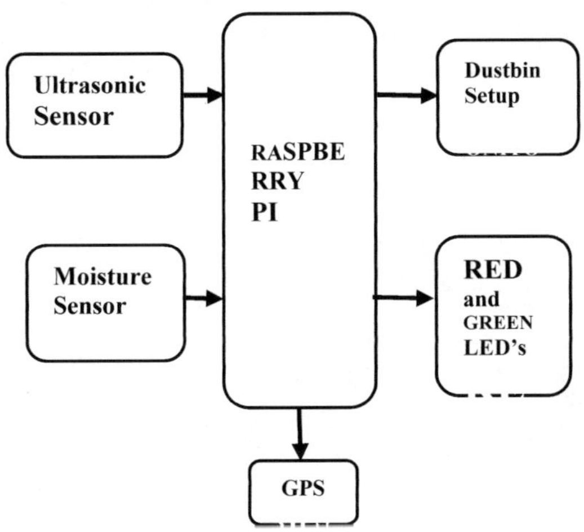

Fig.1. Architecture of Smart bin

The detailed process of each sensor and other mechanisms are explained below.

5. IOT BASED SOLID WASTE MANAGEMENT

The proposed work includes a smart dustbin that separates dry and wet waste so that the waste recycling can be done more efficiently. Also it detects over weight through moisture sensor.The Moisture sensor is used to measure the water content (moisture) in garbage.

In wet waste, dielectric permittivity is a positive capacity of the water content. The sensor makes a voltage corresponding to the dielectric permittivity, and in this manner the water substance of the waste. There is a 2 cm zone of impact regarding the level surface of the sensor; however it has almost no affectability at the outrageous edges

Fig. 2. Moisture Sensor

The working conditions of Moisture sensor are,

Working Voltage : 5V
Working Current :<20mA
Interface type : Analog
Working Temperature: 10°C~30°C

The Ultrasonic sensor calculates the interval between the sent sound wave and its reflection.

It calculates the interval by sending a sound wave at a particular recurrence and tuning in for that sound wave to reflect. By recording the slipped by time between the sound wave being created and the sound wave reflecting back, it is easy to compute the separation between the sonar sensor

Fig. 3. Ultrasonic Sensor

Since it is realized that sound goes through air at around 344 m/s, increase time for the sound wave by 344 meters to locate the aggregate round-trip separation of the sound wave. Round-trip implies that the sound wave voyaged 2 times the separation to the protest before it was identified by the sensor. It incorporates the 'outing' from the sonar sensor to the protest the 'outing' from the question the Ultrasonic sensor (after the sound wave skipped off the protest).

$$distance = \frac{speed\ of\ sound \times time\ taken}{2} \quad (1)$$

SMS ALERT

When the garbage reaches the threshold distance, an SMS should be sent to garbage collecting authorities.The SMS includes the location of the filled bin and the type of bin filled (either wet or dry).After the bin reaches threshold distance, red LED will glow indicating the filled bin. After the timer gets started, for each interval of 1 hour, the bin is rechecked. If the waste is collected within the threshold time, a green LED will glow indicating the empty bin.

STATUS OF SMART BIN

A timer will be set to a threshold time. The waste must be collected within the stipulated time. Otherwise, a complaint SMS will be generated to higher officials regarding their negligence. The SMS includes the location of the filled bin and the type of bin filled (either wet or dry).

IOP Conf. Series: Materials Science and Engineering **590** (2019) 012020 doi:10.1088/1757-899X/590/1/012020

6. SHORTEST ROUTE FOR BIN COLLECTION: DIJKSTRA'S ALGORITHM

Dijkstra's algorithm[ ] is used to find the effective garbage collecting routes. The Dijkstra's algorithm uses min-priority queue which is based on three operations:

add_with_priority(), decrease _priority() and extract_min ().

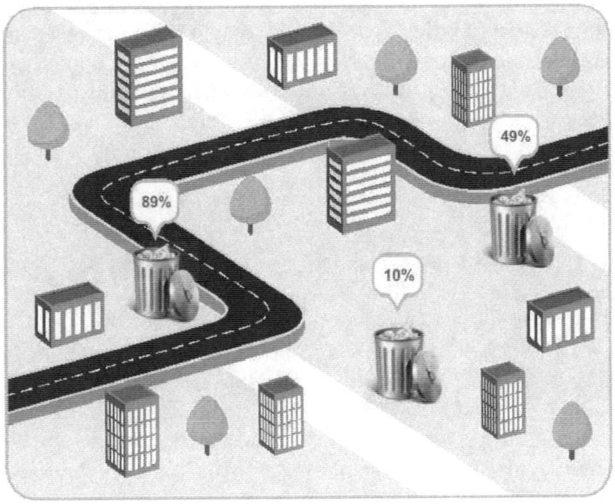

Fig: 4. Com2mon situation of garbage cans' filling status in todays' cities[3]

The Dijkstra's algorithm is as follows [11]:

```
function Dijkstra(Graph, source):
    dist[source] ← 0

    create vertex set Q

    for each vertex v in Graph:
        if v ≠ source
            dist[v] ← INFINITY
            prev[v] ← UNDEFINED

        Q.add_with_priority(v, dist[v])


    while Q is not empty:
        u ← Q.extract_min()
        for each neighbor v of u:
            alt ← dist[u] + length(u, v)
            if alt < dist[v]
                dist[v] ← alt
                prev[v] ← u
                Q.decrease_priority(v, alt)

    return dist, prev
```

Fig. 5 Program

Dijkstra's single source shortest path algorithm is used for finding shortest paths between nodes in a graph. It is used to monitor efficient garbage truck routes and finds shortest route from one node (source node) to every other node.

7. CONCLUSION

Theproposed systemhelps the smart cities to collect the dustbins on time over the cities. It helps the people in the city to live in hygienic conditions. It also makes the cities free from unpleasant odour from dustbins. It is used to separate wet and dry garbage. It helps to trace the filled garbage bins over the city with an SMS alert. If the filled bins are not collected within the stipulated time, another SMS alert will be sent to govt. officials regarding their negligence.

REFERENCES

[1] M. A. Mammon, M. A. Hanna, A. Hussein, and H. Basra, "Wireless Sensor Network Prototype for Solid Waste Bin Monitoring with Energy Efficient Sensing Algorithm", IEEE International Conference on Computational Science and Engineering (CSE), 2013, pp. 321-380.

[2] P. Jabez J, Gowri S, Vigneshwari S, Albert Mayan J and Srinivasulu S (2018)," Anomaly Detection by Using CFS Subset and Neural Network with WEKA Tools", Smart Innovation, Systems and Technologies,Vol.107, pp. 675–682.

[3] A. Alkali, L. Anthopoulos, and S. Kroch, "Smart Cities Data Streams Integration: experimenting with Internet of Things and social data flows", In the Proceedings of the 4th ACM International Conference, 2014, pp. 10-25.

[4] Kamalesh M. D., Albert Mayan J., Felix Y., Sumanth B. S., & Sai Tej B. (2018), " Magrisys: A Smart And Ubiquitous Controlled – Environment Agriculture System " , 2nd International Conference on I-SMAC (IoT in Social, Mobile, Analytics and Cloud) , pp: 1-5 , 2018

[5] Asha P, Albert Mayan J, Canessane A (2018),"Efficient Mining of Positive and Negative Itemsets Using K-Means Clustering to Access the Risk of Cancer Patients",Communications in Computer and Information Science ,ICSCS 2018, Kollam, 2018,pp.373-382..

[6] A. Gallardo, M. Carlos, M. Perils and F.J. Calmer, Methodology to design a municipal solid waste pre-collection system. A case study, Waste Management (0956-053X), Vol. 36, p. 1–11. 2015.

[7] M. Minogue and D. Koninis, Optimizing the treatment and disposal of municipal solid wastes using mathematical programming Vol. 70, p. 42–59. 2013.

[8] Mayan J.A, Arifa S, Pavithra R,"Semantic based multi lexical ranking technique for an effective search in protected cloud,2016 International Conference on Control, Instrumentation, Communication and Computational Technologies (ICCICCT), Kumaracoil, 2016, pp. 570-576.

[9] K. Mahajan and J.S. Chitode, International Journal of Innovative Research in Science, Engineering and Technology (ISSN: 2089-8893), Vol. 2, Issue 5,2014.

[10] Y. Glouche and P. Couderc,Waste Bin Monitoring System Using Integrated Technologies in: 2nd International Conference on Smart Systems, Devices and Technologies (SMART13). Rome, Italy, 2013.

[11]Dhanalakshmi R, and Muthukumar B.(2017),"An efficient image watermarking protocol based on composite image repudiation",in proceeding of IEEE Third International Conference on Science Technology Engineering and Management (ICONSTEM -2017),pp.419 – 422.

[12]Albert Mayan J, Anto Praveena M D , Telkar Bharath Rao , Uravakonda Uday Sagar (2018), "Optimized test data generation over suspicious implementation of oracle problem", IEEE International Conference on Power, Control, Signals and Instrumentation Engineering , pp.2559-2563.

[13]Sankari.A,AlbertMayan.J,"Retrieving call logs and SMS by messaging services",International Journal of Pharmacy &Technology,Vol. 8 , No.4 , pp.22951-22958,Dec 2016.

[14] Muthukumar B, Dhanalakshmi R, and Aroul canessane R.(2017),"Data Recovery and Reconstruction Using Business Intelligence Methodology Erasure Code Implementation Using Cloud Computing",International Journal of Advanced Research in Electrical, Electronics and Instrumentation Engineering,vol. 6, special issue 3, Nov 2017, pp.77-82.

Effective Credit Card Forgery Prevention Using Multilevel Authentication

[1] **M.AJAY CHOWDARY**
U.G Student, Department of Computer Science & Engineering
Sathyabama Institute of Science and Technology, Chennai, India
ajayazu9@gmail.com

[2] **M. KUNDAN**
U.G Student, Department of Computer Science & Engineering
Sathyabama Institute of Science and Technology
Chennai, India
pranjaldhar234@gmail.com

[3] **DR. A. VIJI AMUTHA MARY**
Assistant Professor, Department of Computer Science & Engineering
Sathyabama Institute of Science and Technology
Chennai, India
vijiamumar@gmail.com

Abstract. In the present system, credit card forgery is frequently happening, there is no big security is implemented at all. In the proposing system, Multi level security is implemented. It includes the Hex Keypad, GPS, and RFID. The secure information inside can be further secured with secondary system that consisting of formal fingerprint scanner. Modifying process is our Implementation, rather having multiple sensors integrated output in one process, which is actually time consuming, and we implement RFID with Keypad Matrix. POS based payment system. OTP is also verified from the Android application of the user so as to track the location of original user. Payment is made to the receiver only if both the User's Card & Mobile are in the same location. This system will avoid payment made by the attackers when Credit card is lost or when your details are hacked.

Keywords. Credit card, Hex Keypad, RFID, GPS, Mobile, OTP.

1. Introduction

Credit card fraud happens every day in various places in a variety of ways. But we can create some obstacles and make it tougher for someone to forgery your cards and card numbers. In this paper, we are proposing Mastercard imitation counteractive action by utilizing multilevel verification by utilizing keypad network, OTP alongside location and RFID. The client should enroll his/her points of interest like Visa number, account number, name, and so on which is put away on the server. While the exchange is in the process the OTP is sent alongside location if the exchange location is coordinated with it then the exchange will be fruitful. This can keep the Visa phony by unapproved people. The figure Fig 1 shows the fraud complaints by the numbers in recent years.

Fig.1.Fraud complaints in recent years

2. Existing System

You will register our details to the bank, when a transaction takes place with our card it connects to the bank server checks details and sends back to the merchant receiver then the transaction will be successful/unsuccessful depending upon the card information sent by the bank server. American Express and Discover are the installments organize and the Visa backer, so they favour charge card exchanges themselves. Visa and MasterCard, nonetheless, don't issue charge cards and should contact the Mastercard backer. The charge card guarantor sends back an approval code for the exchange. On the off chance that your charge card is declined, you won't get a reason for the purpose of an offer, only a message that the card was declined. You'll need to contact your card guarantor specifically to discover why your card was declined. The store's bank sends their interchanges electronically either through the telephone line or through the web. You may have been to a store or eatery and heard the shrieking and static from the charge card terminal speaking with the trader bank. Presently you realize what's happening.

3. Literature Survey

Because of the improved ability of foes, electronic frameworks are presently progressively powerless against falsifying and theft. The larger part of fake frame works today are of a cloned write, which has been on the ascent in the current years. Guaranteeing the security of such frameworks is of extraordinary worry as a foe can make a secondary passage or embed a malware to sidestep security modules. The unwavering quality of such frameworks could likewise be sketchy as the parts utilized as a part of these frameworks might be fake and additionally of mediocre quality. It is of prime significance to create arrangements that can keep a foe from making these non-real frameworks. In this paper, we show a novel framework level common validation approach for both the equipment and firmware. The equipment confirms the firmware by checking the checksum amid the catalyst. Then again, firmware checks the personality of the equipment and can't create amend comes about unless it gets a novel equipment unique mark, which we call as framework ID. We propose two secure conventions, TIDP and TIDS, to develop the framework ID and verify the framework by utilizing this novel ID. We demonstrate that our approach is impervious to different known assaults. [1]

A particular administration of testing, discovery, and shirking must be made to handle the overall episode of fake ICs. The electronic part inventory network has been incredibly influenced by across the board fake occurrences. We introduce a few techniques against fake measures to keep this broad falsifying, and we additionally consider the adequacy and confinements of these hostile to forging systems. [2]

IOP Conf. Series: Materials Science and Engineering **590** (2019) 012021 doi:10.1088/1757-899X/590/1/012021

As electronic gadgets wind up omnipresent and interconnected, individuals are progressively depending on coordinated circuits (ICs) for performing security-touchy errands and additionally dealing with delicate data. Another PUF circuit configuration in light of ring oscillators, which has preferences in the simplicity of usage and dependability over beforehand proposed outlines. [3]

4. Proposed System

In the proposing framework, Multi-level security is actualized. These incorporate with the HEX KEYPAD, OTP WITH LOCATION, and RFID. The resources in the internal vault are additionally secured with an optional framework totally isolated from the essential, comprising of a unique finger impression scanner.

Numerous sensors coordinated yield in one process, which is really tedious, we actualize RFID with Keypad Matrix or POS based installment framework. OTP is likewise confirmed from the Android utilization of the client to track the location of the first client. The installment is made to the beneficiary just if both the User's Card and Mobile are in a similar location. This framework will maintain a strategic distance from installment made by aggressors when a Credit card is lost, copied or hacked.

4.1 User Registration

Users have to register their details in login page. Users have to give their name, account details and credit card number in login page. User registration is for authentication purpose that is to verify whether he is the original user or the other person. This module is shown in Fig 2.

Fig.2.User Registration of Credit Card

4.2 Main Server

A server is a computer program running to serve the requests of other programs, the "clients". Thus, the "server" performs some computational task on behalf of "clients". The clients either run on the same computer or connect through the network. Here the server will store all the user's information in the database. Also the server will monitor all the user access. The server will likewise store the client get to subtle elements in the database. It will monitor the user movement and transaction.

4.3 Android Application

The android application is for verifying the user whether he/she is the authorized person. User has to give the OTP on mobile application, then the current location will be taken from application and send it to the server along with OTP.

4.4 POS- Hardware Setup:

POS is Point Of Sale which means where the purchase is made by user. Here the user will give their pin number machine (keypad matrix) by showing RFID card on machine. Key pad id is verified by the server then after verification OTP is send to the user mobile. If it is not correct transaction will be unsuccessful. The setup is shown in Fig 3.

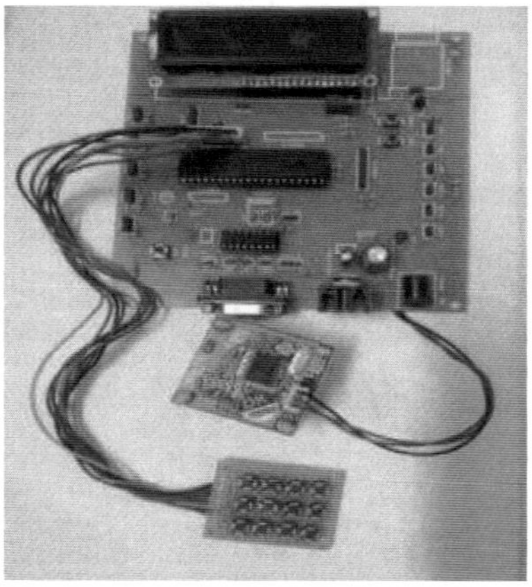

Fig.3 Hardware Components

4.5 Hardware and Software/Verification &Transaction

Here both the hardware and software are verified by the server to check whether the user is authorized person. Once the location sent from the mobile and location of purchase made by the person is matched, then the transaction is successfully made. The flow diagram is shown in Fig 4 and the architecture diagram of Credit card forgery is shown in Fig 5.

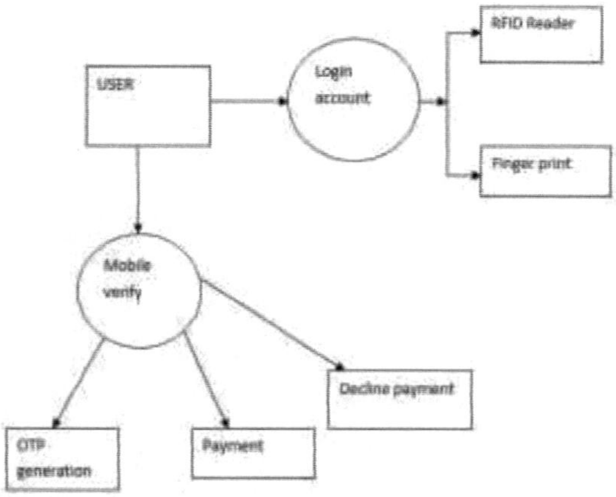

Fig.4.Flow Diagram of Multilevel Authentication verification

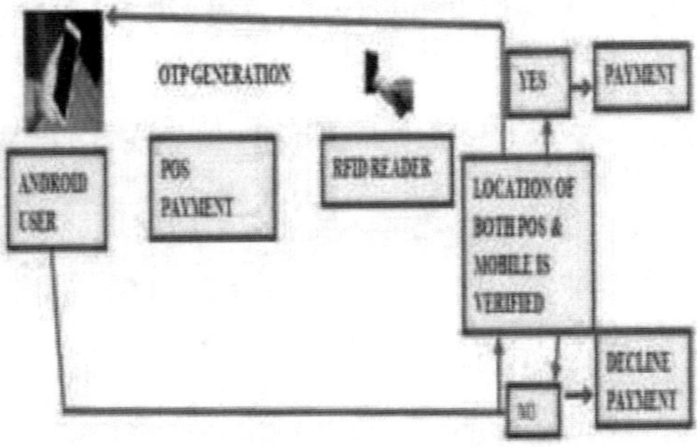

Fig.5 Architecture Diagram of Credit Card Forgery

5. Result

By this procedure when the exchange happens both the versatile and exchange place ought to be same then the exchange will be triumphs. In the event that the two areas are diverse, exchange will be a disappointment. While entering OTP from keypad network, the shrouded tag should coordinate alongside the area to finish an exchange.

6. Conclusion

Thus the paper represents that maximum level of card forgery is reduced by this method. Point of Sale (POS) process will verify the card transaction from where the sale was made. Through this an effective method was identified by the application for credit card forgery.

References

[1] SMA: A System-Level Mutual Authentication for Protecting Electronic Hardware and Firmware Ujjwal Guin, Member, IEEE, Swarup Bhunia, Senior Member, IEEE, Domenic Forte, Member, IEEE, and Mark M. Tehrani poor, Senior Member

[2] Asha P, Albert Mayan J, Canessane A (2018),"Efficient Mining of Positive and Negative Itemsets Using K-Means Clustering to Access the Risk of Cancer Patients",Communications in Computer and Information Science ,ICSCS 2018, Kollam, 2018,pp.373-382.

[3] Albert Mayan.J,Surya.B,Pranoy Prabhakar,Princekumar,"Department–Student Library Using Twig Pattern Query Processing Over Admin-User Login Privilege", Pak. J. Biotechnol. Vol. 13 , Pp. 489 - 493,(2016)

[4] Physical Unclonable Functions for Device Authentication and Secret Key Generation G. Edward Suh, Srinivas Devadas2007, San Diego,

[5] Anti Theft Control System Design Using Embedded System Department of Instrumentation And Control Sri Sai Ram Engineering College Chennai India.

[6] Ubiquitous smart home system using android application Shiu Kumar, Department of Information Electronics Engineering, Mokpo National University,534-729, Mokpo, South Korea

[7] Albert Mayan J, Anto Praveena M D , Telkar Bharath Rao , Uravakonda Uday Sagar(2018),"Optimized test data generation over suspicious implementation of oracle problem",IEEE International Conference on Power, Control, Signals and Instrumentation Engineering (ICPCSI-2017),pp.2559-2563.

[8] "Blood donation and life saver app",2017 2nd International Conference on Communication and Electronics Systems (ICCES), Annish Brislin M R, J Albert Mayan, R Aroul Canessane, M R Anish Hamlin,pp. 446 - 451,2017

[9] Online Toll Gate Payment System using RFID, Viji Amutha Mary A, ARPN Journal of Engineering and Applied Sciences, Vol. 13, No.13, 2018.

IOP Conf. Series: Materials Science and Engineering **590** (2019) 012021 doi:10.1088/1757-899X/590/1/012021

[10] Jabez J, Gowri S, Vigneshwari S, Albert Mayan J and Srinivasulu S (2018)," Anomaly Detection by Using CFS Subset and Neural Network with WEKA Tools", Smart Innovation, Systems and Technologies,Vol.107, pp. 675–682.
[11] Kamalesh M. D., Albert Mayan J., Felix Y., Sumanth B. S., & Sai Tej B. (2018), " Magrisys: A Smart And Ubiquitous Controlled – Environment Agriculture System " , 2nd International Conference on I-SMAC (IoT in Social, Mobile, Analytics and Cloud) , pp: 1-5 , 2018

IOP Conf. Series: Materials Science and Engineering **590** (2019) 012022 doi:10.1088/1757-899X/590/1/012022

DESIGN AND ANALYSIS OF WEARABLE ANTENNA FOR WIRELESS BODY AREA NETWORK

S. Karthikeyan[1],
Associate professor, Dept of ECE,
Sathyabama Institute of Science and Technology
OMR Road, Chennai, India
skarthi1879@gmail.com

Yedduri. Venu Gopal[2],
Final year B.E, Dept of ECE,
Sathyabama Institute of Science and Technology
OMR Road, Chennai, India

V. Giri Narendra Kumar[3],
Final year B.E, Dept of ECE,
Sathyabama Institute of Science and Technology
OMR Road,Chennai, India

T. Ravi[4],
Associate Professor, Dept of ECE,
Sathyabama Institute of Science and Technology
OMR Road,Chennai, India

Abstract: The fast development of wearable low-power devices has increased the requirement of solutions for WBAN implementation. In addition to the need, to reduce the device's dimensions and improve power consumption, it is also important to work on developing flexible antenna's that may be integrated into a wide variety of applications. This paper is a study of wearable antenna design for wireless body area network applications. A literature review of existing wearable antennas is studied. A wearable antenna is designed with size optimization. The antenna is designed using ANSYS High Frequency Structural Simulator (HFSS) software. The designed antenna works at operating frequency of 2.45 GHz. A rectangular patch made up of copper is used as radiating patch and the ground is also made with copper. The substrate material used is bed sheet cotton with a relative permittivity of 3.27 and the loss tangent is 0.00786. The size of the antenna is 40mm*34mm*1.26mm. The properties such as gain directivity and VSWR are studied for the designed antenna. The properties are studied using the ANSYS HFSS software.

Keywords: - Flexible, Garment, Radiation Pattern, Textile, Wearable, WBAN

1. INTRODUCTION

Portable electronic devices have become a major part of everyday human life. Mobile phones which we carry throughout the day not only allow us to just make telephone calls but also provide internet access, multimedia, personal digital assistant and GPS functionality. This form of continuous connection allows a forward step in the computer paradigm [1]. In the upcoming generations human beings are likely to carry a range of devices and sensors, including medical sensors which constantly communicate with each other and the outside world and these are very easy to carry. It is of major importance to provide this functionality as the technologies are upgrading every day[3,4]. A key technology to achieve this goal is wearable

electronics and antennas. Microstrip patch is a well suited for any wearable application, as it can be made conformal for integration into clothing [2–10].A wearable antenna is mainly used for biomedical application. Body centric communication systems like paramedics, fire fighters, and military applications are utilized by different specialized occupation segments [11, 12].

The wearable antennas can be applied to all types of aged people and also for athletes for the purpose of monitoring. Wearable antennas should contain the following properties they should be light in weight, very low cost, almost maintenance-free and with no installation [11]. They should be comfortable and conformal to the body shape, but they should be highly reliable and efficient. There are many problems with the applications of Wearable antenna. The major problems are they should be integrated onto a garment and increase the comfort of wearable, the antenna should be of minimum size. Furthermore, it is difficult to keep the antenna flat and operating when the people move, so the antenna under different bending and crumpling conditions has been studied in [6]. Finally, the influence of human body to the antenna should be taken into account as well .Now in this paper we will study about the design of a wearable antenna by optimizing its size. And we will study the analysis of the antenna with different parameters. In this we also study about the interaction between a wearable antenna and human body and the analysis.

2. Literature Survey:

A planar inverted-F antenna is designed to operate on WBAN it is preferred because of its low profile and Omnidirectional properties. The patch and the ground layer perform a more important task of isolating and decoupling the wearable antenna from the human body. The patch size of the antenna is increased to reduce the return loss and to improve efficiency even the SAR value is reduced when the patch size increases. Polydimethylsiloxane (PDMS) as the. The design is done using CADFEKO 5.5 and the results are also verified with the same software [4]. A compact planar dipole antenna is built by using fully flexible materials such as nitrile butadiene rubber polymer composition is used as substrate. The patch is made up of copper. The performance of the antenna is analyzed by using three layer human tissue. Design and analysis of the antenna are carried using the full wave electromagnetic simulation software XFDTD which is based on the finite-difference time-domain method a non-uniform mesh technique is used for calculating the SAR [15]. UWB wearable textile antenna for body area networks with Hexagonal patch with a partial ground plane is proposed [20,21]. The ultra-wide bandwidth proposed can be obtained by reducing its size, and by introducing a square notch in the ground and a novel slot in the patch. Dacron fabric with a relative permittivity of 3 is used as a substrate. Commercial electromagnetic simulation package CST Microwave Studio is used for the design and simulation of the antenna [3,7].

3. Interaction Between Wearable Antenna And Human Body

Wearable antennas are very close with the human body. The close proximity of human body with an antenna with a high dielectric constant and loss are known to have some detrimental effect on input impedance and efficiency of the antenna. The body affects some of the major properties of the antenna like the mismatch which is caused by characteristic impedance of transmission lines, electrical length changes, and continuously increasing losses, thus damaging its original operation. The interaction between the body and wearable antenna can be studied under two categories:

3.1 Effects of Antenna on Human Body

The non-ionizing radiations like the sound waves, visible light and microwave may not have enough energy to ionize atoms or molecules but this energy is enough to move atoms or even to make them vibrate. Therefore, the non-ionizing radiation can move human cells and also increase the cell temperature. This rise in temperature affects the human tissues, the most common of this type is dielectric heating. This occurs when a dielectric material is heated by rotations of polar molecules induced by the electromagnetic field. Specific Absorption Rate (SAR) is the parameter that is used to measure the rate at which energy is absorbed by human tissues. It can also be defined as the rate at which RF electromagnetic energy is imparted to unit mass of biological body. SAR is usually measured average either over the whole body, or over a small patch sample volume typically 1 g or 10 g of tissue [14, 15]. The value thus obtained is the maximum level measured in the body part studied over the stated volume or mass. The formula used to calculate SAR is given as : $SAR=\sigma E/\rho$ Where, σ stands for electrical conductivity, E stands for RMS electric field and ρ stands for sample density. There are certain rules and regulations to be followed in the world regarding the SAR limit of electromagnetic devices because the high SAR values may have severe consequences on the human body. The SAR limit is set to 1.6 W/kg averaged over 1 g of actual tissue by the Federal Communication Commission (FCC), while the limit is set to 2 W/kg averaged over 10 g of actual tissue by the Council of European Union [14, 15]. The limit of temperature increase in head tissues is 1 K .This increasing in the temperature in the head tissue may affect the behavior, functioning and memory of the people besides causing anatomical injuries [4,13].

3.2 Effect of Human Body on Wearable Antenna

The human body also has some effects on the antenna as it very closely located to the human body. The human body is lossy and disturbs the communication link between antenna and outside world. The human body impact on the antenna is of different types

3.3 Human Body-Induced Gain

The human body induced gain is the ratio of gains (in dB) between body-worn antenna and that of the antenna in free space. Human body has various tissues with various dielectric properties. Also the electrical properties on different frequencies have different values, the gain of the antenna is affected [13]. So the gain of the wearable antenna differs from that of a normal antenna

3.4 Human Body-Worn Efficiency

The human body-worn efficiency is the ratio of total radiated power when antenna is worn in the body to the total radiated power in free space isolation. The overall power loss in a human body can be represented by this [6,13].

Human Body Effect on Impedance

The input impedance of the antenna will be low whenthe user is too close to the antenna. Also the input impedance is dependent on the moisture conditions of the human body. And the position of the placement of the antenna [13].

4. ANALYSIS REQUIRED FOR WEARABLE ANTENNAS

Generally, the measurements required for normal antenna design are return loss, radiation pattern, gain and efficiency. However, these planar antennas always remain flat, so it is not necessary to measure its bending characteristic. But we have to measure the bending characteristics of a wearable antenna and some other factors should be taken into careful consideration to guarantee the performance of the antenna in a body-worn context. They are as follows

4.1 SAR Modeling

As the wearable antennas are attached very close to the human body the SAR model values must be taken into serious consideration because they may affect the human body. The SAR and its effective values to be maintained are studied in the earlier sections [14, 15].

4.2 Measurement with Different Bending

IOP Conf. Series: Materials Science and Engineering **590** (2019) 012022 doi:10.1088/1757-899X/590/1/012022

As we are studying about the wearable antenna we have to study the properties of the antenna under different bending conditions. Because the human body does not remain flat always. We can study the bending affects by attaching the antenna to a cylinder like structure or other different things with different shapes. By studying all the measurements and if the appropriate end efficient only then we have to apply them to the human body otherwise we would get false values. The antenna had to be designed with a wide frequency bandwidth for effective bending effects. The complete bending effects like bending it to 90° and 180° must be studied [16].

4.3 On Body Measurements

Beside stand-alone antenna measurements, where the antenna was measured without presence of human body, onbody measurements have to be carried out as well, in order to ascertain the performance of the antenna at different on-body positions. Positions of wearable antennas will differ, depending on the application of the antenna. Wearable antennas might be designed to be placed on the chest, arm, back of the body and etc. In [17] researchers also went as far as including human body for the measurement. From these previous investigations, it was found out that the antenna placed on the back of the body is the most stable location that will reduce the change of body orientation compared other parts such as the arm

4.4 PROPOSED SYSTEM:

The main drawback of the existing system is they contain narrow bands and large in size. The proposed antenna overcome all these drawbacks of the existing system. The dimensions of the antenna are 40*34*1.26mm which is compact in size.

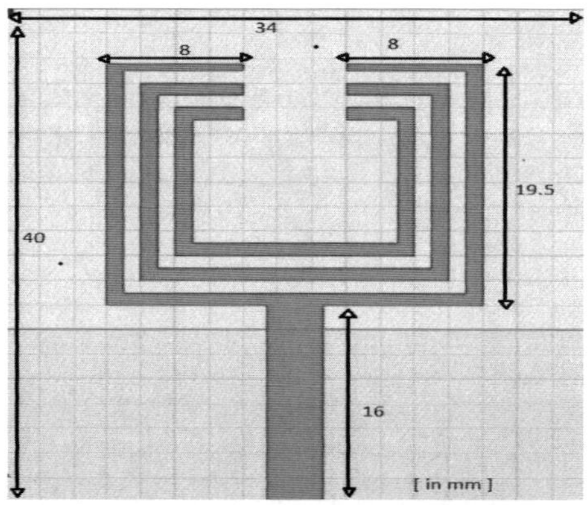

Fig 1: patch of the designed antenna

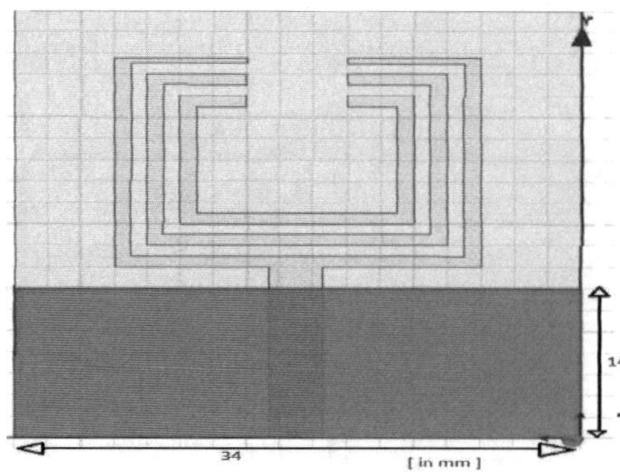

Fig 2: Ground of the antenna

IOP Conf. Series: Materials Science and Engineering **590** (2019) 012022 doi:10.1088/1757-899X/590/1/012022

Multiple cuts are made on the rectangular patch antenna to make the antenna resonate at 2.4GHz frequency. The size is made compact to make it a wearable antenna. The antenna is designed using bed sheet cotton with a relative permittivity of 3.2 and a loss tangent of 0.0027 as the substrate. The designed antenna is of thickness 1.26mm this makes it even more comfortable for on body applications. A partial ground is used to get good directivity and wider bandwidth

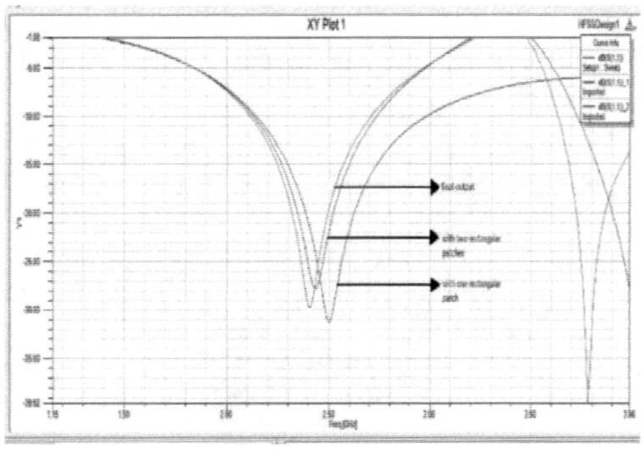

Fig 3: gain obtained for the antenna

ANALYSIS:

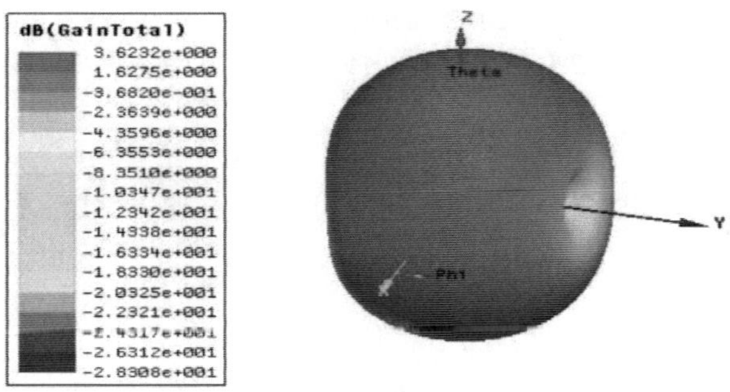

Fig 4: compared return loss and VSWR for the designed antenna

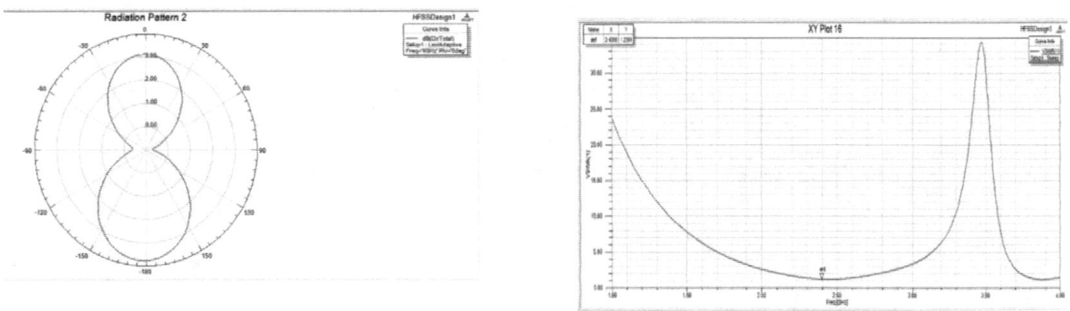

Fig 5: a) Radiation pattern of the antenna and b) gain measurement

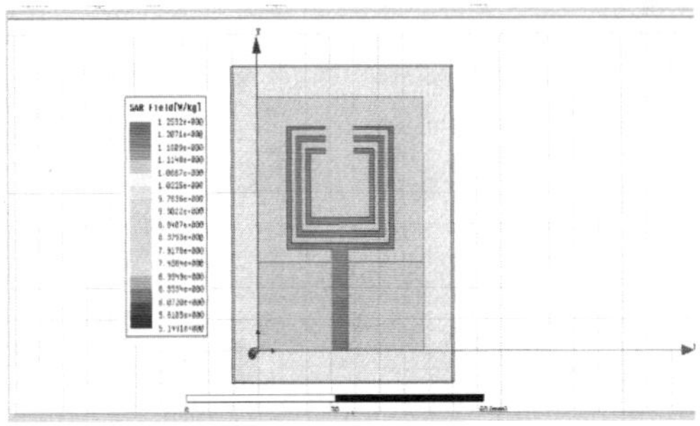

Fig 7: SAR value for the antenna

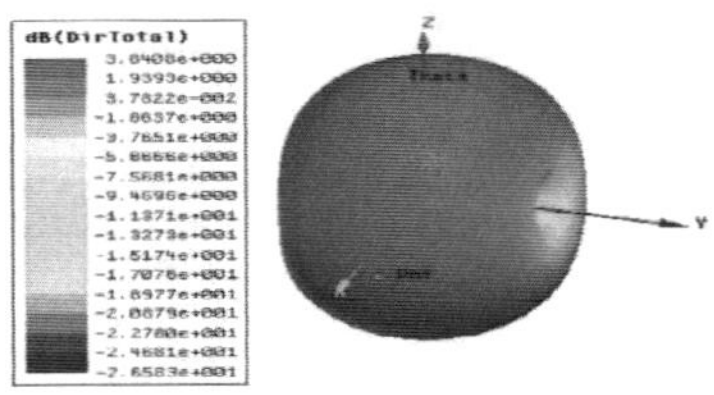

Fig 8: directivity of the designed antenna

(A) From fig 3, it can be inferred that the designed antenna has very good return loss. The designed antenna acts as a multiband antenna. The antenna has a good return loss with in the frequency range of 2.1GHz to 2.7GHz and 3.6 GHz to 4.3GHz which is <-10dB. The designed antenna has good bandwidth. The designed antenna works in the ISM band which is best suited for wireless body area network.

(B) From fig 4, it can be inferred that the designed antenna has good VSWR which is 1.2391.

(C) From fig 5, it can be inferred that the designed antenna has good radiation pattern. The radiation pattern is good along both the axis.

(D) From fig 6, it can be inferred that the designed antenna has good gain. The gain of this antenna shows that the antenna is well suited for body area network applications. The efficiency of the antenna obtained is also high.

(E) From fig 7, it can be inferred that the designed has a good SAR value compared to the previous works. The SAR value obtained for the designed antenna is 1.25W/kg. This is below the required SAR value range which is 1.6W/kg. So the designed antenna can be well suited for body area network applications.

(F) From fig 8, it can be inferred that the designed antenna has good directivity. The designed antenna has a directivity of 3.840. The efficiency of the antenna can be calculated by using the gain and directivity of the antenna.

The efficiency of an antenna is gain divided with directivity and multiplied with hundred.

$$\text{Efficiency} = \frac{Gain}{Directivity} * 100$$

6. Applications On WBAN

Wearable antennas are mainly designed for the wireless body area networks that is to for body parameters measurement. Like heartbeat and disease detections these are also mainly used by the firefighters and in military operations also. These wearable antennas can be used by the sports persons to measure their heartbeat, blood pressure fatigue measurement and other parameters. Wearable antenna has a future scope also they can be used for IOT applications by using them with the combination of sensors.

7. CONCLUSION:

The design of wearable antenna for wireless body area network applications is studied in this paper. The designed antenna is compact in size compared to the previously designed antennas. The designed antenna has given good result in all the properties. As the antenna size is 40*34*1.26mm this can be well suited for wearable applications. Also the designed antenna acts as a multiband antenna and it has wide range of operating frequency. This antenna provides a wider bandwidth with in the frequency range of 2.1GHz to 2.7GHz and 3.6 GHz to 4.3GHz.The antenna has an efficiency of 94%. The SAR value of the designed antenna is 1.25 which is less than the SAR limitation so this antenna will not cause radiation problem when placed on a human body. So based on all the analysis made the designed antenna can be well suited for wireless body area network applications.

REFERENCES:

[1].S. Sankaralingam and B. Gupta Development Of Textile Antennas For Body Wearable Applications And Investigations On Their Performance Under Bent Conditions Progress In Electromagnetics Research B, Vol. 22, 53–71, 2010

[2]. Salomen, P. and H. Hurme, “Modeling of a fabric GPS antenna for wearable applications,” Proceedings of IASTED International Conference Modeling and Simulation, Vol. 1, 18–23, 2003.

[3]. T.Ravi, “ Design And Performance Analysis Of Ultra Low Power RISC Processor Using Hybrid Drowsy Logic In CMOS Technologies” International Journal of Applied Engineering Research, Vol 10, No. 2 (2015), pp. 4287-4296.

[4].M.Manoranjani and T.Ravi, “ “Multithreshold CMOS sleep stack and logic stack technique for digital circuit design” ARPN Journal of Engineering and Applied Sciences, Vol. 10, No. 10, June 2015. pp: 4550-4556.

[5].M R Annish Brislin, J Albert Mayan, R Aroul Canessane, M R Anish Hamlin,"Blood donation and life saver app",2017 2nd International Conference on Communication and Electronics Systems (ICCES), pp. 446 - 451,2017

[6]. S.Karthikeyan, S.Jayashri, “Energy Utilization Strategies using Novel Approach in Wireless Sensor Networks”, International Conference on Software Engineering and Mobile Application Modelling and Development, Dec 2012, pp: 1-5.

[7]S.Karthikeyan, S.Jayashri, “Energy Efficient System for Heterogeneous Wireless sensor Networks”, European Journal of Scientific Research, Volume 72,No 4 march 2012, PP:599-607. ISSN: 1450-216X.

[8] G.Savitha and S.Karthikeyan,"Architecture for diverse modulation techniques in SDR", concurrency computation practice and experience,John wiley & sons, Ltd, February 2019,PP: 1-10.

[9]. Kamalesh M. D., Albert Mayan J., Felix Y., Sumanth B. S., & Sai Tej B. (2018), " Magrisys: A Smart And Ubiquitous Controlled – Environment Agriculture System " , 2nd International Conference on I-SMAC (IoT in Social, Mobile, Analytics and Cloud) , pp: 1-5 , 2018.

[10]. Albert Mayan J, Kuldeep Anand D.S, Neha Sadhvi(2017),"Efficient and secure server migration on cloud storage with VSM and dropbox services", International Conference on Information Communication and Embedded Systems (ICICES),Chennai pp. 1-5..

[11]. Merli. F, L. Bolomey, Meurville. E, and Skrivervik. A.K, “Implanted Antenna for Biomedical Applications”, IEEE Antennas and Propagation Society International Symposium, San Diego, CA,1-4, 2008.

[12]. Albert Sabban “Wearable Antennas for Medical Applications Lab Chip, vol. 11, no. 19, pp. 386–397.

[13]. Mayan J.A, Arifa S, Pavithra R,"Semantic based multi lexical ranking technique for an effective search in protected cloud,2016 International Conference on Control, Instrumentation, Communication and Computational Technologies (ICCICCT), Kumaracoil, 2016, pp. 570-576.

[14International Commission on Non-Ionizing Radiation Protection, “Guidelines for limiting exposure to time-varying electric magnetic, and electromagnetic fields (up to 300GHz),” Health Phys., vol.74, pp. 494-522, 1998.

[15]Julian Menezes.R, Albert Mayan .J, M. Breezely George,” Development of a Functionality Testing Tool for Windows Phones”, Indian Journal of Science and Technology,Vol:8,No:22,pp: 1-7, 2015, ISSN : 0974-6846

[16]. I. Locher, M. Klemm, T. Kirstein, G. Troster, “Design and Characterization of Purely Textile Patch Antennas”, Advanced Packaging, IEEE Transactions on, 29(4), 777-788, 2006.

[17]. Ankita Priya Brajlata Chauhan "A Review of Textile and Cloth Fabric Wearable Antennas". International Journal of Computer Applications (0975 – 8887) Volume 116 – No. 17, April 2015.
[18]. J.G. Santas, A. Alomainy, H. Yang, "Textile Antennas for OnBody Communications: Techniques and Properties". The Antennas and Propagation, 2007. EuCAP 2007.
[19] S.Karthikeyan, S.Jayashri, "Proficient Data Gathering by Mobile Agent using Jade",i-manager's Journal on wireless communication networks,Vol.1, No. 2, 2012, pp. 32-40.
[20] Gopika. S. Kumar, S. Karthikeyan and N. Ambily, "A Metaphorical Scrutiny in Banishing Salt and Pepper Noise",Biosciences Biotechnology Research Asia,December 2014 Vol. 11(3).
[21] Yovan Felix A , Jesudoss A ,Albert Mayan J (2017) ,"Entry and exit monitoring using license plate recognition ", IEEE International Conference on Smart Technologies and Management for Computing, Communication, Controls, Energy and Materials (ICSTM), pp.227-231.
[22] G. Karudaiyar, S.Karthikeyan and B. Sainath, "Encryption and Decryption Schemeby Using Finite State Machine", Biosciences Biotechnology Research Asia, December 2014 Vol. 11(3), 1867-1872.

IOP Conf. Series: Materials Science and Engineering **590** (2019) 012023 doi:10.1088/1757-899X/590/1/012023

DESIGN AND DEVELOPMENT OF SENSOR BASED INTELLIGENT AUTO IRRIGATION SYSTEM

Dr.P.Asha, Anusha Ambati, R.Anuhya

Department of Computer Science and Engineering, Sathyabama Institute of Science and Technology, Chennai.

ashapandian225@gmail.com

Abstract: The quick improvement and scaling down of sensor gadgets, current advances sensor innovations, are enabling researchers and specialists to grow little sensors that can be utilized to ceaselessly screen the wellbeing and steadiness of the Plants Environment. Sensors with figuring, handling and correspondence abilities that can persistently detect and Triggers Solenoid esteem, where naturally can be prepared and seen progressively. This venture gives a nitty gritty investigation and usage of a sensor for checking of Green House parameters like moistness, water levels, etc.... A Sensor network was implemented using an Aurdino based AVR the microcontroller.

1. Introduction

The point of the framework is to create and execute the horticultural yields and to make ordinary cultivating into savvy cultivating. The sensor-based water system framework is principally conceptualized such that the agriculturist can gain higher benefit utilizing the framework. It will be worked naturally. More tasks can be performed utilizing robotization. Naturally, control ecological conditions. The proposed framework depends on Arduino based smaller scale controller. Sensors are associated with the Arduino board. Arduino UNO can be controlled by USB link from PC. The dirt nourishment sensor is utilized to distinguish the sustenance level of dirt. Water level sensor is to perceive the water level in the developing field. Dampness sensor is utilized to distinguish the air dampness level if the dampness level is low at that point water sprinkler is on. An Ultrasonic sensor measures the separation of target objects. The sensor faculties and sends the data to the cloud and it gives an alarm message to the agriculturist.

2. Literature Survey

The authors proposed a framework that utilizes correspondence innovations, enabling the client to dissect the data acquired by various sensors from any gadget in a simple way. The proposed design depends on different remote hubs outfitted with the GPRS network. Diverse business sensors are utilized for estimating the extensive variety of parameters of the dirt, plant and environment are associated with the hubs. The data is sent to remote server, where the data is dealt with and stores the information of the sensors in a database. An electronic board is a fundamental board which is in charge of sensors information. It is spoken with whatever remains of the sheets. The Sensor-Board gives the interface the associated sensors. The outline of the arduino is the best use for agrarian observing, permits the working association of a typical exactness horticulture sensors [1]. By methods for a GSM/GPRS convention the GPRS-Board sets up the correspondence in the versatile system and sends the information to the remote stockpiling server for facilitate counsel. When it is controlled up, the small scale controller secures the game plan information from the SD-Card. This card contains the case time allotment, transmission rate, GPRS and remote server outlines information which are related with the sensor stacks up.The information from the sensors related with the contraption is discontinuously sent to a data server and data server techniques and stores in a social database [2]. To make the gadget vivaciously self-ruling, a comprehensive lively examination of every situation was done to decide the most reasonable battery and sunlight based board in order to dissect the power utilization of the gadget. This depicts the outline, task and advancement of a functional application to streamline water assets in inundated horticulture by checking soil water status and the water system water. To evaluate the power usage of the contraption the testing rate of the sensors and sending rate using a GPRS affiliation is urgent [3].

The target of the work is to screen soil dampness progression in the best soil in detail, in both space and time. The proposed checking framework depends on absolutely autonomous sensor hubs, which permit both ongoing and noteworthy information administration and are associated through an information/yield interface to a WSN stage. In the present work, first describes about the deployment of a WSN, successively the calibration of the relationship between sensor reading and soil water content as a function of the voltage supply is described with reference to the EC-5 capacitance sensors, and finally an example of application concerning the optimization of the irrigation schedule is presented [4]. The soil moisture sensors provide an output reading in mV, which must be converted in soil volumetric water content through a calibration curve. The proposed framework is created by 27 WSN hubs, every one furnished with 5 soil dampness sensors and a subsoil thermometer. Such sensors are associated through an info/yield interface to a Tiny Node 584 WSN stage. The WSN observing framework is conveyed and tried for the ability to give spatially dispersed hydrological information applicable for eco-hydrological examines. In general, this proposes water utilization can be diminished without bargaining the quality and amount of apple generation by presenting a water system plan in view of the data gave by WSN [5].

Authors mechanized a water system sensor that can really scale well with rural development. The sensor utilizes an advanced mobile phone to catch and process computerized pictures of the dirt adjacent the root zone of the yield, and gauges optically the water substance. The water system sensor depends on an installed camera with an advanced mobile phone, encased in a waterproof and light-tight covered chamber. The sensor is kept in a chamber under controlled brightening and covered at the root level of plants [6]. An Android App was produced in the PDA to work specifically the figuring and availability parts, for example, the advanced camera and Wi-Fi organize. The portable App awakens the advanced cell, enacting the gadget with a client characterized parameters. At that point, the implicit camera takes a photo of the dirt through a hostile to the intelligent glass window and an RGB to the dull procedure is proficient to assess the extent among the wet and dry region of the photo. After the Wi-Fi association is empowered, the proportion is transmitted through a switching hub to an entryway for control a water system water pump. At long last, the App sets the advanced cell into the rest mode to save its vitality. The sensor is powered by rechargeable batteries, charged by a photovoltaic panel.The advanced cell water system sensor was assessed in a pumpkin edit field along 45 days. The exploratory outcomes demonstrate that the utilization of advanced mobile phones as a water system sensor could turn into a down to earth device for farming [7-10].

Proposed procedure is installed in the system portal making the framework a really keen and self-ruling remote choice emotionally supportive network. This has been planned keeping in mind the end goal to help the end-client to decide the harvest require water system water, how much water is required and how to flood the product. In this, a remote choice emotionally supportive network for the upgraded administration of the water system in horticulture has been exhibited. The properties of the WSAN innovation have been abused to gain heterogeneous natural parameters and to control the working of the water system framework. The FL-based system has been planned and aligned with the signs of the ranchers with a specific end goal to emulate the human experience and to appropriately comprehend the status of the product [11-13].

The authors supported with various advancements related to the Internet of Things have the phenomenal potential for application in the region of sustenance and cultivation, especially in the context of the societal and normal troubles looked by this fragment. From ranch to fork, IOT advancements could change the part, adding to sustenance security, and the lessening of farming sources of info and nourishment squander. A principle advance for the more prominent take-up of these advances will be the execution of IOT-based substantial scale pilots (LSPs) in the whole inventory network. This article diagrams the difficulties and requirements that a LSP sending of IOT in this space must consider. Innovative difficulties are portrayed so as to recognize an arrangement of mechanical and food prerequisites. An engineering in light of an arrangement of frameworks approach is quickly exhibited, the significance of tending to the interoperability challenges looked by this area is featured, and we expand on prerequisites for new plans of action, security, protection, and information administration [14].

Authors explained that India's people are come to past 1.2 billion and the masses rate is extending well ordered, there will be completely serious an issue of sustenance, so it is vital to creating cultivating. Today, the farmers are encountering the nonattendance of deluges and lack of water. The goal is to develop a programmed water system framework to spare the time, cash and power. Manual intercession is required for the customary homestead arrive water system strategies. With the computerized innovation of the water system, the human intercession can be limited. The temperature and dampness sensors sense the temperature and stickiness of the surroundings advertisement gives an interference with the flag to the micro controller. The Discourage weeds spare water and time, measurable information can be utilized to control infections and contagious development, least complex model. The principal weakness is this framework is simply restricted to the mechanization of the water system framework and needs in additional common highlights. The proposed framework is to underpin forceful water administration for the rural land [15].

3. Existing System

The effectively existing we need to Monitor every last procedure and may prompt some product is going to spoiling and we don't know why the yield is decaying some little misstep that makes part of economic issues there is the real disservice in the current framework. And in some of the smart agriculture they placed some sensor, but they configure into the internet. So each and every time human have to monitor, the factors like water level, humidity level in farming.

4. Proposed System

The given proposed framework is finished by utilizing Arduino based miniaturized scale controller. Furthermore, we need to interface the sensor to Arduino board, the sensor is the dampness sensor, soil sustenance sensor and water level sensor (Fig. 1).

IOP Conf. Series: Materials Science and Engineering **590** (2019) 012023 doi:10.1088/1757-899X/590/1/012023

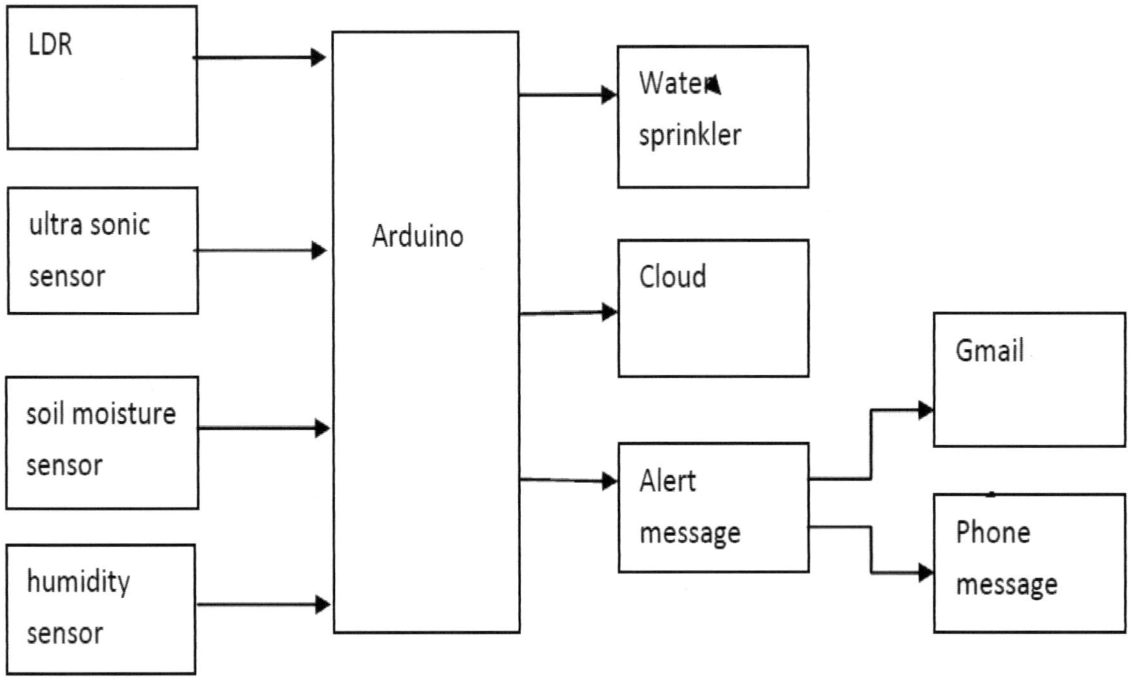

Fig.1. Architecture Diagram

The dirt nourishment level sensor is utilized to distinguish the sustenance level of a sensor. Water level sensor is utilized to recognize the water level in the cultivating field. The Moistness sensor is utilized to distinguish the air dampness level if the dampness level is low means water sprinkler is on. These all parameter spared in the cloud for future reference to recognize the harvest yielding. Also, give the alarm message to the rancher if the water level is the low or sustenance level of a dirt.

Sensors connected to the board are LDR, ultrasonic sensor, humidity sensor (Fig. 2) and soil moisture sensor (Fig. 3). The USB association is likewise how you will stack the code onto your Arduino board. Humidity sensors specifically measures the temperature along with the relative humidity and delivers completely calibrated digital outputs. Dampness sensor detects, measures and reports the relative moistness noticeable all around. LDR sensor is utilized to quantify the force of daylight in the earth. Ultrasonic measures remove between the articles. Soil dampness sensor measures the water level in the dirt. An Irrigation sprinkler is a gadget used to water farming harvests, gardens, scenes, and different territories. All the data subsequent to detecting will be put away in the cloud. Any unusual condition happens then it will alarm a mail or telephone message.

IOP Conf. Series: Materials Science and Engineering **590** (2019) 012023 doi:10.1088/1757-899X/590/1/012023

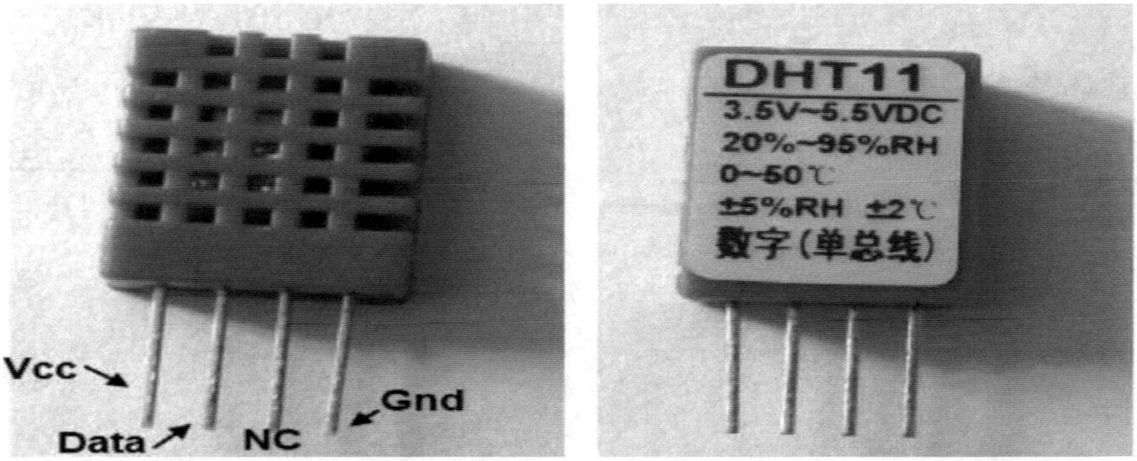

Fig.2. Humidity Sensor

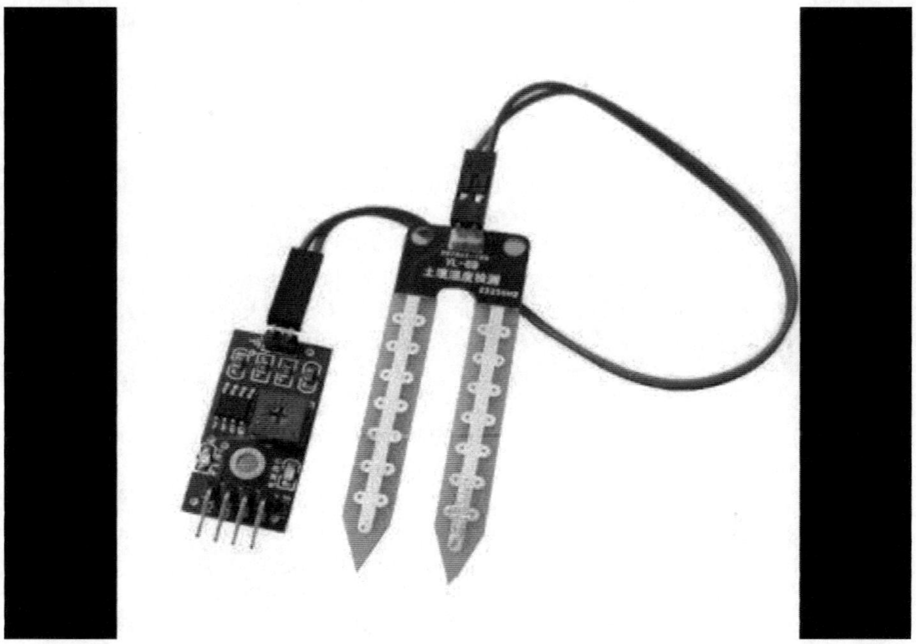

Fig. 3. Soil Moisture Sensor

Disadvantages of the existing system

1.It cannot be operated automatically.
2.Less operations can be performed using the robot.

Advantages of the proposed system

1.It can be operated automatically.
2.More operations can be performed using automation.
3.Has intelligence to avoid flooding of the field.

5. Conclusion

The system is used widely in industries and household but mainly suitable for agriculture. Finally, it is most usable and profitable for farmers. It is less time consuming with good power saving capability.

References

[1] B. Majone, F. Viani, E. Filippi, A. Bellin, A. Massa, G. Toller, F. Robol, M. Salucci, “Wireless sensor network deployment for monitoring soil moisture dynamics at the field scale,” Procedia Environmental Sciences, vol. 19, pp. 426-435, 2013.

[2] O. Kaiwartya, A. H. Abdullah, Y. Cao, R. S. Raw, S. Kumar, D. K. Lobiyal, I. F. Isnin, X. Liu, and R. R. Shah, “T-MQM: Testbed-based multi-metric quality measurement of sensor deployment for precision agriculture - a case study,” IEEE Sensors J., vol. 16, no. 23, pp. 86498664, Dec. 2016.

[3] J. G. Jaguey, J. F. Villa-Medina, A. Lopez-Guzman, and M. A. PortaGandara, “Smartphone irrigation sensor,” IEEE Sensors J., vol. 15, no. 9, pp. 5122-5127, Sep. 2015.

[4] Mayan J.A., Ravi. T ,"Test Case Optimization Using Hybrid Search Technique",ACM International Conference Proceeding Series ,ACM New York, NY, USA ©2014, ISBN: 978-1-4503-2908-8 ,doi:10.1145/2660859.2660906

[5] P.Asha, Dr.S.Srinivasan, “Analyzing the Associations between Infected Genes using Data Mining Techniques”, International Journal of Data Mining and Bioinformatics, Inderscience Publishers, 15(3), pp.250-271, Mar 2016.

[6] J.H. Ziegeldorf et al., “Privacy in the Internet of Things: Threats and Challenges,” Security and Commun. Networks, vol. 7, no. 12, 2014, pp. 2728–42.

[7] Albert Mayan J, Anto Praveena M D , Telkar Bharath Rao , Uravakonda Uday Sagar(2018),"Optimized test data generation over suspicious implementation of oracle problem",IEEE International Conference on Power, Control, Signals and Instrumentation Engineering (ICPCSI-2017),pp.2559-2563.

[8] L.D. Xu et al., “Internet of Things in Industries: A Survey,” IEEE Trans. Industrial Informatics, vol. 10, no. 4, 2014, pp. 2233–43.

[9] O. Vermesan and P. Friess, Building the Hyperconnected Society: IoT Research and Innovation Value Chains, Ecosystems and Markets, River Publishers, 2015.

[10]EC, “Call — Internet of Things, in HORIZON 2020-Work Programme 2016-2017 Cross-Cutting Activities (Focus Areas)”; http://ec.europa.eu/research/participants/data/ref/h2020/wp/2016_2017/main/h2020-wp1617-focus_en.pdf, July 2016, accessed Nov. 25, 2016.

[11]Asha P, Albert Mayan J, Canessane A (2018),"Efficient Mining of Positive and Negative Itemsets Using K-Means Clustering to Access the Risk of Cancer Patients",Communications in Computer and Information Science ,ICSCS 2018, Kollam, 2018,pp.373-382.

[12]J. Dymond, A.-G. Ausseil, R. Parfitt, A. Herzig, and R. McDowell, "Nitrate and phosphorus leaching in New Zealand: a national perspective," New Zealand Journal of Agricultural Research, vol. 56, pp. 49-59, 2013.

[13]S. Ancy, R. Kumar, R. Asokan and R. Subhashini, "Prediction of onset of south west monsoon using multiple regression," Proceedings of IEEE International Conference on Computer Communication and Systems ICCCS14, Chennai, 2014, pp. 170-175.

[14]Prem Jacob, T., Pravin, A., & Asha, P. (2018). Arduino Object Follower with Augmented Reality. *International Journal of Engineering & Technology, 7*(3.27), 108-110.

[15]Pravin, A., Prem Jacob, T., & Asha, P. (2018). Enhancement of Plant Monitoring Using IoT. *International Journal of Engineering & Technology, 7*(3.27), 53-55.

Detecting phishing website using Pattern Mining

Sreelekha B, Harika B,Mrs.L.Sujihelen

Department of Computer Science & Engineering,
School of Computing,

Sathyabama Institute of Science and Technology, Chennai

sujihelen@gmail.com

Abstract. Nowadays users are purchasing products and payments through online. While using these websites they will ask for the information of the users like username and password etc. There are some websites which are used to hack the details of the user. This type of websites is called phishing websites. For detecting these websites, they are proving an effective, intelligent data mining algorithm is used. In this paper, they are surveying the detection of phishing websites by using pattern mining technique.

Keywords— Novel algorithm, NLP technique, Supervised learning algorithm, streaming analytics, SVM, Moby fish.

1. Introduction

To hack and obtain the personal information from the user phishing technique is used. To overcome these phishing techniques there are many anti-phishing websites are used. Every day the usage of internet is increasing rapidly. Nowadays all the users are using internet for sending the messages, emails, and also online banking etc. From this, some hackers may ask for username and the password to collect the sensitive information of the users. Sometimes the users use the fake websites and give their information; they will target the main user of the particular bank or the financial organisation for collecting the information. For this purpose, they are providing the phishing technique to detect these types of websites and emails. In this without the use of internet connection we are going to detect the phishing websites. The pattern mining technique is used for detecting.

2. Related Work

"PHISHSTORM: DETECTING PHISHING WITH STREAMING ANALYTICS"

In this technique, we are going to use Phish Storm, which is a phishing detection used in real time for any URL to identify phishing sites. The experimental evidence will support this paper and is used to detect phishing websites. To evaluate this, we are going to use this concept which is Intra-URL related process which are extracted from words that compose the URL. Machine -learning-based classification features are usedto detect Phishing URLs from real data set. In this paper,They are using efficient patterns which allow real-time analytics by using Big data diagrams such as STORM and advanced data structures are used.

A HYBRID MODEL TO DETECT PHISHING -SITES USING SUPERVISED LEARNING ALGORITHM

Because of this, they can catch the credential details such as bank account number, password etc. To overcome this threat, They propose a hybrid algorithm. In this hybrid algorithm, select the best classifier in terms of high accuracy and less error rate. Then combine the best classifiers with others and develop the hybrid model

IN THIS, THEY USE DATA FROM UCI REPOSITORY. DATASETS CONSIST OF ATTRIBUTES AND INSTANCES INCLUDING

- DOMAIN REGISTRATION-LENGTH
- USING POP-UP-WINDOW
- URL-OF-ANCHOR

Data splitting criteria

In this, the processed data is classified into training and testing categories. Some sets are used for training and others are used for testing.

Data mining classification technique

THEYUSE DATAMINING CLASSIFICATION TECHNIQUES SUCH AS/ RANDOM FOREST, DECISION TREE.

Ensemble technique

- Boosting
- Bagging
- Stacking

"DETECTING SPAM AND PHISHING EMAILS USING SVM AND OBFUSCATION URL DETECTION ALGORITHM"

They propose a hybrid model and evaluate their performance. In this, They proposed a system that uses SVM along with map reduce technique to overcome the problems related to SVM. They use BURL detection algorithm which provides multilayer security. Since/ the attackers use a number of ways to obfuscate the URL. OBURL detect the maximum number of attacks by using tests which are performing for checking

- NS test
- IP address test

SVM algorithm-support vector machine

Support vector machine can be applied toa large amount of input data. It finds the separating hyper plane such that error for the unseen pattern is minimized.

A NOVEL ALGORITHM TO DETECT PHISHING URLS

As they know that the use of the internet is increasing every day. Due to this many online attacks are increasing. The well-known attack is phishing. Phishing is the technique in which it is used to extract the delicate data of the users by providing the source in a web transaction. In many cases, phisher chooses HTTP URLs for the attack. For this purpose, they are proposing a Novel algorithm which will detect whether a given HTTP URL is a phishing site or not. The alert message will be displayed, if the URL found to be phishing, otherwise it will display the safe message. The algorithm enhances the performance when displaying with known/old phishing URLs.

In this, we are proposing phishing URL detection algorithm to find out maximum URL phishing attacks because it executes multiple tests such also URL features test. It also uses top Google a search engine results. The extent of our algorithm is limited to identify phishing URLs rather than detecting an extensive variety of malicious URL, rather than detecting an extensive variety of malicious URLs. In this algorithm, we are using URL hyperlink as input. The output will display the alert message whether the URL is phishing or not.

DC SCANNER: DETECTING PHISHING ATTACK

The proposed system uses the html data which consists of emails and also web pages. Also, domain and domain related authority to web pages are analyzed. There are two problems one is, a normal URL shown it in the address bar which appears like address of any website. Defining the work is meant for an email scanner to filter phishing by using the contents of web pages, also verifying the web domain and details.

Web page scanning

The deceive will come true when phishing web page having its normal appearance and URL same of the genuine web page, ask for credentials from users.

Script scanning

They can associate script codes. It is also checked whether the tags which follow the definitions are tortured.

DETECTION AND PREVENTION OF POSSIBLE LOGIN ATTEMPTS THROUGH STOLEN CREDENTIALS FROM A PHISHING ATTACK IN AN ONLINE BANKING SYSTEM

In this they detect and prevent unauthorized login attempts using behavioural based analysis, IP and device identification technologies.

Online banking facility

In order to access an online banking facility, usually a customer should first register with online banking service provide by the bank and setup username and a password.

a) Phishing attacks on online banking systems:

The concept of phishing is related to deceiving the customers to reveal their login credentials to attackers by masquerading as a trustworthy entity and the term first used by the spam people to describe about stealing of America Online (AOL) accounts.

b) *Currently available methods to safeguard online banking:*

The main action a financial institute is taking to protect their online customer is to educate the customer on online banking services security. The bank sometimes uses commercial products like Guardian Analytics, Threat Matrix, Actimize and the Ver safe Tot All to protect their customers from attackers.

c) *Anomaly- based detection of frauds*

Statistical methods-Detects the abnormalities of users or systems by monitoring the behaviour over a time period by measuring certain variables.

Distance-based-methods-More advanced than statistical methods and uses algorithms to detect outliers by computing distances among points.

"EFFECTIVE DEFENSE SCHEMES FOR PHISHING ATTACKS ON MOBILE COMPUTING PLATFORMS"

A lightweight anti-phishing scheme which is used for mobile platforms. Mobifish verifies the validity of applications, webpages and accounts by comparing with the actual identity. The performance of Moby fish is experimentally evaluated with phishing URLs and also legitimate URLs, also phishing applications. Hence the result is that Moby fish is very effective way for detecting Phishing attacks which are used on mobile phones.

A COMPUTER VISION TECHNOLOGY TO DETECT PHISHING ATTACKS"

The proposed approach is a combination of visual similarity-based techniques and whitelist. SURF detector which is the computer vision technique is used to extract features from both harmful and normal websites.

Some phishing has same visual appearance as the legitimate site such that we cannot differentiate with normal eyes. In proposed system we are working by detecting and extracting unique key points from both screenshots and it will finds the similarity ratio to identify phishing status.

This basic of the proposed solution is given below

- Image database consisting of all URLs and popular websites screenshots has to be maintained.
- After obtaining the accessed URL then compare with a whitelist of URLs.
- URL is considered as innocent if the comparison is successful and no checking required.
- In the whitelist if the given URL is not found then then apply SURF algorithm website screenshot and legitimate image database.
- From both suspicious website screenshot, image database extract the SURF features and compare for similarity check.

1. Web page will be considered as suspected, if the similarity score is greater than the threshold
2. URL is considered to be innocent, if the similarity score is less than threshold. In next update the domain will be the part of whitelist.

3. A COMPARATIVE ANALYSIS AND AWARENESS SECURITY OF PHISHING DETECTION TOOLS

In this paper comparison of eight phishing detection tools has been done to find the best one by testing each tool on the data set, and further an awareness survey was carried out about these tools. After testing on the data set it is found that Antiphishing toolbar did a very good job at identifying 94.32% of phishing as well as legitimate websites from the data set.

Table 1. Study of existing work

1	Detecting phishing attacks from URL by using NLP techniques (IEEE 2017)	Natural Language Processing (NLP) techniques. The system was implemented by examining URLs used in Phishing Attacks before opening them with using some extracted features	It is not predicted accurately
2.	Detecting spam and phishing mails using SVM and obfuscation URL detection algorithm	a system that uses SVM technique along with map-reduce paradigm to achieve a higher accuracy in detection of the spam	Perhaps the biggest limitation of the support vector approach lies in the choice of the kernel

	(IEEE 2017)	email	
3.	Phish Storm: Detecting Phishing with Streaming Analytics (IEEE 2014)	Phish Storm can interface with any email server or HTTP proxy. argue that phishing URLs usually have few relationships between the part of the URL that must be registered and the remaining part of the URL	they can be abysmally slow in test phase
4.	A Hybrid Model to Detect Phishing-Sites Using Supervised Learning Algorithms (IEEE 2016)	To identify such features is a classification task and can be solved using data mining techniques	You can scam your emails or pop ups
5.	A novel algorithm to detect phishing URLs (IEEE 2016)	a novel algorithm which will detect whether a given http URL is of phishing a site or not	phishers can use the data to access a victim's account and withdraw money or purchase merchandise or services

4.CONCLUSION

This paper focuses on various literature survey done to identify phishing websites. In this we are using pattern mining technique which is more effective when compared to the classification algorithm.

REFERENCES

[1] Ebubekir Buber, Banu Diri, Ozgur Koray Sahingoz, "Detecting phishing attacks from URL by using NLP techniques" IEEE 2017.
[2] Albert Mayan.J,Surya.B,Pranoy Prabhakar,Princekumar,"Department–Student Library Using Twig Pattern Query Processing Over Admin-User Login Privilege", Pak. J. Biotechnol. Vol. 13 , Pp. 489 - 493,(2016)
[3] PrajaktaPatil,Rashmi Rane, MadhuriBhalekar ,"Detecting spam and phishing emails using SVM and obfuscation URL detection algorithm" 2017.
[4] Samuel Marchal, Jerome Francois, Radu State, Thomas Engel, "Phish Storm: Detecting Phishing with Streaming Analytics", 4 dec 2014.
[5] Julian Menezes .R, Albert Mayan .J, M. Breezely George," Development of a Functionality Testing Tool for Windows Phones", Indian Journal of Science and Technology,Vol:8,Issue:22,pp: 1-7,September 2015, ISSN : 0974-6846.
[6] Varsharani Ramdas Hawanna, V.Y.Kulkami, RA.Rane, "A novel algorithm to detect phishing URLs", 2016.
[7] Binay Kumar, Pankaj Kumar, Ankit Mundra, Shikha Kabra, "DC scanner: Detecting Phishing Attack", IEEE 2015.
[8] Shammi Ishara Hewamadduma, "Detection and prevention of possible login attempts through stolen credentials", IEEE 2017.
[9] Longfei, Xiaojiang DU,"Effective defense schemes on mobile computingplatforms", IEEE 2015.
[10] Routhu Srinivasa Rao, Syed Taqi Ali, "A computer vision technology to detect phishing attacks",IEEE 2015.
[11] Himani Sharma, Er.Meenakshi,Dr.SandeepKaur Bhatia,"A comparative analysis and awareness survey of phishing detection tools", IEEE 2017

[12] Albert Mayan J , Kumar Akash Deep , Mukesh Kumar, Livingston Alvin Siva Prasad Reddy (2016) , "Number plate recognition using template comparison for various fonts in MATLAB ", InternationalConference on Computational Intelligence and Computing Research, pp: 1-6.
[13]Sujihelen, L., Jayakumar, C.,"Inclusive Elliptical Curve Cryptography (IECC) for Wireless Sensor Network Efficient Operations "Wireless Personal Communications, pp.1-22, 2018.

IOP Conf. Series: Materials Science and Engineering **590** (2019) 012025 doi:10.1088/1757-899X/590/1/012025

NAVIGATION ASSISTANCE USING AN UMBRELLA

Dr. S. Prince Mary[1], Dr. B. Bharathi[2], Bayana Venkata Sai Ujwala[3], Ravi Laxmipriya[4],

[1] Associate Professor, Department of Computer Science and Engineering, Sathyabama Institute of Science & Engineering, Chennai
[2] Professor, Department of Computer Science and Engineering, Sathyabama Institute of Science & Engineering, Chennai
[3] [4] Student, Department of Computer Science and Engineering, Sathyabama Institute of Science & Engineering, Chennai

princemary26@gmail.com

Abstract. Internet of Things (IOT) is one of the fastest growing technologies which allows versatile creation of worldly applications with devices that communicate via computer based systems. It has made way for exceptional innovations which has made our lives assisted and easy. One such example is smart umbrella, which not only acts a shield from rain but also consists of various other features such as alerts of weather and virtual assistance for the elderly. This paper proposes an umbrella which acts as a navigation assistant to pedestrians to reach their respective destinations.

1. INTRODUCTION

The primary usage of an umbrella is to protect its users from rain and sun. And as days change, smart umbrellas now redefine its traditional purpose. The new smart umbrella is now a navigation assistant for pedestrians. Other significant features include theft prevention using SMS alerts, speech assistance, and weather forecasting. The expansion of usage of electronic devices while driving and walking on the roads have had an increased effect on the road accidents that occur annually. This proposed system reduces this risk of accidents by eliminating the need to look at mobile phones and assists in navigation. The umbrella is made of 8 LED bulbs. These are attached to tip of umbrella ribs. These LED bulbs assist the users by pointing to the direction they are supposed to take based on the destination. Raspberry Pi interacts with the LED bulbs by providing direction information which is obtained by the Google Maps Direction API. A developed Android application is used that sends real time co-ordinates to Raspberry pi through Bluetooth. This helps track the location pinned. The module helps pedestrians travel with ease and safety that doesn't require them to look at smart phone screens

Figure 1. System of navigation umbrella

2. LITERATURE SURVEY

PradnyaBattin, Dr.S.D.Marakandehave effectuated an android application that alerts and reminds the users to perform a particular task based on location instead of time. The application is run by Google map services API. This decreases the possibility of forgetting the task to be performed in a particular location which is independent of time. But unfortunately this cannot be implemented for indoor locations [1].

Shigeru HAGA et al have researched about the ramifications of using smart phone devices while walking on the road. They conducted various experiments such as sending messages, watching videos, playing video games on undergraduates while walking, which proved that pedestrians using cell phones on road are prone to accidents [2].

The indoor light system was proposed to be controlled using web technology and raspberry pi. Cesar Cheuque et al proposed this system. The interoperability between mobile and system did not fail. A more durable platform should be created that is user friendly. The two problems that weren't considered were safety and serving larger purpose.

SathyaNarayanan E et al invented a cane that is used by the visually challenged. Which provides services like detecting obstacles, guidance for level crossing, indicates current location by using sensors and GPS. The user is provided instructions using text and speech converter. This is transmitted through headphone. This module is relatively inexpensive and weighs only about 2 kilograms. But again unfortunately this module can't detect obstacles with heights of 1.5 m and above (from the ground).

Yohei Iwaski, Nobuo Kawaguchi, Yasuyoshi Inagaki proposed a system that uses both location and direction. Its biggest advantage is that it provides an extensive range of application services.

3. PROPOSED SYSTEM

The model is a smart umbrella paired with mobile applications that act as a navigation assistant by using the Bluetooth technology and provide additional features to improve the already existing system. Main concept that we use is the LED intimation for navigating the direction and mobile application for the select the root map of the destination. The proposed umbrella should continue to point in the same direction sent from smart phone navigator overcoming a variety of disturbing behavior such as rotating the umbrella. If it does not point in the correct direction, the customer would not be able to arrive at the correct location. To solve this problem, we continue to update the direction using 9DOF sensor calculation module. It updates the direction periodically using a three-axis electronic compass sensor (Sparkfun 9DOF).

The additional features are stated below: When the umbrella goes out of range, an SMS will be sent to the user's mobile number stating the same. When the user reaches the destination, the listed contacts will receive an SMS stating that he/she has reached the destination. The weather forecast updates are sent as an SMS alert to the user, intimating them about the weather. Voice instructions of the directions will be provided using a speaker; the user can enable or disable this feature and also the one which controls the LED bulbs.

The aim of our model is to reduce accidents caused by excessive usage of smart phones while walking on roads. The primary purpose is to eliminate the need for using smart phones by replacing them with this model for navigation. This smart umbrella will guide us our way using the LEDs attached on the ribs of the umbrella. This helps in minimizing the risk of accidents considerably. The weather forecasting in our model helps the users to decide and take precautionary measures based on it. Loss prevention is also provided for the users by notifying them via SMS alerts when the umbrella goes out of range. Speech assistance is present so that the users can navigate easily in case they go in wrong direction.

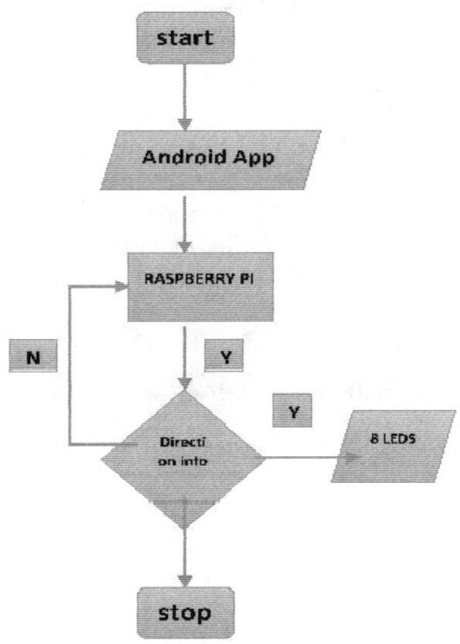

Figure 2. Workflow of Model

The below sub contents provide the whole system architecture which provides the information about the inter connections of the LEDs and the Raspberry Pi .It also provides us the information about the connections between the HC-05, 9DoF and Raspberry Pi.The architecture diagram in Figure 3.3 shows 8 LED bulbs which are connected in series with resistors each of 330 ohms resistance and their connections to the microcontroller i.e. raspberry pi 3 B+. The LED bulbs used are of 3 different colors, namely red, green and yellow. The cathodes of LED bulbs are all connected to the ground pin of raspberry pi 3.The anodes obtain power supply from GPIO pins of the raspberry pi 3 which passes through resistors and powers the LED on or off. The GPIO pins used for the 8 LED bulbs are GPIO 25,12,16,05,21,18,23,24 representing the bulbs in clockwise order.

HC-05 AND 9DOF WITH RASPBERRY PI

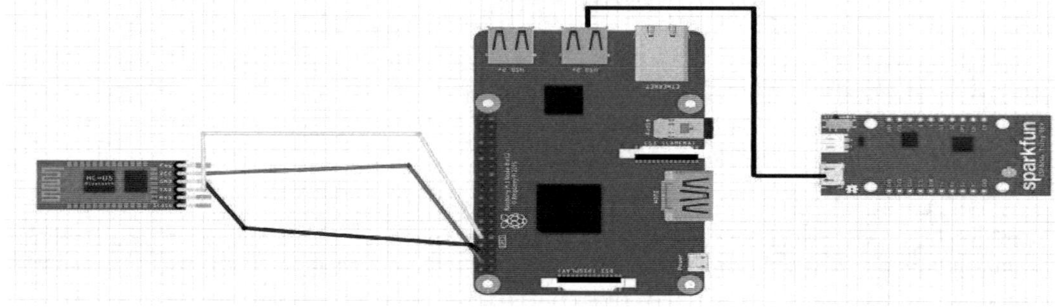

Figure 3. HC-05,9DOF with PI

The above figure consists of raspberry pi 3,Sparkfun 9DoF Inertial Measurement Unit(IMU),HC-05(Bluetooth module).The transmitter of HC-05 is connected to the serial port, UART receiver GPIO pin number 15 of the raspberry pi by a yellow wire as shown in the figure.HC-05's Ground pin is connected to the ground pin number 6 of raspberry pi represented by a black wire in the figure.HC-05 is provided with power of 5 Volts by connecting to the 5V pin of raspberry pi which is numbered 4 shown by a blue coloured wire in the figure. Once, the HC-05 is paired with the mobile phone's Bluetooth, values can be sent from the Android application to the pi via HC-05.The baud rate in the program should be set to 9600 bits per second (bps). To get the current orientation of the device, Sparkfun 9DoF sensor is used. An FTDI basic breakout board is attached to the 9DoF sensor. This is used to connect to the USB port of raspberry pi using a USB connector. The values can be read by writing python code to open the USB port of the raspberry pi.

Table 1. Magnetometer Readings

DIRECTION	START RANGE	END RANGE
North	337.6	22.5
North-East	22.6	67.5
North-West	292.6	337.5
East	67.6	112.5
West	247.6	292.5
South	157.6	202.5
South-East	112.6	157.5
South-West	202.6	247.5

Google Maps Direction API is used to receive the turn by turn guide from the starting point to destination mentioned by the user as inputs. The current orientation of the umbrella is obtained by using a Magnetometer sensor (9DoF). Magnetic declination is also considered to get the accurate heading value. Based on this obtained information the respective LED bulb pointing to that direction glows. The pedestrian needs to follow the direction shown by bulbs to reach the destination. The range of boundary is set for the umbrella. Once it goes out of that range, an SMS alert is sent to the registered users which include the information about the current location of the

IOP Conf. Series: Materials Science and Engineering **590** (2019) 012025 doi:10.1088/1757-899X/590/1/012025

umbrella. This proposed model also gives weather forecasting updates of the destination which can also be sent as SMS alerts based on which the user decides whether to carry the umbrella or not. The directions are provided as text as well as speech which now can be transmitted to the user via headphones or speakers by connecting to Raspberry pi. In table 1 the magnetometer reading is shown.

4. CONCLUSION

This paper proposed the navigation assistance using an umbrella module that exists on the top of the umbrella. Use of smart phones while walking on the roads can cause safety problems to pedestrians due to the dispersion of the field of view. Therefore, we made the smart navigation module such that the user can give the source and destination and the umbrella helps in the navigation to their particular destination using attached LEDs on the ribs of the umbrella instead of smart phone. Due to this the number of accidents decreases. It can provide safety and convenience. The weather forecast updates are sent as an SMS alert to the user, intimating them about the weather. Voice instructions of the directions will be provided using a speaker; the user can enable or disable this feature and also the one which controls the LED bulbs.

REFERENCES:

[1]. Battin, Pradnya, and S. D. Markande. "Location based reminder Android application using Google Maps API." In Automatic Control and Dynamic Optimization Techniques (ICACDOT), International Conference on, pp. 649-652. IEEE, 2016.

[2]. Cheuque, C., Baeza, F., Marquez, G. and Calderon, J., 2015, November. Towards to responsive web services for smart home LED control with Raspberry Pi. A first approach. In Chilean Computer Science Society (SCCC), 2015 34th International Conference of the (pp. 1-4). IEEE.

[3]. SathyaNarayanan, E., Nithin, B.P. and Vidhyasagar, P., 2016, November. IoT based smart walking cane for typhlotic with voice assistance. In Green Engineering and Technologies (IC-GET), 2016 Online International Conference on (pp. 1-6). IEEE.

[4]. Fernandes, Ana Isabel, Miguel Goulão, and Armanda Rodrigues. "A Comparison of Maps Application Programming Interfaces." arXiv preprint arXiv:1305.3485 (2013).

[5]. Finkelstein, Stanley M., Stuart M. Speedie, Xinyu Zhou, Edward Ratner, and Sandra Potthoff. "VALUE: virtual assisted living umbrella for the elderly-user patterns." In Engineering in Medicine and Biology Society, 2006. EMBS'06. 28th Annual International Conference of the IEEE, pp. 3294-3296. IEEE, 2006.

[6]. Asha P, Albert Mayan J, Canessane A (2018),"Efficient Mining of Positive and Negative Itemsets Using K-Means Clustering to Access the Risk of Cancer Patients", Communications in Computer and Information Science ,ICSCS 2018, Kollam, 2018,pp.373-382.

[7]. Sri Harsha S.L.S, Chakrapani Reddy S, Prince Mary S(2017),"Enhanced home automation system using Internet of Things ",Proceedings of the International Conference on IoT in Social, Mobile, Analytics and Cloud, I-SMAC 2017 ,pp.89-93.

IOP Conf. Series: Materials Science and Engineering **590** (2019) 012026 doi:10.1088/1757-899X/590/1/012026

AUTOMATED PAYROLL USING GPS TRACKING AND IMAGE CAPTURE

Dr.K. Mohan Prasad,
Associate Professor
Dept of CSE, School of Computing
Sathyabama Institute of Science and Technology
Chennai 600119, Tamil Nadu, India
mohanaprasad1983@gmail.com

Rajeev Sai Nagendra Goru[1],
Dept of CSE,School of Computing
Sathyabama Institute of Science and Technology
Chennai 600119, Tamil Nadu, India
rajivgoru@gmail.com

Desu Vamsi
Dept of CSE,School of Computing
Sathyabama Institute of Science and Technology
Chennai 600119, Tamil Nadu, India
vamsydesu@gmail.com

Mr.J. Albert Mayan,
Associate Professor
Dept of cse,School of Computing
Sathyabama Institute of Science and Technology
Chennai 600119, Tamil Nadu, India.
albertmayan@gmail.com

***Abstract*—** The main theme of this paper is tracking the employee using an android device and calculating payroll for the working hours he worked. From olden days there are many methods for evaluating the attendance, for example, paper and pen method in this method either the supervisor will take the attendance or under his control, the workers used to sing with their names, but this process had many backlogs and many proxies can be generated. Later on the technology developed to great extent in the similar way the way of taking attendance also developed a lot for example in current days they are using RFID chips, biometric devices, etc. but each of the methods is having a same or different backlogs to eradicate the disadvantages and for accurate result we are introducing attendance method using GPS tracking. Now a day's people or any organization wants their work to be completed fast without taking any time one of the examples payroll allotment so we also include a module for paying payroll according to the number of days that they worked. This project contains two phases one is the mobile phase i.e. android app for field workers for tracking their position for every 5 minutes and there is a web phase where the HR and admin will monitor the employees and for security purpose they are allotted with the employee identification number and password and in the web phase the tracking is done under the control of HR and admin.

***Keywords*—**
GPS tracking, image capture, automated payroll, employee, HR, admin, SQL database.

1. INTRODUCTION

Now a day, monitoring, tracking employees had became a major task for the private and for public institutions and companies. From olden days there are many methods for evaluating the attendance, one of the oldest one is pen and paper system there are many drawbacks and disadvantages mainly eradicating the attendance proxy is the main theme of the project to eradicate the proxy attendance and taking attendance in that way takes more time. Day by day there are many changes in rapid technology as the technology changes the way and systems of taking attendance also gradually changed some of the processes are using RFID sensors, electronic tags, biometric devices like eye scanning, face scanning. All these processes have different issues to eradicate all the issues and disadvantages we introduced software called automated payroll with GPS tracking and image capture. It will track the employee for every **5** minutes the geographical coordinates and help to calculate the payment details.

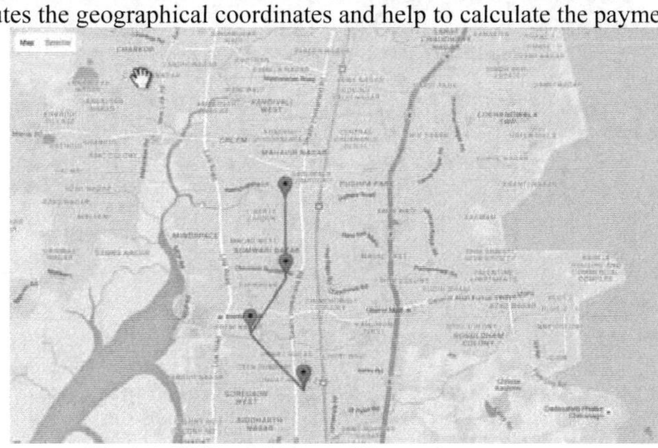

Fig.1.1 GPS tracking and Image capture

1.1 RELATED WORK

Sonal in 2016 developed a project called Employee Tracking and Monitoring System with Android Device. In this paper they provided different security system on same android phone i.e. they provide different logins and passwords to different employees respectively. They used dynamic centralized SQL database utility which retrieves data or information from the database. They provide a mode whenever he/she enters into the company surroundings. Through the android mobile all information about the workers mobile like outgoing call data, SMS, incoming call data, missed call data, employee geographical location coordinates, data usage, unauthorized call data and web browser data, browsing details are tracked. The necessary conditions are that employees should have an android phone where as Manager Functions are also tracked in web phase.

Aparna in 2013 developed software called Smartphone Monitoring System; this is software helps to monitor the employee's office mobile phone and data used by the user. All incoming call details, outgoing call details, text details, emails, images and multimedia messages can be seen and disturbed by the managers and are saved in a centralized SQL database, the manager can also monitor where their employees are, present in the given surroundings or not if anyone is getting out of his given area they receive a text or SMS's or messages, the manager can track whether the employees are receiving texts from unapproved numbers or calls or messages from by the unauthorized people. This software helps managers to monitor their employees with the help of mobile phones.

Priti in 2015 developed software called monitoring employee's smart phone using android application. This software uses android based mobile phones for the software to be executed. The mobile phone with the worker should have android device, and the managers does not have any mobile device and also he will have web inter phase to view the activities of the field workers as the manager is going will receive alert messages from the worker through SMS only. Alert SMS are stored in the database server like the details of incoming call, multimedia messages and text and the location coordinates update of their Employee and their attendance. The manager can log in to the server and see the view of their Employee's mobile usage.

Shermin in 2015 developed a Smart, Location Based Time and Attendance Tracking System Using Android Application. This software is proposed for tracking employees using mobile GPS location rather than the biometrics or any sensors. Using the android phone, the GPS location, the employee geographical location is tracked. GPS location is tracked in our mobile, and it is tracked in the personalized computer by the admin and the HR.

Nirmal in 2016 developed a project called Employee Surveillance System Using Android Smart Phone, Their system is contains of employee tracking and GPS location Tracking System using Android device. All the activities and functions of the Employee will be tracked and viewed using this system. The system works on 3G/4G network or WIFI network communication between the ends. All the functions of a field worker on his android phone and personal computer, like outgoing calls, data usage and all incoming and web browsing, secured document changing and unauthorized transfer of company's information like a blueprint, stocks, projects, contacts, etc. will be tracked by the admin, HR or manager. Not only this, the location of the employee will be tracked using GPS location. Therefore, the organization or institution will be tracked that will stop the unauthorized usage of its resource files by the workers during working hours. The system was beneficial for the growth of the company and allows the HR and admin to check the working activities of the employees.

Ashwini in 2015 developed software called Employee Monitoring System Using Android Smartphone. In this software, all the functions like incoming call data, missed call data, outgoing call data, SMS history, web history, data usage, unauthorized call list/website list are saved in the centralized database server. The manager can monitor the history by logging into the SQL server. The manager can also track employee's geographical location through GPS. If the employees are going outside of the field surroundings then, the manager gets an alert message in SMS format. They analyzed the employee behavior, by checking the unauthorized call data, messages data whether it is good or average or loyal or bad. An Android device is given to the user. The manager doesn't need the Android mobile instead of that he is given with a web inter phase. This software is helpful to the manager to view the activities which are done by the employee.

2. SYSTEM OVERVIEW:

This software automated payroll using GPS tracking and image capture mainly consists of two phases one is application phase and another one is web phase. First one the application part, the APK is installed in the employee's Android mobile and the employee will in by the given credentials i.e. username and password, after logging in the application will track the geographical coordinates and send to the main server for providing more security we are capturing the
person's image and sends to the server there in the server image will be verified and the account will be verified. During the logging of time also, it will track the geographical coordinates and sends to the server and finally for security purpose again an image will be captured and send to the server. While the web inter phase it tracks the employee location coordinates. There are many features like adding and deleting an employee, changing the password of the employee, we can calculate the working days and based on that we can calculate the payroll of the employee.

2.1 System Design:

The automated payroll using GPS location tracking and image capture is client-server based software and follows unique hardware and software architecture. Combining both the hardware and software is the main and important part and the hardware and software work together. The whole software is divided into two parts.

1) Application for android mobile
2) Application for personal computer.

2.2 Software Architecture:

The software architecture contains the centralized SQL database, the application program, and the server.

• Database: The database contains a large number of data tables which store data of the worker such as employee id, employee name, geographical coordinates, images, working hours and salary details. We use the SQL database which is easy, fast and efficient and can the store records and requires a little configuration.

• Android Application: The android application program is developed with Android language using the Eclipse framework. The application program provides the UI (User Interface) to both the employees, hr and admin office server. Programming using Android language is simple, user-friendly and Android offers a good data connectivity.

• Web phase application: The web phase application is developed and deployed on the personalized computer using Apache-Tomcat 7. Tomcat 7 is free, easily deployable and robust.

Fig 2.1 web phase application

2.3 Hardware Architecture:

The basic requirement of the automated payroll using GPS location tracking and image capture system is an Android device, which will run the Android application, with which the employees will log in using their credentials and sends the data for every **5** minutes, mark their attendance and take their login, logout time automatically without any inconvenience. The other basic requirement is computed on the database, which will store the data in the data tables. Using this web phase, the hr and admin able to monitor the employee presents the given location and tells the payroll details of the employee.

3. Methodology:

Employee security and authentication are one among the factors in the current system. Every employee is secured based on their unique user employee identification number. This unique employee identification number is the number which is given in the office to secure their account. The employee identification number along with other information such as location coordinates; images are also saved in the employee's Android device. Firstly the employee has to install the required APK files on their Android mobile. Login by his/her login credentials and takes an image for security purpose. The GPS mobile location service has to be turned turn on when the system was running for tracking the location. If mobile location service is off then the process won't go to further process and a SMS is sent to turn on the GPS location. The GPS mobile location service helps to track the employee longitude and latitude positions. When the employee enters the working area, the Android device of the employee it should be referred to through the internet access or any WIFI and a SMS is sent to the server which is present in the office, with the employee identification number and local time of the android device which is referred as login time of that employee. When an employee leaves the working place, a SMS is sent to the office computer with employee id and local time which is referred as logout time and for security purpose, an image is taken to secure the account.

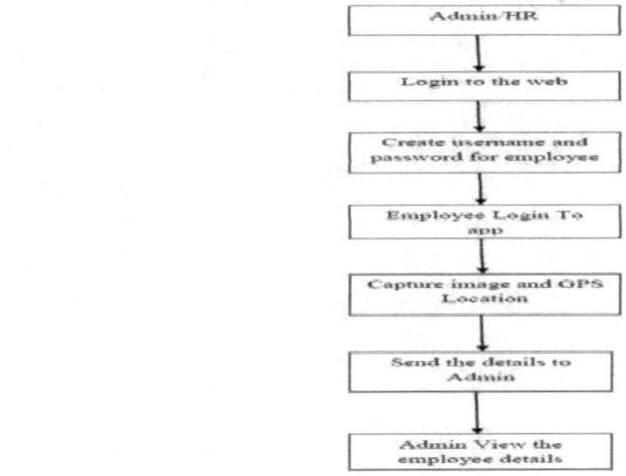

Fig 3.1 Architecture Diagram

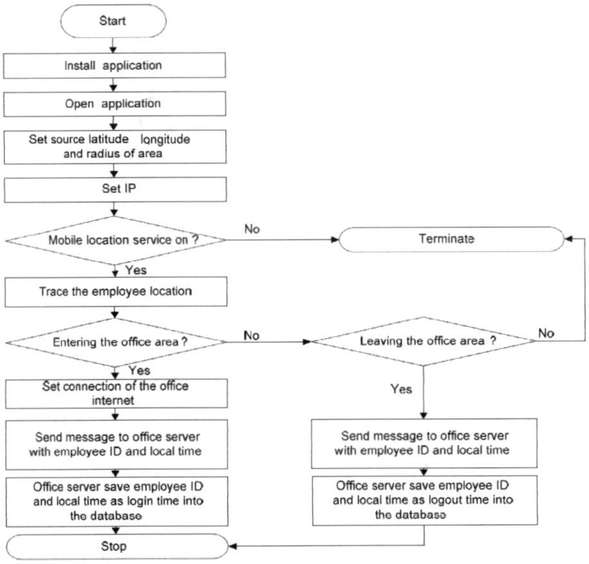

Fig 3.2 Flow chart for trace the employee

4. CONCLUSION

we are introducing automated payroll using GPS location tracking and image capture which uses to view an employee by using the employee's geographical location and locate the longitude and latitude positions and we can view the path the way he travelled throughout the day and calculates the employee salary, it eradicates the proxy attendances and fake salary calculations.

REFERENCES

[1] Sonal Kasliwal, H.D.Gadade and Sushma Kotkar (2016), Employee Tracking and Monitoring System With Android, International Journal of Innovative Research in Advanced Engineering (IJIRAE) SSN: 2349-2763, No03, Vol 3, pp. 1-4

[2] Asha P, Albert Mayan J, Canessane A (2018),"Efficient Mining of Positive and Negative Item sets Using K-Means Clustering to Access the Risk of Cancer Patients", Communications in Computer and Information Science ,ICSCS 2018, Kollam, 2018,pp.373-382.

[3] Aparna Chandran (2013), Smartphone Monitoring System, International Journal of Computer Science & Engineering Technology (IJCSET) ISSN: 2229-3345 Vol. 4 No. 04, page 451-452

[4] Shermin Sultana1, Ishrat Jahan Mouri and Asma Enayet1 (2015), A Smart, Location Based Time And Attendance Tracking System Using Android Application International Journal Of Computer Science, Engineering And Information Technology (Ijcseit), Vol. 5, No.1,

[5] Usha Nandini , Saravanan M , Albert Mayan J , Murari Devakannan Kamalesh , Mohana Prasad K (2018) , " Automatic traffic control system using PCA based approach",International Conference on Energy, Communication, Data Analytics and Soft Computing (ICECDS-2017),pp.2387-2392.

[6] Prajakta Kokare, Ashwini Jaybhaye, Tanmay Kulkarni and Bhakti Toradmal (2015), Employee Monitoring System Using Android Smartphone, International Engineering Research Journal (IERJ) Volume 1 Issue 2 Page 32-35, ISSN 2395-1621

[7] Shamal Bangar , Kalyani Bhagwat Priyanka Salunkhe. (2015), Employee Monitoring System Using Android Smart Phone, International Journal on Recent and Innovation Trends in Computing and Communication ISSN: 2321-8169 Volume: 3 Issue: 2 537 - 541 537 IJRITCC.

[8] Shoewu.O, Amisu, A.A and Makanjuola, N.T (2015) Design and Implementation of An Employee Monitoring System In Lasu Epe Campus, Lagos State University, Journal of Advancement in Engineering and Technology, Volume 4, Issue 1, ISSN: 2348-2931

[9] J. Albert Mayan and S.P. Avinaash Ram (2015), Mobile Attendance Management And Employee Registration Application, Journal Of Engineering And Applied Sciences, Vol. 10, No. 8, Issn 1819-6608, page 3727- 3730

[10] Kanchan A. Patil, Nitin P. Jagtap Nitin S. Ingle and Shaziya Sayyed Shakil (2015), Mobile Activity Monitoring System Using Android Spy, International Journal of Advanced Research in Computer and Communication Engineering, Vol. 4, Issue 2, page 158-162

[11] Dr.R.Sabitha, Dr.K. Mohana Prasad “Meta Physical Algorithmic Representation for Flawless Clustering” Journal of Theoretical and Applied Information Technology , ISSN: 1992-8645, Vol76, NO 1, PP 82-87.

[12] Dr. R. Sabitha, Dr.K.Mohana Prasad “Evolution Of An Algorithm For Formulating Efficient Clusters To Eliminate Limitations” International Journal of Applied Engineering Research, Volume 9, Issue 23, pp. 20111-20118.

[13] Surendar E, Thomas V.M, Posonia Mary A,"Animal tracking using background subtraction on multi threshold segmentation" ,Proceedings of IEEE International Conference on Circuit, Power and Computing Technologies, ICCPCT 2016 ,pp.1-6

[14] Dr. R. Sabitha, Dr.K. Mohana Prasad” Yoking of Algorithms for Effective Clustering”, Indian Journal of Science and Technology, ISSN: 0974-6846 Vol 8(22), IPL0269, September 2015, pp 1-4.

OPTIMAL AND SUBOPTIMAL RESOURCE ALLOCATION TECHNIQUES BY USING VIRTUALMIGRATION IN CLOUD COMPUTING

C.Prasanth Varma[1], L. Infant Cinty[4]
U.G Students, Department of Computer Science & Engineering
Sathyabama Institute of Science and Technology, Chennai, India
prasanth2310195@gmail.com, cintyinfant@gmail.com

Muthukumar B[2]
Professor , Department of Computer Science & Engineering
Chennai Institute of Technology,Chennai, India
muthukumarb@citchennai.net

Albert Mayan J[3]
Assistant Professor , Department of Computer Science & Engineering
Sathyabama Institute of Science and Technology , Chennai, India
albertmayan@gmail.com

Abstract : This paper discusses about data transmission between hot machines and warm machines. We employ a Main Cloud Server (MCS) for this purpose. MCS provides a variety of services. User have to give request to MCS and it will provide necessary service. User will be sends request to MCS for data upload and the request will be sent to CSP and that allocates job to Virtual Machines. We need to use two machines such as HOT MACHINES and WARM MACHINES. Hot machines can handle current job. Warm machines can be idle state until the job is assigned. Every machines will be having three virtual machines. The user can raise the service request CSP to access the resources found in cloud. So, the CSP provides approve or rejection is accept by CSP, then user can immediately access or download the resource from cloud server.

Keywords- Cloud Computing, Optimal and Suboptimal, Virtual Migration.

1. Introduction

Energy can be virtually green for a specific time is abundance is available periodically in shorter time interval along with rest of time frame. The virtualization concept can increase the greenness of energy, rather increasing the amount of green energy.When energy is sufficient to meet the demand, we nullify the degraded interval with the surplus interval[1]. we use virtualization because degraded interval with surplus interval but from the client's or provider's, through the energy is not present instantaneously rather virtually and CLSA extension to validate our proposal. In this heat consumption is reduced when system is idle and also less consumption and system life time is increased. Waiting time is decreased and also Throughput level is increased in this system. To accomplish this execution and cost targets CC suppliers require a persuading asset portion framework that oversees both computational and system assets. Such a productive framework may have a noteworthy money related effect as overabundance assets make an interpretation of specifically into incomes. The perception of our approach. In our Green SLA are checked and assessed per interim time as opposed to demand or application. Despite we parts CSLA the Cloud Service Level dialect permits to characterize SLA in some other dialect for any Cloud administration to help Green SLA it depends on idea of virtualization of environmentally friendly power vitality. In this rendition communicates an administration level target utilizing two limits and backings another class that is character to invalid interim. In this procedure mirrors the instable of efficient power vitality and additionally we present green SLA calculation which the idea of virtualization of environmentally friendly power vitality and CSLA expansion to approve our proposition. In this examination reason genuine work is stack in profile from Planet customer and server control display were followed for direct experiments. We show the Green SLA can be fulfilled without utilizing high cost for add up to vitality use .

2. Related Work

Mohammad Abu Sharkh et all[3] have proposed the issue of asset designation in distributed computing server farms is figured as an improvement issue and explained. In addition, an arrangement of heuristic arrangements are presented and utilized as VM reservation and association planning strategies
S. Venkatesh et all [5] have proposed the three types of system. To start with is HOT, in which the Servers will deal with the Jobs Currently, Second is WARM, in which the Servers are kept in Ideal State, at that point Finally Cold, in which Servers are Turned Off State. Introductory Request is send to HOT – Servers, if those Servers are Busy then the Request is sent to Warm – Servers, at that point at long last if required to Cold – Servers if both the Hot and Warm Servers are Busy.R.Keerthigaet all [6] have proposed the dynamic asset designation procedure in the cloud and the issue of Dynamic Migration of Virtual Machines (DM-VM) is examined. Also executed another BGM-BLA (Binary Graph Matching – based Bucket code Learning Algorithm) calculation for DM-VM issue, this dynamic relocation of VM gives a strategy to appropriate physical asset all the more sensibly without suspending the administrations. Hence the vitality utilization, stack adjusting and activity cost is decreased.

MakhloufHadjiet all [2] have proposed the arrangement of virtual assets isn't sufficient to guarantee framework proficiency. This paper displays a novel and unique on the web direct programming calculation, in view of b-coordinating hypothesis, to ideally tackle the union issue with unimportant SLA infringement. To absolutely dispense with SLA infringement, and to manage dynamic workloads varieties, we decide the ideal sum of an asset pool used to deal with over-utilized servers issue. We infer an answer whose execution is contrasted with a Best-Fit also, a Bin-Packing definition going about as benchmarks.

Neha Sharmaet all [7] have proposed the productive assignment of assets to the suitable client, a proficient undertaking booking calculation is required, which can accomplish least reaction time, high throughput and appropriate asset usage, and so forth. Heuristic based calculation is one such approach to accomplish the ideal or close ideal arrangement of undertaking planning for the cloud condition. Heuristic strategies are the subset of meta-heuristic approach.

3. Proposed Work

We split MCS into two separate service like CSP 1, CSP 2, CSP 3, CSP 4 and for every CSP there is an multiple virtual servers. User will give request to the MCS for data upload and request will send to CSP and allocate job in VM. 1. Hot Machines can handle the current job. 2. Warm Machines are kept idle state until job is assigned. We deploy three Virtual servers for every machine. 1 st Job is assign to the Hot machine 1st Virtual machine and same way following jobs are assigned to other VMs. Now jobs are assigning to Warm machines once all the VMs of Hot category have occupied with the jobs. Automatic migration of job is implemented, so as to transfer the load to the Hot VM from Warm VM once it has completed the job.VM1 CSP1 User Main Cloud Server VM2 VM3 CSP2 VM1 VM2 VM3

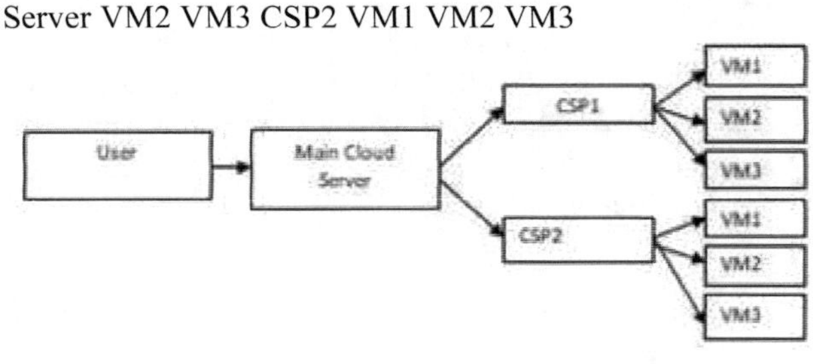

Fig1:Implemented cache Mechanism

In figure 1, we are including either over aminimizing Virtual migration incidents of major goal. Off-line Virtual migration.This does not demand client environment every 9's availability is become an expectation. We need to load with more copying is required. In this challenge associated with virtual migration it causes the cloud computing solution to welcome some solution it does not include migration . In this session, will be using six modules with help of figures such as:

a. User Registration:
we need to user application to allowed to operates the data from the Server to the CSP. We need to create a account and automatically allowed to access the Network. After login into their account and request the Job from the CSP. Based on the User'srequest, the CSP will process requested Job and respond tothem. All the User details will be storing the Database of the CSP. we will design the User Interface to Communicate with Cloud Server through Network Coding using the program Languages like Java. After sending the request to CSP user can access the requested data they authenticated by CSP.

b. Cloud Server Deployment:
It contain the large amount of data in their Data Storage. It will maintain all User information to authenticate user and then we need to login into their account. User need to get the information will be and put away in Database from Cloud Service Provider. Likewise Cloud Server will guide the User asking for employment to the Resource Assigning Module to process the User asked for Job and every one of the Users will process by the Resource Assigning Module. We have to speak with the Client and the with alternate modules from Cloud Network the Cloud Server will interface between them. For this Purpose we have to make a User Interface Frame. Likewise CSP will sends Job ask for to the Resource Assign Module Fist in First out way.

c. Intermediate server deployment:
We can process the User Requested Job and keeps up the Resources the Cloud Server. Along these lines, we need to spare energy when they are not process the Job.

d. Green Computing Setup:
We have to mean proficient utilization of assets in figuring is Green IT . we will be process the User asked for Job.It will divert to RAM from Cloud Server. Slam will contain two Types of the HOT Server and WARM Server. It contain 'n' number of virtual Server to process the User asked for Job. With the goal that the Job can be effectively process.

e. Migration Virtual Server:
In this separate principle utilization of migration to move the activity starting with one server then onto the next server, thus, vitality can be diminishing and work load will be adjusted, by utilizing Migration we have to move the procedure starting with one machine then onto the next machines without loss of information.

f. Cache server Implementation:
Reserve Memory need to land the User asked for position it stores the period time. In the event that the another User asks for a similar Job to the Server of the CSP, the Server will check in the Cache Memory first . With the goal that we can lessen the Job Processing Time. On the off chance that the demand Data is introduced, at that point the Server will give the Data from User quickly. In the event that the demand Data isn't in the Cache Memory, at that point the Server procedure the User asked for Job by exchanging it to the RAM.

Virtual Migration

Using this algorithm, we optimised dynamic virtual migration resource from first machine and another machine. Once hot machine had completed its job then dynamic work is migrated from the warm machine to the corresponding hot machine[3]. It is used for load balancing and physical machine fault tolerant. It can also be used to reduce power consumption in cloud data centres.

4. Optimal and Suboptimal

We utilizing a N-hub coordinate with 2 servers and 9 joins associates with them, the quantity of factors can be N factors when number of variables that have a place with 5 machines. To move to the subsequent stage which is finds a strategy that accomplishes an imperfect arrangement. we presented in view of a decay system[4].

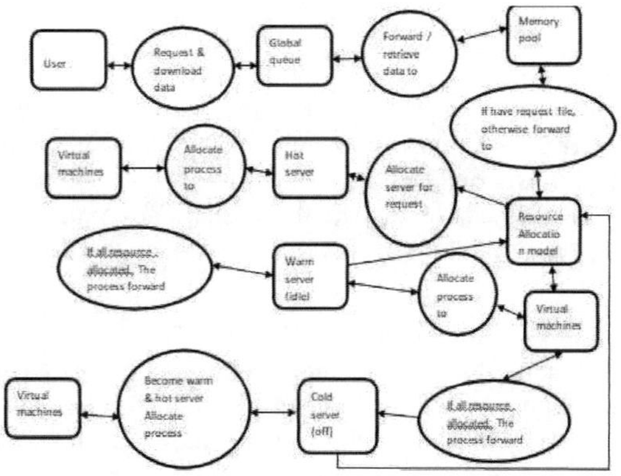

Fig2:Architecture of Optimal and Sub Optimal

In figure 2, In Step 1, an arrangement of known association demands are preprocessed to create reliance of at least two thing on each different estimations. This is figure ascertain the recurrence of correspondences between every two focuses in the system. The recurrence of association asks for between each VM I and VM j is ascertained and also the recurrence of association asks for between VM I and hub n which speak to a private cloud.

In stage 2, an utility capacity is developed in light of the association recurrence esteems created in stage 1. The utility capacity fills in as the target capacity of the some issue that assigns VMs on has. The recurrence work is an esteem that will speak to reliance of at least two thing on each other between two VMs or between a VM and a private cloud. Next, we handle the assignment of VMs from servers and connects to specific paths without generated. The minimum produced from the problem is the objective value are looking for. As in any decomposition based optimization, the success of the decomposition technique depends on the way the solution master problem is chosen. We formulate the problem and the subproblem as follows.

5. Model description:

In figure 3 ,We acquaint a model with the asset portion issue for a gathering of cloud client demands. This incorporates the cloud supplier acknowledges a demand from a client. The model comprises of a system of server farms hubs (open mists) and customer hubs (private mists)[8].These hubs are situated in various urban areas as in Fig. 8. We likewise expect that every datum focus contains various servers associated through Ethernet associations. Cloud supplier offers various kinds of machines for customers to pick diverse composes fluctuate in the particular of each registering asset like memory, CPU and capacity. We will utilize these three kinds of assets in our trial. We have to get the asked for VMs is allotted on a server in one of the server farms. Customers sends the demand to server association. For each demand, the customer characterizes the source, goal, begin time and length of the association. In this way, the goal progresses toward becoming to limit the normal lateness of all association demands. An example of customer demands.

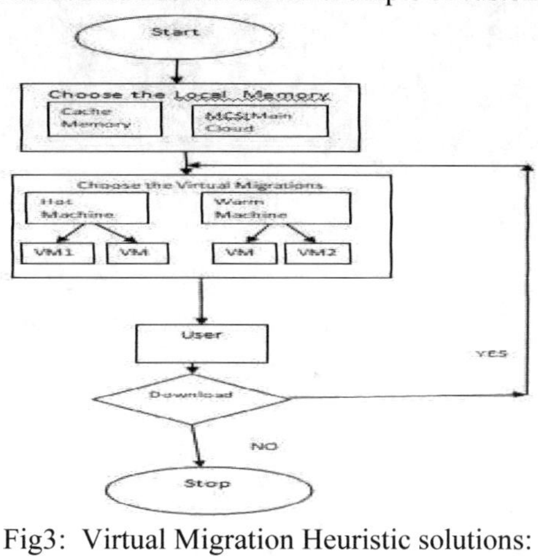

Fig3: Virtual Migration Heuristic solutions:

IOP Conf. Series: Materials Science and Engineering **590** (2019) 012027 doi:10.1088/1757-899X/590/1/012027

6. Heuristic model:

A focal controller deals with the solicitations with the target of limiting normal late and demand blocking[11]. The answer for settle the suppliers cost challenges and the cloud applications execution issues. It apportions the asked for machines from servers request upon the technique or strategy is utilized. In this target will achieved less way is utilized. In this methods getting great, in spite of the fact that not ideal, execution esteems while giving the achievable arrangement in a satisfactory measures of time.

a. Heuristic Techniques:

In this process will divided into some consecutive steps: We are using two types of database in the Green Networking. Firstly, we need to open the MYSQL Query Browser. After Opened the browser need to login the system password should be “ADMIN”. We have to copy the client program and the paste the program in Mysql Query Browser. At the end program line to execute the program thecursor should be at the end of the line, this process repeatedly to each and every line. After executing the program, we can see the right side of the browser Schemata. In that different types of schemata like Info Schema, Mysql, Prediction Server: a)File table b)List c)User details, Server and test and also open the notepad Server and copy the program and paste the program into the browser. At the end of the program line to execute the program the cursor should be at the end of the line, this process repeatedly to each and every line. After executing the program, we can see the right side of the browser Schemata. In that Server file two types: a)File table b)list. Secondly, we need to open code file in that file sub-files are present and then open the cloud file. In cloud file we can see types of files in that we need to open is Compiler.bat and compile the code and then Server.bat run the code and it will show u box as Cloud: Server as shown in figure 4 ,5 and 6.

Fig4:Compile.bat

Fig5:Server.bat

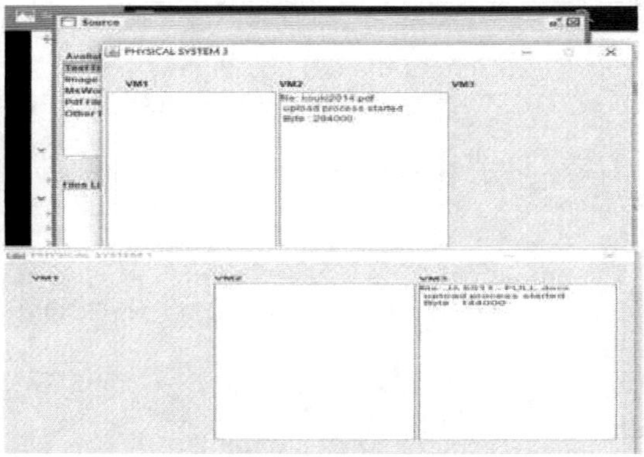

Fig6 :Virtual Migration

7. Performance Analysis:

In figure [7] shows X-axis represent time(m/s) and Y-axisrepresent Byte Transfer.

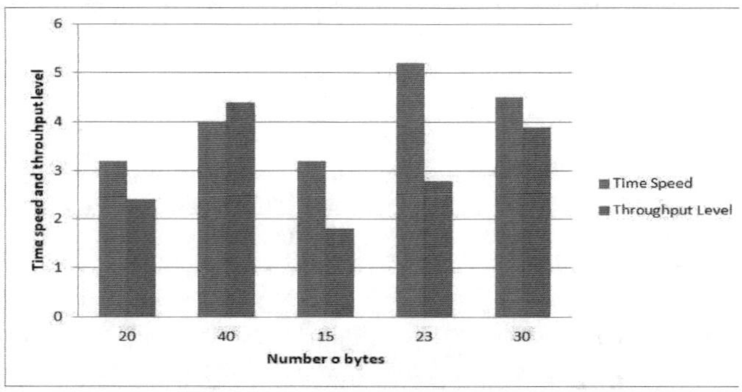

Fig7:Analysis between Time speed,Byte Transfer and Throughput Level.

In Figure. 7, blocking rate is appeared for the two investigation for various issue sizes of time and throughput level. Issue measure here is spoken to by the quantity of solicitations submitted to the focal controller per cycle situations as the permitted lateness level increments. It performs reliably superior to Green booking calculation[9]. It additionally demonstrates that RB-DP performs at an indistinguishable level from for low permitted before demonstrating leverage for high permitted lateness.

8. Conclusion:

A new scheme for energy management based Green SLA. Due to the nature of renewable sources, Green SLA was not introduced specific based on energy before to overcome this difficulty, our proposal included three novel features: Concept of virtualization of energy and Extension of CSLA language that supports Green SLA and Green SLA algorithm, which concept of virtualization of energy.Results demonstrate that, by smartly exploiting the renewable sources, Green SLA can be a possible reality in Cloud Computing. Throughout the paper we argue that, Green SLA should be established based in the presence of green energy, thus carbon footprint can be reduced andsocial responsibility for being Green can be achieved. Moreover, any large power consuming infrastructure similar to Cloud computing, which has large power demand variation over time and has aspiration to reduce carbon footprint by using green energy, can easily extend our proposed research work.

References:

[1] M. Hague, K. Le, I. Goiri, R. Bianchini, and T. Nguyen, "Providing green SLAs in high performance computing clouds," in Proc. Int. Green Comput. Conf., 2013, pp. 1–11.
[2] Makhlouf Hadji, D. Wang, B. Urgaonkar, and A. Sivasubramaniam, "Carbon-aware energy capacity planning for data centers," in Proc. IEEE 20th Int. Symp. Model., Anal., Simul. Comput. Telecommun. Syst.,

Aug. 2012, pp. 391–400.
[3] Mayan J.A., Ravi. T ,"Test Case Optimization Using Hybrid Search Technique",ACM International Conference Proceeding Series ,ACM New York, NY, USA ©2014, ISBN: 978-1-4503-2908-8 ,doi:10.1145/2660859.2660906
[4] Anish Hamlin M R, Albert Mayan J,"Blood donation and life saver-blood donation app",International Conference on Control, Instrumentation, Communication and Computational Technologies (ICCICCT),pp. 625 - 628, 2016.
[5] S. Venkatesh, Z. Zeng, X. Liu, and P. R. Kumar, "The answer is blowing in the wind: Analysis of powering internet data centers with wind energy," in Proc. IEEE INFOCOM (Mini Conf.), Apr. 2013, pp. 520– 524.
[6] R.Keerthiga, K. Le, T. Nguyen, J. Guitart, J. Torres, and R. Bianchini, "Green Hadoop: Leveraging green energy in data-processing frameworks," in Proc. 7th ACM Eur. Conf. Comput. Syst., Apr. 2012, pp. 57–70.
[7] Kamalesh M. D., Albert Mayan J., Felix Y., Sumanth B. S., & Sai Tej B. (2018), " Magrisys: A Smart And Ubiquitous Controlled – Environment Agriculture System " , 2nd International Conference on I-SMAC (IoT in Social, Mobile, Analytics and Cloud) , pp: 1-5 , 2018.
[8] C. Stewart and K. Shen, "Some joules are more precious than others: Managing renewable energy in the data center," in Proc. Workshop Power Aware Comput. Syst., 2009.
[9] Asha P, Albert Mayan J, Canessane A (2018),"Efficient Mining of Positive and Negative Itemsets Using K-Means Clustering to Access the Risk of Cancer Patients",Communications in Computer and Information Science ,ICSCS 2018, pp.373-382.
[10] Albert Mayan J, AntoPraveena M D ,Telkar Bharath Rao , Uravakonda Uday Sagar (2018), "Optimized test data generation over suspicious implementation of oracle problem", IEEE International Conference on Power, Control, Signals and Instrumentation Engineering , pp.2559-2563.
[11] Yovan Felix A ,Jesudoss A ,Albert Mayan J (2017) , "Entry and exit monitoring using license plate recognition ", IEEE International Conference on Smart Technologies and Management for Computing, Communication, Controls, Energy and Materials (ICSTM), pp.227-231.
[12]B.Bharathi and Mahesh kumar (2016), 'Non invasive BG scrutinizer system', Global Journal of Pure and Applied Mathematics, vol.12, No: 8, pp. 5123 – 5125
[13] Muthukumar B, and Praveen Kumar R.(2015)," Intelligent Intrusion Detection System for Private Cloud Environment",***Springer*** *,Communications in computer and Information Science Series(CCIS)*, vol 356, pp 54-65
[14]Sankari.A,AlbertMayan.J,"Retrieving call logs and SMS by messaging services",International Journal of Pharmacy &Technology,Vol. 8 , No.4 , pp.22951-22958,Dec 2016.
[15] Muthukumar B, Praveen Kumar R, and Nagarajan G.(2015),"Hybrid Intrusion Detection System for Private Cloud: A Systematic Approach",***Elsevier****, Procedia in Computer Science*, vol 48, pp. 325-329

IOP Conf. Series: Materials Science and Engineering **590** (2019) 012028 doi:10.1088/1757-899X/590/1/012028

ANALYSIS OF KNN ALGORITHM WITH MAPREDUCE TECHNIQUE ON BIG DATA

Tatikonda.Bhavana[1], J.padmavathy[2] ,R.Sethuraman[3]J.K Jeevitha[4]

[1,2]UG Scholar,[3]Assistant Professor Department of Computer Science and Engineering,

Sathyabama Institute of Science and Technology,Chennai.[4]Assistant Professor,*PSNA College of Engineering and Technology*

[1]bhavanachowdary424@gmail.com,
[2]keerjey@gmail.com,[3]srssethuraman@gmail.com[4]selvajeeva31@gmail.com

Abstract: Due the fast growth of new technology application like social media analysis, web data analysis and medical information network analysis, here the various types of data are processed frequently. The large amount of effective data management and analysis is very vital goal. To reduce the data processing complexity, time complexity, and space complexity in Big Data, the paper going to propose the k-nearest neighbor join (KNN) operation. KNN is used to find the K nearest points in S. It is a computational task that will handle the large range of applications such as knowledge discovery or data mining. When the volume and the dimension of data increases, then only distributed approaches can perform the big operations in a given time .Recent works have done on implementing the efficient solutions using the map reduce programming model because it is used for distributing the large scale data processing. Although these works provide different solutions for the same problem, each one has particular constraints and properties. This paper compares the existing of different computation of KNN on MapReduce. First the paper compares the solutions in to three steps for KNN computation on MapReduce: 1) Data processing, 2) Data partitioning and 3) Computation. The Experiment in this paper explains the variety of different data sets, and analyzes the data volume, data dimension and the value of k from many perspectives like time and space complexity, and accuracy.

1. Introduction:

As we have to get the relevant information or details from large amount of data we use KNN algorithm and Map Reduce algorithm [1][2]. We will merge both the algorithms. KNN algorithm means it will find or search the data nearest to it. MapReduce means it will process large amount of data. Suppose if we take a query like (1990, 2002) employee name. It will give the all the employee details in single form or code. From the huge amount of data we can easily get the information what we required. Here we will use Hadoop frame work that provides a distributed platform for executing Map Reduce jobs.

Hadoop frame work is an open source network that will provide the large amount of data sets. It consists of K nearest neighbor that helps in finding the nearest points in a dimensional space. MapReduce is a

IOP Conf. Series: Materials Science and Engineering **590** (2019) 012028 doi:10.1088/1757-899X/590/1/012028

parallel programming model and this consists is based on three concepts 1) it will represent the data as key -value pairs 2) it will define a map function 3) it will define reduce function. The map function will take the key value pairs as input, and it produce zero or more key value pairs. Outputs with the same key are gathered together, and that key pairs are given to reducers. Data preprocessing will transform the original data that will benefit the particular properties.

This step is done before the partitioning of data in to two different goals.1) either to reduce the dimension of data.2) to select central points of data clusters. Data partitioning and selection process is also used. MapReduce is a shared –platform in order to process data. We have to divide the data sets into independent pieces called partitions. It can be applied to a large number of fields such as medical imagery, bio-information. The primary application of a KNN join is k-nearest neighbor classification. Some data points are given for training and some new unlabeled data is given for testing. The aim is to find the class label for the new points. For each unlabeled data; a KNN query on the training set is to be performed to estimate the class membership. This process will be considered as a KNN join for testing set with training set[3][4][5].

2. Related work:

The author proposed the MapReduce method in cloud computing for improving the performance of data processing and information retrieval; additionally they concentrated the KNN algorithm in Hadoop platform for the analysis of time consumption [6].

Pradesh [7] implemented the data mining concepts in KNN on distributed environment with apache server. The apache server ran in the distributed computing environment, which used Hadoop platform and MapReduce technique for huge amount of data processing.

The author implemented the MapReduce with KNN algorithm in large scale data set up to 30 dimensions for improving the efficiency in data processing [8].

3. Proposed Work:

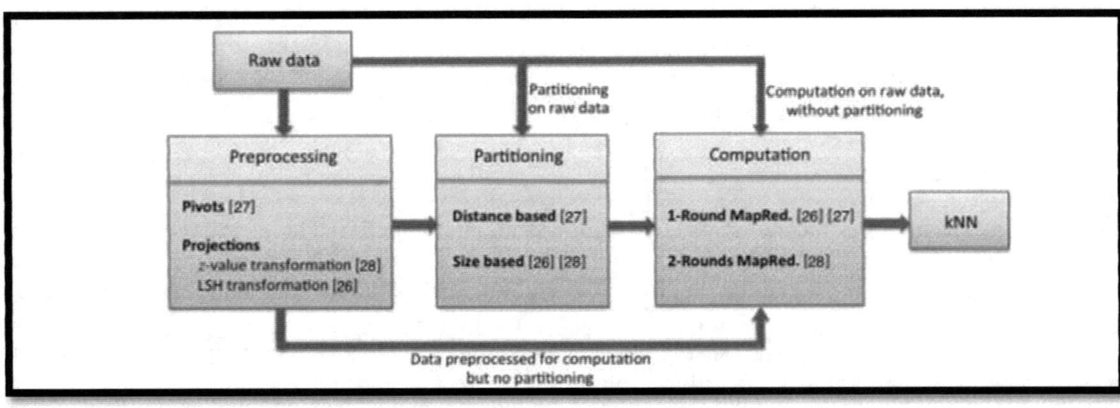

Figure-1: Work flow diagram for MapReduce and KNN.

Figure 1 summarizes the work flow of MapReduce and KNN computation. The work flow of MapReduce and KNN computation architecture processed in three steps.

Step1: First step had done the data processing; moreover it performs the selections of data projection (dominating points) from maximum to minimum dimension.
Step2: This step is used to Splitting and segregating the processed data. This step makes simplifies the KNN computation process
Step3: Third step did one or two function of MapReduce, according to the distance calculation and sorting.

KNN algorithm is very useful and uncomplicated algorithm. Here KNN used the whole dataset in the phase of preparation. While the prediction is demanded the unknown data inputs , that time it seeks out the prepared data set for K-nearly related input and the data with most suitable input. At the final stage the most suitable input is considered as prediction.

4. Result and Discussion:

In this we will use both KNN algorithm and Map Reduce algorithm .KNN algorithm is used to find the nearest point's .Map Reduce is used for process the large amount of data.KNN can be measured by using distance function formulas. The distance functions are implemented in mathematical way [9][10].

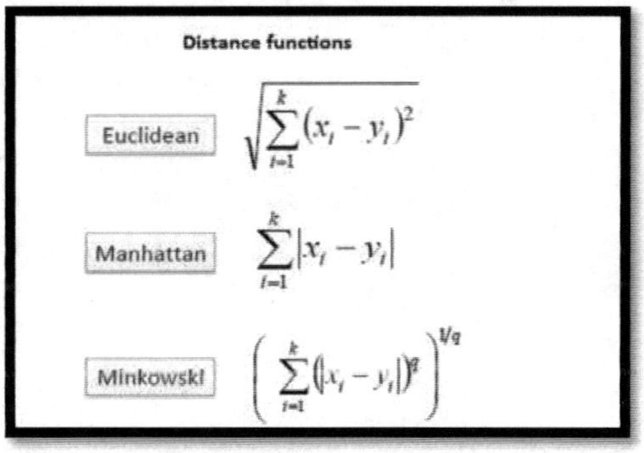

Figure-2: Distance Function.

It will bring the issues of using the variables that starts with 0 and 1.

Figure 2 shows the graph between loan and age. This points can be represented in default and non – default. By using Euclidean distance formula, this paper measured the accurate value.

In the table 1 suppose if we consider k=5 then Y=4 default and N=1 defaultIf we want to find the default value we have to find distance for all the persons in the graph and then we have to take the smallest values from the given table and then from that take the smallest value and we have to find the value.

IOP Conf. Series: Materials Science and Engineering **590** (2019) 012028 doi:10.1088/1757-899X/590/1/012028

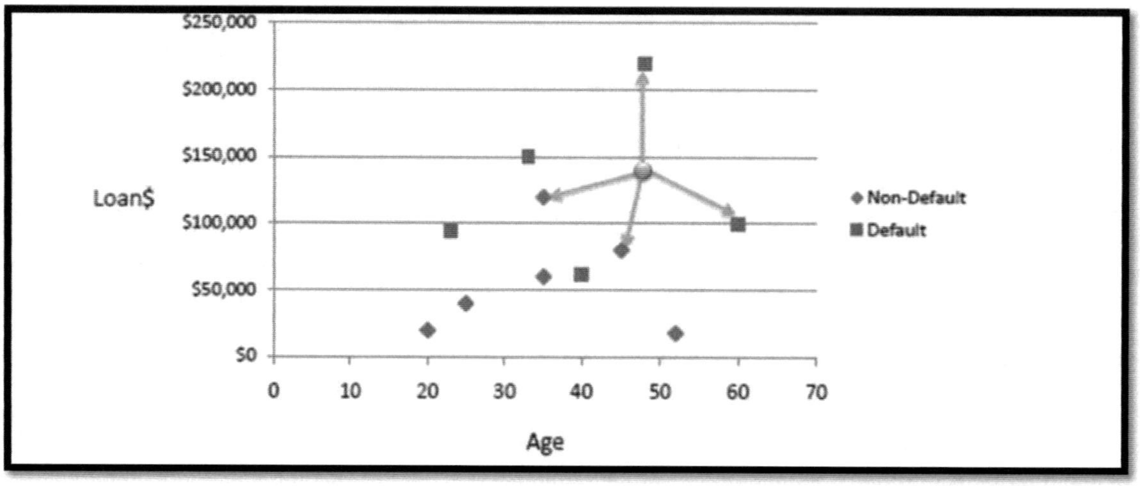

Figure-3 Loan and Age Comparison

Age	Loan	Default	Distance	
25	$40,000	N	102000	
35	$60,000	N	82000	
45	$80,000	N	62000	
20	$20,000	N	122000	
35	$120,000	N	22000	2
52	$18,000	N	124000	
23	$95,000	Y	47000	
40	$62,000	Y	80000	
60	$100,000	Y	42000	3
48	$220,000	Y	78000	
33	$150,000	Y	8000	1
48	$142,000	?		

Euclidean Distance

$$D = \sqrt{(x_1 - y_1)^2 + (x_2 - y_2)^2}$$

Table-1- Euclidean Distance

MapReduce can be done by the following steps:
Step1) The input for the map has to be prepared
Step2) Run the Map code provided
Step3) Shuffle the outputs of the map to the Reducer Preprocessors
Step4) Run the code
Step5) Product the final output

IOP Conf. Series: Materials Science and Engineering **590** (2019) 012028 doi:10.1088/1757-899X/590/1/012028

In this MapReduce Data Flow can be happened in following steps:
1) Input Reader: The input reader will divide the input in to different sizes and read the data thatis provided by the user
2) Map Function: It will take the input of key value pairs and provides output of zero or more key value pairs
3) Partition Function: Provide the key and number of reducers and return the index of the reducer
4) Comparison Function: Compare the inputs of the map and reduce functions
5) Reduce Function: It will take the input values and provide the single output
6) Output Writer: It will give the output for the reduce function

5. Conclusion:

In this paper had analysed the performance of KNN algorithm with the process of MapReduce .This approach measured the solution in to three different steps for computing the accuracy of the space and time complexity in Big Data with the combination of the Map Reduce function and KNN computation. And this process had done in three steps data partitioning, pre-processing and computation. In our future work, we decided to improve the speed of the data processing.

References:

[1] D. Li, Q. Chen, and C.-K. Tang, "Motion-aware knn laplacian for video matting," in ICCV'13, 2013.
[2] K. Inthajak, C. Duanggate, B. Uyyanonvara, S. Makhanov, andS. Barman, "Medical image blob detection with feature stability and knn classification," in Computer Science and Software Engineering, 2011.
[3] C. Yu, R. Zhang, Y. Huang, and H. Xiong, "High-dimensional knnjoins with incremental updates," GeoInformatica, 2010.
[4] C. Ji, T. Dong, Y. Li, Y. Shen, K. Li, W. Qiu, W. Qu,and M. Guo, "Inverted grid-based knn query processing withmapreduce," in Proceedings of the 2012 Seventh Grid AnnualConference, ser. CHINAGRID '12. Washington, DC, USA:IEEE Computer Society, 2012, pp. 25–32. [Online]. Available: http://dx.doi.org/10.1109/ChinaGrid.2012.19.
[5] Apexa B. Kamdar , Ishan K. Rajani,"Improved Adaptive K Nearest Neighboralgorithm using MapReduce",International Journal of Science, Engineering and Technology Research (IJSETR), Volume 4, Issue 6, June 2015.
[6] Prajesh P Anchalia, Kaushik Roy," The k-Nearest Neighbor Algorithm Using MapReduce Paradigm", 2014 Fifth International Conference on Intelligent Systems, Modeling and Simulation.
[7] Chi Zhang, Feifei Li, Jeffrey Jestes," Efficient Parallel kNN Joins for Large Data in MapReduce",March 26-30, 2012, Berlin, Germany, ACM.
[8] J. Dean and S. Ghemawat, "Mapreduce: Simplified data processingon large clusters," Commun. ACM, 2008.
[9] G. Song, J. Rochas, F. Huet, and F. Magoul`es, "Solutions forProcessing K Nearest Neighbor Joins for Massive Data on MapReduce,"in 23rd Euromicro International Conference on Parallel, Distributed and Network-based Processing, Turku, Finland, Mar. 2015.
[10] M. I. Andreica and N. T. pus, "Sequential and mapreduce-basedalgorithms for constructing an in-place multidimensional quadtreeindex for answering fixed-radius nearest neighbor queries," 2013.

IOP Conf. Series: Materials Science and Engineering **590** (2019) 012029 doi:10.1088/1757-899X/590/1/012029

ANALYSIS OF ROAD ACCIDENT

P. Sundar Kumar[1], V. Viswanadham[2], Dr. B. Bharathi[3]
Department of Computer Science and Engineering
Sathyabama Institute of Science and Technology, Chennai
{sundarkumarp69.SK[1], vetsavissu[2], bharathivaradhu[3]}@gmail.com
Chennai, India

Abstract - The paper focused the main idea of this project the reduce road accidents by dissecting the accident prone areas and taking security methods to diminish them. Generally road Accidents frequently occur, due to two main reasons, one is drivers mistake due to over speed or due to avoiding signals and may be due to ignoring traffic rules and the second case is external factors like road conditions, atmospheric conditions bringing this into mind we made application contains various limits and to inspect the standard reason of mischances. The created application stores a huge data of earlier years in the HADOOP to dissect why the mishaps occur in one region consistently. For exploring this colossal data we are using Big data for the faster examination. As Big data can store, control, separation and besides can mine tremendous fender bender dataset. The include is utilized Map reducing for organizing information and for the examination of unstructured information Spark is being utilized. By analyzing this we can take some safety measures to avoid the accidents which are being taken place in one area for to the similar reason.

Keywords - Fetching, Connector, Analysis of Query Language, Analysis Latin Script, Processing, Map Reduce, HADOOP.

1. INTRODUCTION

A mishap, otherwise called unexpected damage, is a bothersome, accidental, and impromptu occasion. Street mischances are incessant and by and large, the reason for the most harm. The purposes behind this are the to a great degree thick street activity and the moderately awesome opportunity of development given to drivers. Mishaps strike the vivaciously supplied vehicles in view of the avoiding the interstate code, and furthermore the dedication for drivers to alter their speed, which impacts stopping partitions, to the movement and atmosphere conditions (Rain, ice, fog, et cetera.). The neutralizing activity of road setbacks is moreover basic and will be ensured by strict laws, in particular, and police. To diminish the incidents, we are separating the best data to know the most astonishing accident zones locales and the reason for the purpose behind the disaster.

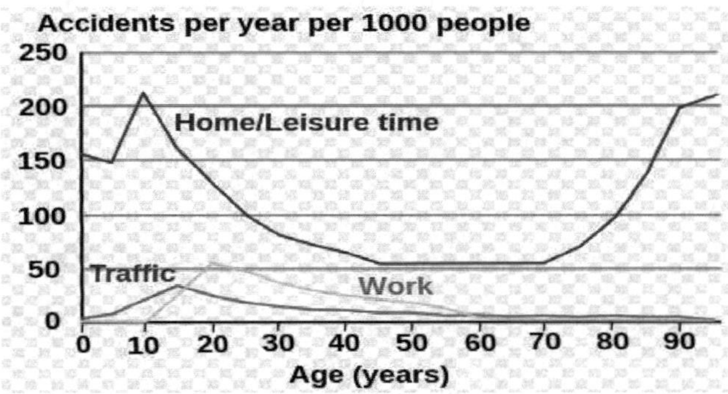

Fig 1: Frequently happened accidents analysis

All through this, we can get an idea how to decrease the accidents. To depict what kind of setbacks is of times happened, everything considered over through Fig.One. Subsequently here aggregate wrongdoing is relative to the sort of mishap.

2. LITERATURE SURVEY

The paper "Applying Mining Algorithms Rules for the Accidents in Dhaki" was proposed by Visas Pareek et al. Information mining utilizes assorted systems and checks to find picking up from information and see sensible cases from the information. It is considered as an achievement among the essential causes in data advancement in the earlier decade [1]. Alliance oversees mining estimations are used to find all principles in the database, fulfilling some base help and also least certainty requirements [2]. Order administers mining plans to locate few standards in the database to shape a precise classifier [3]. A little arrangement of principles is produced by standard order calculations to shape a classifier, however, these calculations utilize space autonomous inclinations and heuristics [3]. An Apriori Algorithm is computed for burrowing progressive thing sets of Boolean alliance rules [4]. Two separate advances are followed for the calculation to produce affiliation rules; applying least help to locate all successive item set in a database and utilizing these regular item sets and the base certainty limitation to create rules [4]. On the other hand, support and sureness restrictions are joined to a singular measure known as exactness which is utilized to make the Apriori alliance represents in the Apriori estimation [5]. Of course, support and sureness restrictions are joined to alone measure known as accuracy which is utilized to make the Apriori Association oversees the Apriori figuring [5].

The paper "An Overview Analysis of the Traffic Accident by using the Data Mining Technique"proposed by S. Krishnaveni and M. Hemalatha. This proposed work researches the utilization of Naive Bayes, AdaBoostM1, PART, J48, for various Forest Classifier and looks at these calculations executions for damage seriousness. As indicated by the Hong Kong's records of 2008 dataset, it does exclude travelers' data. It incorporates the names of seriousness, area board locale, attempt at manslaughter, climate, rain, normal light, intersection control, street arrangement, vehicle developments, sort of impact, number of vehicles included, number of individuals harmed, their age, their sex, area of damage, part of loss, driver Age, driver sex, year of make, seriousness of mischance and vehicle class.

C Chandrasekar et al proposed "A Detailed Description of Data Mining on the Road Accident Analysis". This discussses road prosperity authorities and examiners oversee enormous volumes of quantitative information and assembled estimations, with a particular true objective to understand and measure the social and money related cost of the setbacks and to have familiarized security outlines all together with the thwart or reduce the occasions of accidents. The street activity and mischance measurements must be displayed in such an approach which is less demanding to be both perceived and deciphered by a human administrator. The past takes investigation included measurable strategies and formal procedures. Insights tables and common graphing methods are not adequate for exhibit day necessities and this causes challenges in the powerful perception of results and examples. Another weakness is that customary strategies constrain human association in the investigation assignments. In this paper, we have broken down different examinations done by different creators in mischance databases and furthermore the measures proposed for the mishap and its demise rate.

Sachin Kumar and Dura Toshniwal in their paper analyse road accident using the data mining techniques. India has the second biggest street organize on the planet. Street mischances happen often, and they guarantee excessively numerous lives each year. It is basic for the primary driver of road setbacks with a particular ultimate objective to keep up a vital separation from them. Reasonable information mining approach must be connected to gathering the data sets speaking to happened street mishaps to recognize conceivable concealed connections and associations between different components influencing street mischances with lethal results. The outcomes got from mining methodology can help comprehend the difficult variables or regularly rehashing designs. The present example demonstrates the perilous street as far as Road Accidents and the required various measures to be taken to minimize the Accidents.

3. INFERENCES FROM EXISTING SYSTEM

The inferences from existing system are: .

- Consequently, these data are needed to be preprocessed for getting suitable data for analyzing purposes.
- Optimization techniques, intelligence, processing, integrating Data are limitless.
- The trouble of breaking down for the large volumes due to the information engaged with the street activity dataset.
- The Capacity for data mining is not sufficient processing speed.

- The betterment of resources can't be used maximum speed.

4. EXISTING SYSTEM

MySQL is a database administration framework. RDBMS utilizes relations or tables to store Traffic mishap as a lattice of lines by segments with essential keys and remote keys. With MySQL language, an accident in tables can be collected, stored, processed, retrieved, extracted and manipulated mostly for business purpose. Existing idea manages to give back by utilizing MySQL which has a lot of disadvantages i.e. information constraint is that preparing them is high when the information is enormous and once information is lost we can't recuperate. So, we proposed concept by using HADOOP tool.

5. PROCESS

A. Preprocessing Road accident Database:

In this module, analyzing the data with different kinds of fields in Microsoft Excel, then it converted into the comma delimited format which is said to be CSV (The comma separated value.) filed and moved to MySQL backup through the database. The clearest explanation is shown in below in the fig. 2

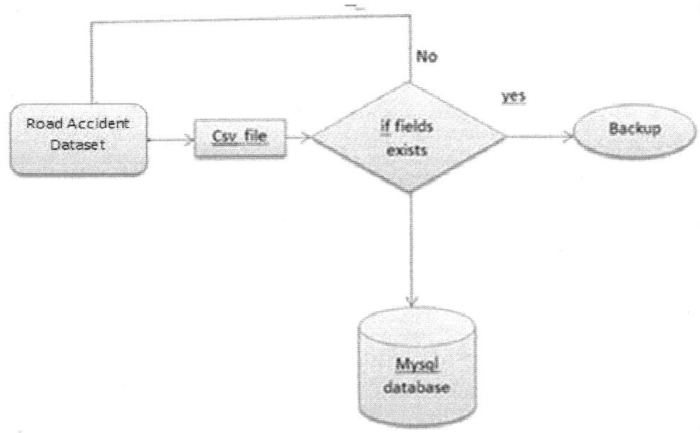

Fig.2.Steps of Preprocessing

B. Storage:

The user takes MySQL data into the software tool by fetching the data through a scope and store in HDFS. The clearest explanation is given in below fig. 3

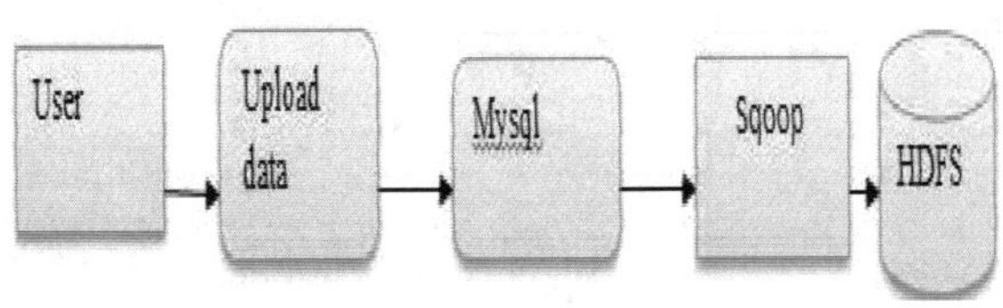

Fig.3.Steps of storage mechanism

C. Data integration with Sqoop:

The SQOOP is a summon line interface application for exchanging Traffic mischance between social databases (MySQL) and Hadoop. We have Traffic accident tables in a MySQL database, and we have to import it to HDFS using SQOOP. The online Traffic accident can be moved into HDFS/Hive from MySQL and then it will generate the Java classes. In past cases, the stream of information was from RDBMS to HDFS. Utilizing "trade" instrument, we can import information from HDFS to RDBMS. Before sending, SQOOP separates the metadata from the MySQL database. So for this, we have to make a metadata at first. And the connections are clearly shown in the fig.4

IOP Conf. Series: Materials Science and Engineering **590** (2019) 012029 doi:10.1088/1757-899X/590/1/012029

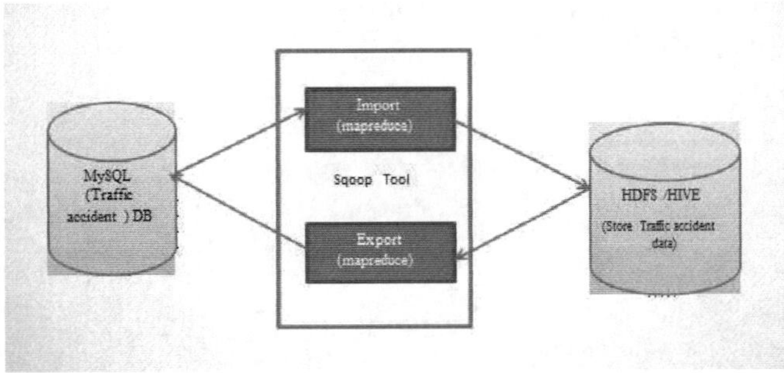

Fig.4.Data integration mechanism

D. Data Analyze, using HIVE:

Using hive query language, it can be analyzed in different functions like partitioning, bucketing in the structured table and produce in console output according to queries and analyzing the road accident database based on different types of incidents that are being taken from the database. Which is being more clearly shown in the fig. 5

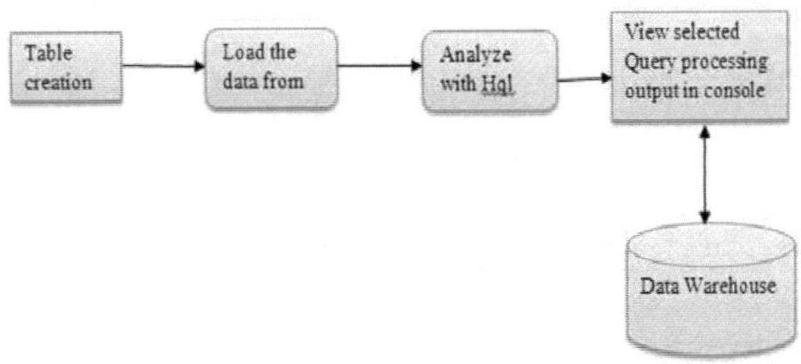

Fig.5.Data analysis

In Hive, Traffic accident tables and databases are created first and then the data is loaded into these tables. A hive composes Traffic mischance tables into segments. It is a method for the part a table into comparative parts in light of the benefits of partial sections. The separate structures are being made to the table data by using the concept of bucketing. Bucketing work depends on the estimation of a hash capacity of some segment of a table as shown below in the fig.6.

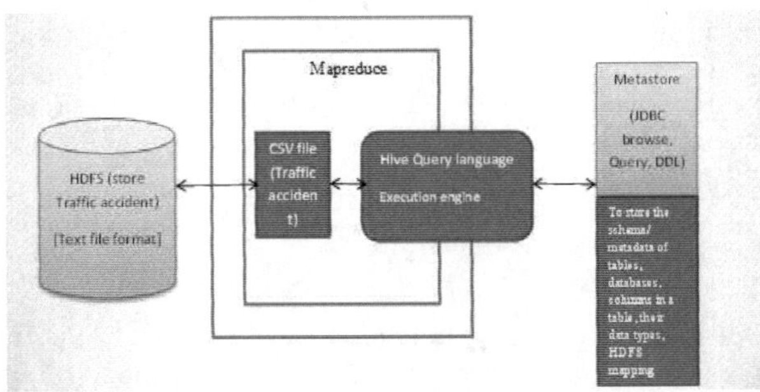

Fig.6. Bucketing

E. Data Analyze, using Pig:

Using Pig we can explore both Structured and besides unstructured data and moreover, it can summon code in various tongues like Java, J Ruby, et cetera. A decent case of a Pig application is the ETC exchange demonstrates that portrays how a procedure will remove information from a source, change it as per a govern sets and after that heap it into a data center. Pig can ingest information from documents, streams or different sources utilizing the User Defined Functions(UDF). When it has the information, it can perform to select, emphasis and different changes over the information. Again the UDF highlight enables passing the information to more perplexing calculations for the change. Pig substance is changed over into a movement of Map Reduce occupations that continue running on the Apache HADOOP gathering. As a feature of the interpretation, the Pig mediator performs improvements to speed execution on Apache HADOOP. We will compose a Pig content that will do our information examination assignment.

F. Data Analytic With Mapreduce:

The Map-Reduce programming model is being formed by two strategies Map and joining. The information which is being available in the database is first being broken down and later the information is being coordinated and appeared as required. In addition, there region unit limits exist in the Map-Reduce execution appear, for instance, set up and kind, for dealing with widely appealing information. On the Map viewpoint the setup ability is associated, and to boot execute information exchange by the key once Map() In these lines, information a closely resembling key are conveyed to a singular scale back () work. The kind ability is moved on the scale back viewpoint later.
Than information exchange. By using key information advancing in the kind field to the social occasion every last one of the sets by routes for a practically equivalent to a key for any making prepared.

Algorithm.1. Map Reduce Execution

```
Class Mapper
Method Mapping (doc_id a, doc d)
For all term t ∈ doc d do
Emit (term t, count 1)
Class Reducing
Method Reducing (term t, counts [c1, c2, . . .])
Sum ← 0
For all count c ∈ counts [c1, c2, . . .] Do
Sum ← sum + c
Emit (term t, count_sum)
```

6. CONCLUSION

In this paper, we exhibited an investigation on Traffic mishap and expectation in informal communication administrations, for example, a smaller scale blog application. For analyzing traffic data in the HADOOP ecosystem. The Hadoop environment is hived, pig, Map Reduce, which is utilized to locate some profound investigation and dynamic cooperations among Traffic mischances. In future, the beginning is introduced which is 100 times Speedier than HADOOP and its examination snappier.

7. FUTURE ENHANCEMENT

Apache Spark is an open-source preparing motor worked around speed, the instance of utilization and examination. If you have a considerable measure of data that requires low idleness taking care of that a regular Map Reduce program can't give by then Spark is the choice. Begin bunches the give data along these lines, that it works 100 times speedier than HADOOP and besides reinforces Java and Python APIs for the straightforwardness of change.

REFERENCES

[1]. Sachin Kumar, Durga Toshniwal, "Investigating Road Accident Data Using Association Rule Mining," Computing, Communication, and Security (ICCCS), 2015 International Conference on 4-5 Dec. 2015, DOI:10.1109/CCCS.2015.7374211.

[2]. B.Bharathi and Mahesh kumar (2016), 'Non invasive BG scrutinizer system', Global Journal of Pure and Applied Mathematics, vol.12, No: 8, pp. 5123 – 5125

[3]. Sachin Kumar, Durga Toshniwal, "An information mining Approach to portray street mishap areas," J. Mod. Transport. (2016), DOI 10.1007/s40534-016-0095-5, pp.62-72.

[4]. Suwarna Gothane, Dr. M. V. Sarode, "Examining Factors, Construction of Dataset, Estimating the significance of the factor and age of affiliation rules for Indian street Accident," IEEE International Conference on Advanced Computing, 2016, DOI 10.1109/IACC, pp. 15-16.

[5]. S.Krishnaveni, Dr.M.Hemalatha, "A Perspective Analysis of Traffic Accident utilizing Data Mining Techniques," International Wireless Communications 2017 International Conference on Innovations in data Embedded and Communication Systems (ICIIECS) and Mobile Computing Conference (IWCMC) 2016, DOI: 10.1109/IWCMC.2016.7577067, pp. 256-261.

IOP Conf. Series: Materials Science and Engineering **590** (2019) 012029 doi:10.1088/1757-899X/590/1/012029

[6]. Sachin Kumar, Durga Toshniwal, "Examination of hourly street mischance tallies utilizing progressive grouping and cophenetic relationship coefficient," Springer, 2016, DOI 10.1186/s40537-016-0046-3, pp. 2-11
[7]. Mayan, J.A., Priya, K.L.," Novel approach to reuse unused test cases in a GUI based application", IEEE International Conference on Circuit, Power and Computing Technologies, ICCPCT 2015, pp.1-7
[8]. Sinan Kaplan, Mehmet Amac Guvensan, Ali Gokhan Yavuz, and Yasin Karalurt, "Driver Behavior Analysis for Safe Driving: A Survey,"IEEE Transaction on Intelligent Transportation System (2015), DOI:10.1109/TITS.2015.2462084, pp. 3017-3032
[9]. Kamalesh M.D., Bharathi B,"Slicing an efficient transaction data publication and for data publishing ",Indian Journal of Science and Technology, Vol 8(S8), pp.306-309, 2015.
[10]. Youthful Sung Cho, Keun Ho Ryu, "Prescient Pattern Analysis utilizing the SOM in therapeutic, informational collections for Medical Treatment Service," Conference on Computational Intelligence in Bioinformatics and Computational, Biology, (IEEE 2014), DOI:10.1109/CIBCB.2014.6845512, pp. 1-5.
[11]. Ahmed Alsayat, Hoda El-Sayed, Haidong, "Effective Genetic K-Means Clustering for Health Care Knowledge Discovery," fourteenth International Conference on Software Engineering Research, Management and Applications (IEEE 2016), pp. 45-52.
[12]. Ahamed Al Malki, Mohamed M. Rizk, A. A. Mousa, "Half and half Genetic Algorithm with K-Means for Clustering Problems," Open Journal of Optimization, (2016), volume:5, pp.71-8
[13]. Asha P, Albert Mayan J, Canessane A (2018),"Efficient Mining of Positive and Negative Itemsets Using K-Means Clustering to Access the Risk of Cancer Patients",Communications in Computer and Information Science ,ICSCS 2018, Kollam, 2018,pp.373-382..
[14]. Madhusmita Mishra, H.S. Behera, "Kohonen Self Organizing Map with Modified K-implies bunching For High Dimensional Data Set," International Journal of Applied Information Systems, Volume 2– No.3, May 2012, pp. 2249-0868.
[15]. Morth (2014) Road Accidents in India 2013. New Delhi: Ministry of Road Transport and Highways, Transport Research Wing, Government of India, August 2014.http://morth.nic.in/showfile.asp?lid=1465.

IOP Conf. Series: Materials Science and Engineering **590** (2019) 012030 doi:10.1088/1757-899X/590/1/012030

Review on Non Invasive Glucose and Cholesterol Measurement System

V. Joseph Devakumar[1]
ME Embedded System,
Sathyabama Institute of Science and Technology,
Chennai.
josephdevsamue@gmail.com

T. Ravi[3]
Associate Professor,
School of Electrical and Electronics Engineering,
Sathyabama Institute of Science and Technology,
Chennai
ravivlsi123@gmail.com

S.Karthikeyan[4]
Associate Professor,
School of Electrical and Electronics Engineering,
Sathyabama Institute of Science and Technology,
Chennai
skarthi1879@gmail.com

Abstract- Diabetes and Cholesterol has become a common disease all over the world. Continuous monitoring of the glucose and cholesterol levels is needed in case of physically disabled and unconscious people where it is hectic to take measurements invasively or through blood samples. The proposed model brings out an idea of measuring glucose and cholesterol levels without the need of blood samples or involving contact. This is done with the help of simple IR sensors and Arduino UNO. The patient's fingers are placed on the IR sensor and power LED sockets with one involving in the measurement of cholesterol and other in measurement of glucose. For the diabetic patients and pregnant ladies have to monitor the glucose and cholesterol level frequency. The minimum level of cholesterol in blood is causes to the cardiovascular disease, that leads to stoke and heart attack. Based on the reflection of power LED light through the blood cells, the values are fed to the Arduino UNO. With the help of program dumped in the Arduino UNO, glucose and cholesterol values are calculated. In addition to that Hemoglobin and also LDL and HDL values are also calculated which is a further advantage. The proposed model costs less and saves more time. Further advancements can be done by employing this project by using IoT which can feed the measured data online. Non invasive measurement methodology of measuring glucose and cholesterol is not implemented anywhere which describes the uniqueness and innovation involved in this method. The model is not affected by external interrupts such as external sources of light etc and works efficiently under every conditions.

Keywords: ***Arduino Uno, LCD, Photo-diode, Power LED***

1. INTRODUCTION

Diabetic is most killer disease in the world. It is also the most common disease nowadays, to finding the blood glucose level by taking the blood sample for the person. While not drawing blood, puncturing the skin, or inflicting pain or trauma. It is not possible to finding the level of glucose measurement. It is more irritating to the patients ex-specially for children it gives more painful for them. For the diabetics and pregnant ladies has frequently a change the blood glucose level. For that there have to puncturing the skin for every time. Then the major disease next to glucose is cholesterol. For cholesterol also need the blood sample to measure the total cholesterol and LDL, HDL cholesterol measurement. The minimum level of cholesterol in blood is causes to the cardiovascular disease, that leads to stoke and heart attack. This project is solving the major glucose and cholesterol measurement level by the non invasive method. This method requires the no need of blood sample for measuring the level of glucose and cholesterol measurement. To finding the blood glucose and cholesterol measurement by using the power LED and photodiode. This process the finger tip has to place over the sensor. Power LED and photodiode sensor calculating the density and thickness of the blood by passing the light spectrum to it. By passing the light to finger the small most of current will generate and they will capture by the photodiode sensor. The current generate from the photodiode, to convert the current to voltage pass though the sensor output. The outputs of the sensor are connected to the ardiuno uno. This method is continuous measurement and pain free glucose and cholesterol measurement system.

2. LITERATURE REVIEW

Gilwon Yoon and Kye Jin Jeon (2012) described about the blood substances such as total hemoglobin, glucose, protein, cholesterol was investigated based on optical spectroscopy. Measurement was made either non-invasively or from blood or serum without using a test ship or wet chemistry. Non-invasive technique is available for measuring blood flow, SpO, bilirubin, muscle composition. Hemoglobin measurement, we have selected wavelengths between 500 and 1000nm.A custom made LED array composed of several wavelengths is used. Wavelengths were selected so that totals hemoglobin, SpO, pulse rate could be measured at the same time. Only pulsatile portion of detected signal was used to avoid other tissue effect as used in the case of pulse oximetry. For non-invasive determination of hemoglobin, a LED array of 569, 660, 805, 940, 975nm were used. Light was irradiated into a finger nail and transmitted light was measured. Digital filtering and outlier detection algorithm were applied to reduce the noise induced by the electronic circuit and unwanted effect generated by body movement. Prediction accuracy with serum is higher in case of the FIR band measurement except for cholesterol. For hemoglobin related measurement such as total hemoglobin, non-invasive measurement is possible. Reagent less determination with blood or serum appears to be feasible in terms of glucose, cholesterol, albumin and globulin. [7]

Mohd Norzaliman and Mohd Zain (2014) proposed really sensitive optical polarimetry measuring technique victimization ratio-metric gauge boson enumeration detection has been introduced and tested for a variety of best-known aldohexose concentrations that mimic the amount of aldohexose in human blood. A 785nm diode optical device that emits weak coherent optical signal onto aldohexose concentration samples in binary compound. The result shows a linear proportional of different glucose concentration and successfully detected 10 mg/dl to 260 mg/dl glucose samples. This indicates a potential improvement method for non-invasive glucose measurement by a sensitive polarimetry based optical sensor in single photon level for biomedical applications. Non-invasive measurement of glucose concentration based on photon counting polarimetry measurement. Low limit of measured detection obtained at 10 mg/dl with interval of 20 mg/dl until 260 mg/dl. This indicates the setup is highly sensitive and able to detect very low concentration of glucose. Therefore, this photon counting polarimetry technique will make a very sensitive and a good optical glucose sensor. [1,6]

Brince Paul and Melvin P Manuel (2016) proposed the invasive methods cause pain, time consumption, high cost and potential risk of spreading infection diseases. Therefore there is a great demand to have reliable cost effective and comfortable non invasive system for the detection of blood glucose level continuously. The planned methodology is predicated on the direct impact of aldohexose on the scattering properties of the organ are shown in figure 2.1. Aldohexose decreases the mate in index of refraction between scatterers and their close media, resulting in a smaller scattering constant and, consequently, a shorter optical path. The reduction in scattering is due to an increase in glucose concentration. As a result, with the growing concentration of glucose, fewer photons are absorbed and the light intensity increases. In the present work, we have used PPG technique [8]. An algorithm was developed from the PPG data for monitoring blood glucose are shown in the figure 2.1. It can be observed that the output from the glucometer shows an increase in glucose concentrations from both Glucocard and the developed setup. It is possible in future to improve the performance of a specially designed non invasive glucometer by including different IR wavelengths ranging from 700nm to 2500nm. [5]

R Periyasamy and Sneh Anand (2016) focused on the continuous blood glucose monitoring can help and prevent hyperglycemia level in diabetic patients. In the present study we aimed to estimate the blood glucose concentration using cost effective non-invasive forearm capacitance measurement technique. In this technique is based on the fact that glucose level affects the dielectric properties of blood and underlying tissue using a designed parallel plate capacitor circuit. Forearm part act as a dielectric medium of parallel plate capacitor and 555 timer circuits is used to produce pulse waveform based on measured capacitance. In first phase, capacitance was measured after consumption of sucrose (table sugar). In second phase, capacitance was measured after consumption of artificial sweetener. The trend of measured capacitance obtained after sucrose consumption is similar to the trend which was already established in oral glucose tolerance (OGT) test [6]. The potential of this technique in predicting invasive blood glucose concentration. An attempt has been made to non invasively measure blood glucose level using capacitance measurement in forearm. Capacitance measurement is a direct measure of blood glucose. It is well known that skin parameters. The fat thickness at forearm varies for different people, so individual calibration may require for this method. Finally, a deeper study of stability is still required, which should be done in parallel with improving sensitivity of the technique. The process was future work to connect the LCD display via microcontroller for direct display of capacitance value. [14]

V.Ashok and A.Nirmalkumar (2011) proposed a method and equipment for non invasive measuring of blood sugar concentration supported transilluminated irradiation to the finger has been reported during this paper. This methodology depends on atomic gas (He-Ne) optical device operational at 632.8nm wavelength. Throughout measuring, the finger is inserted into the aldohexose sensing unit, the transilluminated optical signal into electrical signal, compared with the reference electrical signal, and therefore the obtained distinction signal is processed by signal process unit that presents the ends up in the shape of blood sugar concentration. The maximum scattered signal at 290 & 3370 and transilluminated signal at 1670 was obtained according to anisotropy factor that was selected for further study [10]. The transillumination and blood glucose concentration at fasting and post Parandial conditions of individual subjects are compared and found that transillumination is inversely proportional to the concentration of blood glucose. The difference in glucose concentration between fasting and post Parandial is correlated with transillumination. [4]

Megha C. Pande (2015) analyze the non-invasive blood sugar measuring exploitation the infrared region that is that the most fitted region for blood sugar measuring. During this technique that is analogous to pulse oximetry supported close to infrared spectrum analysis. AN infrared emission of specific wavelength is passed through tip containing an blood vessel pulse element. AN optical sensing element is developed for blood sugar exploitation wavelength of 940nm and 1450nm.Output is obtained on scope correlation is established between actual blood glucose concentration and measured output voltage. The resultant wavelengths are used for non-invasive measure since its linear relationship with aldose concentration provided various error inflicting elements unit of measurement reduced. A sensing element developed with terribly high signal to noise quantitative relation will be designed. [9]

Praful P. Pai (2016) purposed a photo acoustic (PA) measurements were made on glucose solutions at multiple NIR excitation wavelengths. A variety of time and frequency domain features, including amplitude and area based features, were extracted from the PA measurements. These features were observed to be proportional to the glucose concentration of the sample. PA measurements from samples of whole blood at different glucose concentrations showed similar results. The results were compared to reference glucose concentrations made using a regular blood glucose meter. A comparison of 196 measurement pairs of predicted and reference glucose concentrations using a Clarke Error Grid gave a point distribution of 87.24% and 12.76% over zones A and B of the grid, with no measurement pairs falling in unacceptable zones C-E of the error grid. The predicted measurements had a mean absolute difference (MAD) and a mean absolute relative difference (MARD) The use of NIR PAS for continuous non-invasive glucose monitoring was demonstrated on glucose solutions and whole blood. An apparatus for making PA measurements was constructed, and time and frequency domain features of PA measurements at different glucose concentrations were examined for calibration. The features were found to increase smoothly with the glucose concentration and saturated at high concentration values. In PA measurements were performed on individuals and a quadratic fit was used for calibration. The calibration accuracy obtained was superior to earlier studies using PAS. [15]

Roopa.G and K.Rajanna (2011) focused on metabolic process disorders like respiratory disease by analyzing the recorded respiratory pattern. though there square measure devices like measuring device to diagnose respiratory illness, they measure terribly inconvenient for patient's use as a result of patients are created to exhale air through mouth forcefully. Presently developed sensor will overcome this limitation and is helpful in the diagnosis of respiratory related abnormalities. In the figure 2.3 are shown the Polyvinylidene fluoride (PVDF) film in cantilever configuration is used as a sensing element to form the breath sensor. Two identical sensors area unit mounted on a spectacle frame, such the stream of indrawn and exhale air can collide with sensing element, for sensing the respiration patterns. These patterns area unit recorded, filtered, analyzed and displayed victimisation CRO. Further the sensor is calibrated using a U-tube water manometer. The added advantage of piezoelectric type sensing element is that it is self powered without the need of any external power source.

The PVDF sensing element used in the present work is coated with gold at top and bottom surfaces for electrode lead attachment. Double enameled copper wire (0.07mm diameter) is attached to the top and bottom surfaces of PVDF sensing element. Two similar PVDF sensing elements are mounted on a spectacle frame. The spectacle frame was worn by the themes and their several respiration patterns square measure recorded. The recorded signals square measure filtered victimization low pass filter with a cut-off frequency five cycles/second. The respiration rate of asthmatic patient is higher compared to normal person. Though there are devices like spirometer to diagnose asthma, it basically measures the lung capacity. And they are very inconvenient for the patient to use because patients are made to exhale air through mouth forcefully. The presently developed sensor overcomes this limitation, as the patient is made to inhale and exhale air through nose in a normal way. [16]

Satoshi Shimawaki (2014) proposed a non-invasive cholesterol level measurement using near-infrared light. 6-week-old male rats were fed either control diet alone or control diet for 8 weeks [19, 20]. A cuff was subsequently placed on the proximal part of the tail, the tail distal to the compression site was exposed to near-infrared light (wavelength, 850 nm), and transmitted light was photographed. Near-infrared transmission pictures were no inheritable before and once cuff inflation and hemoprotein and haematocrit levels in veins distal to the compression web site were determined. Accrued dietary cholesterin levels resulted in accrued total body fluid cholesterin levels. Before cuff inflation, rats that was fed the two. 0% cholesterin diet had considerably lower hemoprotein levels than people who fed the management diet. There have been no important variations in hemoprotein and haematocrit levels at five min when cuff inflation. From the association between total humour cholesterin levels and variations in hemoprotein levels, the variations in hemoprotein levels were nearly constant once total humour cholesterin levels hyperbolic to close a hundred and fifty mg/dL. The correlation between total cholesterin levels and relative changes in near-infrared lightweight transmission considerably adhered thereto between total serum cholesterin levels and variations in hemoprotein levels. Few changes within the variations in hemoprotein levels were ascertained at total humour cholesterin levels 150 mg/dL, the variations in rumpled with a rise in total humour cholesterin levels. Changes in NIR light transmission depended on the differences in hemoglobin levels, namely, the non-invasive technique of blood vessel visualization was considered to enable the determination of serum cholesterol levels. The findings of this study indicate that the use of NIR light transmission could be highly beneficial for simple, non-invasive diagnosis of hypercholesterolemia, thereby reducing the incidence of atherosclerosis. [17]

I.M.M.Yusoff (2015) proposed a methods for non-invasive blood analysis were disclosed during which blood was light at variety of separate wavelengths elite from the NIR spectrum, supported the acquisition of data from the analysis of skin parts by employing a easy take a look at intensity of mirrored or transmitted light-weight at such wavelengths were measured, and a comparison between non-invasive sterol meter and invasive sterol meter was performed. Infrared transmitted the signal, to the patient skin, absorb to the blood stream, calculate by software package (Arduino), and result show on the liquid crystal display monitor. Experiment was conducted to match the information from invasive sterol meter with the non-invasive sterol meter that has been created. The information from invasive sterol meter is that the most correct because it is taken directly from the blood; meantime the information from the non-invasive sterol meter depends on the reading through NIR crystal rectifier. 5 samples are taken to prove the thought. it's clearly ascertained that the information variations between the 2 instrumentality ar tiny and also the tolerance is 0 to 0.3. [13]

3. CONCLUSION

The non invasive glucose and cholesterol measurement are using power LED sensor has more efficient than the invasive method. Power LED sensor has low cost, highly sensitive and able to detect very low concentration of glucose and cholesterol measurement. This non-invasive method, measure the parameters without blood samples. This method is more comfortable for continuous monitoring of glucose for the diabetic patient and children admitted in ICU and this system is more comfortable for cholesterol measurement than invasive method. Using this system public can measure glucose and cholesterol by their own without ant hospital or testing lab support without any discomfort. This system has less expensive than the existing invasive methods.

REFERENCES

[1] Afiqah Musa, Mohd Hafizulfika Hisham, Mohd Zain "Photon counting polarimetry measurement towards non-invasive biomedical glucose monitoring" International Conference on Photonics (ICP), IEEE EMBS Conference on. IEEE, 2014, pp. 156–158.

[2] Akesh Govada, Ch Renumadhavi , K B Ramesh , " Non-Invasive Blood Glucose Measurement", International Journal of Advanced Research in Computer and Communication Engineering- 2014.

[3] Anuj Srivastava, Md Koushik Chowdhury "Blood Glucose Monitoring Using Non Invasive Optical Method: Design Limitations and Challenges" International Journal of Advanced Research in Electrical, Electronics and Instrumentation Engineering -2013.

[4] Ashok V, Nirmalkumar A, "A Novel Method for Blood Glucose Measurement by Non invasive Technique Using Laser" International Journal of Medical – 2011.

[5] Brince Paul, Melvin P Manuel, "Design and Development of Non invasive Glucose Measurement System" Electronics and Instrumentation Engineering -2013.

[6] M.Manoranjani and T.Ravi, " "Multithreshold CMOS sleep stack and logic stack technique for digital circuit design" ARPN Journal of Engineering and Applied Sciences, Vol. 10, No. 10, June 2015. pp: 4550-4556.

[7] Gilwon Yoon, Kye Jin Jeon, "Reagentless / Non-invasive Diagnosis of Blood Substances" IEEE Conference on Biomedical Engineering and Sciences, IEEE, 2013, pp.226- 227.

[8] Samuel Lazar.M and Ravi.T, "Li-Fi design for high speed data transmission" ARPN Journal of Engineering and Applied Sciences, ARPN Journal of Engineering and Applied Sciences, Vol. 10, No. 14, August 2015, pp: 5999-6003.

[9] Megha C. Pande, Prof. A. K. Joshi, "Non-Invasive Blood Glucose Measurement" International Journal of Computational Engineering Research (IJCER), April – 2015,pp. 26-28.

[10] Ravi.T and Kannan.V "Effect of N-type cntfet on double edge triggered D flip-flop based PISO shift register" IEEE Proceedings of the International Conference On Emerging Trends in Science Engineering and Technology: Recent Advancements on Science and Engineering Innovation, INCOSET 2012, Dec-2012, pp. 344-349. .

[11] Naoki Wadamori, Ryou Shinohara and Yasutoshi Ishihara, "Photoacoustic depth profiling of a skin model for non-invasive glucose measurement" International IEEE EMBS Conference, IEEE, 2010, pp. 5644-5647.

[12] Ogawa M, Yamakoshi Y, "A New Non-invasive Method for Measuring Blood Glucose Using Instantaneous Differential Near Infrared Spectrophotometers" International Conference of the IEEE EMBS, 2007, pp. 2964-2967.

[13] Omar W.R.W, Yahya R, Yusoff I.M.M, , "Non invasive cholesterol meter using Near Infrared sensor" Innovation & Commercialization of Medical Electronic Technology Conference (ICMET),IEEE ,2015,pp. 100-104.

[14] Periyasamy R, Sneh Anand, "A study on Non-invasive blood glucose estimation an approach using capacitance measurement technique" International conference on Signal Processing, Communication, Power and Embedded System (SCOPES) IEEE, 2016, pp. 847-850.

[15] Praful P. Pai, Pradyut Kumar Sanki, "NIR Photoacoustic Spectroscopy for Non-Invasive Glucose Measurement" IEEE Conference, IEEE, 2015.

[16] Roopa.G, K.Rajanna, M.M. Nayak, "Non-Invasive Human Breath Sensor", IEEE Conference on Biomedical Engineering and Sciences, IEEE, 2011.

[17] Satoshi Shimawaki, Yohei Kobayashi, "Non-invasive Serum Cholesterol Detection Using Near-infrared Light Transmission" Biomedical Engineering Research Sept. 2014, Vol. 3 Iss. 3, PP. 80-87.

[18] Sri haindavi K, Swathi Kamasahayam "Non Invasive Estimation of Blood Glucose using Near Infra red Spectroscopy and Double Regression Analysis" Seventh International Conference on Sensing Technology, IEEE EMBS Conference on. IEEE, 2013, pp. 627-631.

[19] S.Karthikeyan, S.Jayashri, "Proficient Data Gathering by Mobile Agent using Jade",i-manager's Journal on wireless communication networks,Vol.1, No. 2, 2012, pp. 32-40.

[20] S.Karthikeyan, S.Jayashri, "Energy Efficient System for Heterogeneous Wireless sensor Networks", European Journal of Scientific Research, Volume 72,No 4 march 2012, PP:599-607. ISSN: 1450-216X.

[21] S.Karthikeyan, S.Shanmugapriya and S.Jayashri, "Sub-head Transmission of Heterogeneous Data by Cloned Agent to Android Mobile",Research Journal of Applied Sciences, Engineering and Technology, 8(1):pp.24-34, 2014.

[22] R. M. Joany, E. Logashanmugam, "Design and Analysis of Compressive Sensing on MRI", Journal of Medical Imaging and Health Informatics, Volume 8, Number 3, March 2018, pp. 596-601, American Scientific Publishers.

[23] R. M. Joany, E. Logashanmugam, M.Ooha, "Algorithmic Approach on Video Compression", International Journal Control Theory Applications, 9(7), 2016, pp. 339-343,ISSN : 0974-5572 , International Science Press.

IOP Conf. Series: Materials Science and Engineering **590** (2019) 012031 doi:10.1088/1757-899X/590/1/012031

WIRELESS POWER TRANSMISSION

V. Bhavya Keerthi[1]
Student, Dept of ECE,
Sathyabama Institute of Science and Technology,
OMR Road, Chennai, India
bhavyakeerthi98@gmail.com

V. Swetha[2]
Student, Dept of ECE,
Sathyabama Institute of Science and Technology,
OMR Road, Chennai, India
vutukuriswetha747@gmail.com

E. Annadevi[3]
Assistant Professor, Dept of ECE,
Sathyabama Institute of Science and Technology,
OMR Road, Chennai, India
annadevimit@gmail.com

Abstract - In recent days, Wireless Power Transfer (WPT) and harvesting energy received lots of attentions due to the necessity of having a flexible, suitable and secure source of energy. With the explosion of wireless technologies including wireless sensor network (WSN) like internet of things, such intelligent systems had highlighted this need. Wireless power transfer is a key technology to enable any device operation without replacing their batteries, often challenging due to cost of replacement and the difficulty to reach inaccessible areas. Anyhow wireless power transfer is being considered as eco-friendly. This also removes the need to dispose dead batteries. Inductive coupling can be used to charge the low power devices quickly and effectively without the help of wires. This method is done by using charging a resonant coil from AC and then transmitting subsequent power to the resistive load. In this case, oscillating circuit is used in the conversion of DC to AC in order to transmit the magnetic field by passing frequency and inducing the receiver coil. The system is designed and constructed as two parts. They are Transmitter part and Receiver part. To calculate the inductive coupling between the transmitter oil and receiver coil, Ampere's Law, Biot-Savart's Law and Faraday's Law are used as per the requirement.

Keywords - Rectenna, Microwave, Charging, Power transmission, Resonance, Current.

1. INTRODUCTION:

The main objective of their project is to develop a device for wireless power transfer. At first, the concept of wireless power transfer was evolved by Nikola Tesla. This can make a remarkable change in the field of the electrical engineering which eliminates the use of conventional copper cables and current carrying wires. This concept tells about the development of transferring power within a small range. It is built upon using an electronic circuit which converts AC 230V 50Hz to AC 12V, high frequency. The secondary coil develops voltage of HF 12 volts. The power transfer is done from the primary to the secondary within a distance of 5cm (approx). Then the transfer is seen as the primary transmits. The secondary receives the power to run load.

In the transmitter side AC to DC filter, HF inverter, HF transformer and primary copper coil is being used. The transmitter coil has 7 turns and the secondary one with 36 turns. The HF transformer is used to convert 230V 50Hz AC into 12V 25KHz AC. In the same way, in the receiver side, receiving coil, bridge rectifier, filter, voltage regulator and load is used. The rectifier is used to convert AC received by the secondary oil

IOP Conf. Series: Materials Science and Engineering **590** (2019) 012031 doi:10.1088/1757-899X/590/1/012031

into DC. The filtering is used to remove the unwanted AC ripples from the rectifier. To avoid the damage of the component the supply to the load should be regulated. Hence, LM7805 voltage regulator is used to regulate the voltage given to the load. This technique is able to use in many applications such as to charge a mobile phone, ipod, laptop battery, propeller clock etc

2. EXISTING SYSTEM:

The existing methodology for wireless power transfer is QI STANDARD. This method is used for low power applications by using an inductive coupling. In this method power transfer efficiency depends on quality factor of an induction coil, coupling coefficient, distance between transmitter and receiver, impact of core used for the coil. In wireless power transfer, according to coupling it is classified as near field and far field power transmission i.e Inductive and Magnetic resonance coupling respectively.

Qi Standard method is primarily used on inductive coupling where coupling between transmitter and receiver in short range. The power transferred from transmitter to receiver by exciting dc on a transmitter coil, it generates magnetic field which induces a voltage in receiver circuit. This voltage is used to charge any device or battery. In Qi standard, it can transfer low power of 5W, operating frequency lies between 100-200KHZ and distance between transmitter and receiver is 5mm.

In this method, the communication between transmission and receiver is done in the form of packets. This method is done in accordance with the Qi Standard. This communication process can be known as data structure. The power receiver uses to communicate a message to the power transmitter. The packet used here consists of a preamble, a header byte, a message and a checksum. After the transmitter is powered it sends analogy pings every 420ms to check the presence of receiver.

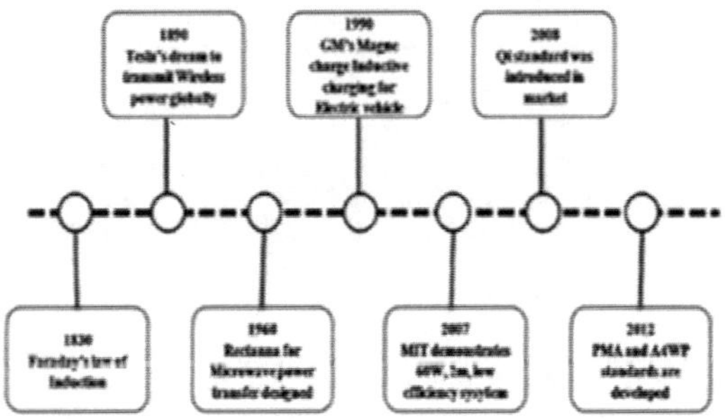

Fig 1: Evolution of wireless power transmission in various periods of the previous two centuries

3. PROPOSED SYSTEM:

There is several wireless charging techniques. Two of them are OI WIRELESS CHARGING and A4WP WIRELESS CHARGING. The Qi Wireless Charging is used to transfer energy supply over a distance of 1.6 inches. It has two types Low Power and Medium Power. Low Power can deliver 0.5 watts power and frequency is about 110 to 205KHZ. It is mostly used in domestic portable devices like mobile phones, music players, and Bluetooth earpieces. High Power can deliver up to 120 Watts and frequency is about 80 to 300KHZ.It consists of two elements Base Station and Mobile Devices. Base Station provides wireless transmission using inductive power. It has a flat surface. The device is placed on the top of this surface. Mobile Devices consumes the transmitted power. It is used to charge the battery in mobile phones. If the coupling between transmitter and receiver is as high as possible then efficiency is as high as possible. If the device is placed on a particular position to charge is called GUIDED POSITION. If the device is placed at anywhere in a wide area by using multiple transmitter coils is called PLACEMENT ANYWHERE. A4WP Wireless Charging is used to deliver power over a large area in a wireless charging. It is unnecessary to place device in a particular position. By using single power transmitter it charge many devices at a time. It uses the same magnetic inductive technique using this we can charge devices with in the inductor ring. It also allows

IOP Conf. Series: Materials Science and Engineering **590** (2019) 012031 doi:10.1088/1757-899X/590/1/012031

the Z-axis charging placement i.e. the chargers can be built deeper in to other objects. Some of features of Wireless Power Transfer is, it is used in high frequency range of 6.78MHZ. It avoids inductive heating using lower frequencies. CONTROL/MANAGEMENT PROTOCOL is uses a frequencies of 2.4GHZ ISM (Industrial Scientific and Medial) band. It is used in smart phones and other electronic items. POWER RECEIVING UNIT is used to transfer the power from transmitter to receiver.

INDUCTIVE POWER TRANSMISSION uses inductive coupling between two circuits. Its operation is similar to its name. It couple between transmitters to receiver by using inductive coupling. The transformer has 2 paths primary and secondary. Primary coil is in power source and Secondary coil is in batteries. Primary and Secondary circuits are coupled in inductive coils to increase the magnetic field. Inductive Power Transmission efficiency depends on Inductor size, Inductor shape, Distance between coils, Coil resistance.

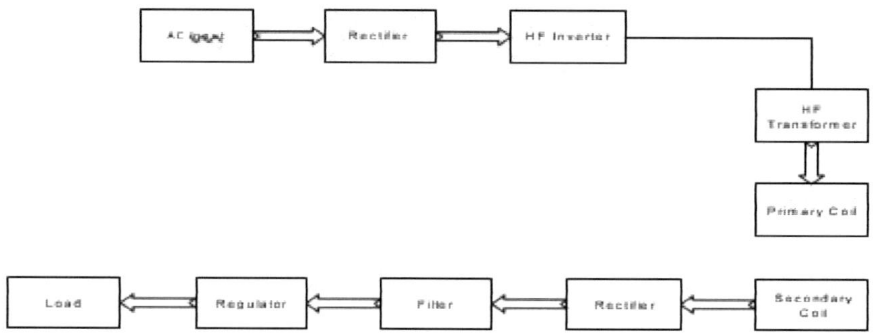

Fig 2: Block Diagram

3.1. *RECTIFIER:*

Rectifier is an electronic device. It convert alternating current to direct current. Bridge rectifier is essential for this method because it has good stability and it shows full wave rectification.

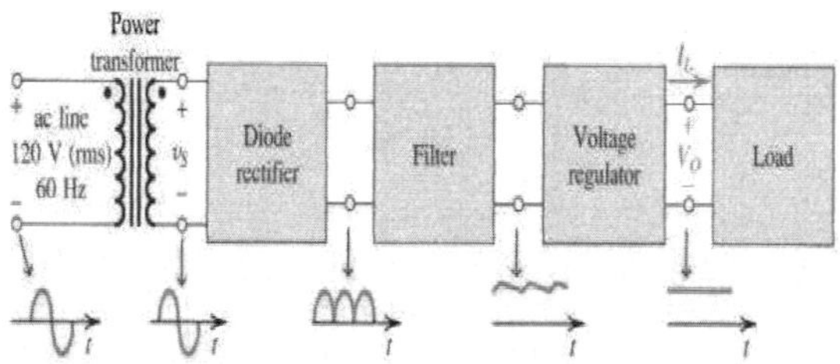

Fig 3: DC Voltage Rectification

3.2. *HF TRANSFORMER:*

Transformer is an electrical device it transfer energy from one circuit to another by magnetic coupling. High frequency transformers transfer electric power with frequency 20 and 100KHZ. It is normally used in high voltage power transmission to transmit power over large distance. It is available in thumbnail sized coupling transformer in a microphone.

IOP Conf. Series: Materials Science and Engineering **590** (2019) 012031 doi:10.1088/1757-899X/590/1/012031

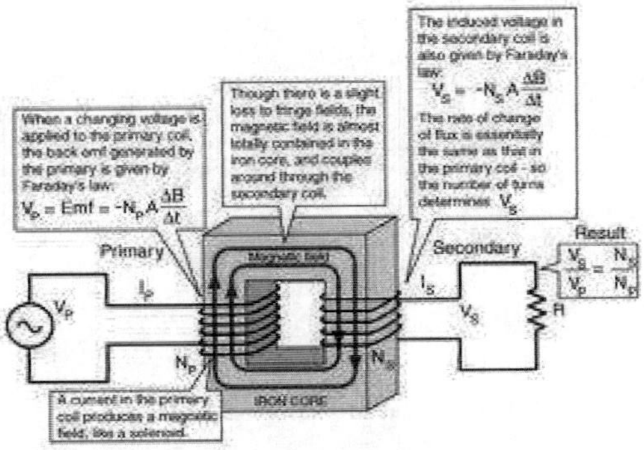

Fig 4: HF Transformer

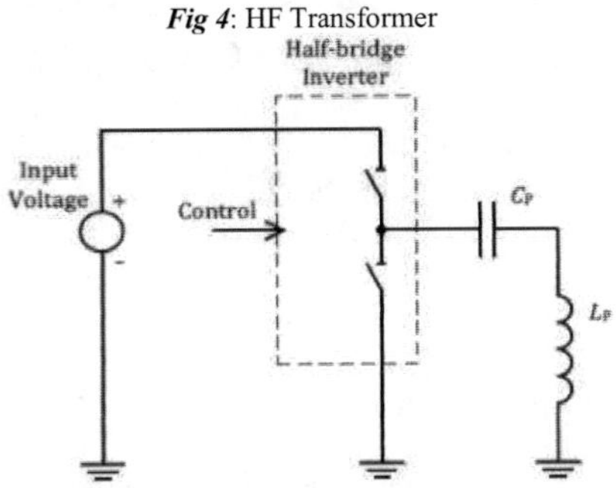

Fig 5: Transmitter circuit

3.3. TRANSMITTER CIRCUIT:

In wireless power transfer transmitter or base station has either 12V or 19V DC that is given to the half bridge rectifier circuit which has two IGBT's to convert DC to AC. The AC effects the emf in primary coil in the secondary coil of magnetic induction.

3.4. RECEIVER CIRCUIT:

In receiver side AC to DC is connected. It convert DC-DC which convert regulated DC from unregulated. Proper regulated is involved in battery charger which act as bridge between wireless power receiver circuit and the battery of the device.

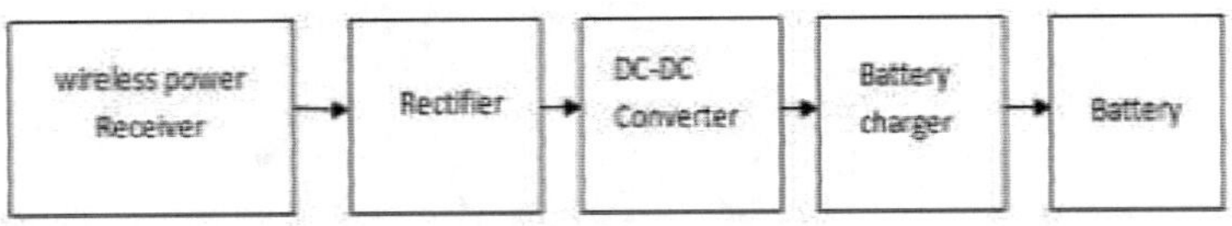

Fig 6: Receiver circuit

3.5. FILTERS:

Filters are used to remove unwanted frequency from the signal and allow wanted frequency. In this filters capacitance filter is very useful method. Because it is used in high voltage and low current power supplies.

3.6. VOLTAGE REGULATOR:

Voltage Regulator is automatically used to maintain a constant voltage level and it is also regulate one or more AC or DC voltages. It can protect from short circuiting the device. By using external components it can adjust both current and voltage. It can produce 1A output current.

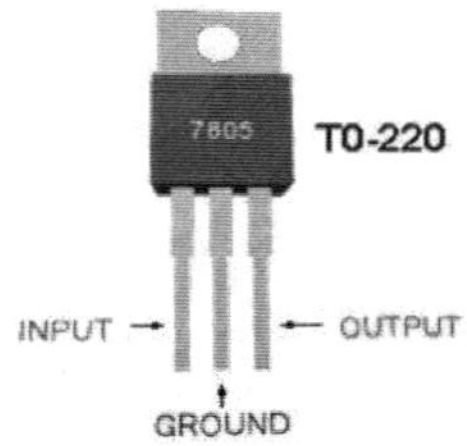

Fig 7: Voltage regulator

4. OPERATION:

Transformer works on the basis of half bridge and double line frequency. Half bridge rectifier converts the given AC input signal in to DC through resistor. During half cycle capacitor gets charged and Q1 starts conducting. Q1 transmitter is biased by the one of the feedback winding coil of the transformer. The current passes through the primary coil of capacitor to the ground. During another half cycle Q2 transistor starts conducting and it is biased by another feedback winding coil of the transformer. The current flows from the capacitor to the primary coil that is connected to the negative of the Q2 transistor. Thus in both half cycle the current direction is opposite and the biasing of transformer feedback winding is done automatically. So we don't know which coil is biasing. In primary coil the current produced during both half cycles generate AC in secondary coil. It produces 25 KHz of frequency because the transistors are fast switching devices. This feds to secondary of transformer through copper winding L1. By using EMF (principle of transformer) L1 transfers 25 KHz to L2. The voltage in L2 is connected to bridge rectifier and it converts into DC. It is filtered by electrolytic capacitor of 1000 micro farads. This DC voltage is regulated by a regulator of IC LM7805 to get 5volt constant. Again it is filtered through a capacitor of 10 micro farads. The output generated by the circuit is used for battery charging.

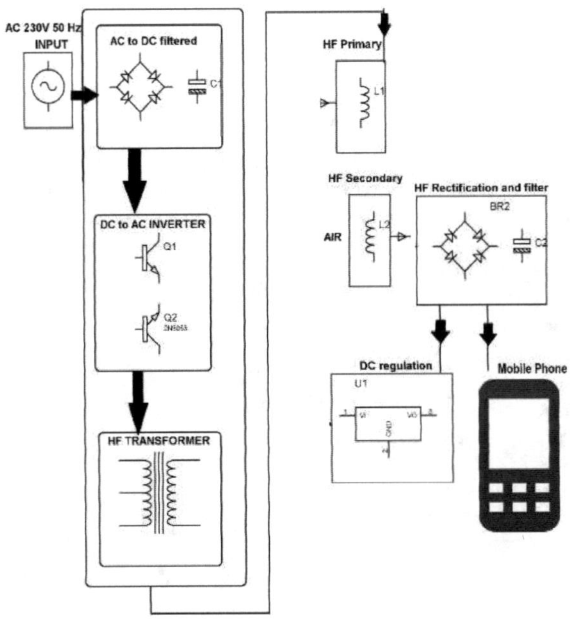

Fig 8: WPT Circuit diagram

5. ADVANTAGES:

1) Wireless Power Transfer gives the convenience of charging.
2) It is effective way to transfer power in any medium convenient.
3) It doesn't require bunch of wires.
4) It is eco-friendly and removes the need to dispose dead batteries.
5) It does not reproduce any sparkle and electric shock.
6) It is rust free, safe and effortless.
7) It has less consumption of electricity.
8) It is easy way to charge the device.

6. CONCLUSION:

As per the survey done on Automatic Wireless Charger by using an Inductive coupling method low power devices can be charged at a short distance (few CM). There are various types of methods to transfer power wirelessly that is Inductive coupling, Magnetic resonance coupling and Microwave with rectenna. Among all these methods Inductive coupling is one of the best methods to transfer power safely and efficiently. Inductive coupling emphasizes that by varying the magnetic flux between two inductive coils to transfer energy from the source to load. Inductive coupling is working under the principle of Electromagnetism, where magnetic flux from one coil interlinks the second coil. To transfer energy to moderate distance, "Resonance" is added in the inductive coupling. Resonance is where primary and secondary is operated at the same resonant frequency to transfer energy to moderate distances. This is achieved by the adding the resonant circuit in the primary and secondary side. The energy transfer is based on the range between primary and secondary coil, operating frequency and the amount of transmitted power.

REFERENCES:

[1]. Mohamed Hassan, Amr El Zawawi, "Wireless Power Transfer (Wireless Lightining), proceeding of IEEE.
[2]. Abishek Gupta, Ankush Tandon, Akhilesh Jeriya, Naveen Tanwar, Ramesh Kumar, Dinesh Kumar,"Automatic Wireless Mobile Charger", Vol 4, No.1, pp:113-119, 2016.
[3]. Omkar Singh, "Wireless Mobile Charger",Vol 5, No. 6 ,2016
[4]. Vikash Choudary, Satendar Pal singh Vikash Kumar and Deepak Prashkar, 2011, "Wireless Power Transmission: An Innovation Idea" Volume 1, Number 3(2011).
[5]. Gum HKawng Tu Raw, Wai Phyo Ei, 2017, "Static Contactless Charging System Using Inductive Coupling Method" Volume 6, Issue 4, April 2017.
[6]. Grant A. Covic, Senior Member IEEE, and John T.Boys, 2013, "Inductive Power Transfer".
[7]. Maged Abdou, Ayman El-Tager, 2016, "Wireless Power Transmission Enhancement using Magnetic Resonance Coupling", International Conference on Signal Processing, Communication, Power and Embedded System(SCOPUS)-2016
[8]. Julian Menezes .R, Albert Mayan .J, M. Breezely George," Development of a Functionality Testing Tool for Windows Phones", Indian Journal of Science and Technology,Vol:8, No:22,pp: 1-7,2015, ISSN : 0974-6846
[9]. T.S.Chandrasekhar Rao and K.Geetha,Categories, Standards and Recent Trends in Wireless Power Transfer: A survey", Indian Journal of Science and Technology, Vol 9, 2016.
[10]. R.Ramya and T.Ravi, "Design of cache memory mapping techniques for low power processor" ARPN Journal of Engineering and Applied Sciences, VOL. 10, NO. 11, JUNE 2015, pp:4783-4788.
[11]. Brij Bhushan Ojha, S.Karthikeyan ,"Design of an interconnection network using VLSI photonics: Issues and challenges " International Journal of Pure and Applied Mathematics, Volume 118 No. 17 2018, PP: 211-224.
[12]. S. Vimalsree and S. Karthikeyan, "A Low Power Multibit Flip Flop Merging Technique Using WSN Nodes", ARPN, vol. no:11, No.15, August 2016, Page no: 9358-9363.
[13]. Gopika. S. Kumar, S. Karthikeyan and N. Ambily, "A Metaphorical Scrutiny in Banishing Salt and Pepper Noise ", Biosciences Biotechnology Research Asia,December 2014 Vol. 11(3)

An Improved Forgery Detection Method for Images

Tiss Tom[1], D.Usha Nandini[2], S.Princemary[3], B.Ankayarkanni[4]

School of Computing, Sathyabama Institute of Science and Technology, Chennai, India
tisskayathinkara@gmail.com, ushaduraisamy@yahoo.co.in, princemary26@gmail.com,
ankayarkanni@gmail.com

Abstract: A forgery image detection system is introduced to detect the forgery in images using global, local and pixel based features. Global features will consider the whole images which include luminance and chrominance features. These global features can be extracted by using Zernike moments. The local feature consists of descriptors of multiple interest points. Image authentication is done by using hash method. The set of images will be trained in the database earlier itself. Then the test image and reference image will be compared based on pixels. By analyzing the hash distance the system can identify the test image is forged or not. As an improvement, pixel based feature extraction is used. Pixel based feature extraction is carried out by using supervised classification algorithm.

Keywords: forgery detection, global, local and pixel features, supervised classification algorithm

1. Introduction

We are indubitably having life in an age where we are exposed to a eminent array of visual imagery. In earlier days images may have trustworthy. But now a days the powerful digital image editing software undermine our trust inimages. Even though the image editing software becomes powerful the more becomes the forgery detection systems. The image editing software will carry out the manipulations such as deleting some objects from the images, inserting some objects into the images or replacing some objects. In this system introducing global,local and pixel based features.

The system [1] proposes mapping of an image function onto a set of orthogonal basis functions over the unit circle has been developed on the basis of Zernike moments. To make the image reconstruction from its moment simple the orthogonal property of Zernike moments used. The image representation ability of each order moment and in addition its contribution to the reconstruction process can be enabled by using Zernike moments. The highest order moment is used to yields a reconstructed image which will be close to the original one. The features are tested with series of experiments on two different data sets. And also Zernike features can be performed well in moderate level of noise.

In this system [2], proposed two contrast enhancement based forensic algorithms through histogram peak/gap artifacts analysis. First, to detect the global contrast enhancement for both uncompressed and previously JPEG-compressed images extends it. To identify features, exploit zero-height gap bins in gray level histograms. This contrast enhancement detector accomplished high performance, i.e., Pd = 100% at Pfa = 1%. Second, the system proposed a novel method to expose source-enhanced composite image, which vitiated the prior detection methods. The composition boundary was precisely located by detecting the disproportionateness between detected block wise peak/gap positional distributions.

The system[3]proposed a method for embedding and detecting chaotic watermarks for large images. Adaptive clustering method is used in order to obtain robust region representation of original image. Ellipsoids are used to approximate robust regions. To avoid annoying artifacts from the embedded watermark a visual masking technique is used.

The images and videos on the social networks which would be edited and misused will affect trustworthiness of online information. To estimate the parameters of geometric transforms and to detect local tampering Radon transform and scale theory is used in [4] this system. The alignment component of forensic hash can provide correct calculation of parameters of geometric transforms.

In a wireless multimedia sensor network (WMSN)for secure image transmission and storageImage authentication is critical. A perceptual hashing-based robust image authentication scheme issued in the system [5]which uses the distributed processing strategy for perceptual image hashes and can provide visual fragility, compactness, security in digital image authentication and perceptual robustnessfor WMSN. A secure pseudorandom chaotic sequence with keys will be generated by cluster head node and dispatch it to the image capturing node. By using this received sequence, several overlapping rectangles will be derived and then calculate the gravity centers and binary distance of two gravity centersin each of the general cluster member node.

The system [6] analyze the log spectrum of input image then extract spectral residual of the image in spectral domain and constructing saliency map in spatial domain. But it can apply only to static images.

In the system [7] hash method is based on Fourier transform features and controlled randomization. It specifies the security issues of existing image hashing scheme. Multimedia hashes will use in content authentication and content based retrieval from databases.

IOP Conf. Series: Materials Science and Engineering **590** (2019) 012032 doi:10.1088/1757-899X/590/1/012032

2. Existing System

The global features which include luminance and chrominance extract using Zernike moments representing the characteristics of the images as a whole. Position and texture information of salient regions in the image includes the local features. Image authentication carried out using hash method. Disadvantage of the existing system is that it analyzes only local and global features and also the secret key is complicated.

3. Proposed System

Initially the global and local features of the given input image are extracted. The extracted global features are group to form the hash. The hash is constructed by using Zernike moments. Zernike moment features from order 1 to order 5. Magnitudes of the Zernike moments are rounded and used to form a global vector. Thus the global features are extracted from the image. The local features are extracted by using k-element vector values.
The extracted images are compared with the trained image which is already stored in the database. During the comparison, the image needs not to have the identical pixel values. However, either both image or one of the images may have small modification like contrast, brightness, etc. In addition, to find the similarity our system calculates the distance between the hashes. To compute the distance both the global features vector values and local features vector values are combined by using Euclidean distance calculation.

If the tested image in fake or mismatched with the trained image then it said to be forged image. Four types of forgery can be identified. They are removal, insertion and replacement of objects, and unusual color changes. The performance of proposed system is very high in detecting the forged image. This is because of using hash methods.
Finally the pixel features are extracted from the image. Supervised classification algorithms are applied in pixel-based classification. This technique is used find the exact location of forgery and also the type of forgery. It also includes the features of local and global from the image.

The performance of the proposed system will be high. This is due to the pixel based feature extraction which is more accurate to identify the location and type of forgery.

It efficiently identifies the forgery image which is inserted or removed or replaced and can be used in the real-time image detection applications.

IOP Conf. Series: Materials Science and Engineering **590** (2019) 012032 doi:10.1088/1757-899X/590/1/012032

3.1Implemenation

Trained Dataset

The number of images can be trained and stored in the database. The images may be taken from the internet or can be captured using cameras.

Preprocessing

Rescaled the image to a fixed size and converted from RGB values to YCbCr representation.

Feature Extraction

Global Feature Extraction

The Zernike moment of order 'n' and repetition 'm' of an image f(x,y) can explain as follows:

$$A_{nm}=\frac{n+1}{\pi}\sum_x \sum_y f(x,y)\, V_{nm}(x,y) \qquad (1)$$

Where$x^2 \mid y^2 <= 1$

The image $V(n,m,x,y)$ is the Zernike basis images of order 'n' and repetition 'm'. These basis images are complex and orthogonal. The Zernike moments are essentially the projections of the input image onto these basis images.

Zernike moments has been employed for reconstruction of the original image. The N-th order approximation is given by

$$f(x,y) = \sum_{n=0}^{N}\sum_{m} A(n,m)V(n,m,x,y) \qquad (2)$$

The information content of the nth order moments is

$$I(x,y,n) = \sum_{m} A(n,m)V(n,m,x,y) \qquad (3)$$

The moments are returned in the order of (0,0)(1,1)(2,0)(2,2)(3,1)(3,3) and so on, where m<=n and n =0

Local Feature Extraction

Local features will be extracted using an image descriptor called surf that combines interest points with visual words.

Pixel Based Feature Extraction

For extracting pixel based features the system uses supervised classification algorithm.

In supervised classification algorithm, it can differentiate one class of other, usually by providing samples of pixels already knows that should be assigned to a specific class. To classify the other pixels on the image, the algorithm use the information already provided. The task is to identify, for each pixel in one image, which class should be assigned to that pixel.

IOP Conf. Series: Materials Science and Engineering **590** (2019) 012032 doi:10.1088/1757-899X/590/1/012032

Pseudo code:

Input : Image A and Image B

```
Steps 1: Select File Reff_Image=new File("Image A");
Steps 2: Select File Test_Image=new File("Image B");
 Steps 3: File gh[]=Reff_Image.listFiles();
 Steps 4: File f9 = Test_Image.listFiles();
 Steps 5: for(inti=0;i<f9.length();i++)
            {
            int width=cimge.getWidth();
            int height=cimge.getHeight();
            for(int w=0;w<width;w++)
              {
            for(int h=0;h<height;h++)
                 {
      if(inimge.getRGB(pixelshash(w,h), h)!=pixelshash(cimge.getRGB(w,h), h))
            {
            Flag=false;
                 endif;
            }
       }
      }
 Steps 6: Stop
```

Output: Comparison Result – Image A and Image B (Matched or Fraud will be detected)

Hash Value Generation

For the above features hash values are generated using randomkey.

Forgery Identification

The calculated pixel values of reference image and tested are compared. If there is no difference with pixels then the reference image and tested image are same. If the reference hash key value is less than test hash key value then the forged type is inserted. Otherwise forged type is removed.

4. Conclusion & Future Enhancement

The proposed system is implemented to find the forged image which is inserted or removed or replaced with new objects. This is achieved by using pixel based features extraction which includes the supervised classification algorithm. The pixel based features are extracted and then compared with reference image which is already trained in the database. By decomposing the hash values used to find the type of forgery and location of forgery. As an enhancement the pixel based feature extraction is used along with local and global feature extraction method in order to identify the forgery image in an effective way. In Future, the study is desired to find features that better represent the image contents so as to enhance the hash's sensitivity to small area tampering while maintaining short hash length and good robustness against normal image processing.

References

[1] AlirezaKhotanzad and Yawhua Hong," Invariant Image Recognition by Zernike Moments", IEEE Transactions On Pattern Analysis And Machine Intelligence.Vol. 12. No. 5. May 1990

[2] RemyaS,"Digital Image Forgery Detection by contrast Enhancement",IOSR Journal of Computer Engineering(IOSR-JCE) e-ISSN:2278-8727, volume 16, issue 5, Ver.IX(Sep-Oct. 2014), PP 01-07

[3] AthanasoisNikolaidis and IoannisPitas,"Region Based Image Watermarking", IEEE Transactions on Image Processing, vol.10, No.11, November 2001

[4] Wenjun Lu, Avinash L Varna and MinWu,"Forensic Hash for Multimedia Information", IEEE Transactionsof Information Forensics and Security, vol. 1, No.
2, pp. 93-116, 2009

[5] HongxiaWang and Bangxu Yin, "Perceptual Hashing-Based Robust Image Authentication Scheme for Wireless Multimedia Sensor Networks", International Journal of Distributed Sensor Networks, Volume 2013, Article ID 791814, 9 pages.

[6] XiaodiHou and Liqing Zhang," Saliency Detection: A Spectral Residual Approach", IEEE Transactions on Pattern Analysis and Machine Intelligence,20(11):1255–1259, 1998. 1, 5

[7] A. Swaminathan, Y. Mao, and M. Wu, "Robust and secure imagehashing," IEEE Transactions of InformationForensics and Security, vol. 1, no. 2, pp.216–230, Jun. 2006.

[8] Ankayarkanni, B., and A. Ezil Sam Leni. "A new statistical rule model for image retrieval system of remote sensing images." In Circuit, Power and Computing Technologies (ICCPCT), 2016 International Conference on, pp. 1-5. IEEE, 2016.

[9] D.Usha Nandini, Dr.Ezil Sam Leni, "Shadow Identification using Ant Colony Optimization", Journal of Theoretical and Applied Information Technology, vol. 78,no.2, pp.195-200, 2015.

[10] S.Prince Mary, Dr.E.Baburaj, "Performance Enhancement in Session Identification", Conducted by ICCICCT, Noorul Islam University, on July 10 &11,2014.Published in IEEE Xplore,pgs: 837 – 840, ISBN: 978-1-4799-4191-9.

SECURE AND TRUSTABLE ROUTING IN WIRELESS SENSOR NETWORKS USING ACTIVE DETECTION ROUTING PROTOCOL

S. Karthikeyan [1],
Associate professor, Dept. of ECE
Sathyabama Institute of Science and Technology,Chennai
skarthi1879@gmail.com

Katikareddy Venkata Suhas Reddy[2],
Associate professor, Dept. of ECE
Sathyabama Institute of Science and Technology,Chennai

Katasani Venkata Bypu Reddy [3],
Associate professor, Dept. of ECE
Sathyabama Institute of Science and Technology,Chennai

T. Ravi [4]
Associate professor, Dept. of ECE
Sathyabama Institute of Science and Technology,Chennai

Abstract: A Wireless Sensor Network (WSN) contains gadgets that are self-sufficiently conveyed in space used to direct the physical and natural conditions with the assistance of sensors. They are primarily utilized as a part of security basic applications, for example, checking nature, following the environment and controlling regions. As a result of the asset impediment and computational limitations the WSNs are straightforward to different security assaults. Dark opening assault is a genuine assault which seriously influences the bundle exchange from source to goal. Here the Active Trust conspire is utilized to stay away from the Black opening assault. The Active Trust conspire keeps away from the dark openings through Active discovery directing and information steering which enhances the course security. Both complete hypothetical investigation and exploratory outcomes demonstrate that the execution of the Active Trust conspire is superior to anything that of past examinations. Dynamic Trust can fundamentally enhance the information course achievement likelihood and capacity against dark gap assaults and can enhance arrange lifetime.

Keywords: — Security, BHL, NL- Network lifetime, Trust, Wireless sensor networks(WSN).

IOP Conf. Series: Materials Science and Engineering **590** (2019) 012033 doi:10.1088/1757-899X/590/1/012033

1. INTRODUCTION

Wireless Sensor Networks are rising as a promising innovation as a result of their extensive variety of utilizations in modern, ecological checking, military and regular citizen areas. A remote sensor arrange is a self-sorting out system comprising of sensor hubs which can differ from hundreds to thousands in numbers. Every sensor hub has confined preparing, stockpiling limit, computational control. In remote sensor arranges, each sensor hub speaks with other condition to think about its neighborhood condition and the information will send to any hub if any solicitations are coming. In remote sensor systems, it isn't much secure when vast zone of system contrasted with little zone of system [3]. Due to the innate qualities like memory constraints, open condition, control confinements and unattended nature, the security of a remote sensor arrange is traded off [11]. These feeble attributes makes the system effortlessly traded off by a foe to make assaults bringing about deplorable results. Dark Hole assault is one of the perilous assault which abuses a Dependability of a system by promising steering of information bundles to the goal realizing that it has a Most brief way yet as a general rule it drops all parcels and in addition specifically drops the parcels, and therefore undermines unwavering quality. As remote sensor systems are causing numerous security dangers, in this paper we are keeping away from the dark opening hub by proposing a procedure of information directing with no come up short. The dark gap assault gives the circumstance where an aggressor attempting to trade off some of hubs to track the data and hinder with the ordinary working of the WSN by consistently evolving, irritating, or breaking the usefulness of the hubs in the framework. This assaults will brings about producing the dark openings: zones inside which the enemy can either inactively catch or effectively piece data conveyance. The assailants can give numerous dark gaps because of the unattended idea of WSNs. By this assault enemy can upset typical information conveyance between sensor hubs and the sink, or even parcel the topology.

Be that as it may, the present trust-based course methodologies confront some trying issues [24]. (1) The center of a trust course lies in getting trust. In any case, getting the trust of a hub is exceptionally troublesome, and how it should be possible is as yet vague. (2) Energy productivity. Since vitality is extremely restricted in WSNs, in most research, the trust procurement and dispersion have high vitality utilization, which truly influences the system lifetime. (3) Security. Since it is hard to find noxious hubs, the security course is as yet a testing issue. In this way, there are still issues deserving of further investigation. Security and trust directing through a dynamic recognition course convention is proposed in this paper. The principle developments are as per the following.

(1) The ActiveTrust conspire is the principal directing plan that utilizations dynamic identification steering to address BLA. The most noteworthy distinction amongst ActiveTrust and past research is that we make numerous identification courses in locales with buildup vitality; in light of the fact that the aggressor doesn't know about location courses, it will assault these courses and, in this manner, be uncovered. Thusly, the assailant's conduct and area, and nodal trust, can be gotten and used to maintain a strategic distance from dark openings when preparing genuine information courses. To the best of our insight, this is the primary proposed dynamic discovery system in WSNs.

(2) The ActiveTrust course convention has better vitality proficiency. Vitality is valuable in WSNs, and there will be more vitality utilization if dynamic discovery is handled. Thusly, in past research, it was difficult to envision embracing such high-vitality utilization dynamic discovery courses. Be that as it may, we think that its conceivable after deliberately examining the vitality utilization in WSNs. Research has noticed that there is still up to 90% buildup vitality in WSNs when the system has kicked the bucket due to the "vitality opening" marvel. Subsequently, the ActiveTrust plot takes full preferred standpoint of the buildup vitality to make location courses. Those recognition courses can recognize the nodal trust without diminishing lifetime and along these lines enhance the

IOP Conf. Series: Materials Science and Engineering **590** (2019) 012033 doi:10.1088/1757-899X/590/1/012033

system security. As indicated by hypothetical investigation and test comes about, the vitality proficiency of the ActiveTrust conspire is enhanced in excess of 2 times contrasted with past directing plans, including most brief steering, multi-way directing.

(3) The ActiveTrust plot has better security execution. Contrasted and past research, nodal trust can be gotten in ActiveTrust. The course is made by the accompanying standard. To begin with, pick hubs with high trust to maintain a strategic distance from potential assault, and after that course along a fruitful discovery course. Through the above approach, the system security can be moved forward.

(4) Through our broad hypothetical examination and reproduction consider, the ActiveTrust directing plan proposed in this paper can enhance the achievement steering likelihood by 1.5 times to 6 times and the vitality productivity by in excess of 2 times contrasted and that of past.

2. RELATED WORKS

Anbuchelian. S et al proposed [1] a vitality sparing grouping calculation for identifying the assaults on bunch heads and therefore prompts the better vitality utilization in the remote sensor systems. Vipul Sharma et al [9] proposed a strategy for the discovery and concealment of dark opening assault in Leach based sensor systems. The go for this examination work is to propel an instrument that can distinguish and defeat the impact of dark opening assault in sensor arrange. The bad mark in this paper was it won't identify the sensor hubs as a dark gap hub.

Barleen Shinh proposed [2] a system to recognize and segregate the dark gap assault. The proposed instrument will recognize the vindictive hub and stop, it from the system. The system depends on the throughput of the system. At the point when the throughput of the system, will abatements to certain limit esteem, hubs in the system will go to head hub and identify the vindictive hub. Single-way steering is a simple directing convention however is immediately caught by the assailant. Regardless of whether there is an assault in one course, the bundle can safely achieve the goal.

Tao Shu propose [7] a randomized multi-way directing calculation to distinguish dark opening assault. Under the outline, the courses taken by the "offers" of various bundles change after some time. So the assailant can't track the courses passed by every parcel if the steering calculation ends up known to the aggressor. Wenjing Lou propose and explore a novel plan, [4] Security Protocol for REliable dAta Delivery (SPREAD), to improve the information privacy benefit in a system. The proposed SPREAD plan expects to give encourage security to mystery messages from being traded off when they are conveyed over the uncertain system D Loganathan likewise propose [5] a crossover multipath plot (H-SPREAD) to enhance both security and unwavering quality of this errand in a conceivably antagonistic and inconsistent remote sensor arrange. The new plan depends on a N-to-1 multipath convention which finds various ways from each hub to in one course revelation process. There are diverse multipath course development techniques.

H.- M. Sun [8] proposes a multi dataflow topologies (MDT) way to deal with oppose the specific sending assault. In the event that one topology neglected to send the information, the goal will get the information through other topology.

Yuxin Liu, [10] proposed a dynamic put stock in conspire for secure and trustable directing in remote sensor systems. The paper proposes a dynamic identification directing of information for better security and trust. The primary objective of the plan is to guarantee that the nodal information securely achieve the sink and are not hindered by the dark gap. The identification course discovers high put stock in hub and in information directing, it will chooses the course without dark gap hub and in this way enhance the achievement proportion of information achieving the sink. Indeed, even there is numerous examination on dark hub assault and evasion, there is still parts for additionally ponder. Single-way directing is a basic steering convention [12] however is effectively hindered by the aggressor. In this way, the most characteristic approach is by means of multi-way directing to the sink. Regardless of whether there is an assault in some course, the information can even now securely achieve the sink [9]. Multi-way directing conventions can be grouped into two classes relying upon whether the information bundle is isolated. One is multi-way steering without share division. The other is multi-way steering with share division, i.e., the parcel is separated into shares, and diverse offers achieve the goal by means of various courses [9].

(1) Non-share-based multi-way steering. There are diverse multi-way course development strategies. Ref. [25] proposes a multi dataflow topologies (MDT) way to deal with oppose the specific sending assault. In the MDT approach, the system is isolated into two dataflow topologies. Regardless of whether one topology has a pernicious hub, the sink can in any case acquire bundles from the other topology. In such conventions, the insufficiency is that if the parcel is steered through n courses all the while, the vitality utilization will be n times that of a solitary way course, which will truly influence the system lifetime; comparable research can be seen in multi-way DSR [25], the AOMDV [18] and AODMV [26].

(2) Share-based multi-way directing conventions. The SPREAD calculation in [27] is a commonplace offer based multi-way directing convention. The essential thought of the SPREAD calculation is to change a mystery message into different offers, which is known as a (T, M) edge mystery sharing plan [28]. The M shares are conveyed by different free ways to the sink with the end goal that, regardless of whether few offers are dropped, the mystery message overall can in any case be recuperated [9, 16, 28]. The benefit of this calculation is that through multi-way directing, every way courses just a single offer, and the aggressor must catch in any event T offers to reestablish nodal data, which builds the assault trouble [9]. Subsequently, the protection and security can be progressed. In the above research, the multi-way steering calculations are deterministic with the end goal that the arrangement of course ways is predefined under a similar system topology [9]. This shortcoming opens the entryway for different assaults if the directing calculation is acquired by the foe [9]. For the shortcoming said above, Ref. [29] proposed four irregular engendering methodologies: arbitrary spread (PRP), coordinated irregular proliferation (DRP), non-dreary arbitrary spread (NRRP), and multicast tree helped arbitrary spread (MTRP). The general procedure is as per the following. To begin with, partition the message into M shares, and the course way of each offer isn't foreordained. Along these lines, regardless of whether the enemy procures the directing calculation, it is hard to dispatch a pinpointed hub trade off or sticking assault. Since it is hard to catch more than T shares, the security is additionally made strides. In multi-to-one information gathering WSNs, we contend that for exemplary "cutting and collecting" or multi-way directing strategies, cut offers will converge in a similar way with high likelihood, and this way can be effectively assaulted by dark openings. Subsequently, in [16], a Security-and Energy-productive Disjoint Route (SEDR) plot is proposed to course cut offers to the sink with randomized disjoint multipath courses by using the accessible surplus vitality of sensor hubs. The creators exhibit that the security is expanded without decreasing the lifetime in the SEDR convention. Another strategy to keep away from assault and enhance course achievement likelihood is trust steering. Trust administration [20] is turning into another main impetus for explaining challenges in specially appointed systems [21], distributed systems [22], and WSNs [23, 24].

Zhan et al proposed a trust-mindful directing structure convention (TARF), utilizing trust and vitality cost for course choices, to keep vindictive hubs from deceiving system movement [30]. Ref. [31] proposes the Sec-CBSN calculation, which creates distinctive trust computation strategies in light of nodal parts. Ref. [32] builds up an assault safe and lightweight trust administration convention named ReTrust, which can oppose assaults through a trust administration approach for medicinal sensor systems (MSNs). Ref. [33] presents a proposition named TRIP, which means to rapidly and precisely distinguish malevolent or childish hubs spreading false data in vehicular impromptu systems (VANETs). Ref. [34] additionally proposes a flexible confide in display, SensorTrust, for various leveled WSNs. Ref. [24] presents the idea of property closeness in finding possibly agreeable hubs among outsiders.

3. EXISTING SYSTEM

Blockhole assaults is a standout amongst the most run of the mill assaults and fills in as takes after. The enemy bargains a hub and drops all parcels that are directed by means of this hub, bringing about touchy information being disposed of or unfit to be sent to the sink. Since the system settles on choices relying upon the hubs' detected information, the outcome is that the system will totally come up short and, all the more truly, settle on erroneous choices. In this manner, how to distinguish and maintain a strategic distance from BLA is of awesome criticalness for security in WSNs. There is much research on dark opening assaults. In any case, the present trust-based course systems confront some trying issues. (1) The center of a trust course lies in getting trust. Be that as it may, getting the trust of a hub is exceptionally troublesome, and how it should be possible is as yet vague. (2) Energy proficiency. Since vitality is exceptionally restricted in WSNs, in most research, the trust obtaining and dissemination have high vitality utilization, which genuinely influences the system lifetime. (3) Security. Since it is hard to find malignant hubs, the security course is as yet a testing issue.

3.1 DISADVANTAGES OF EXISTING SYSTEM

The arrange settles on choices relying upon the hubs' detected information, the result is that the system will totally come up short and, all the more truly, settle on mistaken choices

4. PROPOSED SYSTEM

In this paper we propose security and trust directing through a dynamic recognition course convention. The most gigantic refinement among ActiveTrust and past research is that we make various area courses in zones with store essentialness; in light of the way that the attacker doesn't think about recognizable proof courses, it will ambush these courses and, in this way, be revealed. Thusly, the attacker's direct and territory, and furthermore nodal trust, can be procured and used to avoid dull openings while getting ready honest to goodness data courses. To the best of our understanding, this is the foremost proposed dynamic acknowledgment segment in WSNs.

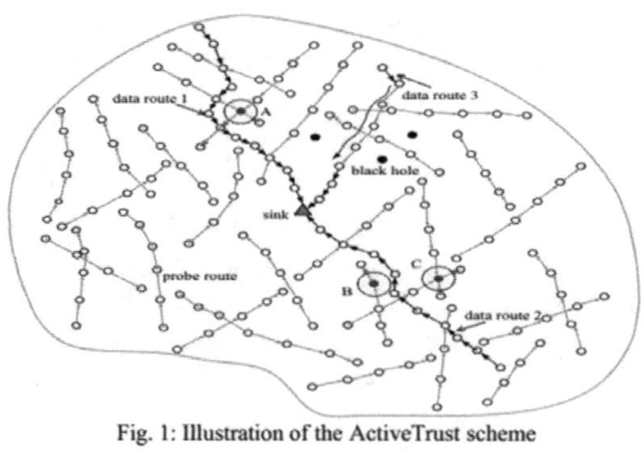

Fig. 1: Illustration of the ActiveTrust scheme

4.1 ADVANTAGES OF PROPOSED SYSTEM:

- The ActiveTrust conspire is the main directing plan that utilizations dynamic identification steering to address Blockhole assaults (BLA)

- The ActiveTrust course convention has better vitality effectiveness
- The ActiveTrust plot has better security execution

4.2 MODULES DESCRIPTION:

4.21. DETECTION ROUTE

An identification course alludes to a course without information parcels whose objective is to persuade the enemy to dispatch an assault so the framework can distinguish the assault conduct and after that stamp the dark opening area.

4.2.2 DATA ROUTING

The information directing alludes to the procedure of nodal information steering to the sink. The directing convention is like regular steering conventions in WSNs; the distinction is that the course will choose a hub with high trust for the following jump to keep away from dark openings and in this manner enhance the achievement proportion of achieving the sink.

5. WORKING PRINCIPLE

In this paper, we have proposed a novel security and trust directing plan in light of dynamic discovery, and it has the accompanying superb properties High fruitful steering,

□ The Active Trust plan can rapidly identify the nodal trust and afterward maintain a strategic distance from suspicious hubs to rapidly accomplish an almost 100% fruitful directing likelihood High vitality proficiency.

□ The Active Trust plot completely utilizes buildup vitality to develop different discovery courses.

□ The hypothetical examination and exploratory outcomes have demonstrated that our plan enhances the effective steering likelihood by in excess of 3 times, up to 10 times at times.

Be that as it may, the present trust-based course methodologies look in getting trust. Be that as it may, acquiring the trust of a hub is exceptionally troublesome, and how it should be possible is as yet hazy. Vitality proficiency. Since vitality is extremely restricted in WSNs, in most research, the trust securing and dispersion have high vitality utilization, which truly influences the system lifetime. (3) Security. Since it is hard to find noxious hubs, the security course is as yet a testing issue. Accordingly, there are still issues deserving of further investigation. Security and trust steering through a dynamic discovery course convention is proposed in this paper.

The essential advancements are according to the accompanying. The Active Trust plot is showed up in the principle coordinating arrangement that uses dynamic acknowledgment directing to address BLA. The most immense differentiation between Active Trust and past research is that we make distinctive disclosure courses in districts with development imperativeness; in light of the fact that the attacker doesn't think about acknowledgment courses, it will ambush these courses and, in this way, be revealed. Thusly, the assailant's direct and territory, and furthermore nodal trust, can be gained and used to avoid dim openings while getting ready certified data courses. To the best of our understanding, this is the central proposed dynamic acknowledgment instrument in WSNs.

The Active Trust course convention has better vitality proficiency. Vitality is valuable in WSNs, and there will be more vitality utilization if dynamic identification is prepared. Accordingly, in past research, it was difficult to envision embracing such high-vitality utilization dynamic location courses. Nonetheless, we think that its conceivable after precisely examining the vitality utilization in WSNs. Research has noticed that there is still up to 90% buildup vitality in WSNs when the system has passed on due to the "vitality gap" wonder. Thusly, the Active Trust plot takes full favorable position of the deposit vitality to make identification courses and endeavors to

IOP Conf. Series: Materials Science and Engineering **590** (2019) 012033 doi:10.1088/1757-899X/590/1/012033

diminish vitality utilization in hotspots (to enhance arrange lifetime). Those location courses can identify the nodal trust without diminishing lifetime and consequently enhance the system security. As indicated by hypothetical examination and test comes about, the vitality effectiveness of the Active Trust conspire is enhanced in excess of 2 times contrasted with past directing plans, including briefest steering, multi-way directing. The Active Trust conspire has better security execution. Contrasted and past research, nodal trust can be acquired in Active Trust. The course is made by the accompanying standard.

Table 1. Performance of algorihm

Algorithm	Malicious node is in the middle of the route	Destination node is a malicious node	Malicious node is identified but with major loss of packets	Malicious node is identified and change of route occurs(proposed system)
Packets sent	155	155	155	155
Packets Received	57	0	95	134
Packet Loss	98	155	57	21
PDF	36.77	0.00	63.22	86.45
End to End Delay(s)	1.290	2.025	0.861	0.637

To start with, pick hubs with high trust to stay away from potential assault, and afterward course along an effective identification course. Through the above approach, the system security can be made strides. Through our broad hypothetical investigation and recreation examine, the Active Trust steering plan proposed in this paper can enhance the achievement directing likelihood by 1.5 times to 6 times and the vitality productivity by in excess of 2 times contrasted and that of past looks into.

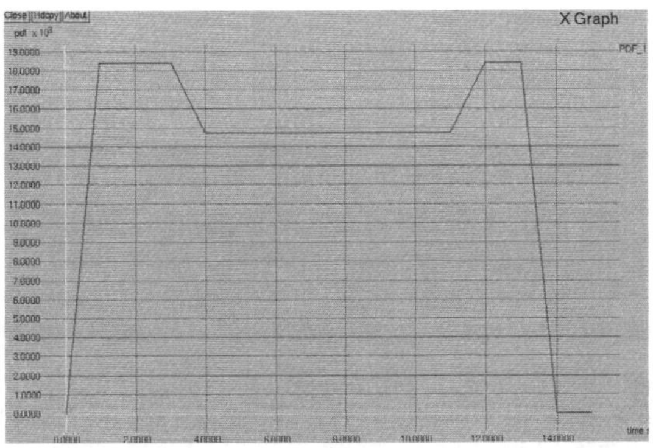

Figure 2. Malicious node

The above graph refers to the first row of the table, in this case, the malicious node is in between the source and destination nodes. The malicious node drops packets very frequently and sometimes send to its neighbour node after the storage of the node capacity.

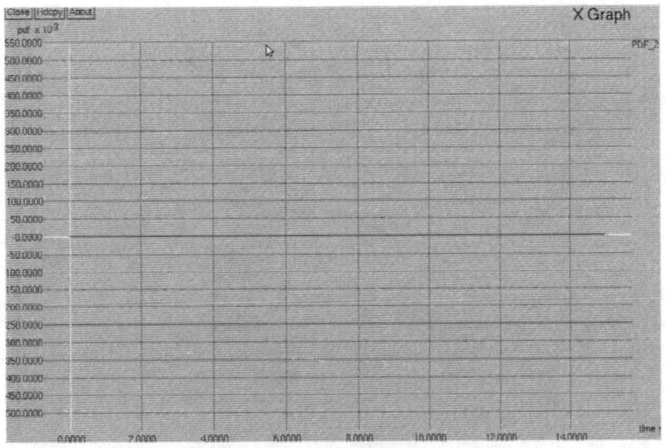

Figure 3. Malicious Send

The above graph refers to the second row of the table, in this case, the malicious sends the request to the source node and source node sends all the data packets to the fake destination.

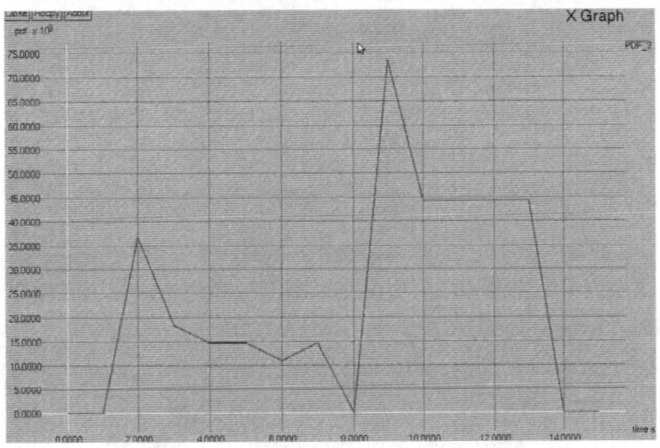

Figure 4. Malicious source node

The above graph refers to the third row of the table, in this case, the malicious node is identified by the source node and takes another path to the destination node, but there will be so much of delay in the route selection, so some of the packets will be dropped.

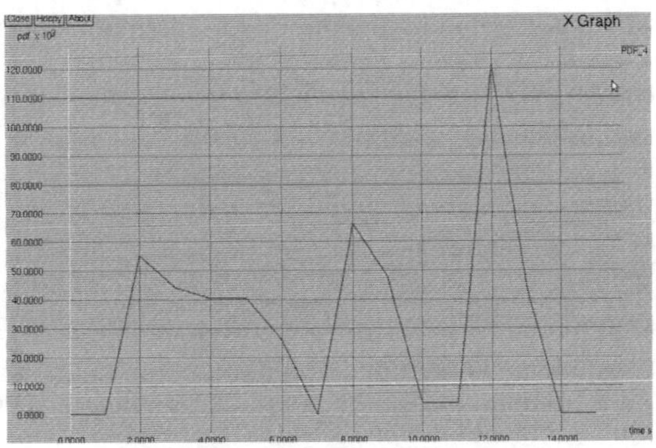

Figure 5. Malicious source identify

The above graph refers to the fourth row of the table, In this case, the malicious node is identified by the source node and takes another path to the destination node, it also checks the trust on the neighbour node to which it has to send the data packets.

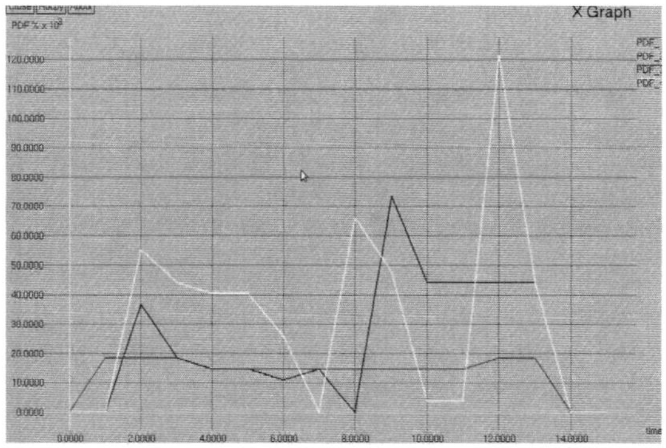

Figure 6. Comparison of nodes

The above graph gives the comparision between the existing systems and our proposed system. The yellow color represents the proposed system, it shows that the packet delivery fraction is high and the end to end delay is very less i.e.,0.637sec(approx.).

6. CONCLUSION

In this paper, we have proposed a novel security and trust directing plan in view of dynamic identification, andit has the accompanying superb properties: High fruitful steering likelihood, security and adaptability. The Active Trust plan can rapidly distinguish the nodal trust and afterward maintain a strategic distance from suspicious hubs to rapidly accomplish an almost 100% fruitful directing likelihood. High vitality proficiency. The Active Trust conspire completely utilizes deposit vitality to develop different identification courses. The hypothetical examination and trial comes about have demonstrated that our plan enhances the fruitful steering likelihood by in excess of 3 times, up to 10 times sometimes. Further, our plan enhances both the vitality productivity and the system security execution. It has imperative importance for remote sensor arrange security.

REFERENCES

[1] Anbuchelian, Selvamani. K, Chandarasekar. A "An EnergyEfficient Multipath Routing Scheme by Preventing Threatsin Wireless Sensor Networks", Electrical and ComputerEngineering (CCECE), IEEE 27th Canadian Conference,2014.

[2] Kumar, M.K., Surya, J., Kumar, V.G.S., Karthikeyan, S."Online prediction of driver distraction based on brain waves activity patterns", International Journal of Pure and Applied Mathematics, Volume 117 No. 21 2017, PP: 139-146

[3] Larisha Rayen, B., Monica, B., Karthikeyan, S., Sivakumar, V.G.," Multi-band circularly polarized antenna for 2.4/5.3/5.8 ghz WLAN and 3.5 ghz WI-MAX applications", Journal of Advanced Research in Dynamical and Control Systems, 2017,vol. No:16, PP: 1311-1322

[3] Liu A, M. Dong, K. Ota, and J. Long, "PHACK: An efficientscheme for selective forwarding attack detection in WSNs,"Sensors, vol. 15, no. 12, pp. 30942–30963, 2015.

[4] Lou W, W. Liu, Y. Fang, "SPREAD: Enhancing dataconfidentiality in mobile ad hoc networks, IEEE INFOCOM 2004, HongKong, China,March 2004

[5] Savitha.G and S.karthikeyan, "Effective cellular transmission in software defined radio",ARPN journal of Engineering and Applied Science, Vol.11,no.22 , November 2016, PP: 12675-12680.

[6] Merin Mary Koshy and S.karthikeyan, "A survey on advanced technology for communication between deaf/dumb people using eye blink sensor & flex sensor", 2017 International Conference on Innovations in Information, Embedded and Communication Systems (ICIIECS),IEEE xplorer conference PP:

[7] Balaji and S.Karthikeyan, "A survey on moving object tracking system using Image Processing", 11th international conference on Intelligent systems and control, IEEE explorer conference, vol. , Jan 2017, PP: 11-16.

[8] Mayan J.A, Arifa S, Pavithra R,"Semantic based multi lexical ranking technique for an effective search in protected cloud,2016 International Conference on Control, Instrumentation, Communication and Computational Technologies (ICCICCT), Kumaracoil, 2016, pp. 570-576..

[9] Gopika. S. Kumar, S. Karthikeyan and N. Ambily, "A Metaphorical Scrutiny in Banishing Salt and Pepper Noise",BIOSCIENCES BIOTECHNOLOGY RESEARCH ASIA,December 2014 Vol. 11(3).

[10] G. Karudaiyar, S.Karthikeyan and B. Sainath, "Encryption and Decryption Scheme by Using Finite State Machine", BIOSCIENCES BIOTECHNOLOGY RESEARCH ASIA, December 2014 Vol. 11(3), 1867-1872.

[11] S.Karthikeyan, S.Jayashri, "Energy Utilization Strategies using Novel Approach in Wireless Sensor Networks", International Conference on Software Engineering and Mobile Application Modelling and Development, Dec 2012, pp: 1-5.

[12] S.Karthikeyan, S.Jayashri, "Energy Efficient System for Heterogeneous Wireless sensor Networks", European Journal of Scientific Research, Volume 72,No 4 march 2012, PP:599-607. ISSN: 1450-216X.

[13] S.Karthikeyan, S.Shanmugapriya and S.Jayashri, "Sub-head Transmission of Heterogeneous Data by Cloned Agent to Android Mobile",Research Journal of Applied Sciences, Engineering and Technology, 8(1):pp.24-34, 2014

[14] S.Pavithra, S.Karthikeyan, V.J.K.KishorSonti, S.Jayashri, "Competent Realization of Co-operative spectrum sensing in cognitive radio systems", International journal of Engineering systems modeling and simulation, Inder science publishers, Vol. 7, No. 2,pp 103-110 2015.(FREE JOURNAL)

[15] K. Bharathi,S.Karthikeyan, "A Novel Implementation of Image Segmentation for Extracting Abnormal Images in Medical Image Applications", Indian Journal of Science and Technology, Vol. 8(S8), 333–340, April 2015.

[16] Asha P, Albert Mayan J, Canessane A (2018),"Efficient Mining of Positive and Negative Itemsets Using K-Means Clustering to Access the Risk of Cancer Patients",Communications in Computer and Information Science , ICSCS 2018, Kollam, 2018,pp.373-382.

[17] S.Karthikeyan, "Ontogeny Smart Bulletin Board",ARPN Journal of Engineering and Applied Sciences, ISSN 1819-6608,Vol. 10, NO. 5, pp. 2148-2151, March 2015.

[18] R. Thenmozhi, S. Karthikeyan, "An Efficient Parallel Architecture in Design and Implementation of Adaptive LMS Algorithm",ARPN Journal of Engineering and Applied Sciences, ISSN 1819-6608,VOL. 10, NO. 7, pp 3150-3153, April 2015.

[19] J.Nandhini, K.Shabatini,S.Karthikeyan, "Wireless Colour Sensing Arm Robot", International Conference on Robotics, Automation, Control and Embedded Systems – RACE, ISBN: 978–81–925974–3–0, PP:1065-1071,February 2015.

[20] Albert Mayan J, Anto Praveena M D , Telkar Bharath Rao , Uravakonda Uday Sagar(2018),"Optimized test data generation over suspicious implementation of oracle problem",IEEE International Conference on Power, Control, Signals and Instrumentation Engineering (ICPCSI-2017),pp.2559-2563.

[21] D. Angeline Deborah Monica, S. Karthikeyan and V. Brindini, "Source and destination anonymity using end-to-end anonymous addressing scheme in wsn", ARPN Journal of Engineering and Applied Sciences, August 2015,Vol 10, No 14, PP:5778-5783.

[22] R.Ananth ,S. Karthikeyan ,"Dual Cluster Head Algorithm For Proficient Routing In Wireless Sensor Networks", Indian Journal of Science and Technology, Vol 9(43), November 2016,PP: 1-7.

[23] M. Tamilselvi , S.Karthikeyan , " A Literature Survey In Face Recognition Techniques", International Journal of Pure and Applied Mathematics Volume 118 No. 16 2018, PP: 831-849

[24] Bagavath sri and S. Karthikeyan, "Secure Elliptic Curve Cryptography Based RFID Schemes for Internet of Things", Indian Journal of Science and Technology Vol 9(42), November 2016 PP: 1-5.
[25] Brij Bhushan Ojha, S.Karthikeyan ,"Design of an interconnection network using VLSI photonics: Issues and challenges " International Journal of Pure and Applied Mathematics, Volume 118 No. 17 2018, PP: 211-224
[26] S. Vimalsree and S. Karthikeyan, "A Low Power Multibit Flip Flop Merging Technique Using WSN Nodes", ARPN, vol. no:11, No.15, August 2016, Page no: 9358-9363.
[27] Usha Nandini , Saravanan M , Albert Mayan J , Murari Devakannan Kamalesh , Mohana Prasad K (2018) , " Automatic traffic control system using PCA based approach", International Conference on Energy, Communication, Data Analytics and Soft Computing (ICECDS-2017),pp.2387-2392.

Weather Prediction Using Multi Linear Regression Algorithm

N Anusha*[1]
Assistant Professor
Department of Computer Science and Engineering
Sathyabama Institute of Science and Technology
OMR Road,Chennai, India

M Sai Chaithanya[2]
Under Graduate Student
Deaprtment of Computer Science and Engineering
Sathyabama Institute of Science and Technology
OMR Road,Chennai, India

Guru Jithendranath Reddy[3]
Under Graduate Student
Department of Computer Science and Engineering
Sathyabama Institute of Science and Technology
OMR Road,Chennai,India

anusha.nallapareddy@gmail.com*[1]

Abstract: Weather forecasting is one of the applications of science and technology, used to predict the weather condition depending on the input attributes. Most of the existing systems are implemented using statistical approaches for Support Vector Machine (SVM), which are incapable of giving the accurate prediction as they cannot capture sudden changes in weather conditions. The proposed technique uses the concept of Multi- Linear regression which can produce better results than existing methods.

Keywords: Weather forecasting, Statistical approaches, Support Vector Machine (SVM), Multi-Linear regression

1. Introduction

Weather forecasting is one of the areas that involve a complex process in meteorology. Ancient weather forecasting methods used by our ancestors usually relied on observing patterns of events. For example, if the sunset in the day looks brighter, assumption was made that the following day will have fair weather. However, not all these predictions proved reliable. In literature, there are (°N. Anusha et.al, Asha P,et.al)different algorithms which can be used for rainfall prediction and are classified into the different types like cloudy, partially cloudy, full cloudy, etc., Out of which some of the methods predicts numerical values but each method as their own merits and demerits.

IOP Conf. Series: Materials Science and Engineering **590** (2019) 012034 doi:10.1088/1757-899X/590/1/012034

More precision in predicted values can be obtained by taking the weather attributes periodically with a very high level of accuracy, in the controlled environment and mainly with the knowledge of current weather condition over the wide area. The small error in initial stage can lead to drastic change in outcome. Accuracy in prediction is more important in many sectors like Marine, Agriculture, Transportation, Aircraft, ([a]N. Anusha,et.al, [b]N.Anusha,et.al) disaster management, defense etc.,

(Wenying Zhang, et.al) compared the predicted with the actual weather and detected fault rate using Support Vector Machine (SVM). Failed to do numerical weather prediction and also requires a large amount of data for training in order to get more accuracy.(Serkan Buhan, et.al) Correlated between Numerical Weather Prediction (NWP) and Reference Wind Mast(RWM) models to localize the data used to get wind patterns by using Support Vector Machine (SVM). Failed to get more accuracy and not able to apply correlation to all data sets. (C. R. Rivero, et.al) Bayesian Enhanced Modified Approach (BEMA) is combined with Relative Entropy in order to issue short rainfall series prediction. Classified the data into clusters and not able to predict numerical value. (Swati Bhomia, et.al) The Dynamic-model-selection-based-multi-model-ensemble technique is developed for monsoon rain prediction. Calculates Root Mean Square Error and proceed according to the rate of estimation for next hour prediction. Not able to apply at different levels.

(S. Laboret, et.al) Altered the parameters using Bayesian and Artificial Neural Network filter methods for cumulative rainfall forecasting. Fair distribution is obtained from the selected data set and computational cost is high. (Julian Pucheta, et.al) Forecaster method was implemented to eradicate noisy data and Mackay-Glass chaotic time series was used to predict numerical value. Failed to maintain consistency with predicted values due to the eradication of noisy data. (Andrew Kusiak, et.al) Broyden–Fletcher– Goldfarb–Shanno (BFGS) used various data mining techniques (SVM, Simple Linear Regression, Random Forest) and compared the results on the basis of error rate. High cost is involved to estimate rainfall rate in real time scenario. (Niya Chen, et.al) Weather prediction was done using Hydrodynamic and Thermodynamic models based on boundary conditions of the atmosphere. Not able to predict rainfall rate more than four hours and model was purely based on wind rate.

(José L. Aznarte and Nils Siebert) Naïve predictor, Trivial predictor, Multivariate adaptive Regression, the generalized linear models and random forests were the used techniques for prediction. Unable to apply on datasets and high cost because of using different models.(Hagit Messer and Omry Sendik) Wireless processing was used to exploit diversity in data. Due to Wireless communication data transmission rate was slow.(Wenyu Gong, et.al) The Atmospheric filtering approach that combines lagged values of the computed rating series, measuring Meteorological variables and NWP values. Dealing with statistical data and rate of significance was varying from 0.67 to 0.77.(Sheikh Nurunnahar, et.al) Weather Forecasting was done by Data Mining Techniques, Classification and Regression Tree (CART). Not able to predict numerical values.Application of Multi-Linear regression on the available data gives a more accurate prediction than the contemporary methods that are in practice.On the basis of literature survey, Temperature, Wind Speed, Wind Direction, Humidity, Atmospheric Pressure and Rainfall rate, are found to be the parameters that influence prediction the most. Data on these parameters on an hourly basis, over five years (9-08-2011 to 19-10-2016) has been acquired from the India Meteorological Department (IMD).

2. About study area and used data set

The Figure.1 shows the study area. This Prediction is mainly concerned with Uttar Pradesh (UP) which is located in the North-Eastern part of India. (Uttar Pradesh) The latitude and longitude of Uttar Pradesh are 26.8467° N, 80.9462° E. (Climate of Uttar Pradesh) Variations exist in different parts of the state because of Indo-Gangetic plain forming a single parametric pattern throughout the state. Indo-Gangetic plain, also known as Indus-Ganga plain, is the main reason for a tropical monsoon in Uttar Pradesh. Uttar Pradesh has a climate of extremes with cyclic climatic conditions like temperature fluctuating from 0^0 to 46 ^{0}C, droughts, and floods due to unpredictable rains.

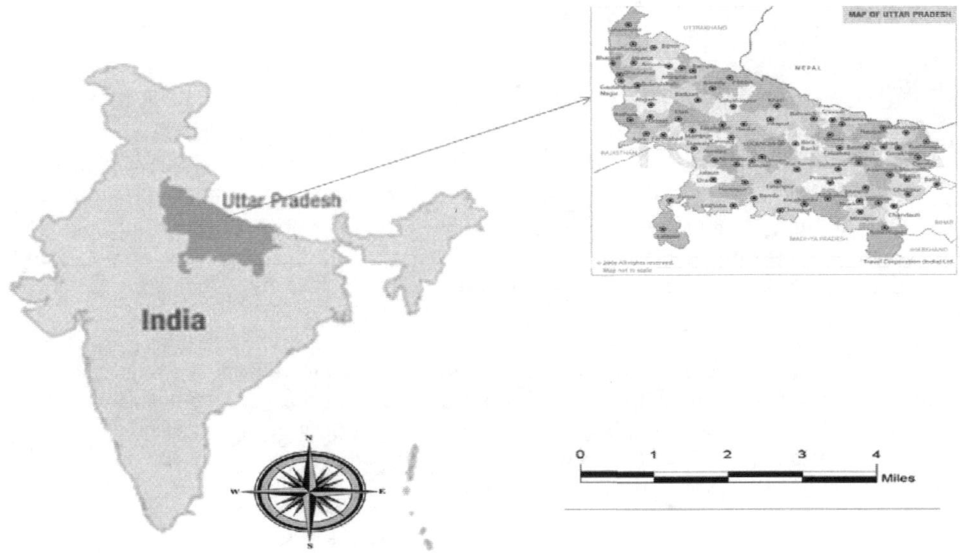

Figure 1. Location of the study area

Meteorological data of Uttar Pradesh over a period of four years starting from 09-08-2011 and up to 19-10-2016 is considered for this study. This data consists of parameters like station, latitude, longitude, altitude, time (IST), date (IST), air temperature, wind speed, wind direction, humidity, atmospheric pressure, rainfall recorded at random intervals during each day. The partial data set used for rainfall prediction is given in the Table 1. The total data set is divided into two categories, namely, training Data and testing Data. 70 percent of the total data is for training algorithm and 30 percent of the total data for the testing.

Table 1. Partial data set used for rainfall prediction

Date(IST)	Temperature in C°	Wind Speed in m/s	Wind Direction deg	Atmospheric Pressure(hecto Pascal)	Humidity	Rainfall(mm) (Output)
09-08-2011 (5:30)	32.9	0.4	93.8	928.3	88	0
10-08-2011 (2:30)	32.4	0.4	71.8	940.2	0	20
04-12-2011 (10:30)	30.1	0.9	229.2	951.5	75	0
27-06-2011 (13:30)	31.7	1.2	294.2	949	99	46

3. Methodology

3.1. Architecture

The proposed system consists of a specific way of dealing with Meteorological data by applying technical analysis and Multi-Linear Regression to get the numerical values about rainfall. During the technical analysis phase, the system analyzes the given input data by pre-processing and then the results are converted into a particular scale. This processed data is further fed as an input to the Multi-Linear regression algorithm and system predicts the rainfall rate by considering attributes as mentioned in the Table 1.

The total data set is divided into two categories, namely, Training Data and Testing Data. Training Data comprises 70 percent of the total data and it is fed to the algorithm to derive a relation between independent and dependent variables, Testing data, which comprises of the remaining 30 percent, is applied to the algorithm and value of dependent variable is calculated the calculated or predicted value is then compared with actual value of rainfall and the error rate is determined.

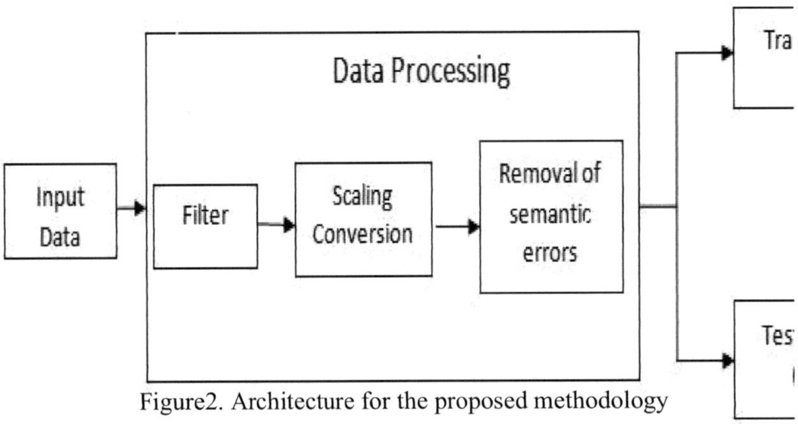

Figure2. Architecture for the proposed methodology

Figure 2 shows the architecture for the proposed methodology using the Multi-Linear regression algorithm. Data is collected from the Indian Meteorological Department (IMD). In the pre-processing phase, data is processed by applying filtering to remove the noisy data which affects the prediction rate, then the scaling conversion is done to form uniformity in data and then rectification of semantic errors is done to rule out values that are not practically possible. After pre-processing it is observed that cyclic patterns appear in the data, based on this, data is divided into three parts namely, summer, rainy and winter data. Then, the processed data is split into two data sets of 7:3 ratio i.e., 70% is taken as training data and 30% as testing data. The Multi-Linear regression algorithm is first trained using the training data to form a relation between various parameters and rainfall. This relationship makes the system capable of predicting the numerical value of dependent attributes.

$$Y = b_0 + \sum_{n=1}^{5}(b_n * X_n) \quad (1)$$

Then above given Eq. (1) is applied on testing data to calculate dependent variable, rainfall rate. Then the calculated values are compared with the actual rainfall data and finally margin of error is calculated.

3.2. Flow diagram

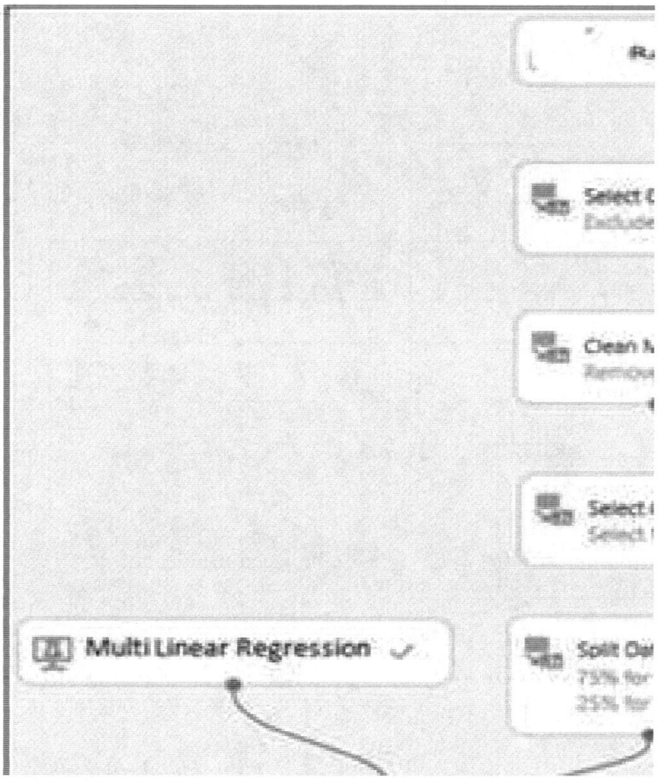

Figure 3. Activity diagram of Multi-Linear model

Weather data, for processing, is collected from the Indian Meteorological Department (IMD). The various attributes that influence the prediction of rainfall most significantly are listed in Table1 mentioned above. In cleaning phase the data is first filtered to remove noises and then data is converted into a single scale for homogeneity. Finally the data is checked for semantic errors like negatives wind speed or negative pressure etc. The processed data is further segregated into categories like summer, monsoon and winter on the basis of rainfall. Each group's data is processed independently and three different relations are generated using multi linear regression. When the test data is fed into the system it gets processed by one of the equations based on the date of recording. The system predicts the rain fall and this value is compared with the recorded value and the difference is calculated and the whole process is depicted in Figure 3.

IOP Conf. Series: Materials Science and Engineering **590** (2019) 012034 doi:10.1088/1757-899X/590/1/012034

3.3. Correlation between the attributes

Correlation quantifies the degree to which attributes are related and computes a correlation coefficient(r) that tells how much one attributes mentioned in Table1 tends to change with respect to the other variable. Mainly correlation coefficient is found between independent attributes mentioned in Table 1 and the dependent attribute is rainfall. Correlation between the attributes is shown in Figure 4.

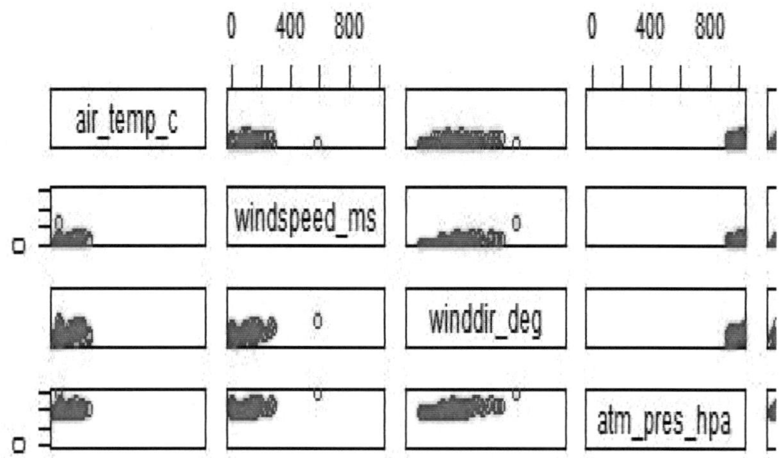

Figure 4. Correlation between the attributes

4. Result and Discussions

Chart 1 describes the projected values of actual rainfall data and predicted rainfall numerical data in mm. actual values are obtained from test data sets and predicted values are outcomes of the trained algorithm and the error rate is 12.16.

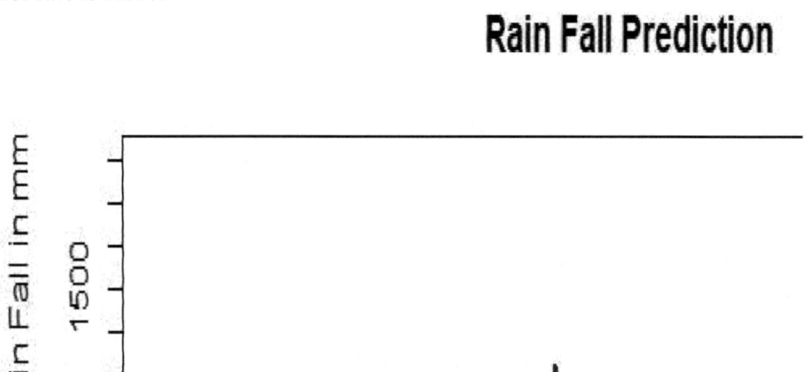

Chart 1. Projected actual and predicted of test data

IOP Conf. Series: Materials Science and Engineering **590** (2019) 012034 doi:10.1088/1757-899X/590/1/012034

Chart 2 Shows the percentage of accuracy among different approaches Support Vector Machine(SVM) 75% accuracy, Bayesian Enhanced Modified Approach (BEMA) approach has 80% accuracy, Multi-Linear has 88% accuracy, Ranys method has 81% of accuracy, BFGS has 85% accuracy, Multi-Variant method has 85% accuracy among proposed technique Multi-Linear method has high rate of accuracy.

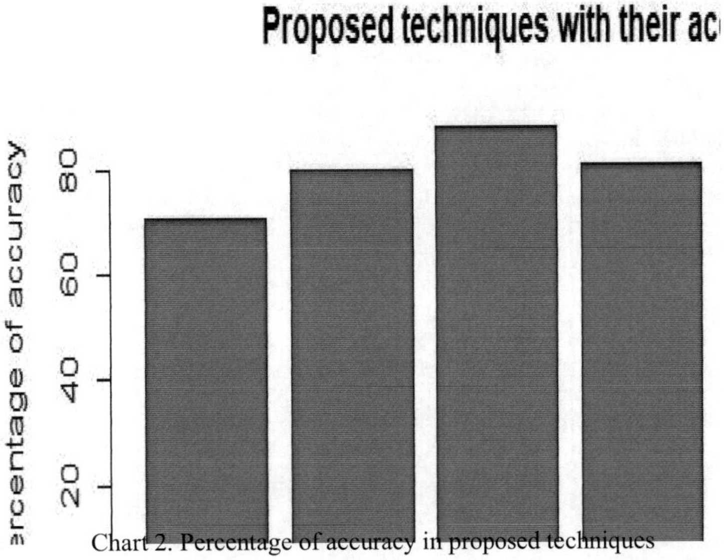

Chart 2. Percentage of accuracy in proposed techniques

5. Conclusion

Most of the existing systems are implemented using statistical approaches, the above module was implemented using Multi-Linear Regression and made the system more accurate than previous prediction methods. The problem with the existing modules lies in their inability to consider the effect of every value of each parameter, on the relation. In other words, the effect of every single value is not considered and a more generalized equation is produced rather than a unique relation. For example, in the SVM methodology, a plane that contains the data is generated and the equation of this plane is used as a relation for prediction. This poses a problem as data points that are far apart from each other in terms of magnitude end up on the same plane and their individual effect gets overlooked. The use of multi linear regression deals with this problem and optimizes the results. Historical data used for this module has independent attributes like temperature, wind speed, wind direction, humidity, and atmosphere pressure. They are used to calculate the amount of rainfall. This module gives an accurate prediction 88% of the time.

References

[1] Wenying Zhang, Huaguang Zhang, Jinhai Liu, Kai Li, Dongsheng Yang,Hui Tian "Weather Prediction With Multiclass Support Vector Machines in the Fault Detection of PhotovoltaicSystem"inIEEE/CAA Journal of Automatica Sinica, Vol. 4, No. 3, July 2017.

[2] Niya Chen, Zheng Qian, Ian T. Nabney, Xiaofeng Meng " WindPowerForecastsUsingGaussian Processes and NumericalWeatherPrediction",IEEE Transactions on Power Systems,Vol.29,No.2,March 2014.

[3] N. Anusha, P. Radhika Priyanka, P. Sai Harini, Segmentation of Multi-temporal Images Using Gaussian Mixture Model (GMM), ARPN Journal of Engineering and Applied Sciences, ARPN, Vol. 13, Issue. 16, pp. 4633-4640, 2018.

[4] Andrew Kusiak, Xiupeng Wei, Anoop Prakash Verma, Evan Roz" Modeling and Prediction of Rainfall Using Radar Reflectivity Data: A Data-Mining Approach", IEEE Transactions on Geoscience and Remote Sensing, Vol. 51, No. 4, April 2013.

[5] James N. K. Liu, Bavy N. L. Li, Tharam S. Dillon "An Improved Naïve Bayesian Classifier Technique Coupled With a Novel Input Solution Method", IEEE Transactions on Systems, Man, and Cybernetics—Part C: Applications and Revoews, Vol. 31, No. 2, May 2011.

[6] C. R. Rivero, J. Pucheta, S. Laboret, M. Herrera, V Sauchelli", Time Series Forecasting Using Bayesian Method: Application to Cumulative Rainfall ", IEEE Latin America Transactions, Vol. 11, No. 1, February 2013.

[7] N. Anusha, B. Bharathi, Flood Detection and Flood Mapping Using Multi-temporal Synthetic Aperture Radar and Optical Data, Egyptian Journal of Remote Sensing and Space Sciences,2019 (Article in Press)

[8] Swati Bhomia, Neeru Jaiswal, C. M. Kishtawal, Raj Kumar "Multimodel Prediction of Monsoon Rain Using Dynamical Model Selection", IEEE Transactions on Geoscience and Remote Sensing,Vol.54,No.5,pp.2911-2917,May 2016.

[9] Abebe S. Gebregiorgis,Faisal Hossain," Understanding the Dependence of Satellite Rainfall Uncertainty on Topography and Climate for Hydrologic Model Simulation", IEEE Transactions on Geoscience and Remote Sensing, Vol. 51, No. 1, January 2013.

[10] N. Anusha, K. Bhavana Sai, K. Srujana, Image Segmentation Using ToZero Method and ToZero Inverse Methods, ARPN Journal of Engineering and Applied Sciences, ARPN, Vol. 13, Issue. 16, pp. 4678-4682, (2018).

[11] C. R. Rivero, J. A. Pucheta, J. S. Baumgartner, S. O. Laboret, V.H. Sauchelli, H. D. Patiño," Short-series Prediction with BEMA Approach: Application to Short Rainfall Series", IEEE Latin America Transactions, Vol. 14, No. 8, August 2016.

[12] Jos´e L. Aznarte, Nils Siebert "Dynamic Line Rating Using Numerical Weather Predictions and Machine Learning: a Case Study" in IEEE/CAA Journal of Automatica Sinica, Vol.32,No.10,pp. 335 – 343, January 2017.

[13] Asha P, Albert Mayan J, Canessane A (2018),"Efficient Mining of Positive and Negative Itemsets Using K-Means Clustering to Access the Risk of Cancer Patients",Communications in Computer and Information Science ,ICSCS 2018, Kollam,pp.373-382, 2018.

[14] Serkan Buhan,Yakup Ozkazanc, Isik Cairic "Wind Pattern Recognition and Reference Wind Mast Data Correlations With NWP for Improved Wind-Electric Power Forecasts" in IEEE Transactions on Industrial Informatics, Vol. 12, No.3, pp.991-1004, September 2016.

[15] Uttar Pradesh, India.https://www.latlong.net/place/uttar-pradesh-india-6809.html (accessed on 15 February 2018)

[16] Climate of Uttar Pradesh. https://en.wikipedia.org/wiki/Climate_of_Uttar_Pradesh (accessed on 16 February 2018).

Smart Irrigation and Tank Monitoring System

Kumar Kunal
U.G Student ,Department of CSE, Sathyabama Institute of Science and Technology, Chennai, India
kr.kunal145@gmail.com

Md. Azhar Hussain
U.G Student ,Department of CSE, Sathyabama Institute of Science and Technology, Chennai, India
azharhussain2u@gmail.com

Dr. N Srinivasan
Professor, Department of CSE,Sathyabama Institute of Science and Technology, Chennai, India
professorsrini@gmail.com

J.Albert Mayan
Assistant Professor, Department of CSE,Sathyabama Institute of Science and Technology, Chennai, India
albertmayan@gmail.com

Abstract: Water, which is regarded to be one of the mostimportant part for the growth of plants in agriculture should be utilized properly. The paper is focused on the smart irrigation of farms as well as monitoring the tank automatically so as to make better irrigation system and tank monitoring which will be beneficial in conserving water and irrigating fields in a smart way. For this project purpose we are integrating different sensors like DHT11, Soil Moisture sensor Ultra sonic sensor, etc. These sensors communicate with the micro controller. Micro-controller perform its action as per its instructions. The proposed framework may prompt upgrade the cultivating rehearses, overcoming the water crises and developingan upgraded agricultural system for the country.

Keywords : Arduino, GSM, Soil Moisture Sensor, Ultra Sonic Sensor, DHT11 Sensor etc.

1. Introduction

Agriculture is considered as the back-bone of Indian economy. Around 70 % of the population still depends on Agriculture. 33% of the National economy comes from Agriculture. The development in agrarian creation has been still for as long as quite a long while. The importance of farming is: 1) Contribution to National Income, 2) Main wellspring of Food,3)Agriculture and Industrial advancement, 4) Sources of Revenue, 5) Source of Foreign exchange. 6) Transport, 7) Source of sparing, 8) Capital arrangement, 9) International significance, 10) Way of life, 11) Effect on costs, 12) Source of work supply, 13) Economic improvement. Our property was losing its fruitfulness being put to development persistently for a considerable length of time together. So we have perused all the current framework and their working and we have discovered that there is no framework that uses the small scale controller, various sensors, information mining and mobile phone to communicate with gsm modem all together. So we are consolidating all the current framework to get the half and half framework. Sensor-Based water and irrigation system framework has been contemplated in much application. These sensors send constant esteems to microcontroller and microcontroller send these qualities to PC by means of serial channel. The system suggests an economical and easy-to-use Arduino-based automated irrigation system and tank monitoring system that utilizes the

GSM modem to impart and send and get instant message as status of the incorporated framework. The data received as a text message from the Arduino is displayed on the User mobile as well as on the LCD screen. The tank is monitored based on the water level of the tank and the system also proposed of preventing the motor from damage in case of continuous running due to leaking.

2. METHODOLOGY AND COMPONENT

The Smart Irrigation and Tank Monitoring System provides a smart way to monitor the water tank and irrigate the crop field automatically. This proposed system reduces the man power of farmers like supplying water to plants and controlling the motor to fill tank. This system uses different components like DHT11 sensor, Soil Moisture sensor, Gsm Modem, ultrasonic sensor etc. and according to this sensor parameters farmer are provided an automated way to irrigate their fields and monitor the tank. By the help of this proposed system we can even get the current status of the soil, motor, tank as well as the status of the different components.

Hardware:

1. **Sensors**: We use sensors like
   - DHT11:
     DHT11 is a Humidity and Temperature Sensor, which creates aligned advanced yield. DHT11 can be port with any microcontroller like Arduino, Raspberry Pi, etc. and get sudden results.
   - Soil Moisture Sensor:
     The Soil Moisture Sensor uses capacitance to measure the moisture content of soil (by evaluating the dielectric permittivity of the soil, which is a part of the water content). Basically embed this rough sensor into the soil to be tested, and the volumetric water substance of the soil is accounted for in percent.
   - Ultrasonic Sensor:
     An Ultrasonic sensor is a device that can quantify the distance to a body or object by utilizing sound waves. It evaluates and measure distance by sending a sound wave at a particular recurrence and tuning in for that sound wave to skip back. By recording the past time between the sound wave being produced and the sound wave bobbing back, it is conceivable to figure the distance between the sonar sensor and the body.

2. **Microcontroller:**
   It is heart of system, implies it control all task of framework. For this task we are utilizing Arduino Uno. The Arduino Uno is a microcontroller board in light of the ATmega328 (datasheet). It contains 14 digital Input/yield pins (of which 6 can be utilized as PWM yields), 6 Analog sources of info, a ceramic Resonator of 16MHz, a power jack, a USB association, an ICSP header, and a reset catch. It contains everything which is required to help the Microcontroller basically interface it to a PC with a USB link or power it with an AC-to-DC connector or battery to begin.

3. **Gsm:**
   A GSM Module is fundamentally a GSM Modem (like SIM 800A) associated with a PCB with various sorts of yield taken from the board – say TTL Output (for Arduino, or any other microcontroller) and RS232 Output to connect specifically with a system or PC.The board will likewise have sticks or arrangements to append mic and speaker, to take out +5V or different estimations of energy and ground associations.
4. **LCD:**

   LCD stands for Liquid Crystal Display. LCD screen is an electronic display module and locate an extensive variety of uses. We are using a 16x2 **LCD**, it can display 16 characters in a line andthere are 2 rows of lines. It has two registers which are Command and Data. The command register stores the command instructions given to the LCD. A command is an instruction fetched to LCD to do a defined task like initializing it, clear the screen, sets the cursor position, controls the display etc. Data are stored in data register which are to be displayed on the LCD. The data is the ASCII representation of the character to be displayed on the LCD.

5. **Relay Switch:**

Arelay switch is a switching device used electromagnetically that open or close circuits. Relays consist of armature that is moved by an electromagnet which control one electrical circuit by opening and shutting contacts in another circuit. As shown in fig below there is open contact when a relay contact is normally open and the relay isnot empowered. It provides isolation and amplifications and is straightforward.

Software:

1. **Windows system**

   1 (GHz) gigahertz or above, 32-bit system processor (x86) or 64-bit (x64) system processor.

2. **Linux system requirements**

   700 MHz processor, 512 MB RAM or above, 5 GB of hard-disk space and above.

3. **Arduino IDE:**

   Arduino is a software tool which is an open-source IDE (Integrated Development Environment) platform used for writing Arduino code. It is used to programme physical Arduino circuit board often called microcontroller that runs on the PC, used to write andupload computer code to the Arduino board.

3. PROPOSED SYSTEM

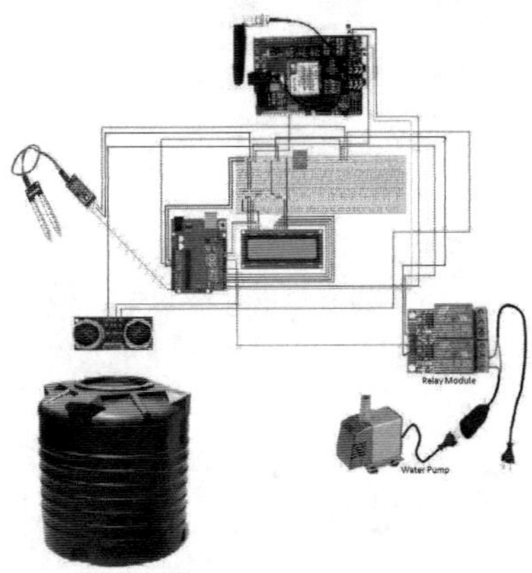

Fig. 1. Proposed System

Working:

This project is about a moisture detecting of soil programmed using Arduino UNO which is used in irrigating the fields. The framework peruses the moisture substance of the dirt utilizing soil moisture sensor and switches ON the motor when the moisture is underneath as far as possible. Right when the clamminess level rises above the set point, the system switches off the pump.

The tank monitoring system uses ultrasonic sensor to measure the level of water inside the tank which is automated and switches on and off the motor comparing to a threshold level which has been set and it also use DHT sensor which prevent the motor in case of leakage as the temperature will increase which switches off the motor going beyond certain level. The system uses GSM modem to communicate with user by sending and receiving text message as status of the integrated system. The status of the moisture, temperature, tank, engine and the dampness level will be shown on a 16×2 LCD show.The logic described behind the algorithm is helpful in understanding the overall working of the system. It initializes the sensors and read the values using Arduino Uno which helps to performs actions.

We uses Relay Circuit board to turn on and off the motor which is used to irrigate the field automatically as soon as the moisture content in the soil goes down a certain fixed level based on the crops requirement and turns off the motor when the field gets irrigated at a defined level. Also the farmer s would be able to get the value of the soil moisture level on their mobile phone through the use of GSM modem which helps in receiving the status of the farm. This helps to remotely monitor the status of the farm.On the other side, the project also focuses on monitoring the tank through the use of sound wave ultrasonic sensor which is integrated with the system that measures the level of water inside the tank. As the water level goes beyond a certain level it switches on the motor using the relay circuit board and turns off as it become full. Also the motor can be prevented by using a DHT11 sensor at the end which can turn off the motor after sensing the high temperature and heat value.

IOP Conf. Series: Materials Science and Engineering **590** (2019) 012035 doi:10.1088/1757-899X/590/1/012035

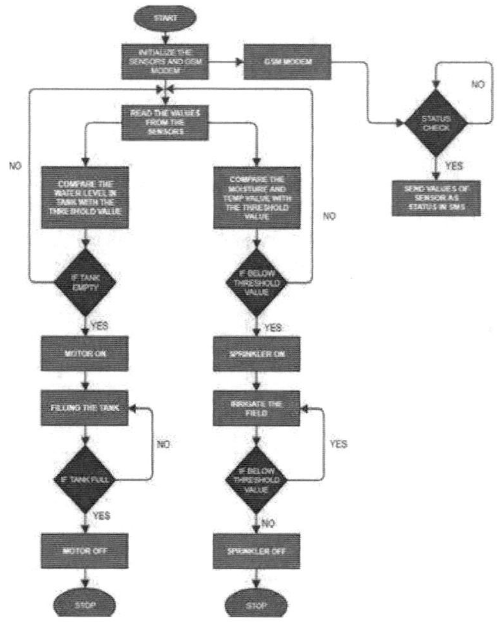

Figure 2. Flow chart

Gradually this system enables to ensure that the moisture content and availability of water in field is well maintained and the tank and motor is monitored and protected respectively.

4. RESULT

The system described in the paper can be accessed anywhere, anytime provided gsm network available for communication between user and the system.

Soil Moisture Sensor:

Data from moisture sensor	Condition of soil
0-300	Soil is dry
300-700	Soil is partially wet
700-1020	Soil is wet

Based on the data provided by the soil moisture sensor its values lies between the ranges of 0-1020 which helps in determining the soil type of field in terms of moisture content. The below tables shows the moisture content and status of the soil and the temperature and humidity along with operating voltage.

IOP Conf. Series: Materials Science and Engineering **590** (2019) 012035 doi:10.1088/1757-899X/590/1/012035

Humidity:

Parameter	Min	Typical	Max	Unit
Accuracy (25°C)	-	±4	-	%RH
Accuracy (0-50°C)	-	-	±5	%RH
Measurement range (25°C)	20	-	95	%RH

Response time: 1/e (63%) 25°C, 1m/s air	6	10	15	S

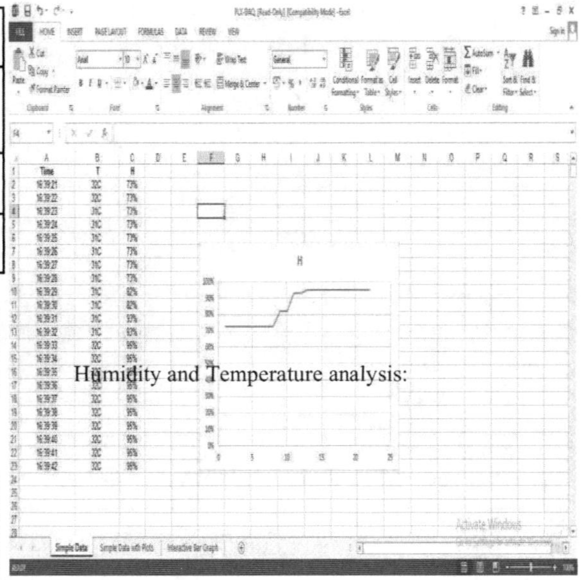

Humidity and Temperature analysis:

Temperature:

Parameter	Min	Typical	Max	Unit
Accuracy (25°C)	±1	-	±2	
Measurement range	0	-	50	
Response time /e (63%)	6		30	S

Figure 3. Performance analysis

The setup provides a better option of knowing the details and status of all the sensors which comprises the complete system. The details like soil moisture content value, water status, relay status and the operation mode can be monitored through the GSM modem. The important thing is that the soil moisture value will be updated to the board for every second. It also enables to supervise the system automatically.

Ultrasonic Sensor:

Power Supply	Quiescent Current	Ranging Distance	Resolution	Frequency
5V DC	<2mA	2cm-500cm	1cm	40k Hz

Operating Voltage:

Parameter	Min	Typical	Max	Unit
Working Voltage	3	5	5.5	VDC
Working Current	0.5	-	2.5	
Sampling interval	1	-	-	S

Figure 4. Operating specs

The person can decide at a certain time if the relay should be ON or OFF and it can be controlled using the GSM itself. The overall system control and the system details are also displayed on a 16*2 LCD.

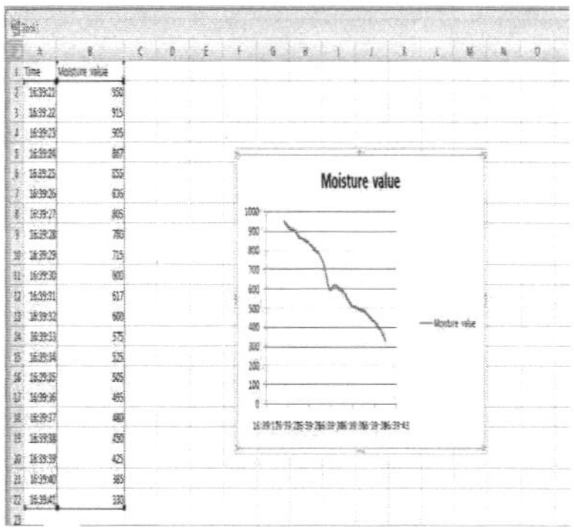

Figure 5. Moisture analysis

5. CONCLUSION

The purpose of this paper is to apply the functionality integrated architecture in the field of agriculture such as Smart Irrigation to provide optimum use of fresh water and better crops production. The Smart Irrigation and Tank monitoring system implemented is feasible and cost effective for using water resources in an optimised way for production of agriculture.The setup was carried out using Arduino UNO board, Soil moisture sensor, ultrasonic sensor, dht11 sensor and relay. The Arduino with soil moisture sensor takes the reading of water content in the soil and perform specific actions based on its value. This proposed system reduces the man power of farmers like supplying water to plants and controlling the motor to fill tank.

The system has successfully completed four objectives but could not complete two objectives i.e., weather forecasting and controlling system through android app in mobile. It uses low cost sensors and other devices which makes the system cheap. As the system is completely automated, it does not require complete attention of farmer's every time, so it saves the time. Overall this system can be used in the place with water scarcity and where human interference is less. The system is incredibly versatile and economical. It doesn't need individuals on duty as it is so easy and reliable.

REFERENCES

[1]X. Wang, W. Yang, A. Wheaton, N. Cooley, and B. Moran, "E_cient registration of optical and IR images for automatic plant water stress assessment", Comput. Electron. Agricult., vol. 74, no. 2, pp

[2]Albert Mayan .J , Dr. T. Ravi, "Optimized Regression Testing using Genetic Algorithm and Dependency Structure Matrix", International Journal of Applied Engineering Research , Vol:9, Issue:20, pp: 7679-7690,Nov 2014 , ISSN: 1087-1090

[3]Prof.B.T.Salokhe, Miss Shilpa G. Gadekar, "A wireless application of drip irrigation automation supported by soil moisture sensors " ,International Journal of Advanced Research in Computer and Communication Engineering, Vol. 4, Issue 4, April 2015

[4] Albert Mayan .J , Sharmila Latha T , Kislay Sinha ," Security Analysis of Three Factor Authentication Schemes for Banking", ARPN Journal of Engineering and Applied Sciences, Vol:10, Issue 8, pp: 3504-3509,May 2015 , ISSN : 1819-6608.

[5]BhagyashreeK.Chate, Prof.J.G.Rana, "Smart Irrigation System Using Raspberry Pi" International Research Journal ofEngineering and Technology (IRJET), vol. 03, Issue 05, pp.1, May-2016.

[6]G. Merlin Suba, Y M Jagadeesh, S Karthik and E Raj Sampath, "Smart Irrigation System Through Wireless Sensor Networks", ARPN Journal of Engineering and Applied Sciences, vol. 10, pp. 1,no. 17, september 2015.

[7]Venkata Narayana Eluri, K. Madhusudhana Rao, A. Srinag, "Wireless Solution for Water Saving In Agriculture Using Embedded System", International Journal of Computer Science and Business Informatics, ISSN: 1694-2108 — Vol. 2, No. 1. JUNE 2013

[8]S. V. Devika, Sk.Khamuruddeen, Sk.Khamurunnisa, Jayanth Thota, KhaleshaShaik,"Arduino Based Automatic Plant Watering System", International Journal of Advanced Research in Computer Science and Software Engineering, Volume 4, Issue 10,pp. 2-3,October 2014.

[9] Asha P, Albert Mayan J, Canessane A (2018),"Efficient Mining of Positive and Negative Itemsets Using K-Means Clustering to Access the Risk of Cancer Patients", Communications in Computer and Information Science , ICSCS 2018, Kollam, 2018,pp.373-382.
[10]ChandankumarSahu, PramiteeBehera, "A Low Cost Smart Irrigation Control System", IEEE Sponsored 2nd International Conference on Electronics and Communication System (ICECS 2015).
[11]P. Jabez J, Gowri S, Vigneshwari S, Albert Mayan J and Srinivasulu S (2018)," Anomaly Detection by Using CFS Subset and Neural Network with WEKA Tools", Smart Innovation, Systems and Technologies,Vol.107, pp. 675–682.
[12] Kamalesh M. D., Albert Mayan J., Felix Y., Sumanth B. S., & Sai Tej B. (2018), " Magrisys: A Smart And Ubiquitous Controlled – Environment Agriculture System " , 2nd International Conference on I-SMAC (IoT in Social, Mobile, Analytics and Cloud) , pp: 1-5 , 2018

BUS TICKET SYSTEM FOR PUBLIC TRANSPORT USING QR CODE

C.UPENDRA REDDY[1], D.L.S.VARA PRASAD REDDY[2]

[1,2] U.G Student , Department of CSE , Sathyabama Institute of Science and Technology, Chennai, India

upendrareddygood30@gmail.com, devireddysaiprasad11@gmail.com

Dr N.SRINIVASAN[3]

Professor , Department of CSE , Sathyabama Institute Of Science and Technology, Chennai, India

professorsrini@gmail.com

ALBERT MAYAN J[3]

Assistant Professor , Department of CSE , Sathyabama Institute Of Science and Technology, Chennai, India

albertmayan@gmail.com

Abstract: In this paper, we are proposing QR reader for a bus ticket. Users can scan QR reader instead of the ticket. In this app, after registration profile, we have to attach our bank details to this app and add money to the wallet. Whenever we go by bus, we have to select from and to location. Then it will generate amount details for per head. Also, we can buy a ticket for extra members and produces the mean count. It Scans the QR code from the conductor of the bus with a particular ID. Then we can scan QR code. So directly money will transfer from our wallet to transport corporation on basis of conductor ID. Then we will get an SMS alert for ticket payment proof. The admin side (Transport Corporation), they calculate amount details of a particular conductor through using web application. Then they can calculate per day amount details of Transport Corporation for buses.

Keywords: QR code, Bus information, Easy ticket process.

1. INTRODUCTION:

In this bus ticket system for public transport using QR code system we are actually trying to make bus ticket process in metropolitan and public transport as an easy way by QR code system. The project works on the basic android application platform. The user will be registered through application on online basis. Once the user gets registered with the system. It helps the passenger to mark the boarding and destination places on the application and check the bus availability in that route at that particular time. Before going to book ticket we have to add the amount into wallet from bank account for travelling. After boarding into the bus. Ticket will be generated by scanning QR code from conductor in the bus. Which deducts the amount from wallet and provides message of ticket details.

2. EXISTING SYSTEM:

For the most part, in the transportation framework, the procedure of the ticket is for the most part taken care of by the conductor. He used to gather the sum from the travelling individual and gives the ticket to them. Fundamentally, the entire procedure is finished by paper-based, the printed papers are issued as tickets. In introducing days the handholding machines are utilized to print the tickets this framework is having different weaknesses. That traveller needs to convey the printed ticket to the last stop of their destination. That the conductor ought to guarantee that each explorer in the transport is having the ticket are most certainly not. The time taken for the printing of ticket of each individual in the transport is too long. Conductors must be prepared for keeping up the handheld machines. For instance, if the traveller wishes to travel by transport he needs to give the cash to the conductor and gather the printed ticket and to complete until loading up from the transport. This procedure will take additional time and the abuse of Human Resource and fundamentals as well. The framework is for the handholding ticket process however in another way, it additionally kept up by the RFID framework, in which tag and per user are utilized output the cards. Subsequent to filtering a card the measure of the ticket cost is charged from the tag naturally. This procedure is likewise having the inconvenience of loosing of RFID tag. That prompts loss of cash. RFID label comprises of cash since it was credited with cash before the travelling itself. The RFID framework is likewise utilized for tracking also.

3. LITERATURE SURVEY:

In this segment, we will review some things up on the various kinds of writing open transport ticket planning process.

3.1. GPS Tracking in Public Transportation

In a large portion of the nation's population is expanding quickly. So the general population, for the most part, utilizes the general population transport than the private transport and they need greater adaptability. To guarantee that people in general transport must be redesigned with introducing innovation that encourages open transport to pull in the general population. Automatic Fare Collection (AFC) System is generally called as Transit Smart Card System that gives benefits on manual fare collection system towards bringing down the work costs and that expands proficiency of fare collection. Bayesian choice tree calculation is created in examining process, the calculation is utilized with the open transportation vehicles that are with GPS system. It expresses that information gathered was the purposes of starting point where the traveller AFC card is filtered, it is pivotal for the procedure of Automatic Fare Collection process [3]. In another paper, the GPS following is utilized towards following of transports. Where that a GPS system of the transport fast transit system is utilized as a part of the transportation for expanding of adaptability for people in general. The BRT System was actualized for expanding the unwavering quality and security on the premise of lessened journey times [1].

3.2. Public transportation with RFID technology

RFID innovation comprises of segments like Tag, Reader and the Middleware which cooperates with the back-end database. Numerous applications utilize RFID innovation that has been proposed lately to make the procedure as simple way. Those endeavours have been tested as to achievability, organization, administration, protection, and security. It demonstrates that RFID innovation made a compelling utilization of these kinds of uses. We should see that in instances of no "observable pathway" between smart cards and per users in a comparative meld with the radiation examples to coordinate them and the situating of receiving wires, there might be changes in designs as these variables are basic for acknowledgment [4]. The exploration was additionally done on how RFID Technology will be utilized to take care of issues looked by open transportation experts in metropolitan urban communities.

By investigating the automatic following of transports that are utilized to give an estimation of transport entry times which improve the traveller comfort. In short run, the RFID procedure may demonstrate the most noteworthy cost. To make productive it must have sifting of information to expel undesirable or off base cost costs and after that ordering helpful information by the total process. Additionally, that RFID procedure is examined on gathering information and that can be used to foresee the transport development that enhances the administration of the general population transport system. The process has the hindrance like when the labels and per users work all the while there is a possibility of happening twofold charges [5]

IOP Conf. Series: Materials Science and Engineering **590** (2019) 012036 doi:10.1088/1757-899X/590/1/012036

3.3. RFID in cashless ticketing system

The cashless system is transcendently observed in different nations which utilize a Smart Card that was set on the ticketing machine introduced inside the transport. The entry will be deducted and issues a ticket by the entire paid, that besides includes date and time of the exchange near to the straggling leftovers of the attributes on the card, the entire framework is defined by the utilization of an Electronic Ticketing Machine. The card can be revived and that can likewise be customized if there should arise an occurrence of misfortune or burglary [6]. Notwithstanding the Smarter Card help procedure of fast and simple ticket process, the capacity to load up a transport and pay for all intents and purposes without hands is acknowledged by the RFID empowered Smarter Card. That was not just recognizable when held close by and furthermore when conveyed anyplace by a voyaging individual. The issued ticket mirrors the voyager and card information accumulated from the filtering and Not withstanding those numerous are additionally assembled like a planned course, transport number, date and time of loading up into the transport. They additionally analysed on protection worries as for accumulated information of individual data. The RFID Cards are utilized as a part of general society transportation which empowers the following of individuals boarded in the transport and those following points of interest are gathered in the database server which is known as the focal server. This enables a person's whole history to be shown as though they required or for the security reason [7]. This paper likewise researched on the proper security and protection necessities for e-ticketing through RFID innovation.

To feature the insufficiency of an existing proposition and present answers for protection saving e-tickets in light of RFID innovation. With this procedure, the known cryptographic strategies are utilized which will debilitate ticket fabrication. E-ticket is generally used to enable the conductor to bring down the procedure of check and exchange to process. That enables the explorer to have a speedy and helpful confirmation process than the paper-based or money based check process

[8]. Notwithstanding that RFID innovation has likewise picked up support in ticketing, by a cashless system utilization. So considering both RFID and GPS advances, they proposed a novel portable based transport following and planning system [2]. It was specified that GPS innovation additionally has a disadvantage where transmission might be hindered when they are crossing through the shut structures, for example, burrow situations or same courses with comparative nature.

3.4. RFID bus ticket for public transport:

The paper-based public transport ticketing system, prevailing in the megacity of Dhaka presents extreme glitch in the system, the malignant contention among the public is corruption and above all else traffic jam. This paper really proposes a substantially more public benevolent, automated system of ticketing and also the credit exchange with the utilization of RFID based tickets. The aggregate system for the most part acts to draw out the consistency among different transport organizations that will finish up in uniform access of travellers in day by day rides through an automated server being refreshed each and every time the travellers go via conveying the RFID based tickets [9].

4. PROPOSED SYSTEM: Architecture

In proposed method, we are introducing QR reader. Here, we will create one android application for selecting the travelling route and generate amount.

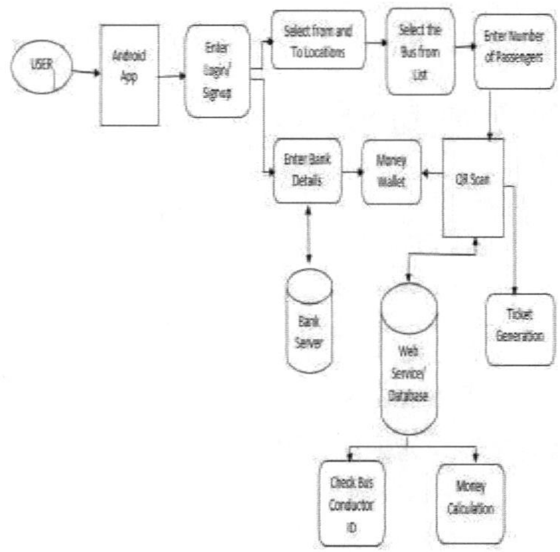

Figure 1. Block diagram

The QR code will be scanned by passenger from the conductor in the bus. Then automatically it will send amount from our wallet to conductor account which is maintained by transport system. Then user will get message of travelling ticket. After that value of money collected by the conductor will be automatically store in database and the details of tickets collected.

4.1. REGISTRATION MODULE:

The user provides personal details accordingly then he provides route details and before that add the amount to wallet from bank. After that it produces QR code scanning page layout in mobile. In this module, the client gives the points of interest like name, make password, mail, telephone number and so forth through enrollment shape and gets enlisted with the vehicle transportation framework.

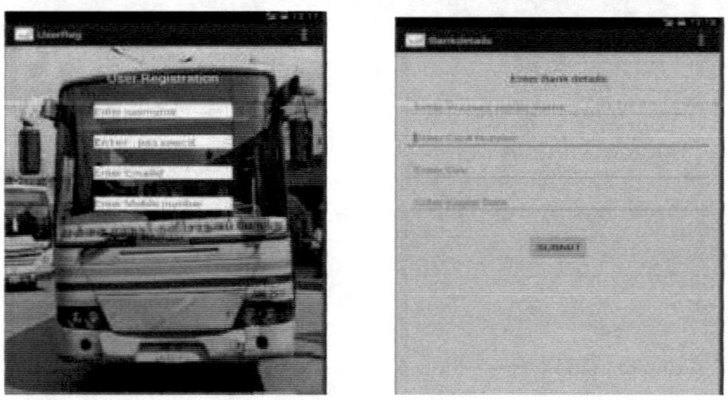

Fig 2. User registration

4.2. USER LOGIN:

The principle page of the application is the client login page that used to get enrolled with the application. If they were selected (who are registered with app) people, they can login by giving the username and password through the user login module

4.3. BANK PAYMENT MODULE:

Once the login is finished it prompts the bank page in which we give the bank subtle elements like card number, name on the card, cvv and so on. In the wake of giving points of

4.4. LOCATION MODULE:

A user needs to choose from and to and it will produce fare details for the journey. At that point we need to enter the check of travellers and we get aggregate sum. From that point onward, we need to utilize QR scanner for versatile installment.

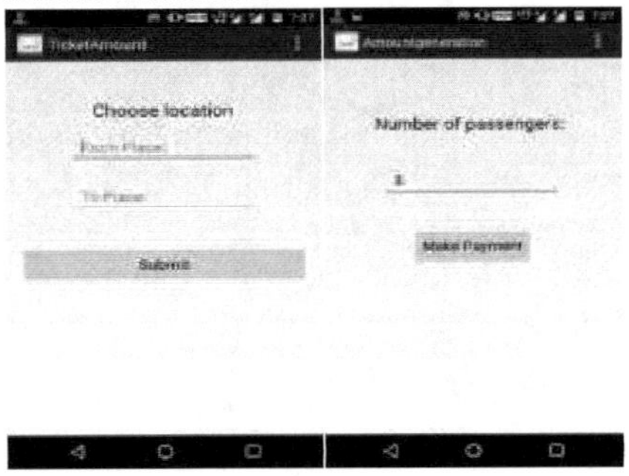

Fig 3. Location and Number of passenger's selection

4.5. QR SCAN MODULE:

In the wake of adding the cash to the wallet. The traveller needs to choose from and to subtle elements of the journey then it creates the cost of travelling per head. At that point we can likewise choose the number of travellers to movement, it figures the aggregate sum and then scans the QR code of the conductor.

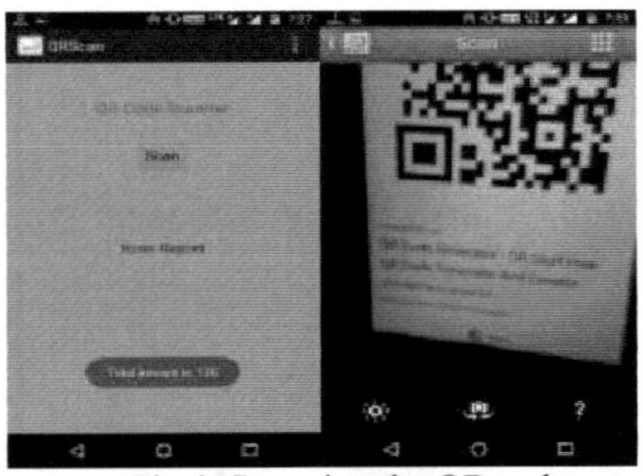

Fig 4. Scanning the QR code

4.6. VERIFICATION MODULE

The QR code generated will be verified by the conductor and the amount will be deducted from the wallet into conductor's wallet and later to the transport corporation from the conductor and tickets collected by conductor will be tracked by admin of the transport corporation.

5. CONCLUSION

The project would be valuable to general society who are looking with the present issues. The Proposed framework would empower the general population to have the ticket for movement in metropolitan and for the most part open transport, in addition, the printed material will be decreased and loosing of the card is additionally wiped out. It would guarantee the minimization of tedious and cash issues like change. Further, the subtle elements of the ticket of each traveller are put away in the database alongside conductor points of interest which are followed by the administrator.

REFERENCE:

[1.] Albert Mayan J, Kuldeep Anand D.S, Neha Sadhvi(2017),"Efficient and secure server migration on cloud storage with VSM and dropbox services", International Conference on Information Communication and Embedded Systems (ICICES),Chennai pp. 1-5.

[2.] Hamilton. P and Suresh. S (2013). "Intelligent Agent based RFID System for On Demand Bus Scheduling and Ticketing", International Journal of Future Computer and Communication, Vol. 2(5), pp.399-406.

[3.] Xiao-Lei, M et al (2012). "Transit Smart Card Data Mining for Passenger Origin Information Extraction", I Journal of Zhenjiang University Science C, Vol. 13(10), pp.750-760

[4.] Surya.V, J. Albert Mayan," A Secure Data Sharing Mechanism In Dynamic Cloud By Using KP-ABE", Research Journal of Pharmacy and Technology , Vol 10 , Issue 1 , pp:83-86,2017.

[5.] Mezghani, M (2008). "Study on Electronic Ticketing in Public Transport", Available from http://www.emta.com/IMG/pdf/EMTA-Ticketing.pdf

[6.] Asha P, Albert Mayan J, Canessane A (2018), "Efficient Mining of Positive and Negative Itemsets Using K-Means Clustering to Access the Risk of Cancer Patients",Communications in Computer and Information Science ,ICSCS 2018, Kollam, 2018,pp.373-382.

[7.] Albert Mayan .J , Dr. T. Ravi," Structural Software Testing: Hybrid Algorithm For Optimal Test Sequence Selection During Regression Testing", International Journal of Engineering and Technology, Vol : 7, Issue:1 , pp: 270-279, March 2015,ISSN : 0975-4024

[8.] Jabez J, Gowri S, Vigneshwari S, Albert Mayan J and Srinivasulu S (2018)," Anomaly Detection by Using CFS Subset and Neural Network with WEKA Tools", Smart Innovation, Systems and Technologies, Vol. 107, pp. 675–682.

[9.] Foisal Mahedi Hasan et al., " RFID-based Ticketing for Public Transport System : Perspective Megacity Dhaka", 3rd IEEE International Conference on Computer Science and Information Technology (ICCSIT), vol. 6, pp. 459-462, 2010

[10] S. Muruganandam , N. Srinivasan ,"Appraisal of Felder - Silverman Learning Style Model with Discrete Data Sets",Indian Journal of Science and Technology, Vol 9(10), DOI: 10.17485/ijst/2016/v9i10/88992, 2016.

[11] Anish Hamlin M R, Albert Mayan J,"Blood donation and life saver-blood donation app",International Conference on Control, Instrumentation, Communication and Computational Technologies (ICCICCT),pp. 625 - 628, 2016.

[12] Kamalesh M. D., Albert Mayan J., Felix Y., Sumanth B. S., & Sai Tej B. (2018), " Magrisys: A Smart And Ubiquitous Controlled – Environment Agriculture System " , 2nd International Conference on I-SMAC (IoT in Social, Mobile, Analytics and Cloud) , pp: 1-5 , 2018

[13] Albert Mayan J , Kumar Akash Deep , Mukesh Kumar , Livingston Alvin , Siva Prasad Reddy , "Number plate recognition using template comparison for various fonts in MATLAB ", IEEE International Conference on Computational Intelligence and Computing Research, 2016 ,PP: 1-6 .

A SURVEY PAPER ON COLOUR DETECTION SYSTEM FOR INDUSTRIAL APPLICATIONS USING ARDUINO

B. Baron Sam[1], Abhilash Ekka[2], M. Ashfaq Hingora[3]

Assistant professor[1], Student[2], Student[3]

School of computing, Sathyabama Institute of Science and Technology, Chennai

baronsam1988@gmail.com[1], abhilashekka2@gmail.com[2], ashhingora@gmail.com[3]

Abstract. A Colour Sensor, as the name recommends, is a gadget that faculties or recognizes hues. A shading sensor makes utilization of an outer wellspring of discharging light (a variety of four light transmitting diodes to be exact) and afterward breaks down the light reflected over from the protest with a specific end goal to decide the question's shading. Shading sensors will give a precise shade of the question. An extensive variety of uses can be done utilizing shading sensors like arranging of items in light of their shading, quality evaluation frameworks, shading improvement in printers and so on. In this project, we intend to design an Arduino Colour Sensing and detection application, which has the capability to detect distinct colours. We have utilized TCS3200 shading sensors for this reason. Prologue to shading sensor, circuit chart and working of the Arduino Colour Sensor venture are clarified further.

keywords: Arduino Mega R3; TCS3200 colour detection sensor; Quality Control; Colour based Sorting.

1. Introduction

The color detection system utilizes Arduino in intriguing and prestigious ventures for techies who might want to consolidate electronics, industries and programming. These sorting machines are utilized mostly for sorting RGB hues. The TCS3200 is utilized with the system. This sensor goes about as an immaculate shading sensor for the proposed system. The TCS3200 PCB module joins a TAOS (4 numbers), LED control circuit with some basic portions. The core of this proposed framework is the Arduino microcontroller. Each one of the relationship in the circuit is made through the Arduino board. This will cut back the affiliation complexities and help in the general execution of machines. The RGB sensors and different parts are related with Arduino and a 9 to 12 V of voltage is used for driving the component. This framework is an ongoing running framework according to the programming code.

Shading is a particular normal for a protest, be it of any sort. Along these lines, it has been used as the establishment for numerous explores and related works. It has been the fundamental aspect of study for various skin pigmentation prospects. M. Wimmer et al.[2] and A. Albiol et al.[3] have worked extensively in this field to present various avenues where color identification can play a significant role. Gang Li et al.[4] and H. Fleyeh[5] have researched its various implementations in traffic control and roadways utility. Here, we intend to utilize the objects' colors for industrial applications such as sorting and quality control.

The premise of the color detection system is based on the fact that various industries are centered on products with their colors as a distinctive feature. Industries manufacturing multi-colored light bulbs, glass bottles, food products etc. Apart from these, shading assumes a critical part in deciding the quality of products in almost every manufacturing industry. Color detection serves as the basis for further processing such as sorting of products and quality assessment.

The proposed system intends to utilize this ability of the color sensor in order to automate further processing tasks such as sorting & arrangement of products and quality assessment. The entire system comprises only of the Arduino Mega R3 and the TCS3200 color sensor. This enables the system to be very efficient and simple to implement.

2. Related Work

P. Sebastian et al [1].This paper contemplates the impact of shading space on the execution of following calculations. The shading spaces that were researched were grayscale, RGB, YCbCr and HSV. The execution of a standardized cross connection following calculation was estimated to decide vigor and exactness in the distinctive shading spaces.

M. Wimmer et al [2].An approach that uses an abnormal state vision module to identify a picture particular skin shading model. This model is then used to adjust parametric skin shading classifiers to the handled picture. This approach is proficient to recognize skin shading from to a great degree comparative hues, for example, lip shading or eyebrow shading.

A. Albiol et al [3].The target of this paper is to demonstrate that for each shading space there exists an ideal skin locator plan with the end goal that the execution of all these skin identifiers plans is the same.

Gang Li et al [4].A yellow license tag area calculation utilizing RGB model of shading picture and dark white surface investigation of the plate specifically to extricate vehicle tag from a confused foundation picture.

H. Fleyeh [5].This paper intends to display three new techniques for shading recognition and division of street signs. The pictures are taken by an advanced camera mounted in an auto. The RGB pictures are changed over into IHLS shading space, and new strategies are connected to remove the shades of the street signs under thought.

2.1 PROBLEM DESCRIPTION

In manufacturing industries, where color plays a significant role, such as glass bottle and light bulb manufacturers, the task of organizing and packaging end products could turn hectic and complicated. Apart from these, industries focused on food products, such as fresh food packaging industries and large scale food products manufacturers often suffer from decayed items spread among the raw materials. This problem may seem trivial in the lower levels but in the broader scheme, it costs a substantial amount of capital to the industries and brings about the loss of the most important resource; time. Although many of these industries have adopted relevant technologies to tackle such

IOP Conf. Series: Materials Science and Engineering **590** (2019) 012037 doi:10.1088/1757-899X/590/1/012037

impediments, these methods usually tend to be very sophisticated and expensive. Here, we propose a generally straightforward and financially savvy answer for this issue. The proposed framework consolidates an Arduino microcontroller board and a TCS3200 shading sensor to adequately recognize hues and mechanize the assignment of association and quality evaluation.

3. IMPLEMENTATION

The implementation of the system is relatively simple. It requires only an Arduino Mega R3 board and a TCS3200 color sensor. The technical description of these hardware are given ahead. The connections between the color sensor and Arduino is made through jumper wires which enable them to communicate. This setting is further installed in a support frame thus unifying the system. An additional LCD screen can be used if required.

The entire functioning of the system is partitioned into 3 phases.

1. Calibration.
2. Color detection.
3. Application.

The entire process follows a linear sequence. The first part deals with the installation and calibration of the components. The initial values are assigned to the components. After each component is installed and tested the system can start its operations. We move to the next stage of the process. In the second stage, viz. color detection, the TCS3200 determines the color of the incoming objects through the use of LEDs and the photodiode. Then the final stage of the process approaches. Here, the RGB values are compared with the standard values stored in the system to carry out further applications such as sorting & organization and quality assessment.

3.1 Calibration

This phase deals with the installation of the components. All the components are connected and incorporated into a frame. The connections are made and the system is booted up. Initially, all the coding has to be fed to the Arduino so it is connected to a computer. Using the Arduino IDE, all the required code is written and uploaded to the board. The results can be viewed through the serial monitor present in the IDE. The key functionality in this phase is provided by the **setup()** function of the Arduino IDE. It is used to initialize the fundamental values of all the components to ensure that everything proceeds seamlessly. The setup() function utilizes two pre-defined functions, that are provided by the IDE. They are

- pinMode() – It is utilized to dole out a usefulness of either info or yield to the pins of the shading sensor.
- digitalWrite() – It is utilized to decide the voltage that must be provided to the pins of the shading sensor.
- This phase is essentially a one-time process. After the calibration is complete, the process moves to the next stage.

3.2 Color detection

This phase deals with the objects that need to be processed. After the calibration is done, the 2nd and 3rd phase are carried out simultaneously over and over again, until all the objects are processed. This phase, along with the 3rd phase, is carried out through the loop() function. In this function, the

IOP Conf. Series: Materials Science and Engineering **590** (2019) 012037 doi:10.1088/1757-899X/590/1/012037

TCS3200 is used to identify the colors of incoming objects and subsequently process them as per the requirement. The TCS3200 comprises of a configurable photodiode which changes over light to electric signs and an extra current to recurrence converter is incorporated which yields square waves. The recurrence of the waves is specifically relative to the light force, in this manner diverse hues with differing powers can be recognized. The loop() function utilizes the following built in functions:

- digitalWrite() - It is utilized to decide the voltage that must be provided to the pins of the shading sensor.
- pulseIn() – It is utilized to decide the shade of the protest by the yield recurrence gave by the shading sensor.
- map() – It is used to normalize the output value provided by the pulseIn() function.
- Right after the identification of the color, the final processing is done within the same loop() function.

3.3 Application

This phase deals with the application part which is usually guided by the primary requirement of the user. The algorithms used in this phase shall differ depending on the purpose. Here, the two main applications are

- Product sorting and arrangement
- Product quality assessment

The operating algorithm for both these purposes is similar. In sorting applications, the value retuned in the 2nd phase will be matched with the respective values and the product will be categorized accordingly. In quality assessment applications, it is checked whether the returned value falls under a certain bracket; if it does then it is accepted, if not then it is discarded.

4. Procedure

1. The components are connected and incorporated in a frame work.
2. The system is then powered up and every component is tested.
3. The initial setup is done; all the initial values are assigned and every component is calibrated.
4. The objects are now fed to the system. The color sensor detects their colors.
5. The color is identified through the reflected light from LEDs and the frequency determines their intensity.
6. The value is returned, which is then normalized to obtain the final value.
7. Thus, the color obtained is used to perform the required task
8. The objects are sorted by comparing the obtained and standard values. Quality assessment is also carried out similarly.

4.1 Architecture Diagram

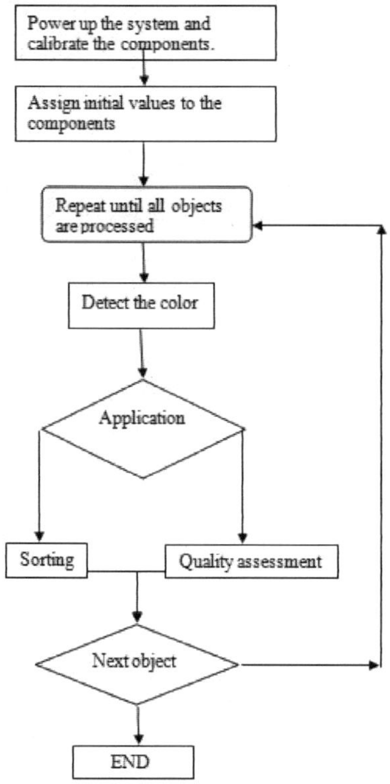

Fig 1. Architecture Diagram

4.2 Circuit Diagram

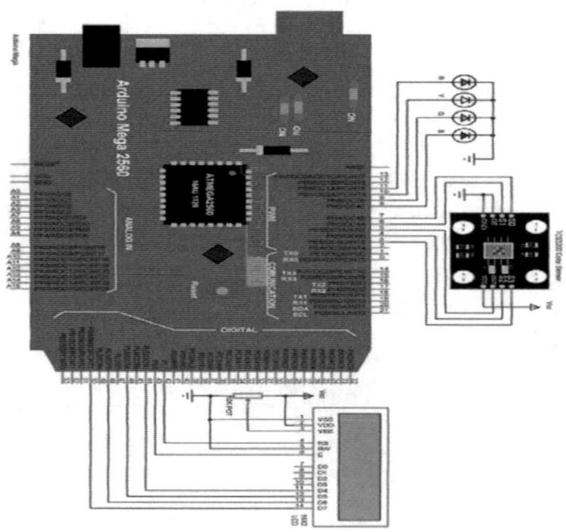

Fig.2 circuit diagram

The key components used in this system are the Arduino Mega R3 microcontroller and the TCS3200 color sensor. The technical details of these components are described below.

4.3 Arduino Mega R3

The Arduino Mega 2560 is a microcontroller board in light of the ATmega2560.It has advanced info/yield pins from 0 to 53 (15 can be used as PWM yields), simple information sources from A0 to A15, 4 UARTs (equipment serial ports), a 16 MHz crystal oscillator, a Universal Serial Bus link, a power line, an ICSP header, and a reset catch. It contains everything expected to help the microcontroller; essentially interface it to a PC with a USB link or power it with an AC-to-DC connector or battery to begin.

4.4 TCS3200 COLOR SENSOR

- High-Resolution Conversion of light intensity to frequency.
- Programmable color to full-scale output frequency.
- Communicates directly with a microcontroller.
- Single supply operation (2.7V to 5.5V).
- Power down feature.
- Nonlinearity Error Typically 0.2% at 50 kHz.
- Stable 200 ppm/°C Temperature Coefficient.
- Low-Profile Lead (Pb) Free and RoHS Compliant Surface-Mount Package.

The TCS3200 is a programmable shading light-to-recurrence converter that consolidates configurable silicon photodiodes and a current-to-recurrence converter on a solitary solid CMOS coordinated circuit. The yield is a square wave (half obligation cycle) with recurrence straightforwardly relative to light force (irradiance). Advanced data sources and computerized yield enable direct interface to a microcontroller or other rationale hardware.

All photodiodes of similar shading are associated in parallel. Pins S2 and S3 are utilized to choose which gathering of photodiodes (red, green, blue, clear) are dynamic. Photodiodes are 110 µm x 110 µm in size and are on 134-µm centers.

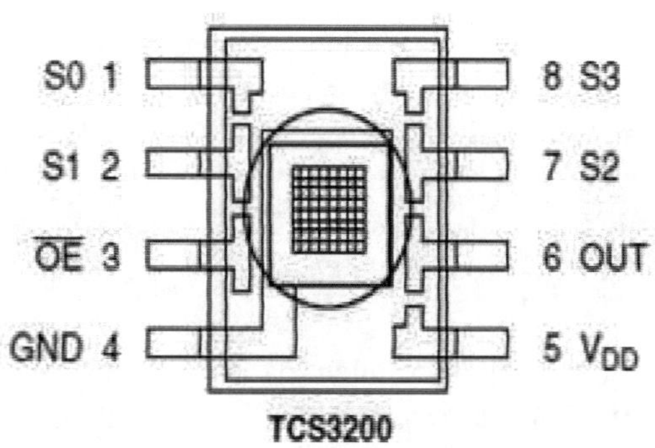

Fig 3.TCS3200

IOP Conf. Series: Materials Science and Engineering **590** (2019) 012037 doi:10.1088/1757-899X/590/1/012037

4.5 Terminal Functions

TERMINAL		**I/O**	**DESCRIPTION**
NAME	**NO**		
GND	4		Power supply ground. All voltages are referenced to GND.
S0, S1	1, 2	I	Output frequency scaling selection inputs.
OE	3	I	Enable for f_o (active low).
VDD	5		Supply voltage.
OUT	6	O	Output frequency (f_o).
S2, S3	7, 8	I	Photodiode type selection inputs

4.6*Selectable Options*

S0	**S1**	**OUTPUT FREQUENCY SCALING (f_o)**	**S2**	**S3**	**PHOTODIODE TYPE**
L	L	Power down	L	L	Red
L	H	2%	L	H	Blue
H	L	20%	H	L	Clear (no filter)
H	H	100%	H	H	Green

5. Conclusion

The proposed framework presents a genuinely straightforward and financially savvy answer for the issue of modern warehousing and bundling and item quality appraisal. As more and more processes are being automated in industries, this system presents a very efficient alternative to the task of colour based sorting and storage. Although some industries have already adopted such technologies, they have to pay a heavy price for it as the systems present currently are very complex and sophisticated. Consequently, the proposed framework is an exceptionally basic approach to complete the insignificant yet critical errand of warehousing The quality assessment process in industries can also be done through this system.

References

[1] P. Sebastian, Y. V. Voon, R. Comley, "The Effect of Color Space on Tracking Robustness", *Industrial Electronics and Applications 2008. ICIEA 2008. 3rd IEEE Conference on*, pp. 2512-2516, June 2008.

[2] M. Wimmer, B. Radig, M. Beetz, "A Person and Context Specific Approach for Skin Color Classification", *Pattern Recognition 2006. ICPR 2006. 18th International Conference on*, vol. 2, pp. 39-42, 2006.

[3] A. Albiol, L. Torres, E. J. Delp, "Optimum Color Spaces For Skin Detection", *Image Processing 2001. Proceedings. 2001 International Conference on*, vol. 1, pp. 122-124, 2001.

[4] G. Li, R. Yuan, Z. Yang, X. Huang, "A Yellow License Plate Location Method Based on RGB Model of Color Image and Texture of Plate", *Digital Media and its Application in Museum & Heritages Second Workshop on*, pp. 42-46, Dec. 2007.

[5] H. Fleyeh, "Color Detection and Segmentation for Road and Traffic Signs", *Cybernetics and Intelligent Systems 2004 IEEE Conference on*, vol. 2, pp. 809-814, 2004.

[6] Asha P, Albert Mayan J, Canessane A (2018),"Efficient Mining of Positive and Negative Itemsets Using K-Means Clustering to Access the Risk of Cancer Patients", Communications in Computer and Information Science ,ICSCS 2018, Kolla

ANONYMOUS DUAL-FACTOR AUTHENTICATION USING SMART CARD

Kambaduru sai Vikas kumar[1]
U.G Student, Dept. Of CSE,
Sathyabama Institute of Science and Technology, Chennai, India
[1]sai355205@gmail.com

Kotte Bhavani sai nikhil[2]
U.G Student, Dept. Of CSE,
Sathyabama Institute of Science and Technology, Chennai, India
[2]sainikhilkotte999@gmail.com,

Albert Mayan J[3]
Assistant Profesor, Dept. Of CSE,
Sathyabama Institute of Science and Technology, Chennai, India
[3]albertmayan@gmail.com

Abstract—Nowadays, IT-enabled service gains more attention due to easy to access resources from a remote place. IT-enabled services extend their service to all kind of business and personal related applications like e-commerce, e-business, e-transactions, e-healthcare, etc.. Such development parallelly brings a need for security concern. The system as part proposes a technique to ensure access security. It uses a two-factor authentication scheme using Elliptic Curve Cryptography thereby providing a secured and multi-level authentication service. Authentication is based on two-factor authentication with smart card and binary code, which provides high security with minimum computational cost. The scheme generates a new session key for every new session with a fresh timestamp and nonce value.

Keyword----Radio Frequency Identification (RFID), smartcard, MYSQL database, Elliptic Curve Cryptography (ECC).

1. Introduction

Anonymous dual-factor authentication that uses passwords and smartcard could preserve user privacy and reduce the risk of using a single authentication factor. Compared to previous security processes this will be more secure because of using the binary code as encryption and decryption. The first password authentication scheme based on a smart card without a confidential verification table stored on the server, which is a key advantage of two-factor schemes compared to confidential password table for this last on the server. Mobile computing devices are of great

IOP Conf. Series: Materials Science and Engineering **590** (2019) 012038 doi:10.1088/1757-899X/590/1/012038

security concernnot only because of the data stored on them, but also for that they may provide access to other services that storeor display non-public data. For almost all these transactions, mutual authentication and user privacy are required in the key exchange before remote servers start providing services to users.

In particular, authentication and privacy play animportant role in applications for industrial networks, wireless sensor networks, distributed networks, as well as RFID systems. Due to the advantages on portability and usability, most proposed authenticated key exchange (AKE) protocols support two factor authentication using passwords and smart cards.Binary code as a key makes authentication more secure. Tapping system will be easy for the user to use or access smart card. Secure level will be very high with the combination of smart card and binary code usage. Extending the security model of AKE to support user anonymity and resist lost-smart-card attack, and then propose a dynamic ID-based Anonymous Two-Factor protocol which satisfies all the following properties: (1) Security against various attacks including de-synchronization attack, lost-smart-card attack and offline dictionary attack; (2) User anonymity and un-traceability;(3) Perfect forward secrecy; (4) No long-term public key;(5) No centralized password storage, and support user password change and smart card revocation; (6)Provable security in extended security model.The rest of the paper is as follows: section II describes the Literature survey of previous papers, section III explains the proposed model for the paper and the dataset, it is processing, section IVdescribes the methodology used and the cryptography algorithm studied, section V presents the results and discussion,lastly section VI concludes the findings.

2. Literature Survey

An Effective 3-factor authentication for Secure Communication using Improved Smart Card Authentication was latently published in 2016. It provides Secure Communication from various attacks and low Communication and Computation cost.
Intercept Behaviour Analysis of Industrial Wireless Sensor Networks in the Presence of Eavesdropping Attack was proposed in 2015. Asymptotic intercept probability analysis is performed to provide an imminent into the impact of the sensor scheduling on the wireless security.
Collaborative Localization and location verification in WSNs was proposed in 2015. It proposed the virtual force model to determine the location by incremental refinement.
Implementation of Single Sign-On Mechanism for Distributed computing was proposed in 2014. Implementation of Single Sign-On for distributed computing using user-id and password along with biometric verification.

Industrial Wireless Sensor Networks, Challenges, Design Principles, and Technical Approaches was proposed in 2014. Wired automation systems require expensive communication cables to be installed and regularity maintained and thus they are not widely implemented in industrial plants because of their high cost.

IOP Conf. Series: Materials Science and Engineering **590** (2019) 012038 doi:10.1088/1757-899X/590/1/012038

3. Proposed Model

Two-Factor Authentication Scheme is combined with smart card and binary code. Elliptic Curve Cryptography is used for encrypting binary code. This algorithm is proposed so that encryption will be more secured. This scheme is verified with Random Oracle Model for real-time security proof. This requires a nominal amount of computational cost and communication cost compare to other related schemes. The binary code is the key using here to encrypt and decrypt. Earlier fingerprint, facial recognition, pin number, one-time password and some other processes are used. Binary code as a private key is a most protected way for user's data. For a fingerprint scanning process, when user's finger got injured, the problem will occur to encrypt and decrypt. Like this for every process got some privacy lag. Binary code accessing is the most secure process for any security process. The components used are RFID reader and RFID tag. Radio Frequency identification reader is used to gather information from an RFID tag. It is used to track the individual objects like tags, electromagnetic fields. The tags contain the electronically stored information. The tags are two types one is Passive tags and the other is an Active tag. Passive tags collect energy from a nearby RFID reader that produces the radio waves. Active tags have a local power source like battery and it may operate hundreds of meters from the RFID reader.

Fig 1: Dual-Factor Authentication System

This RFID technology is similar to the barcodes. The RFID tag can be scanned in any way do not have to be scanned directly or in the line of sight to a reader. The tag must be within the range of 3 to 300 feet to the RFID reader. The tag can be scanned for many times and enables fast identification of a particular product if the RFID reader is surrounded by several items the reader will read the card because the electromagnetic waves are so powerful.

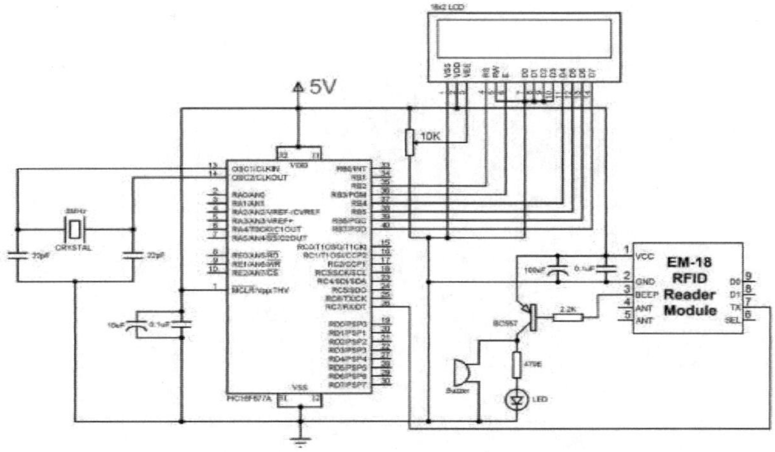

Figure 2. Circuit Diagram

Active RFID and Passive RFID tags are the types of RFID tags. These two types are varied by their owing of power sources. Active RFID tag contains its own power source, often like a battery. Passive RFID tag does not need batteries. From a reading antenna, it gets its power required. These RFID tags identification barcodes cannot be changed because of their cost, each tag has an individual identity for every item.

The RFID technology is used in some applications like Passports, Airplane luggage, Home appliances, Merchandise tags, Telephone and computer networks, Pallet tracking for inventory. RFID tag contains the digital data tracked or easily scanned by the RFID technology. RFID tag is made up of integrated circuits. These contain a tiny antenna for transferring information to an RFID transceiver. The most RFID tags contains an integrated circuit for modulating and demodulating the radio frequency and also an antenna for transmitting and receiving signals. Radiofrequency ranges from low with 125 to 134kHz and 140 to 148.5 kHz, high with 850 to 950 MHz and 2.4 to 2.5GHz. Wavelengths on 2.4 GHz range are very less because they can be absorbed by water. The data in a common RFID tag can be read by anyone easily. User can't be able to know how the data is exposed. This is due to the reason that the RFID tags have a less amount of computing power which is not enough for encryption. In the case of RFID tags used in passports, they have sufficient computer power for the encryption.

There are three pieces of information used in the passport – the passport number, the birth date of the passport holder and passport's expiration date – along with checksum digit for each of the three. This is called Basic Acces Control(BAC). The U.S State Department adopted BAC system in 2007 and added an anti-skimming material to electronic passports to reduce the threat of attempts to steal user's personal information.

4. BLOCK DIAGRAM:

The user will first access the smart card if the smart card is accessible otherwise the access will get denied. Using Elliptical Curve Cryptography the binary code will be generated.

The generated binary code will check the accessibility in the data base provided. If the matching binary code is available in the database, The locking system will be accessed automatically, Otherwise, the control will go back to the user notifying that the password is invalid

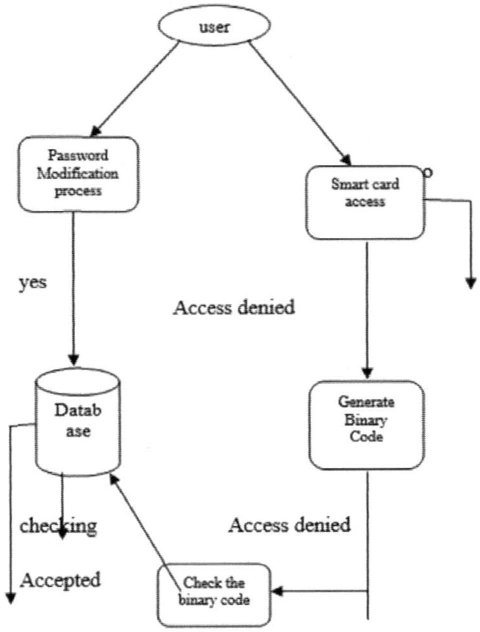

Figure 3. Flow Diagram

. The user can modify the password anytime. Binary System will be more secure than the related schemes. Tapping system makes user easy for access. The password or any information will be hidden under an image and when some people tried to see the information, only the image will be shown. Security level will be increased by using this scheme.

5. ALGORITHM

MD5 Algorithm:

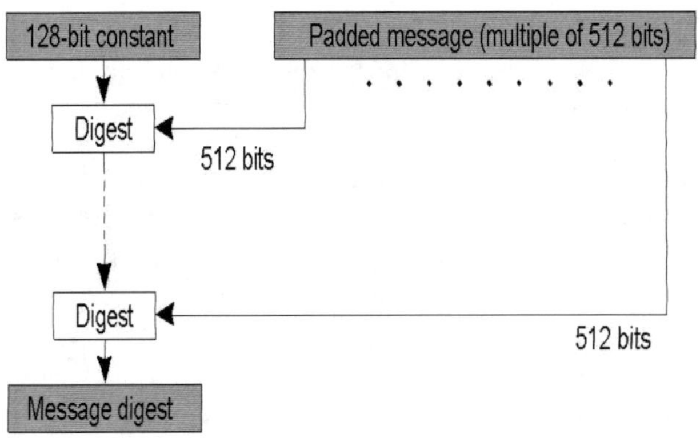

Figure 4. Algorithm

IOP Conf. Series: Materials Science and Engineering **590** (2019) 012038 doi:10.1088/1757-899X/590/1/012038

The MD5 algorithm was developed by Professor Ronald L. Rivest in 1991.MD5 is a message-digest algorithm which produces an output of a message of arbitrary length as a 128-bit message. This is used for digital signature applications where compression takes place for a large file. This is a secure manner beforebeing encrypted with a private key.MD5 is one of the most efficient algorithms currently in use.

Implementation steps:

Step1. **Append padding bits**

The input message is "padded" (extended) so that its length (in bits) equals to 448 mod 512. Padding is always performed, even if the length of the message is already 448 mod 512. Padding is performed as follows: a single "1" bit is appended to the message, and then "0" bits are appended so that the length in bits of the padded message becomes congruent to 448 mod 512. At least one bit and at most 512 bits are appended.

Step 2. **Append length**

A 64-bit representation of the length of the message is appended to the result of step1. If the length of the message is greater than 2^64, only the low-order 64 bits will be used. The resulting message (after padding with bits and with b) has a length that is an exact multiple of 512 bits. The input message will have a length that is an exact multiple of 16 (32-bit) words.

Step 3. **Initialize MD buffer**

A four-word buffer (A, B, C, D) is used to compute the message digest. Each of A, B, C, D is a 32-bit register. These registers are initialized to the following values in hexadecimal, low-order bytes first):

word A: 01 23 45 67
word B: 89 ab cd ef
word C: fe dc ba 98
word D: 76 54 32 10

Step 4. **Process message in 16-word blocks**

Four functions will be defined such that each function takes an input of three 32-bit words and produces a 32-bit word output.

F (X, Y, Z) = XY or not (X) Z
G (X, Y, Z) = XZ or Y not (Z)
H (X, Y, Z) = X xor Y xor Z
I (X, Y, Z) = Y xor (X or not (Z))

Elliptical Curve Cryptography:

ECC offers greater security for a given key size. The smaller key size also makes possible much more compact implementations for a given level of security, which means faster cryptographic operations, running on smaller chips or more compact software. There are extremely efficient, compact hardware implementations available for ECC exponentiation operations, offering potential reductions in implementation footprint even beyond those due to the smaller key length alone.

IOP Conf. Series: Materials Science and Engineering **590** (2019) 012038 doi:10.1088/1757-899X/590/1/012038

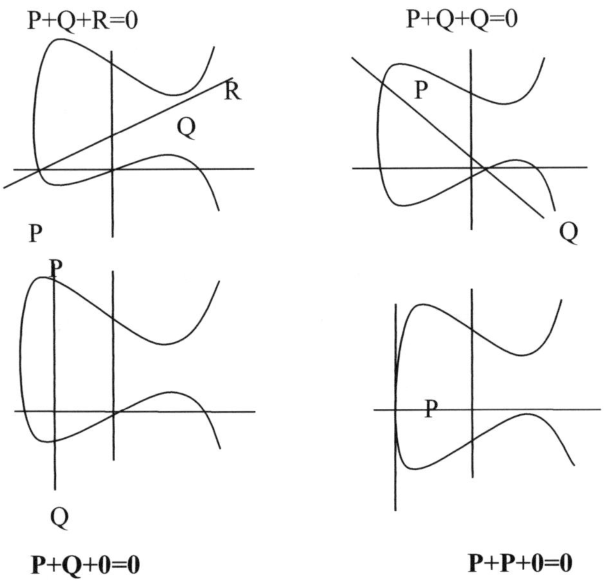

Figure 6. Grouping process

GROUP

A set of objects and an operation on pairs of those object form which a third object is generated. The group must be closed, invertible, the operation must be associative, and there must be an identity element.

Example: integers 0-9 and addition modulo 10

Closed: the sum of 2 numbers from 0-9 modulo 10 is
An integer from 0 to 9.

Identity: 0 since 0+x = x for any x

Invertible: x + (10-x) = 0 mod 10 for any x
So (10-x) is the inverse of x

Associative: (x+y)+z = x+(y+z)

Finite Field:

If p is any prime number and n is any positive integer, then there exists a finite field of size p n. There are no other finite fields.

Example: p = 7, n = 1, operation = * 0 1 2 3 4 5 6

If G = { g0, g1, g2, g3, g4, g5 } 0 0 0 0 0 0 0 0 is a group, then g6 = g0 and 1 0 1 2 3 4 5 6

G is cyclic 2 0 2 4 6 1 3 5 3 0 3 6 2 5 1 4

generator = 3 4 0 4 1 5 2 6 3 3, 2, 6, 4, 5, 1, 3 5 0 5 3 1 6 4 2

inverse: 6 0 6 5 4 3 2 1 1:1, 2:4, 3:5, 4:2, 5:3, 6:6

6. RESULT and DISCUSSIONS :

After entering admin login details, the smart card is to be tapped.

The binary code is generated and one tab is opened for admin to add image and the data that is to be secured.

IOP Conf. Series: Materials Science and Engineering **590** (2019) 012038 doi:10.1088/1757-899X/590/1/012038

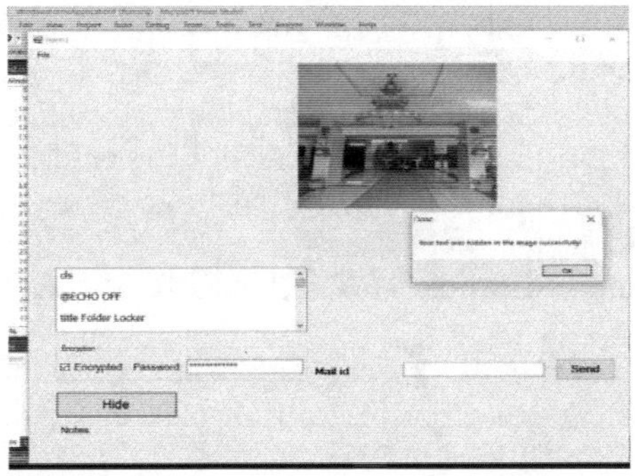

Figure 7. Input image

After adding image and data, the data will be hidden before that Image.
The user after login, cannot see the data directly, he can only see the image. After tapping, the data will be displayed.

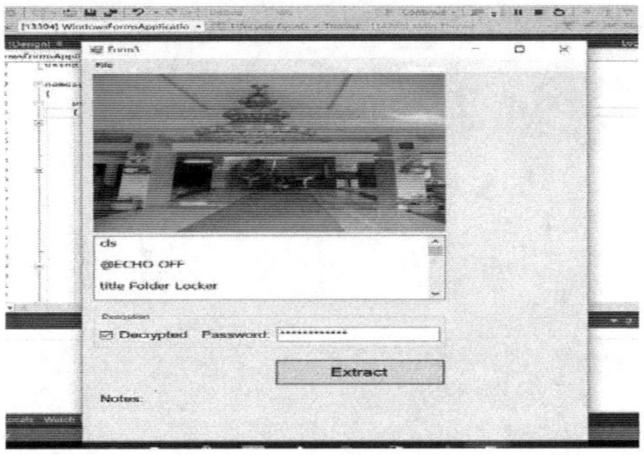

Figure 8. Output image

7. Conclusion:

Smart card based authentication scheme has been widely utilized for various transaction-oriented services such as electronic currency exchange, social insurance payment, and e-commerce payment charge in modern society. To develop a smart card based authentication scheme to support initiator traceability and defend against major security threats for a transaction service user has become a crucial topic for researchers. Recent efforts for developing anonymous authentication scheme with the smart card have failed to provide initiator non-tracing ability for a user or be vulnerable to some security attacks.

References:

[1] A. Valenzano, L. Durante, and M. Cheminod, "Review of security issues in industrial networks," IEEE Trans. Ind. Inf., vol.9, no. 1, pp. 277-293, 2013.

[2] Mayan, J.A., Priya, K.L.," Novel approach to reuse unused test cases in a GUI based application", IEEE International Conference on Circuit, Power and Computing Technologies, ICCPCT 2015,pp.1-7

[3] D. Liu, M. C. Lee, and D. Wu, "A Node-to-Node Location Verification Method," IEEE Trans. Ind. Electron., vol. 57, no. 5, pp. 1526 - 1537, May 2010.

[4] C. Chang and C. Lee, "A secure single sign-on mechanism for distributed computer networks," IEEE Trans. Ind. Electron., vol. 59, no. 1, pp. 629-637, Jan. 2012.

[5] G. Wang, J. Yu, and Q. Xie, "Security analysis of a single sign-On Mechanism for Distributed Computer Networks," IEEE Trans. Ind. Inf., vol. 9, no. 1, pp. 294-302, 2013.

[6]L. Barolli and F. Xhafa, "JXTA-OVERLAY: A P2P platform for distributed, collaborative and ubiquitous computing,"IEEE Trans.Ind. Electron. vol. 58, no. 6, pp. 2163-2172, Oct.2010.
[7]Albert Mayan J, Anto Praveena M D , Telkar Bharath Rao , Uravakonda Uday Sagar(2018),"Optimized test data generation over suspicious implementation of oracle problem",IEEE International Conference on Power, Control, Signals and Instrumentation Engineering (ICPCSI-2017), pp.2559-2563.
[8] B.Wang and M. Ma, "A server independent authentication scheme for RFID systems," IEEE Trans. Ind. Inf., vol. 8, no. 3, pp. 689-696,Aug. 2012
[9] Mayan J.A, Arifa S, Pavithra R,"Semantic based multi lexical ranking technique for an effective search in protected cloud,2016 International Conference on Control, Instrumentation, Communication and Computational Technologies (ICCICCT), Kumaracoil, 2016, pp. 570-576.
[10] Asha P, Albert Mayan J, Canessane A (2018),"Efficient Mining of Positive and Negative Itemsets Using K-Means Clustering to Access the Risk of Cancer Patients",Communications in Computer and Information Science ,ICSCS 2018, Kollam, 2018,pp.373-382.

SIGN SEQUENCE BASED GESTURE RECOGNATITON USING MEMS ACCELEROMETER FOR AUGUMENTAD REALITY APPLICATIONS

Karthick C [1]
Assistant professor, Dept of ECE,
Sathyabama Institute of Science and Technology
OMR Road,Chennai, India
karthivlsi13@gmail.com

VINO T [2]
Assistant professor, Dept of ECE,
Sathyabama Institute of Science and Technology
OMR Road,Chennai, India
vinodhevan@gmail.com

Surendra K [3]
U.G Student, Dept of ECE,
Sathyabama Institute of Science and Technology
OMR Road,Chennai, India

Rewanth kumar Reddy L R [4]
U.G Student, Dept of ECE ,
Sathyabama Institute of Science and Technology
OMR Road,Chennai, India

Abstract- The Increase in Human Machine Interactions in our daily life is getting important day by day. Physical gesture will greatly ease the interaction process and enable humans to more naturally command the machinery like computers. In this hand gestures are mostly used as the control mechanism in virtual reality. To ease these physical gestures MEMS accelerometer can be used which can recognize the several hand gestures. In the real world the physical gesture machinery control is already in existence such as vision system or data glow but these are relatively expensive. By the usage of MEMS, the cost can be reduced and method can be implemented more effectively.

This project deals with an accelerometer-based gesture recognition application. This application 0consists of a tri-axial accelerometer and same is connected to raspberry pi through wire. The tri-axial accelerometer measures the acceleration signals generated by a users hand motions to identify the character. The code for application program in raspberry pi is written using python programming language. The text is displayed on the monitor of the PC.

Keywords: Raspberry pi, Mems accelerometer, Accuracy, gesture recognition, python. Etc

1. INTRODUCTION

Technology has played a very significant role in improving the quality of life. The simplification of work can be done by automating several tasks using complex logic. Much attention from many research communities such as human computer interaction and image processing has been given to gesture recognition. More importance has been given to the interface technology due to the increasing human machine interactions in our daily life. Physical gestures will greatly ease the interaction process and enable humans to more naturally command computers or machines. To convey information a system is created using the primary goal gesture recognition. Vision based and accelerometer based are the two existing types of gesture recognition method. The limitations that we have, while using vision based method are slower dynamic response, unexpected ambient optical noise and relatively large data collections. To overcome these limitations we are using Mems accelerometer [1]. AVR Microcontroller is used in real time hand gesture recognition. In order to get the output object is needed to be presented in front of camera. It limits mobility and cannot use camera in dark area as well as cost is very high [2]. Inertial MEMS present in smart phones are used to tackle the task of symbolic gestures, But Inertial MEMS are more sensitive to feeble movements [3]. First a class of finite state machines (FSM) for gesture recognition were constructed in first phase and then tuned for higher accuracy with the help of some training data and a suitable optimization method in the second phase. In this the power consumption is very high [5].Unsupervised network self-organising map is applied for character recognition and to classify the characters [6].

2. DIFFERENT METHODOLOGIES OF EXISTING

There are three methods in the existing system. They are 1) sign sequence and Hopfield based gesture recognition model 2) velocity increment based gesture recognition model and 3) sign sequence and template matching based gesture recognition mode.

A. Sign sequence and Hopfield based gesture recognition model.

There are two types of gestures one is which has only motions on one axis (up, down, left, right) and the other one is which has 2-D motions (circle, tick, cross). These two are stored separately. Minimum value and maximum value of the accelerations are used to Wright the coding and find the gestures. We get eight numbered coding by combining the x and z axis sign sequences. But the input for Hopfield network can be only 1 or -1. So we will use positive sign, negative sign and zero. We use following rules to get these signs,"1 1" represents positive sign, "-1 -1" represents negative sign, "1 -1" represents zero. Standard patterns for the seven gestures will be pre given and will be compared to get the output. **[1]**

B. Velocity increment based gesture recognition model

Identification of gestures will be done based on velocity increment. An area sequence should be normalized before stored as training data or compared with templates. Normalization is implemented using

$$Anorm = Aoriginal / Amax \qquad \text{eq... } 2.1$$

Where Aoriginal is the original area, Amax is maximum area and Anormal is normalized area in a sequence. The final step is to compare the velocity increment sequence by subtracting two area sequence vectors. The gesture which has minimum value by comparing one area sequence with the training sample can be recognized. [1]

IOP Conf. Series: Materials Science and Engineering **590** (2019) 012039 doi:10.1088/1757-899X/590/1/012039

C. Sign sequence and template matching based gesture recognition mode.

This is very similar to model one. All sign sequences are represented by -1,1and 0 unlike model one. This method is not limited to specific users. In this the algorithm is based on feature of acceleration sign changes. [1]

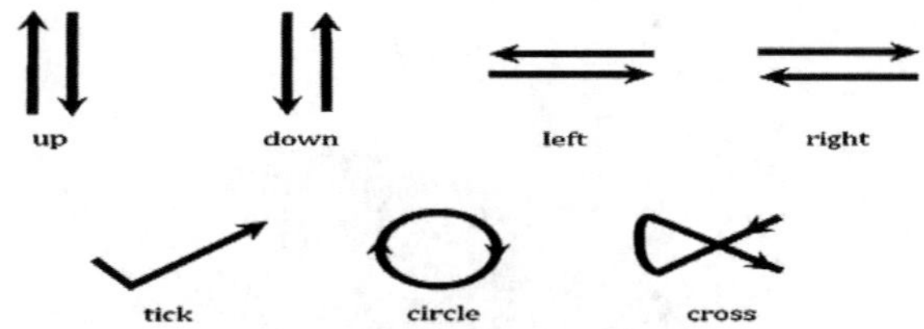

Fig. 1 Motion of Seven Gestures [1]

D. Comparison of gesture recognition accuracy of above three Methods.

	Up	*Down*	*Left*	*Right*	*Tick*	*Circle*	*Cross*	*Mean*
Method I	*95.0*	*86.0*	*91.0*	*84.0*	*64.0*	*75.0*	*61.0*	*79.0*
Method II	*87.0*	*19.0*	*63.0*	*94.0*	*25.0*	*0.0*	*88.0*	*54.0*
Method III	*94.8*	*91.1*	*96.70*	*100*	*94.4*	*97.7*	*94.4*	*95.6*

Table. 1 Comparision of Gesture Recognition Accuracy (%)of Three Methods[1]

3. PROPOSED SYSTEM

LOT measures the similarity between a candidate and the target using locally order less matching, and HSV colour space is used to describe the appearance of each pixel. MEEM uses features extracted in the LAB colour space. In the most recent work, CSK is extended with colour names, and to speed up, the dimension of the original colour names is reduced with an adaptive dimensionality reduction technique.

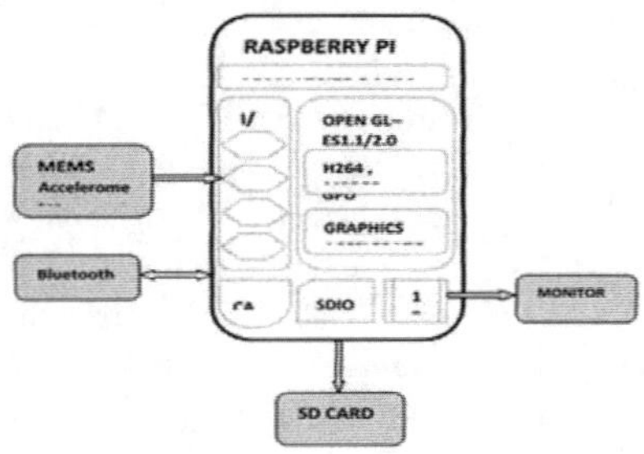

Fig 2 Block diagram of proposed system.

A. Micro-Electro-Mechanical-System (MEMS)

MEMS technology consists of microelectronic elements, actuators, sensors, and mechanical structures built onto a substrate, which is usually silicon. They are developed using micro fabrication techniques deposition, patterning, and etching. The most common forms of production for MEMS are bulk micromachining, surface micromachining, and HAR fabrication. The benefits on this small scale integration bring the technology to a vast number and variety of devices.

Fig 3 MEMS accelerometer

B. Raspberry pi

In this project we are dealing with raspberry pi, which has ARM 11 core architecture with Broadcom BCM2837 Soc chip. It has I GB SDRAM with video and audio outputs. Audio output has 3.5mm jack and HDMI. It has 4 USB 2.0 connectors. It has micro SD slot to store the memory by placing a memory card. Linux operating system is used in this. It has general purpose input output connector (GPIO).

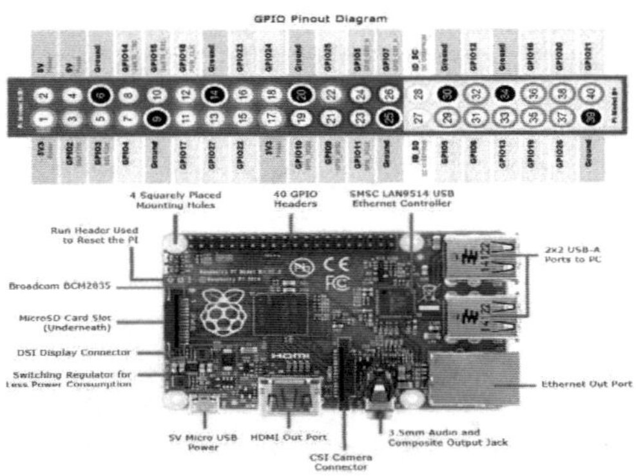

Fig 4. Raspberry pie

C. Accelerometer Sensor

It is an electromechanical device which measures acceleration forces. These forces maybe static which are constant force of gravity pulling at your feet, or they could be dynamic which are caused by moving or vibrating the accelerometer. By moving the MEMS accelerometer gesture will be drawn on the monitor. The drawn shape will be recognized according to the pre given code which has been written using python language. Three switches will be provided for the given functions. First one is for start and stop of the gesture drawing. Second one is to clear the screen and the third one is for getting the output. In this we can get two modes of output. One is in the text form, which is already given in the program for particular gestures and the other one is audio form, where the displayed output text will be read out. There are two ways to get the audio output, one is through HDMI and the other is from audio jack which is present in raspberry pi.

1. RESULT AND DISCUSSION

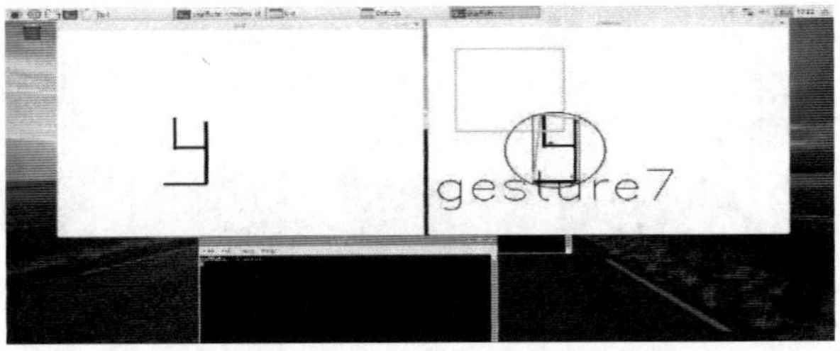

Fig. 5 Gesture Recognition of Alphabet " Y type"

Fig. 6 Gesture Recognition of Alphabet " W type"

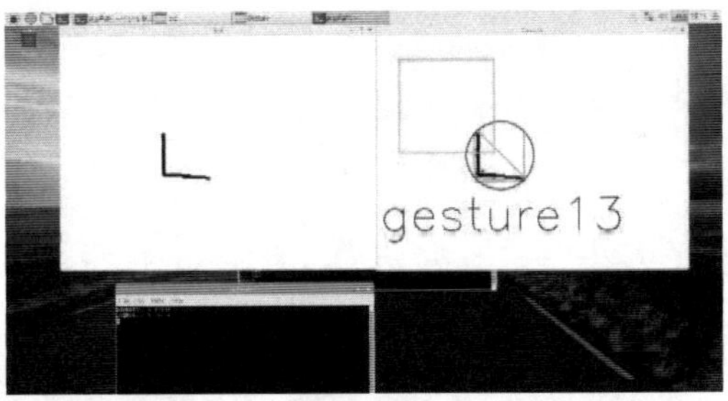

Fig. 7 Gesture Recognition of Alphabet " L type"

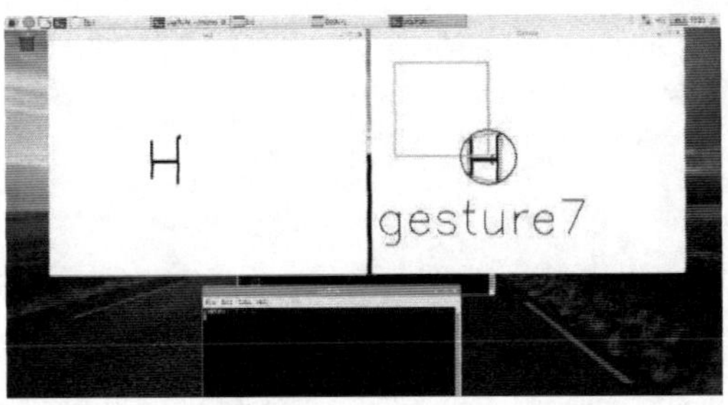

Fig. 8 Gesture Recognition of Alphabet " H type"

IOP Conf. Series: Materials Science and Engineering **590** (2019) 012039 doi:10.1088/1757-899X/590/1/012039

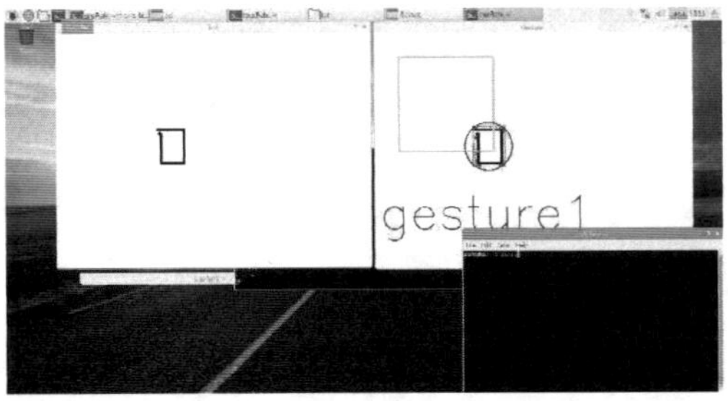

Fig. 9 Gesture Recognition of Alphabet "O type"

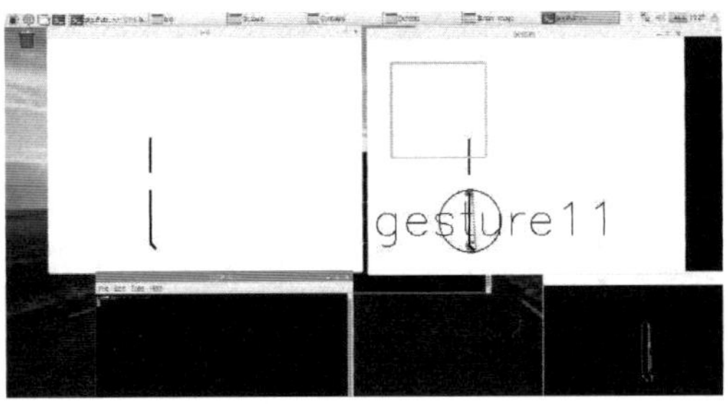

Fig. 10 Gesture Recognition of two separated lines

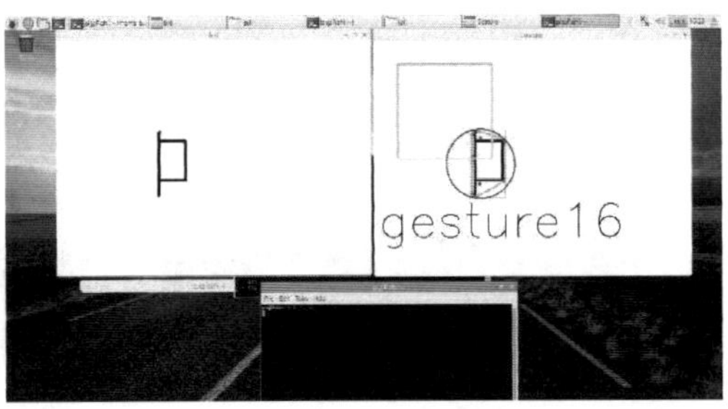

Fig. 11 Gesture Recognition of Alphabet " P type"

Above figures is the output of some alphabetic gestures such as "W, H, L, O and P". In this method we have used angles to differentiate the gestures. So the accuracy is more, when compared to the existing methods. By using MEMS accelerometer the cost can be reduced.

2. CONCLUSION

The gesture can be recognized by MEMS accelerometer sensor. The sensor analog outputs have sent to the ATmega8 microcontroller. In microcontroller, already stored data can compare the input data of gesture and also transmitted to the PC. On running in a PC, the power point presentation has been manipulated. The gesture values are found out for 85 experiments and found that gesture recognition based on sign sequence and template matching gives high (96%) accuracy when compared to other models. The future work will be

IOP Conf. Series: Materials Science and Engineering **590** (2019) 012039 doi:10.1088/1757-899X/590/1/012039

done with the speech output using the speaker with hand gesture input, and also more recognition methods will be investigated.

REFERENCES

1. Ruize Xu, Shengli Zhou, Wen J. Li "MEMS Accelerometer Based Nonspecific-User Hand Gesture Recognition" IEEE sensors council,volume-12,issue 5 2012.
2. M. Shakunthala Devi, R.B. Revathi "Real time hand gesture recognition using AVR Microcontroller", information Communication and Embedded Systems (ICICES), 2014 International Conference on 2014.
3. Samuel Berlemont, Gregoire Lefebere "Siamese neural network based similarity metric for inertial gesture classification and rejection", Automatic Face and Gesture Recognition (FG), 2015 11th IEEE InternationalConference and Workshops on 2015.
4. Nina Rudigkeit, Marion Gebhard, Axel Graser "An analytical approach for head gesture recognition with motion sensors", sensing technology (ICST) 2015 9th international conference .2015.
3. Faisal Waris, Robert G. Reynolds "Human-defined, machine-optimized – gesture recognition using a mixed approach", Swarm/Human Blended Intelligence workshop (SHBI) ,2015.
4. 6 .S.Zhou, Z. Dong ," Hand-written character recognition using MEMS motion sensing technology", technology," in Proc.IEEE/ASME Int. Conf. Advanced Intelligent Mechatronics, 2008, pp. 1418–1423.2008.
5. L. Yun and Z. Peng, "An automatic hand gesture recognition system based on Viola-Jones method and SVMs", in Proc. 2nd Int. Workshop Comput. Sci. Eng., 2009, pp. 72-76
6. A. H. F. Lam, W. J. Li, Y. Liu, and N. Xi, "MIDS: Micro input devices system using MEMS sensors," presented at the IEEE/RSJ Int. Conf. Intelligent Robots and Systems (IROS), Oct. 2002.
7. J. K. Oh, S. J. Cho, and W. C. Bang et al., "Inertial sensor based recognition of 3-D character gestures with an ensemble of classifiers," presented at the 9th Int. Workshop on Frontiers in Handwriting Recognition, 2004.
8. H. Je, J. Kim, and D. Kim, "Hand gesture recognition to understand musical conducting action," presented at the IEEE Int. Conf. Robot & Human Interactive Communication, Aug. 2007.
9. Karthikeyan S., Karthick C., Shibu Prasath S.V. "Human tracking system for victims trapped from collapsed building" ARPN Journal of Engineering and Applied Sciences, 2016
10. Subudhi, J., Karthick C, "Implementation of vedic divider on RSA cryptosystem". ICIIECS 2015 - 2015 IEEE International Conference on Innovations in Information, Embedded and Communication Systems,. 2015
11. Kumaresan, R., Senthilkumar, S., Karthick, C. "Biometric high secure and cost effective finger vein authentication system for ATM" ARPN Journal of Engineering and Applied Sciences, 2015.

IOP Conf. Series: Materials Science and Engineering **590** (2019) 012040 doi:10.1088/1757-899X/590/1/012040

Li-Fi BASED REAL TIME VIDEO TRANSMISSON FOR PATIENT MONITORING SYSTEM

K.Hashwanth Saikumar[1]
BE, Electronics and Communication Engineering,
Sathyabama Institute of Science and Technology,
Chennai
k.hashwanth@gmail.com

K.Akshay Teja[2]
BE, Electronics and Communication Engineering,
Sathyabama Institute of Science and Technology,
Chennai
axaytejachowdary@gmail.com

T. Ravi[3]
Associate Professor,
School of Electrical and Electronics Engineering,
Sathyabama Institute of Science and Technology,
Chennai
ravivlsi123@gmail.com

S.Karthikeyan[4]
Associate Professor,
School of Electrical and Electronics Engineering,
Sathyabama Institute of Science and Technology,
Chennai

Abstract- The aim of this work is to achieve live surveillance of the patient using LI-FI (light fidelity). Li-fi is an advanced and promising technology can be used in many medical applications. In this paper, a prototype of live video broadcast using LEDs is proposed. The prototype presented transfers the data with single LED, and is capable of illumination as well as data broadcast. Whole process is kick started by converting analog signal into digital form in transmitter section. LM359 IC amplifies video signal with two inverting amplifiers which possess high gain bandwidth and allows high frequencies and it is coupled to LED which accelerates the transmission process and it's geared at high speed by receiving the high frequency video signal in digital form by a photo detector. The output from the receiver circuit is analog output given to television module using a RCA cable.***Keywords**:* Li-Fi transmission, Transmitter chip, receiver chip, Android application, prolific cable.

1. INTRODUCTION

In this electronic world, using radio wave frequencies for wireless communication cause a loss in data rate, light fidelity (Li-Fi) modulation technique is used as an alternative. The light emitting diode is a semi-conductor device that emits light when an electric current pass through it. The light can transmit the data in the binary form, which cannot be visualized by a human naked eye. The bandwidths of radio frequencies are limited now-a-days due to increase in the wireless communication technology. Visible light communication is point to point communication (Line Of Sight LOS).Li-Fi has the advantage of being useful in electromagnetic sensitive areas such as in aircraft cabins, hospitals and nuclear power

IOP Conf. Series: Materials Science and Engineering **590** (2019) 012040 doi:10.1088/1757-899X/590/1/012040

plants without causing electromagnetic interference [3]. Li-Fi is expected to be ten times cheaper than Wi-Fi. Different types of photo detectors can be used to detect the transmitting light and convert the binary information into electrical form. A photo detector has a P-N junction which uses reverse bias technique to convert light photons into electrical form. The main objective is to transmit the high frequency video monitoring signal through illumination. Visible light communication is an effective high bandwidth communication system serves point –to- point communication [3]. Visible light communications (VLC) have many advantages over radio links. One of these advantages include the availability of huge bandwidth. The schematic structure of visible light communication system is shown in figure 1. With increase in demand of video and high quality image transmission this paper reviews without any modulation techniques and improvement in channel capacity. In home environment, the combination of illumination and transmission is captivated in specific scenarios which are healthy and secured.

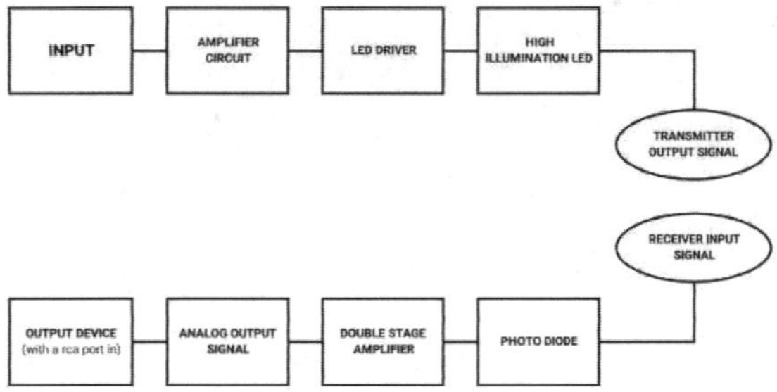

Fig.1 Block Diagram of Li-Fi

The next sections include existing system and proposed work with methodology describing about experimental setup and final result and conclusion.

2. EXISTING WIRELESS COMMUNICATION SYSTEM

The hospital area contains much critical care equipment which is used to monitor patient medical status. For the officials of hospital, to monitor the status of patient they have to use electronic devices such as cell phone or personal computer for the data communication. Most of the hospitals around the world use Wi-Fi (Wireless Fidelity). Wi-Fi is hazardous in such situations as the electromagnetic frequencies radiates in low-gigahertz frequencies. The Electromagnetic interferences due to this, results in potentially harmful events related to medical equipment operations [4,7]. According to researches done 20% patients suffer insomnia due to the effect of electromagnet radiations. Since these routers should work for whole day, patients are effected by electromagnetic frequencies who are being exposed by it 24/7. Some patients' lives are endangered from equipment malfunctioning in intensive care room in such situations. Due to this, the Wi-Fi connection are prohibited completely in some parts of the hospital building. The CCTV in the hospital environment are connected to the Wi-Fi routers to transmit the video broadcast to the doctor or officials to monitor the patient continuously from their rooms. Figure 2 shows wireless communication system used in hospitals [13]. It can overcome such limitations by Li-Fi (light fidelity) [16,18]. The light fidelity can act as illumination as well as data transmission which is an efficient alternative. The main components of this communication system are LED, Which acts as a Transmitter source. Photodiode which shows good response to visible wavelength region serving as the receiving element [9]. This system gives the lower data rates transmission and its cover very less distance System integrate with Embedded Plat form. In the Hospitals the light fidelity system is limited to send only data, audio and images of sensing elements like heartbeat sensors, EEG sensors, temperature, respiration sensor, glucose level, and other related sensors [14].

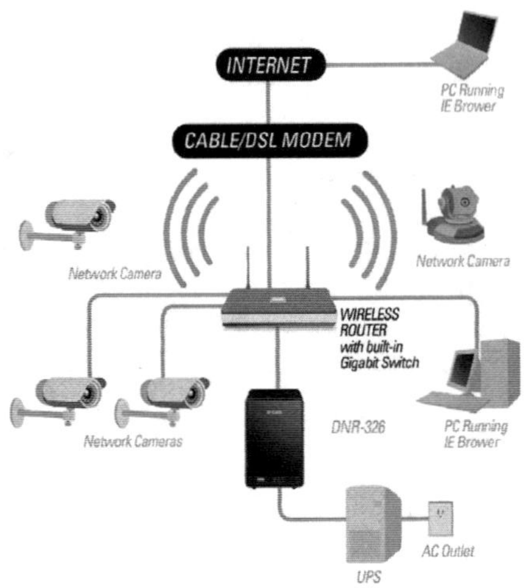

Fig2. Existing Wi-Fi Communication System

3. PROPOSED LI-FI BASED REAL TIME VIDEO TRANSCIEVER SYSTEM

The CCTV surveillance of hospitals especially in ICU's using light fidelity reduces health related issues of patients. The camera installed in hospitals (in intensive care) can communicate each and every information to the doctors monitor with the use of LED. The light fidelity (Li-Fi) technology is one of the promising solutions to increase transmission capacity, one issue in implementing light fidelity is to select the appropriate access technique in the multiuser environment. In this paper we explore the transmission of video through LI-FI, which requires higher frequency to transmit and in equal to that we need such a receiver circuit to hold such a high frequency signal. This receiver circuit is linked to a television with a RCA input jack which display the transmitted video.

The prototype presented below transfers the data with single LED, and is capable of supporting both illumination and data broadcast for monitoring the situation of patient. The following scenarios explain about transmitter section, receiver section and the components which allows the video broadcast through the LED's.

4. PROTOTYPE OF Li-FI BASED REAL TIME VIDEO TRANSMISSION

- **TRANSMITTER**

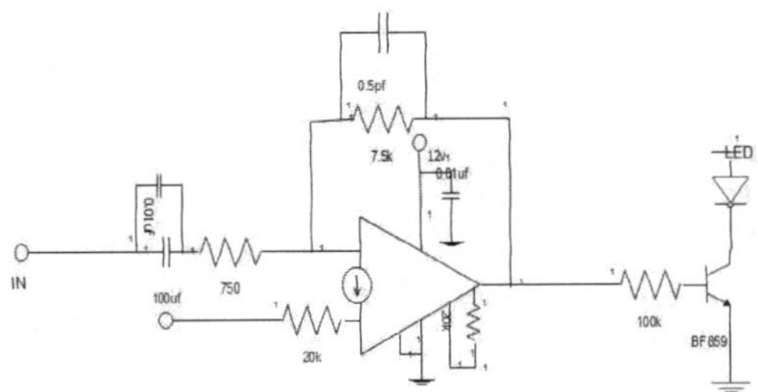

Fig3. Transmitter Circuit Diagram

In the transmitter section, the electric signal comes from CCTV or TV signal using a VGA to RCA converter, the signal received as video/analog signal is converted into electrical signal. LM359 integrated circuit is used to amplify the received signal from the input. The optical source used in this work is a light emitting diode which emits radiation at 650 nm. The maximum output of optical LED is power of 5v. Transistor type BF859 is used to allow the high frequency received from LM359 to the LED and biased it with a suitable resistance and capacitors as shown in figure. The capacitors used in the transmitting section is to reduce the dc components and different ohms of resistors are used to lower the voltage levels, reduce the current and each resistor has capacity to allow maximum frequency of its own.

- **LM359 INTEGRATED CIRCUIT**

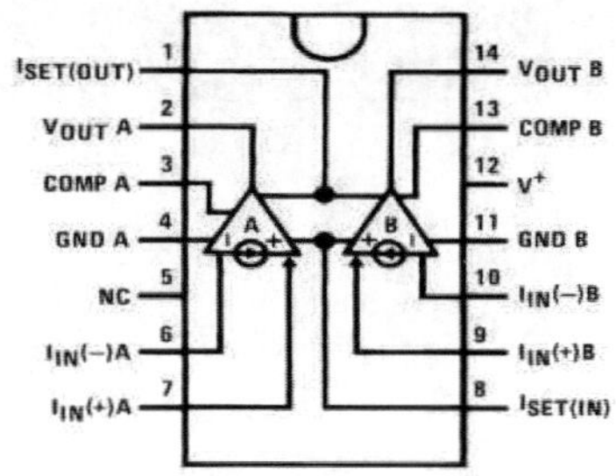

Fig4. LM359 Pin Diagram

The high speed LM359 operational amplifier is used to provide high gain bandwidth. LM359 consist of two inbuilt inverting amplifiers which allows high frequencies. It can be used as general purpose video amplifiers and has wide frequency range waveform generations. The amplifiers in LM359 are designed to operate from a single supply and can accommodate input voltages greater than the supply given. The output of LM359 is amplitude modulated and the amplified signal is converted into optical form through the source connected with driver circuit[17]. The pin diagram of LM359 is shown below.

- **BF859 NPN TRANSISTOR**

BF859 is NPN high voltage transistor. The main application is to use in video output stages of black and white and color television receivers. The first pin emitter is grounded, the second pin collector, connected to mounting base cathode of LED and third pin base connected to output of the LM359. In the transmitter section BF859 is used to allow the high frequency through the LED which should transfer the video broadcast of high frequency.

- **LED DRIVER CIRCUIT**

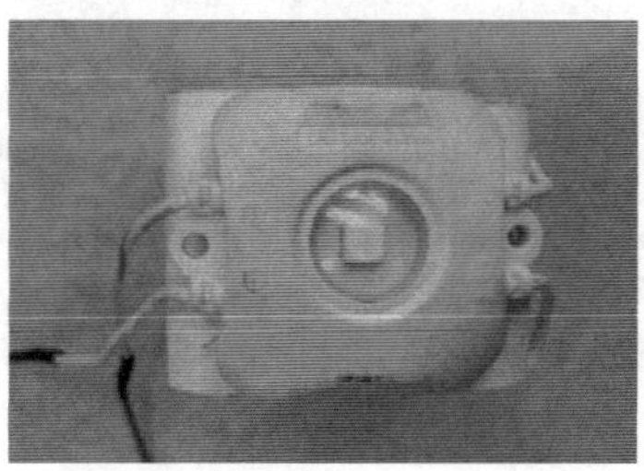

Fig5. Light Emitting Diode

The light emitting diode transfers the information in the digital binary form 1's and o's. The LED flickers 1500 times a second which cannot be observed by a human naked eye. The basic concept of LED is when the light glows it is considered as 1, when it is in off it is considered as 0.

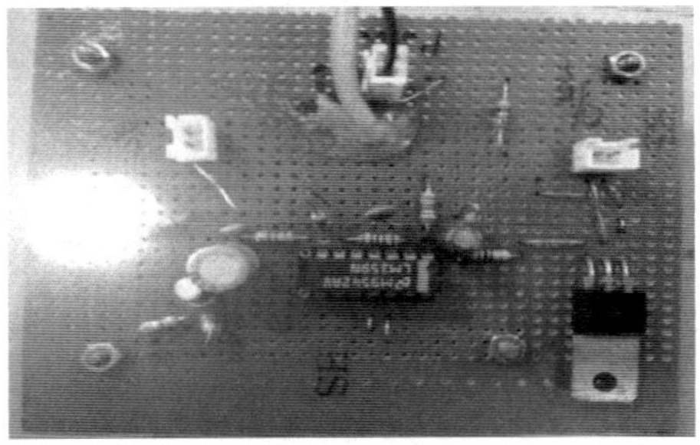

Fig6. Prototype of Transmitter

- **RECEIVER**

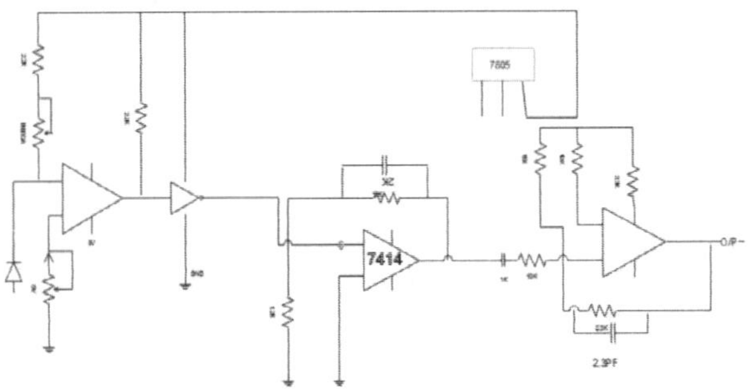

Fig7. Receiver Circuit Diagram

The photodetector receives the information through the light and is responsible for converting the received optical form into electrical form of equivalent pulses. The information received from the photodetector is fed to the amplifier LM359 to amplify the signal received. A dual high speed LM359 preamplifier in the receiver unit amplifies the output signal from the photodetector. The inverting schematic trigger amplifier 7414 is connected to the amplifier LM359. The received signal from the photodetector may not match with frequency or information sent by the transmitter so perfect square wave shaping circuit used. The stages in receiving unit are designed to amplify the signal as much as possible, around of ten times. The regulator integrated circuit 7805 is used to maintain the output voltage. The capacitors used in the receiver section is to reduce the dc components and different ohms of resistors are used to lower the voltage levels, reduce the current and each resistor has capacity to allow maximum frequency of its own. The circuit diagram of receiver section is shown in figure.

- **PHOTODETECTOR**

Photodetectors receive the information from the LED and converts the received the optical information into electrical pulses. A photo detector has a P-N junction which converts light photons into current which uses

IOP Conf. Series: Materials Science and Engineering **590** (2019) 012040 doi:10.1088/1757-899X/590/1/012040

reverse bias technique. The image sensors present in photodetectors transforms an optical image into an electrical signal. The photodetector type BPX 65 with a required rise time and peak responsivity is used. Figure 8 shows the circuit diagram of PhotoDetector.

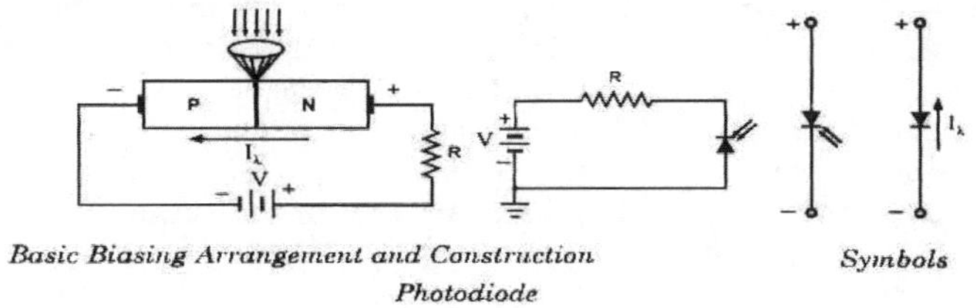

Fig8. Photodetector Circuit Diagram

- **INVERTING AMPLIFIER 7414**

A 7414 inverting amplifier is connected to the amplifier LM359 which has less input impedance, this is because of the feedback resistors. This is used to eliminate the noise near reference point which would cause problems in analog comparators. It operates from very slow edges, it has improved line receiving characteristics and high noise immunity. It can give clean, jitter free output signals.

- **7805 REGULATOR**

The 7805 regulator is connected to the unit to maintain the output voltage at a constant value. The 7805 regulator integrated circuit provides 05 volts as the output with provisions to add heat sink, preferred size of heat sink should be selected.

Heat generated = (input voltage-5)*output current

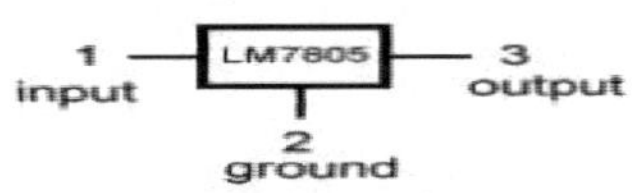

Fig 9. 7805 Regulator

Fig 10. Prototype of Receiver

IOP Conf. Series: Materials Science and Engineering **590** (2019) 012040 doi:10.1088/1757-899X/590/1/012040

6. RESULTS AND DISCUSSION

Li-Fi is advancement in wireless communication systems which possess high security, speed than existing wireless data transmission systems. Each and every LED can act as a transmitting device if it is perfectly implemented in real time. The main advantage of Li-Fi is its unregulated bandwidth unlike radio frequencies which contains a limited bandwidth. Now-a-days increment of radio frequencies usage led to limit some application which can be achieved by Li-Fi. All the equipment by default uses 2.4 Ghz frequency, which makes the equipment very difficult to function as they are interfered. If the medical equipment is interfered with the radio frequency it may fall under faulty condition and can menace patients' lives. The problems of CCTV surveillance in the hospitals are vanquished by the proposed system which is reviewed in this paper. The output of receiver section is the analog form which can be displayed in television module or monitor as shown in the below figure 11.

Fig11. Prototype of Real Time Video Transceiver

The blurriness or noise is due to the conversion of VGA input to RCA which is the input transmitter section. The following figure Figure-13 shows the output of the proposed system. This can be vanquished by using high voltage tuners ant the input side as shown in the figure 12.

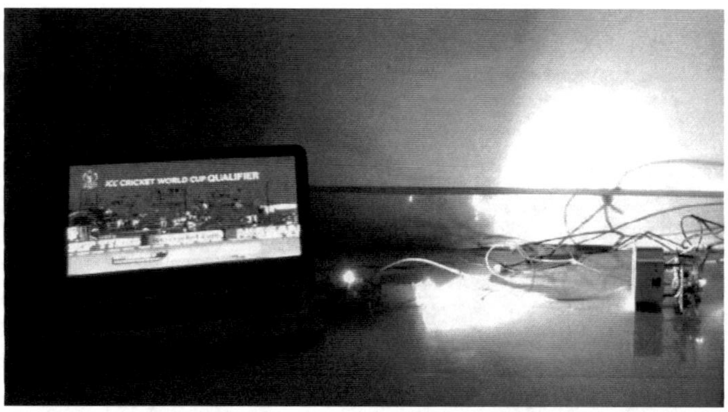

Fig12. Transmitted Real Time Video

Fig13. Received Real Time Video

7. CONCLUSION

This paper concludes that the safe usage of CCTV or camera in the sensitive area like hospital especially in the ICU's or intensive care. The existing systems cause unhealthy effects on the patients' in hospital due to the radio frequency interferences. The proposed system overcomes the effects by replacing the radio frequency components with Li-Fi equipment. The transmission of high frequency live broadcast video transfer is successful. By comparing the original video with the output there are some disturbances due to conversion of VGA to RCA this can be overcome in future work.

REFERENCES

[1] Ankit Navalakha, NeeluMaheshwari, "Data Services of Li- Fi in Hospital Management" , International Journal of Science and Research (IJSR) 2014

[2] Hani J. Kbashi College of Science, Department of Physics, University of Baghdad, Baghdad, Iraq, "Experimental Investigation of a Laser Wireless Video Communication System Using Intensity Modulated /Direct Detection Technique" Iraqi J. Laser, Part A, Vol.9, No.1 pp. 1-6 (2010)

[3] Ravi.T "10-Nanometer carbon nano tube field effect transistor based high celerity transposed polyphase decimation filter" Materials Today: Proceedings, Volume 3, Issue 6, 2016, pp: 1799–1807

[4] Jyoti Rani, PrernaChauhan, RitikaTripathi. 2012. Li-Fi (Light Fidelity)-The future technology In Wireless communication. International Journal of Applied Engineering Research, ISSN 0973-4562

[5] Kahn J. and Barry H. "Wireless infrared communications" Proc. of the IEEE, 85, No. 2, 1997, pp. 265-298.

[6] M. Kavehrad, "Sustainable energy-efficient wireless applications using light," IEEE Communications Magazine, vol. 48, no. 12, pp. 66-73, Dec. 2010, doi: 10.1109/MCOM.2010.5673074

[7] M. Samuel Lazar and T. Ravi, "Li-Fi Design For High Speed Data Transmission", VOL. 10, NO. 14, AUGUST 2015 ISSN 1819-6608, ARPN Journal of Engineering and Applied Sciences

[8] R. Karthika and S. Balakrishnan, "Wireless Communication using Li-Fi Technology," International J. of Electronics and Communication Engineering (SSRG-IJECE), vol. 2, issue 3, pp. 32-40, March 2015

[9] S.Sudha Ms.Indumathy D, Lavanya .A, Nishanthi M " Patient Monitoring In The Hospital Management Using Li-Fi", 2016 IEEE International Conference on Technological Innovations in ICT For Agriculture and Rural Development (TIAR 2016)

[10] T. D. C. Little, P. Dib, K. Shah, N. Barraford and B. Gal-lagher, "Using LED Lighting for Ubiquitous Indoor Wireless Networking," IEEE International Conference on Wireless and Mobile Computing, pp. 373-378, 2008. doi:10.1109/WiMob.2008.57

[11] van der R. Togt, E.J. Lieshout, R. Hensbroek, E. Beinat, J.M. Binne-kade and P.J.M. Bakker, "Electromagnetic Interference From Radio Frequency Identification Inducing Potentially Hazardous Incidents in Critical Care Medical Equipment," J. the American Medical Association, vol. 299, no. 24, pp. 2884-2890, June 2008, doi:100.1001/jama.299.24.2884

[12] Woldeit E., Yong M. "Designed and constructed free space amplitude modulation laser communication system" EE 488-Laser Electronic,1996

[13] Micheal Priyanka.M and Ravi.T, "Survey on role of memristor in electronics "2015 International Conference on Control Instrumentation Communication and Computational Technologies, ICCICCT 2015, 18-19 Dec 2015, Pp:738-744.electronics "2015 International Conference on Control Instrumentation Communication and Computational Technologies, ICCICCT 2015, 18-19 Dec 2015, Pp:738-744..

[14] M.Manoranjani and T.Ravi "Multithreshold CMOS sleep stack and logic stack technique for digital circuit design" ARPN Journal of Engineering and Applied Sciences, Vol. 10, No. 10, June 2015, pp:4550-4556.

[15] Xin LIN, Kenichi IKAWA, Kazutoshi, HIROHASHI, Nakagawa Laboratories, Inc. "High-speed Full-Duplex Multiaccess System for LED-Based Wireless Communi-cations Using Visible Light", the 2nd International Multi- Conference on Engineering and Technological Innova-tion, 2009.

[16] S.Karthikeyan, "Ontogeny Smart Bulletin Board",ARPN Journal of Engineering and Applied Sciences, ISSN 1819-6608,Vol. 10, NO. 5, pp. 2148-2151, March 2015.

[17] R. Thenmozhi, S. Karthikeyan, "An Efficient Parallel Architecture in Design and Implementation of Adaptive LMS Algorithm",ARPN Journal of Engineering and Applied Sciences, ISSN 1819-6608,VOL. 10, NO. 7, pp 3150-3153, April 2015.

[18] Bagavath sri and S. Karthikeyan, "Secure Elliptic Curve Cryptography Based RFID Schemes for Internet of Things", Indian Journal of Science and Technology Vol 9(42), November 2016 PP: 1-5.

Design and Analysis of K-SVD Algorithm for the Noise Removal Using MATLAB

K. Akhil[1]
Department of ECE
Sathyabama Institute of Science and Technology, Chennai –India.
akhil.leo10@gmail.com

K.Prasanth Kumar Reddy[2]
Department of ECE
Sathyabama Institute of Science and Technology, Chennai –India.
prasanthreddykomatla@gmail.com

R.M. Joany[3]
Department of ECE
Sathyabama Institute of Science and Technology, Chennai –India.
mariajoany@gmail.com

Ilayarajaa .K .T[4]
Department of ECE
Sathyabama Institute of science and Technology, Chennai-India.

ABSTRACT - This paper was proposed to developed a dictionary learning algorithm to minimize or remove the noise from the noisy pixels present in the given data sets without distorting the edges and preserving the fine particulars and details of the image for removing the noise there are three stages named INTERPOLATION, OMP(orthogonal matching pursuit)and finally K-SVD dictionary learning matrix algorithm in this algorithm there are different methods for removing different level of noises present in the image by using the masking by using this algorithm noise can be from the pixel range by checking the each and every pixel whether the pixel is a noisy pixel or noise free pixel only noisy pixels are passed the noise removal process so noise free pixels would not get degraded or blurred or changed in their property this K-SVD algorithm can remove the noise Color Images, Black and White Images and Grey Scale Images The algorithm can handle noise level

Index Terms - Dictionary Learning matrix,
Discrete Cosine Transform (DCT), Image De-noising, K-SVD, Orthogonal Matching Pursuit, Redundancy, PSNR (Peak Signal To Noise Ratio),SSIM(Structural Similarity Index)

1. INTRODUCTION

Nowadays digital image processing is used widely in many areas such as medical, face recognition and military and crime investigation etc In Medical it is used for diagnosis of diseases in face recognition for security purpose and In crime investigation when the digital data is corrupted while transmitting and receiving, sometimes when digital evidence is tampered The main fundamental property of image processing is to effectively remove the noise present in a image by preserving the edges and by preserving the details and features of the image intact the nature of the problem depends on the type of noise affected to the image most oftenly the images are corrupted by Gaussian Noise, Salt And Pepper Noise, Speckle Noise, Possession Noise etc; mostly noise is added to the image during the analog to digital conversion ,pre-processing ,compressing of the image and video and while transmitting and receiving the signals with data salt and pepper noise is also known as impulse noise for removing the impulse noise from an image we use median filter and mean filter for removing the Gaussian noise ,for removing the speckle noise wavelet analysis was used and speckle noise is also known as granular noise it mostly occurs in the satellite images such as SAR which makes difficult in extracting the information from the image. Impulse noise is of two types fixed value impulse noise and random valued impulse noise random valued noise does not have any fixed value it corrupts the pixels in the range of 0-255 and it is very difficult to remove the random valued noise fixed value impulse noise is also as salt and pepper noise and this salt and pepper noise effects the image and lowers the image quality makes the difficult for edge detection, segmentation, feature extraction and recognition for the security reasons so the salt and pepper noise present in the images should be removed and the salt and pepper noise can be removed using median filtering method it is a non linear filter and the removes the noise to some extent and preserves the image details to some degree so to overcome this problem many median filtering were proposed such as the centre weighted median filter, adaptive median filter ,adaptive centre weighted median filter

The main disadvantage of these filters is that these methods treat the noisy pixels and noise free pixels present in an image using the same process so the noise free pixels changes after the process of noise removal is completed therefore the images are blurred after the processing median filter can effectively the noise only at the low level only when it comes to the high level of noise its effect drops sharply to overcome this problem first the pixels are divided into noise free pixels and noisy pixels the noise free pixels keep the original grey level and noisy pixels are replaced with their median value in the spatial domain and this method is only theoretical and its can't be archived in practical as we can't set the parameters and thresholds required Gaussian noise can be removed using spatial filter it is caused due to the natural resources such as thermal vibrations of atoms in the communication channel and sometimes this white noise is used as an additive speckle noise is also known as granular noise and it is a inherit noise and it degrades the quality of the image it is mainly occurred in the sar mages and it can be removed using multi look processing and spatial filtering Wavelet analysis transform is able to separate the noise and signal at different scales and orientation hence the original signal at any scale and direction can be recovered and used the above methods change the noise into smooth fine structures and the Gaussian filter and median filter remove the noise in fixed valued impulse noise only it can't remove the random valued noise Edges of the image can be preserved by using the anisotropic diffusion but it will form a mask effect on the uniform regions of the image bilateral filtering has overcome this problem of mask effect but it can only preserve the edges in the smooth images only each pixel of the de-noised result will be the weighted average of its neighbors according to the intensity and distance between them.

2. THE EXISTING METHOD

The existing method was proposed based on noise removal algorithm for salt and pepper noise which can adapt to different noise level it contains a noise pixel detection and noise filtering process this noise pixel detection works by counting the no of closed grey level of pixels in the neighborhood to detect the noisy pixels by this noise detection process archives high correct detection and low false alarm rates and then using the adaptive domain of Euler's distance method by this it can achieve excellent noise removal and good detail preservation using the estimate of the interference rate there is no need to adjust the parameters and thresholds manually in the filtering algorithm median filtering technique is a nonlinear digital filtering technique it is mostly used to remove the noise from images and signal it preserves the edges while removing noise from it also have application in signal processing for removing the noise from the signal without altering the signal or tampering with the data init in median filtering first the pixels in the image is divided into matrix window and the detected noise pixel is replaced by the median value of the matrix median filter is also a smoothing technique like linear Gaussian filtering all the smoothing techniques can remove the noise only at the smooth patches and regions they fail to remove the noise present at the edges so this median filter doesn't provide better results because it can't preserve the edges for small and moderate level of Gaussian noise median filter performs better than the Gaussian filter and Gaussian filter performs well at high level of noise so speckle noise and salt and pepper noise can be removed effectively So, because of this median filtering in mostly used in digital image processing [9].

Bilateral filter is also a nonlinear filter the main advantage of this filter is that it can preserve the edges of the image by removing the noise from it and it works only in smooth images only in this process noisy pixel intensity is replaced by the weighted average of intensity values from the nearby pixels weight is based on Gaussian distribution it can depend on color intensity and depth distance so it preserves the sharp edges

Gaussian filter modifies the input image by using a convolution with the Gaussian function
Anisotropic diffusion is a technique for reducing the image noise by preserving the image content, edges ,lines , other details which are required for the interpretation of the image this method works on

the diffusion process it can remove the noise from the digital images without blurring the edges non-local mean algorithm is a special case of bilateral filter it provides better edge preserving each pixel of the de-noised image from a non local mean is the weighted average of the pixels with surrounding pixels of the noisy image by using Gaussian function as a smoothing function NLM filter removes the noise up to great extent nowadays sparse representation of images has a great attention by using a over complete dictionary that has a prototype image atoms images are represented by sparse linear combination of these atoms this has been already applied in many areas such as compression, feature extraction and image de-noising by combing the NLM and sparse representation algorithm to develop an enhanced image de-noising filter first it trains the dictionary by using K-SVD method from the noisy image then the each pixels surroundings are formed into a patch and it is sparsely represented by the dictionary now the weight of the NLM'S is calculated by these patches by using these weight and with the weight average of a pixel with the surrounding is obtained then we get the de-noise

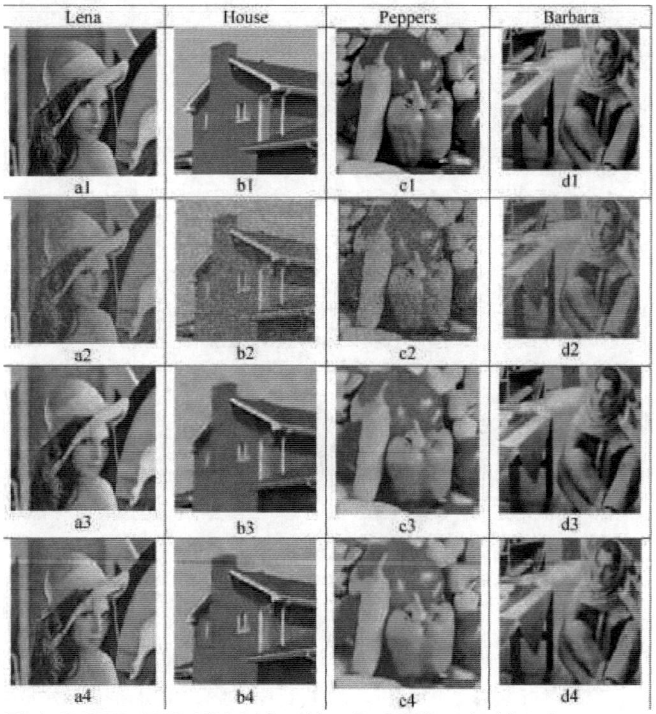

Fig1. K-SVD AND NLM Algorithm for black and white picture

3. PROPOSED METHOD

At present we have many de-noising techniques available for grey level images for the noise removal but most of the techniques perform at removing noise at low level of noise but at high level they fail remove noise effectively in this we will introduce a new method for de-noising the color images, black and white and grey level images we have mainly focus on removing the salt and pepper noise from image by using the K-SVD algorithm with improved PSNR and reduced MSE at both the high density and low density noisy image to 5%to 95% many de-noising techniques proposed are application dependent image processing has two important stages first one is noise detection and the second is that removing the noise and image enhancement [12]. This method is mostly based on K-SVD algorithm and residual error without depending on the low SNR image sparse representation of de-noising algorithm depending on the previous papers methods this paper explains the process and mechanism and analysis of the algorithm PSNR is taken as objective evaluation standard by this we can see the simulation results for different kinds of image de-noising and image sparse decomposition is based on complete atom library which has better effect on de-noising algorithm after this process the

IOP Conf. Series: Materials Science and Engineering **590** (2019) 012041 doi:10.1088/1757-899X/590/1/012041

atoms in the library has the ability to show more characteristic information and effectively extract the image features [11]. In the proposed algorithm image sparse decomposition is used to optimize the dictionary and residual error more than the threshold for accurate division of image information and provide evidence for image de-noising

The proposed method we simulate the output with the help of MATLAB in this process we first we take a color image and then it is subjected to the impulse noise and noise is added to the targeted image after that noisy pixels are detected after the noisy pixels are detected is given to the de-noising process according to the noise density level so the complete de-noising process can be divided into the following steps first stage is interpolation, second stage is orthogonal matching pursuit and finally it is given to the K-SVD algorithm

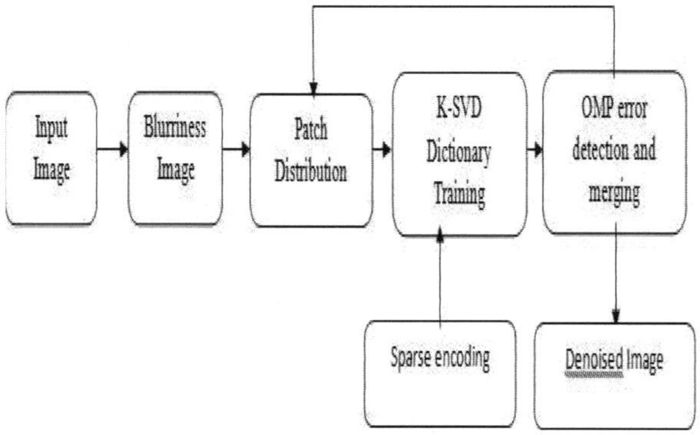

Fig2. Block Diagram of proposed work

According to the block diagram first input image is taken which may be a color or black and white without any noise then after the input stage the image is given to the add the noise then salt and pepper noise is added to the image and it is given to the patch distribution and K-SVD dictionary training with sparse encoding and them it is given to the omp and then we obtain the de noised image

The following steps are taken for removing the noise in an image without damaging the noise free pixel, securing the edges and details of the image noise free pixel are preserved

1. Input Image
2. Interpolation
3. Orthogonal Matching Pursuit
4. K-SVD
5. Output Image

IOP Conf. Series: Materials Science and Engineering **590** (2019) 012041 doi:10.1088/1757-899X/590/1/012041

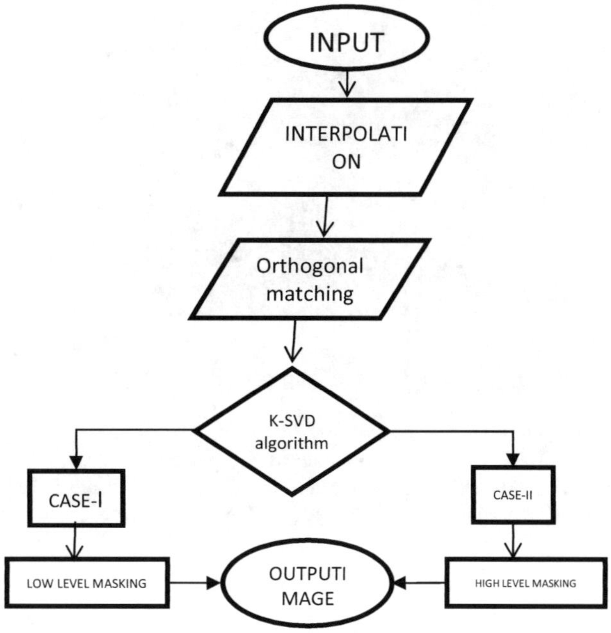

Fig3. Flow chart of proposed work

A. INPUT IMAGE

The Image which is taken input contains the noise that has to be removed from the image to get a clear image Noise in the image that is occurred during the transfer of image from the transmitter to the receiver and conversion of analogue to digital format etc; The image which is taken as input is colour image

Fig4. Image With High Level Of Salt And Pepper Noise

IOP Conf. Series: Materials Science and Engineering **590** (2019) 012041 doi:10.1088/1757-899X/590/1/012041

Fig5. Image With low Level Of Salt And Pepper Noise

B. INTERPOLATION

The input image is given to the interpolation stage at first In this stage input image is divided into blocks (pixel by pixel) to determine the noise present in that block Then the block with the noise is determined and masked up then it is given to the OMP(ORTHOGONAL MATCHING PURSUIT)

C. OMP(ORTHOGONAL MATCHING PURSUIT)

It receives the image after the noisy blocks are masked up. OMP determines the level of noise present in that block depending the level of masking .If the noise in the image is high then the level of masking in that block will be high and then it is given to the K-SVD stage .If the noise in the image is low then the level of masking in that block will be low then it is given to the K-SVD stage.

D. K-SVD

In this K means clustering and SVD means SINGULAR VALUED DECOMPOSITION.K-SVD algorithm searches the image pixel by pixel for any noises present in it and it removes the noise at the pixel range to get the original image K-SVD stage has two different cases for removing two different levels of noises(low or high)It can be determined by the ITER which is given by the OMP stage By knowing the ITER value we can determine the case and the cases can be switched using switch case statement In this the image in the block is converted into the YUV scale by comparing them with the ASV files present in the algorithm Then the block is unmasked to get the original image with the high quality And at the output image with no noise is displayed

E. OUTPUT IMAGE

Fig6. image with high level of salt and pepper noise removed

Fig7. image with low level of saltand pepper noise removed

F. PSNR & SSIM Ratio Table

	21	22	23	24	25	26	27	28	29	30	31	32
1	1.7157e+03	1.7167e+03	1.5806e+03	1.4821e+03	1.4551e+03	1.5369e+03	1.4718e+03	1.5713e+03	1.6718e+03	1.7421e+03	1.7076e+03	1.5800e+03
2	0	0	227.5233	235.4288	230.2248	162.2237	0	0	0	0	0	251.6889
3	0	0	0	0	0	0	0	0	0	141.0761	120.5304	0
4	169.9729	135.2695	0	0	0	0	228.0963	230.4180	159.0975	0	0	0
5	0	0	0	0	-93.8943	0	0	0	0	0	0	0
6	0	0	0	0	0	0	-37.8405	0	0	0	0	-71.7614
7	-61.2655	-72.4394	0	0	0	0	0	-34.8160	0	-68.2077	0	0
8	0	0	0	80.6815	0	0	0	0	0	0	0	0
9	0	0	0	0	0	28.1223	24.0289	0	0	0	0	0
10	0	0	0	0	0	0	0	0	0	0	0	0
11	0	0	0	0	0	0	-45.4619	0	0	58.2829	0	0
12	60.0930	45.1778	0	0	0	0	0	0	0	0	-54.8023	0
13	0	0	0	0	0	0	0	0	0	-33.8606	0	0
14	0	0	0	0	55.8673	0	0	0	0	0	0	0
15	0	0	0	0	0	0	0	0	0	0	0	0
16	0	0	0	0	0	0	0	0	0	0	0	0
17	0	37.9335	0	0	0	0	0	0	0	0	68.8639	99.5640
18	0	0	0	0	0	0	0	0	0	0	0	0

Fig8. Output tabulation of PSNR and SSIM ratios

PSNR = 20 * log10(b/rms)
SSIM(x,y)=[l(x,y)]α·[c(x,y)]β·[s(x,y)]γ

4. CONCLUSION

By using the K-SVD algorithm we can remove noise from an image by preserving the edges and fine details of the image and the salt and pepper noise and Gaussian noise can be removed from 5% to 95%we have calculated the psnr ratio also for the output image

REFERENCES

[1] Julien Mairal, Michael Elad, Guillermo Sapiro," Sparse Representation for Color Image Restoration", IEEE TRANSACTIONS ON IMAGE PROCESSING, VOL. 17, NO. 1, JANUARY 2008
[2] Michael Elad and Michal Aharon," Image Denoising Via Sparse and Redundant Representations Over Learned Dictionaries", IEEE TRANSACTIONS ON IMAGE PROCESSING, VOL. 15, NO. 12, DECEMBER 2006

[3] Qing-Qiang Chen , Mao-Hsiung Hung, Fumin Zou," Effective and adaptive algorithm for pepperand- salt noise removal", ISSN 1751-9659 2nd September 2016 doi: 10.1049/iet-ipr.2016.0692

[4] E.Jebamalar Leavline, D.Asir Antony Gnana Singh," Salt and Pepper Noise Detection and Removal in Gray Scale Images: An Experimental Analysis", International Journal of Signal Processing, Image Processing and Pattern Recognition Vol.6, No.5 (2013), pp.343-352

[5] S.GOPI KRISHNA, T. SREENIVASULU REDDY , G.K.RAJINI," Removal of High Density Salt and Pepper Noise Through Modified Decision Based Unsymmetric Trimmed Median Filter", ISSN: 2248-9622 Vol. 2, Issue 1, Jan-Feb 2012, pp.090-094.

[6] M. Aharon, M. Elad, and A. M. Bruckstein, "The K-SVD: An algorithm for designing of overcomplete dictionaries for sparse representations," IEEE Trans. Image Process., vol. 54, no. 11, pp. 4311–4322, Nov. 2006.

[7] D. L. Donoho, "De-noising by soft thresholding," IEEE Trans. Inf. Theory, vol. 41, no. 3, pp. 613–627, May 1995.

[8] Lu, C.-T., Chou, T.-C.: 'Denoising of salt-and-pepper noise corrupted image using modified directional-weighted-median filter', 2012, 33, pp. 1287–1295

[9] Ravi.T, "10-Nanometer carbon nano tube field effect transistor based high celerity transposed polyphase decimation filter" Materials Today: Proceedings, Proceedings 3 (2016), pp: 1799-1807.

[10] E. Jebamalar Leavline and S. Sutha, "Gaussian noise removal in gray scale images using fast Multiscale Directional Filter Banks", Proceedings of International Conference on Recent Trends in Information Technology, Chennai, India, (2011) June 3-5.

[11] Ravi.T and Kannan.V "Effect of N-type cntfet on double edge triggered D flip-flop based PISO shift register" IEEE Proceedings of the International Conference On Emerging Trends in Science Engineering and Technology: Recent Advancements on Science and Engineering Innovation, INCOSET 2012, Dec-2012, pp. 344-349.

[12] R.Ramya and T.Ravi, "Design of cache memory mapping techniques for low power processor" ARPN Journal of Engineering and Applied Sciences, VOL. 10, NO. 11, JUNE 2015, pp:4783-4788.

[13] S.karthikeyan, Merin Mary Koshy and V.G.Sivakumar, "Smart & automated robotic circular car parking using GSM",International Journal of Pure and Applied Mathematics Volume 117 No. 21 2017,PP: 81-86.

[14] Hari Krishna, K., Swamy, Y., Siva Kumar, V.G., Karthikeyan, S. "Integration of two way communication using gesture control hand movement and voice based text", Journal of Advanced Research in Dynamical and Control Systems, 2017, vol. No:17 , PP: 1124-1135

[15] Mayan, J.A., Priya, K.L.," Novel approach to reuse unused test cases in a GUI based application", IEEE International Conference on Circuit, Power and Computing Technologies, ICCPCT 2015,pp.1-7

[16] Bagavath sri and S. Karthikeyan, "Secure Elliptic Curve Cryptography Based RFID Schemes for Internet of Things", Indian Journal of Science and Technology Vol 9(42), November 2016 PP: 1-5.

[17] Gopika. S. Kumar, S. Karthikeyan and N. Ambily, "A Metaphorical Scrutiny in Banishing Salt and Pepper Noise",BIOSCIENCES BIOTECHNOLOGY RESEARCH ASIA,December 2014 Vol. 11(3).

IoT Based Dual wavelength Non-Invasive Haemoglobin Sensor System

G. Venkata Praveen Kumar[1]
BE, Electronics and Communication Engineering,
Sathyabama Institute of Science and Technology,
Chennai
praveen4149@gmail.com

B. Jagadeesh[2]
BE, Electronics and Communication Engineering,
Sathyabama Institute of Science and Technology,
Chennai
[2]jj71116@gmail.com

T. Ravi[3]
Associate Professor,
School of Electrical and Electronics Engineering,
Sathyabama Institute of Science and Technology,
Chennai
ravivlsi123@gmail.com

Abstract- Haemoglobin is main constituent for characterizing the physiological conditions of a body. For pregnant ladies, haemoglobin is the main factor for prevention of risk factors caused due to anaemia and polycythemiavera. This will used during pre screening test for donating blood and to obtain complete blood count. Hb is calculated from invasive technique in which blood samples are taken using needle which causes pain and discomfort, delay is more in analysing and to receive results, it doesn't allow real time monitoring. But an non-invasive test enable painless and obtain real time monitoring. Red and IR LED are passed through the finger, according to the amount of light absorbed by the tissue, remaining transmitted light detected by a photodiode, the haemoglobin levels i.e. Hbo2and HHb can obtained by interpreting the ratio of red to IR. This method allows pain free online patient monitoring with minimum risk of infection and facilitates real time data monitoring allowing immediate clinical reaction to the measured data.

Keywords: Haemoglobin, LED sensors, NODEMCU, PC display, dual wavelength plethysmography.

1. INTRODUCTION

Haemoglobin is the iron–containing protein molecule in red blood cells that carries oxygen from the lungs to the body's tissues and returns carbon dioxide from the tissues back to the lungs .There are mainly two different types of haemoglobin present in the blood i.e. Oxygenated haemoglobin and de-oxygenated haemoglobin ,Hbo2 is nothing but when oxygen binds to the heme component of the protein haemoglobin in red blood cells molecule whereas for HHb it is not attached. Measuring of Hb is essential for all

IOP Conf. Series: Materials Science and Engineering **590** (2019) 012042 doi:10.1088/1757-899X/590/1/012042

individuals, especially in pregnant ladies, because the low levels leads to anaemia and high levels of haemoglobin leads to polycythemia. The normal haemoglobin range for men is 13.5 to 17.5 and for women 12.0 to 15.5 gm/dl. There are two different ways to measure Hb, one way is invasive and the other is Non-invasive methods. Invasive techniques are Hemocue, Cynameth, copper sulphate gravimetric method .In Cynameth method, blood is mixed with potassium cyanide, potassium ferri cyanide and drabkin solution then haemoglobin converts to cynameth haemoglobin, when a light is passed through that solution some amount of light is absorbed[4,5] , that absorbance is compared with that of the standard haemoglobin cyanide solution by using a formula to obtain the amount of haemoglobin. ,but for this ejection of blood is required. Another way to calculate is Non invasive e.g. spectro photometry, opto-acoustic and transmission spectroscopy. These methods are based on spectrophotometry and beer lambert's law. Spectrophotometry is a method to measure how much a chemical substance absorbs light by measuring the intensity of light as a beam of light passes through sample solution. The principle of Beer-Lambert's law which states that absorption is directly proportional to absorbent and density of the absorbent of an object. Different types of spectrophotometry includes single Wavelength and dual wavelength spectro photometry, multiple spectro photometry. The design uses dual wavelength spectro photometry[6,8].

2. EXISTING INVASIVE METHOD

In invasive techniques hematology analyzers are used to predict haemoglobin in which cells are passed in single stream they are struck by laser which gets scattered. The light scatter at different angles i.e. between 10 and 70 deg is used by VCS instruments. The scattered light gives information about cell surface and granularity [13]. Based on the degree of light scatter, the analyzer obtain a measure of the average haemoglobin content but these analyzers are very costly and patient feel discomfort during ejection of blood.

Fig a :Hb measurement taken on a blood flow model

3. EXISTING NON-INVASIVE METHOD

The recent studies by Ranganathan proposed a Single wavelength spectrophotometry to calculate the HB. It is found that Hbo2 and Hb have different absorption characteristics .The absorption transmission and scattering of light by Hb products are wavelength dependent the light absorption is minimum at wavelength 603 nm for Hbo2 , HB and Hbo2 absorbs equal amount of light at 805nm which is isosbestic point[10] .These optical features are used in the estimation of Hb using Light sources .By using beer-Lambert's[4] law principle

$$OD = ecl = \frac{OUTPUT\ VOLTAGE}{TRANSMITTED\ VOLTAGE}$$

OD-optical density,
I_0-light intensity of incident light,
I-intensity if transmitted light,
e- extinction coefficient of haemoglobin
c- concentration
L-length of light path.

Haemoglobin concentration is calculated. By comparing the readings for both 700nm and 805nm light sources, the output of 700nm shows more linearity than the other one. Also it is observed that thickness of the finger plays a role on the amount of light transmitted through finger. In design part,SD-7BA photodiode range of 320-1150nm is used which has peak sensitivity at 805nm,microcontroller Mcp3551 that uses low power, has single channel, 22-bit delta sigma ADC which is used for conversion from analog ppg signal to the value of Hb[1].

4. PROPOSED NON INVASIVE METHOD

Design system uses dual wavelength spectrophotometry and beer-lambert's law principle. The developed haemoglobin sensor system consist of a number of hardware modules, which include appropriate light sources, constant light intensity circuit, power amplifier, NODEMCU, and PC. The sensor consist of emitter as LEDs, with centre wavelengths of = 660nm, = 940nm. These two wavelengths are selected because at 660 nm wavelength absorbance of de-oxy haemoglobin greatly exceeds the absorbance of oxy-haemoglobin where as at 960nm wavelength absorbance of oxy-haemoglobin greatly exceeds the absorbance of de-oxy haemoglobin (figure 1). These LEDs are installed in the upper shell of a finger clip. And single receiver photodiode is installed in the lower shell of the finger clip. The probe is placed to the patient's body usually on the finger. Red and infrared light is then emitted sequentially through the body tissue. Some light is absorbed by the tissue and remaining light is transmitted which is sensed by photodiode. Output voltage of photodiode increases linearly with light intensity and vice versa.[2]
To operate Led's sequentially a H-bridge circuit is used which consists of two p-n-p and two n-p-n transistors .The output waves that are obtained from the Led's are to be filtered to remove the DC component signal due to bones, tissues, skin. The two AC signals are to be ratio

$$\textbf{Ratio=} \frac{ac\ voltage(red\ LED)}{ac\ voltage(IR\ LED)}.$$

the received ratio is to be calibrated for the resulted Haemoglobin value. To digitalise this analog signal 40 pin Arduino NODEMCU is used which is having inbuilt ten bit analog to digital converter. The software for NODEMCU is used to timely operate the LED's and also for calibration. The nodemcu has inbuilt wifi module which makes it most suitable because the results of Hb are shown directly in a website so, that in an emergency ,doctor can know the patient condition from far distant place.[3]

Fig 1 depicts the absorption spectra of oxy haemoglobin and deoxy haemoglobin, at different wavelengths both LED's absorb different levels.

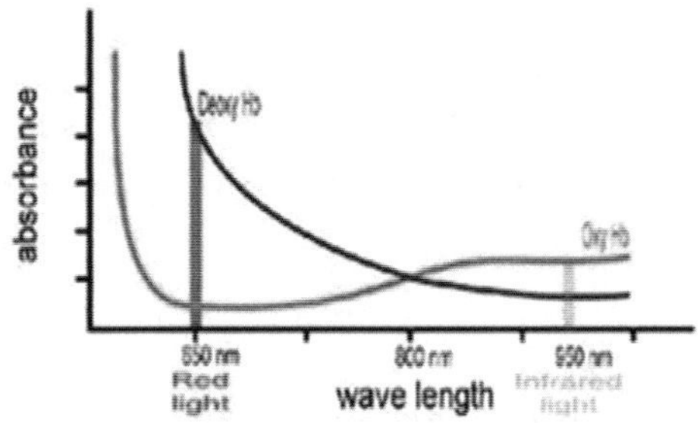

Fig.1: Absorption spectra of de- oxy and oxy-haemoglobin[5,8]

Fig2 tells the overall block diagram of the proposed non invasive haemoglobin sensor system.

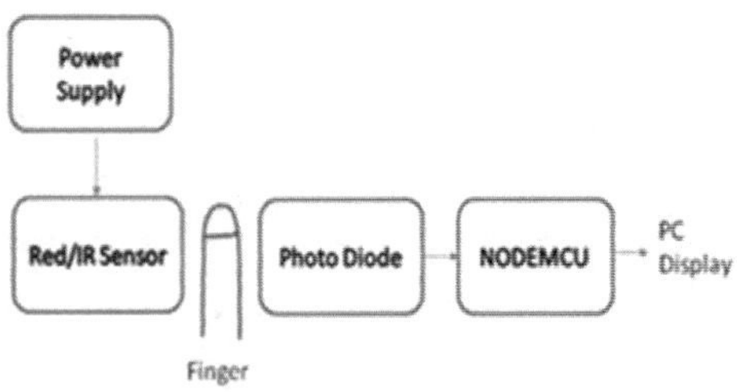

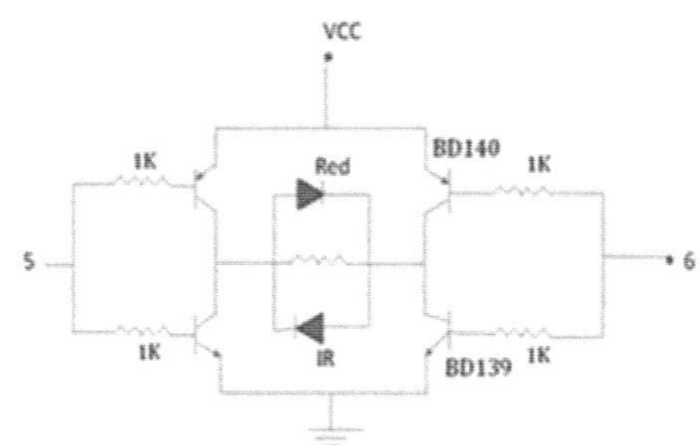

Fig. 2.Block diagram of the proposed system

Fig. 3: H-bridge circuit to operate LED's

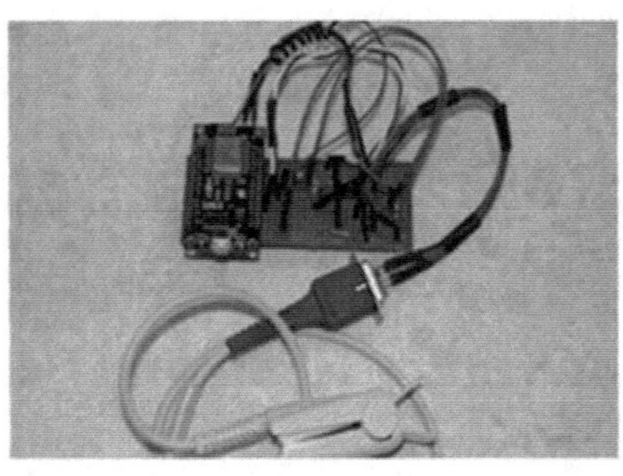

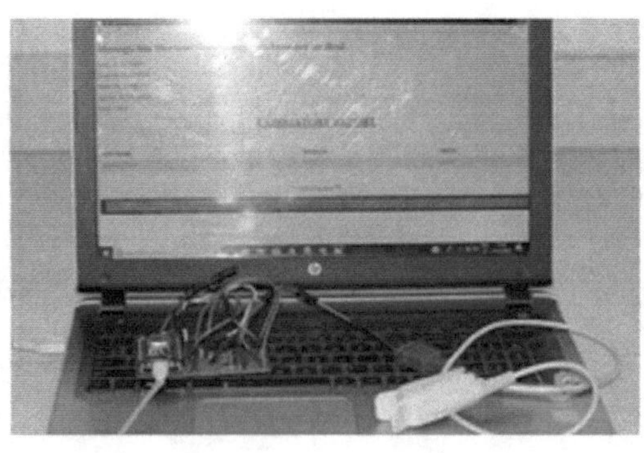

Fig. 4:Circuit design of proposed system

Fig. 5. Proposed prototype overview

IOP Conf. Series: Materials Science and Engineering **590** (2019) 012042 doi:10.1088/1757-899X/590/1/012042

5. RESULT AND DISCUSSION

The design uses spo2 probe which consists 670 nm and 940 nm is tested on various subjects and output ratio is measured using software the ratios are converted to the haemoglobin values and various subjects involves both men and women of age 18-21. The laboratory readings and calculated non-invasive readings are tabulated below. By calibration and analysis developed system readings are close to the laboratory readings.

The prototype output is displayed on the local website that can be checked by the doctor from anywhere in the world, which helps in taking precautions during emergency conditions.

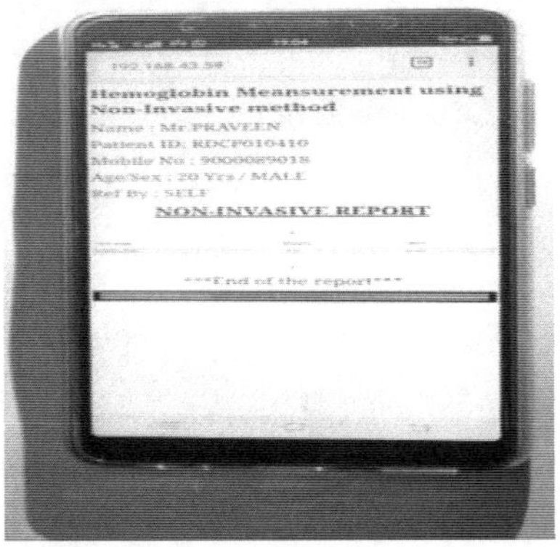

Fig 6. Output on mobile using IoT

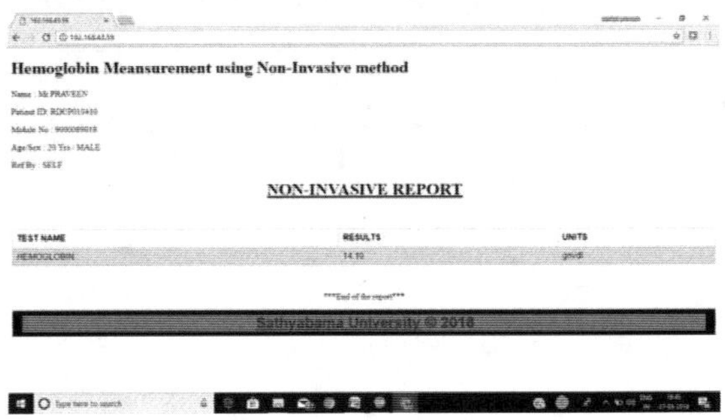

Fig .7: Haemoglobin value output on webpage

Tab.1:**Observed haemoglobin vales from both methods**

s.no	Description	age	sex	Invasive values	NonInvasive values
1	Patient 1	20	M	14.3	14.11
2	Patient 2	21	M	16.7	16.3
3	Patient 3	20	F	11.7	11.3
4	Patient 4	35	F	11.8	11.45
5	Patient 5	42	M	16.5	16.2
6	Patient 6	17	M	14.7	14.53
7	Patient 7	20	M	15.6	15.87
8	Patient 8	20	M	13.9	13.45

5. CONCLUSION

The prototype model of a Non invasive sensor unit for haemoglobin measurement is designed and developed. This paper clearly tells that invasive techniques causes discomfort to patients and delay in measuring haemoglobin, whereas Non-invasive methods are real time and comfort. This paper describes about the measurement of haemoglobin using non-invasive method in order to estimate Hb using a portable homemade device, which is user friendly and used for real time applications.

REFERENCES

[1] Rajashree Doshi, Anagha Panditrao, "Non-Invasive Optical Sensor for Haemoglobin Determination", International Journal of Engineering Research and Applications (IJERA) ISSN: 2248-9622 www.ijera.com Vol. 3, Issue 2, March -April 2013, pp.559-562.

[2] Kumar. R, Ranganathan.H, "embedded system based non invasive device development for haemoglobin estimation", International Journal of Industrial Electronics and Electrical Engineering, ISSN: 2347-6982 Volume-3, Issue-8, Aug.-2015.

[3] U. Timm, D. McGrath and E. Lewis, J. Kraitl and H. Ewald, "Sensor System for Non-Invasive Optical Haemoglobin Determination", IEEE SENSORS 2009 Conference 978-1-4244-5335-1/09/$26.00 ©2009.

[4] Ravi.T "10-Nanometer carbon nano tube field effect transistor based high celerity transposed polyphase decimation filter" Materials Today: Proceedings, Volume 3, Issue 6, 2016, pp: 1799–1807

[5] Modified Beer's Law – historical perspectives and relevance in near-infrared monitoring of optical properties of human tissue. Rammohan V. Maikala*Liberty Mutual Research Institute for Safety, 71 Frankland Road, Hopkinton, MA 01748, USA.

[6] Ravi.T "Design and performance analysis of ultra low power RISC processor using hybrid drowsy logic in CMOS technologies" International Journal of Applied Engineering Research (IJAER) Volume 10, Number 2 (2015) pp. 4287-4296

[7] Khunawat Luangrat1, Prapat Suriyaphol2,3 and Yodchanan Wongsawat.1 Department of Biomedical Engineering, Faculty of Engineering, Mahidol University, Thailan2Faculty of Medicine Siriraj Hospital, Mahidol University, Thailand.3Center for Emerging and Neglected Infectious Diseases, Mahidol University, Thailand.

[8] Head, L.M.; Pierson, R.M., "Functional near infrared spectroscopy for Hb and HbO2 detection using remote sensing," Sensors Applications Symposium (SAS), 2010,pp.166,169, 23-25 Feb. 2010.

[9] Roggan, André, et al. "Optical properties of circulating human blood in the wavelength range 400-2500 nm", Journal of biomedical optics, vol.4, no.1, pp.36-46, 1999.

[10] M.Manoranjani and T.Ravi "Multithreshold CMOS sleep stack and logic stack technique for digital circuit design" ARPN Journal of Engineering and Applied Sciences, Vol. 10, No. 10, June 2015, pp:4550-4556.

[11] Machireddy Amrutha1, K.Arthi2,M.Dhivya3, V.Priya dharshiniISSN: 2393-994xkarpagam Journal Of Engineering Research (KJER)Volume No.: III, Special Issue on IEEE Sponsored International Conference on Innovations in information, Embedded and Communication Systems (ICIIECS'15).

[12] Shah N., Osea E.A., Martinez G.J., 2014, "Accuracy of Non invasive Haemoglobin and Invasive Point of Care Haemoglobin Testing Compared With A Laboratory Analyzer." International journal of laboratory. haematology 36(1), pp. 56-61.

[13] Samuel Lazar.M and Ravi.T, "Li-Fi design for high speed data transmission" ARPN Journal of Engineering and Applied Sciences, Vol. 10, No. 14, August 2015, pp: 5999-6003.

[14] N. Kobayashi, Y. Takamura, T. Usuda, S. Takeda, "Technique of non-invasive measurement for total haemoglobin using pulse-spectro-photometry", The Japanese J. Med. Inst. 73, 2003, pp.587–588

[15] Brij Bhushan Ojha, S.Karthikeyan ,"Design of an interconnection network using VLSI photonics: Issues and challenges " International Journal of Pure and Applied Mathematics, Volume 118 No. 17 2018, PP: 211-224

[16] S. Vimalsree and S. Karthikeyan, "A Low Power Multibit Flip Flop Merging Technique Using WSN Nodes", ARPN, vol. no:11, No.15, August 2016, Page no: 9358-9363.

[17] Balaji and S.Karthikeyan, "A survey on moving object tracking system using Image Processing", 11th international conference on Intelligent systems and control, IEEE explorer conference, vol. , Jan 2017, PP: 11-16.

IOP Conf. Series: Materials Science and Engineering **590** (2019) 012043 doi:10.1088/1757-899X/590/1/012043

REVIEW ON FIR FILTER ARCHITECTURE DESIGN AND ITS COMPUTATIONAL BLOCKS

P.Teja Satyanarayana[1]
BE, Electronics and Communication Engineering,
Sathyabama Institute of Science and Technology,
Chennai
te97ja@gmail.com

P.Lakshman Kumar Reddy[2]
BE, Electronics and Communication Engineering,
Sathyabama Institute of Science and Technology,
Chennai
reddireddylakshman@gmail.com

T. Ravi[3]
Associate Professor,
School of Electrical and Electronics Engineering,
Sathyabama Institute of Science and Technology,
Chennai
ravivlsi123@gmail.com

S.Karthikeyan[4]
Associate Professor,
School of Electrical and Electronics Engineering,
Sathyabama Institute of Science and Technology,
Chennai

Abstract- Low power VLSI circuit employed applications power consumption and high performance, the area is crucial in DSP. A filter is used in almost all electronic devices that function on signals. They are used to extract the useful part from the input signal and the required part of signal reaches the receiver. This paper is mainly focussed on the survey of different architecture design for FIR filter. The different performances such as speed, complexity, PDP, ADP, are reviewed.

Keywords: Re-configurable FIR filter, speed, area, power.

1. INTRODUCTION

In the past, Area, Performance, Cost, and Reliability were only the major concerns of a VLSI designer. whereas the power consumption of the circuit was merely considered. In recent years this has begun to change dramatically. Power consumption is considered and given the importance that has been given to other factors, in some cases or demand of applications, a bit more. Digital signal processing is an important part of electronic devices as signal is an integral part of DSP. A signal is pulse coded with information that needs to be transferred from source to the destination, and a medium is needed to do so, called channel. During the transmission, some of the data is lost due to the noise present in the channel. This information must be retained in the signal in its original form till the transfer is complete to have complete data transfer with minimal loss. A signal has high-frequency range, low-frequency angle, and mid-frequency range. Sometimes the information is contained in only one frequency range to make the extraction and transfer easy. So that the loss induced could also be compromised effectively. To pull-up, the information from the signal filters is used. The filters employed are different based on the signal whether it's digital or analog. We use digital filters in Digital Signal Processing (DSP).

FIR filter is a type of digital filter that is used for linear characteristics applications. It is usually preferred because of its linear phase response and non-recursive property, which makes FIR filter possess stronger resilience to phase shit and higher stability in long-term processing tasks than its FIR counterparts. As a result, to date FIR filters are widely adopted in numerous commercial produces, especially in the base-band processing market. Based on this various apps are implemented to design an efficient FIR filter to reduce power, delay, and area. Minimising area and delay have always been considered important. But, reducing power consumption has been gaining prominence recently as high power operation in small area circuits could rise thermal problems. The customers will take some products which they want to buy and go to the billing counter. The shopkeeper will take the products which are attached with Li-Fi hardware which communicates to the billing server through Li-fi Receiver connected to it. After that, all the product and its details will be shown to the android application and the total billing amount will also show. As the shopkeeper is a visually challenged person, the total billing amount will convert into speech. After that Shopkeeper has to say "BILL" on his mobile application by pressing the power button, then the bill will automatically generate. Then billing will be done successfully.

2. LITERATURE REVIEW

FIR Filter Architecture The FIR filter basically consists of three components, the D flip-flop, multiplier and an adder. The D flip -flop to implement a delay, a multiplier to implement the coefficient, and an adder to sum the nodes at the end of each tap [5]. Figure: A shows the basic block diagram of FIR filter.

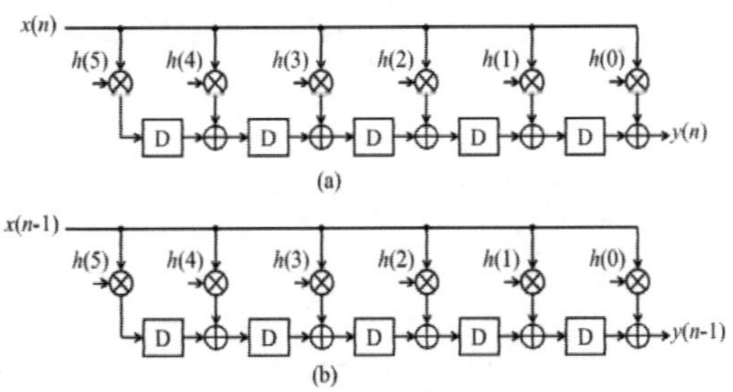

Figure: 1a. – FIR Filter

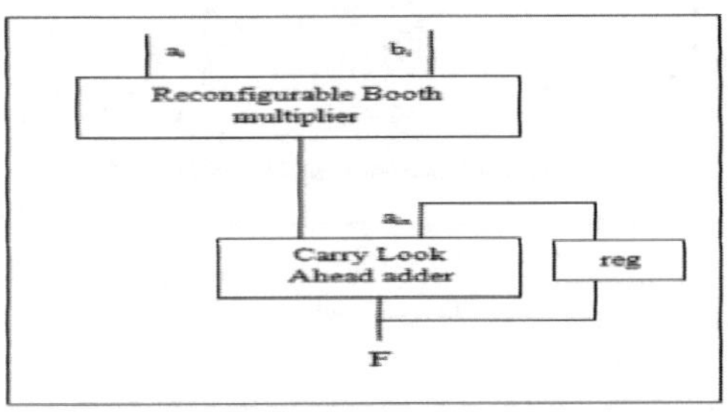

Figure: 1b. –MAC unit in FIR Filter

Deepika etal (2016) described implementation of MAC unit in the design of FIR filter. It is easy when compared to window techniques. MAC unit is multiple accumulation unit. It comprises of multiplier and adder.

IOP Conf. Series: Materials Science and Engineering **590** (2019) 012043 doi:10.1088/1757-899X/590/1/012043

A reconfigurable booth multiplier was selected as it is a high speed multiplier[2,8]. Carry look ahead adder is used for the purpose of final addition and these combinations together make the circuit execute algorithms faster and efficient [10].

Figure 1b, shows architecture of MAC unit. It is a combination of reconfigurable booth multiplier and carry look ahead adder. Due to regular structure and speed by reducing partial product booth multiplier and it can work for n number of bits. A configuration register has been added in booth multiplier to avoid move clock cycle. Register introduced is set for 4-bit,8-bit,16-bit [8]. First of all 2 input's that are given to the MAC unit gets multiplied by using reconfigurable booth multiplier and the multiplication result is added to the previously stored result in the MAC unit. When an input a_1 has been taken which is constant throughout the computation and another input b_1 which is FIR filter co-efficient was taken, if these two inputs get multiplied by using reconfigurable booth multiplier then result is fixed to carry look ahead adder and another variable a_{in} which is set to zero initially works as a second input to the adder later in next cycle, a_{in} is previous output of MAC unit [13].

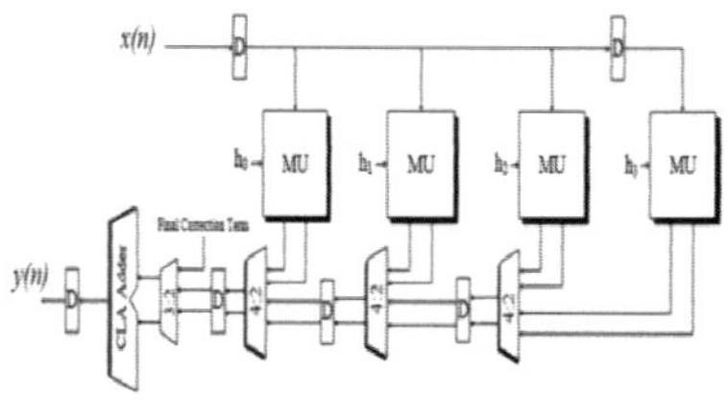

Figure :2 Conventional FIR filter

Evangelos kyritsis et al (2016) explained about the implementation of karatsuba algorithm in the design of efficient and faster FIR filter[16]. The KMA (Karatsuba Multiplication Algorithm) reduces a 2N x 2N bit multiplication to a set of (N+1) x (N+1) and N x N bit multiplication and additions by splitting the 2N bit input operands and generating same auxiliary variable. When karatsuba algorithm is applied to FIR filter equation, a parallel architecture is obtained resulting in increased speed and reduced area especially in case of high dynamic range filter, and also uses carry-save(CS) in order to decrease the critical path and to speed up calculations[13,5].

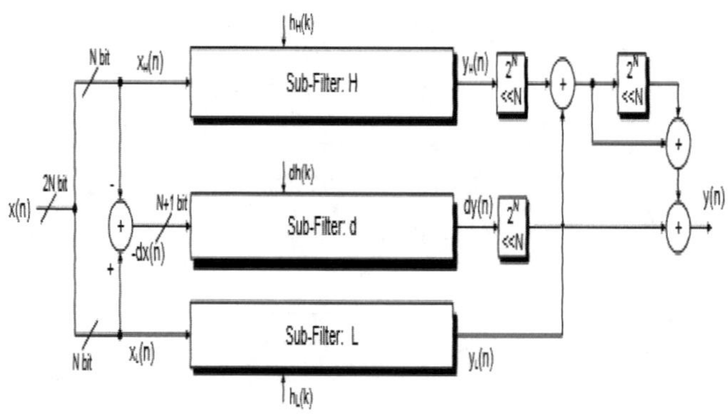

Figure:3 shows the proposed Karatsuba FIR filter

Figure: 2 shows Conventional FIR filter structure in transposed form. It has three sub-filter of reduced dynamic range. A parallel MB pre-encoded, Multiplication Unit(MU) was designed and used as a building block. The

Multiplication Unit contains a Partial Product Generator Unit(PPG) and a Wallace tree for adding the partial products. For the Wallace tree, the Synopsis Design Ware IP (DW02_tree) has been used [4]. CS architecture is used in order to enhance delay savings. Figure:3 shows the proposed Karatsuba FIR filter. The karatsuba design which proposed shows better performance with less circuit area, and lower power consumption.

hamizharan. V et al(2012) described the implementation of computation sharing multiplier in the design of FIR filter. Computation Sharing Multiplier (CHSM) is efficiently used for the low complexity design of FIR filter. Both add and shift operations was represented by multiplications in FIR filtering operations over common computation results using CSHM. CSHM multiplier was implemented by using Carry Select Adder (CSA)[7-9] which is a high-speed adder. In the design by using single ripple carry adder and add one circuit using fast all-one finding circuit and low-delay multiplexers a Carry ahead adder was implemented which will further decrease area and accelerate the speed of CSA. No additional memory is required to share when the common computations are found. A high performance and low power in FIR implementation was approached by computation sharing multiplier [14].

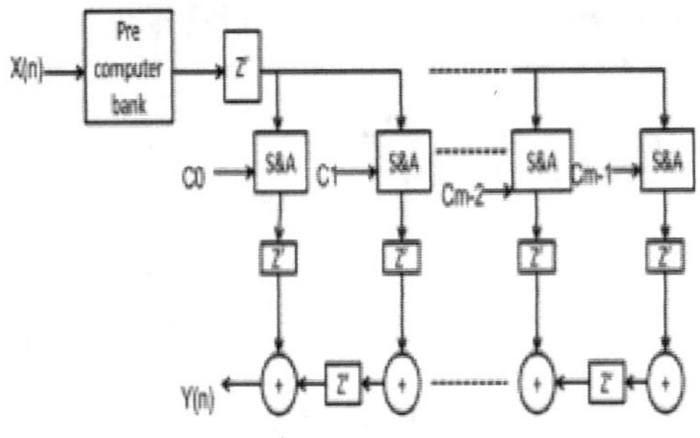

Figure 4 shows FIR filter in transposed circuit form (TDF) fir filter.

Figure 5 points that FIR filter based on CHSM. In transposed direct form (TDF), FIR filter multipliers

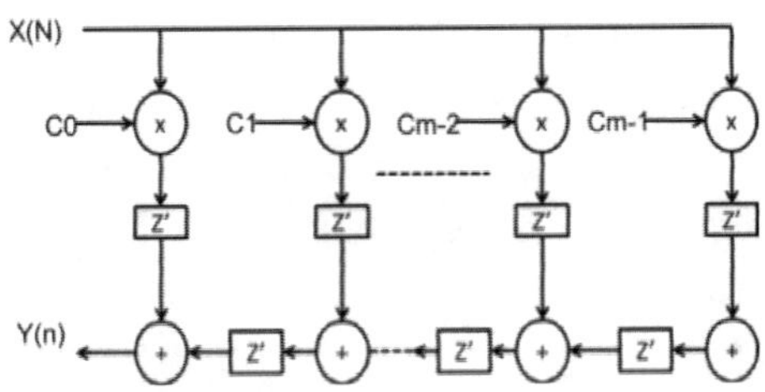

were replaced by S&A's as a pre-computer is connected to input. Therefore, the FIR filter using CHSM consists of one pre-computer and ten S&A's. Then, figure 5, the pre-computer outputs are shared by all the S&M's. The CSHM scheme and circuit-level techniques help to achieve high performance FIR filtering operations.

Bo Yuan et al (2016) introduced the implementation of stochastic multiplier[10-11] in designing of a high accuracy FIR filter. In the proposed edition, it achieved significant reduction in the hardware complexity as compared to the conventional design using the simplicity of stochastic arithmetic unit and the high accuracy non-scaled stochastic adder that has significant increase in the computation accuracy. It avoids the original down scaling effect for large T cases. Based on this method the high accuracy stochastic FIR filter was developed. Here SA refers to stochastic adder and SM to stochastic multiplier. Figure 6 points the hardware architecture of the proposed stochastic adder. Figure 7 shows direct form stochastic FIR filter.

IOP Conf. Series: Materials Science and Engineering **590** (2019) 012043 doi:10.1088/1757-899X/590/1/012043

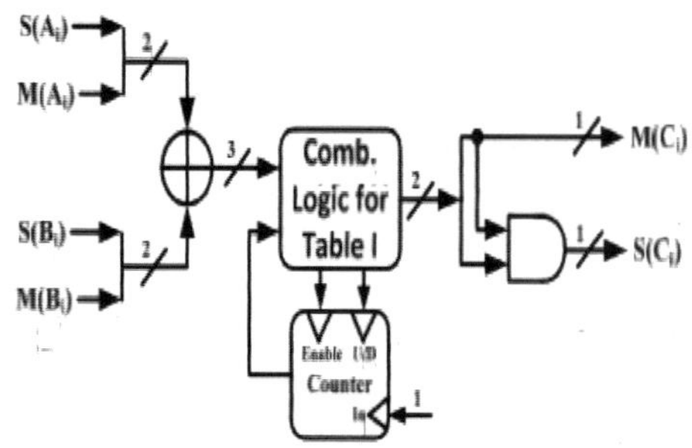

Figure: 6 - Stochastic Adder

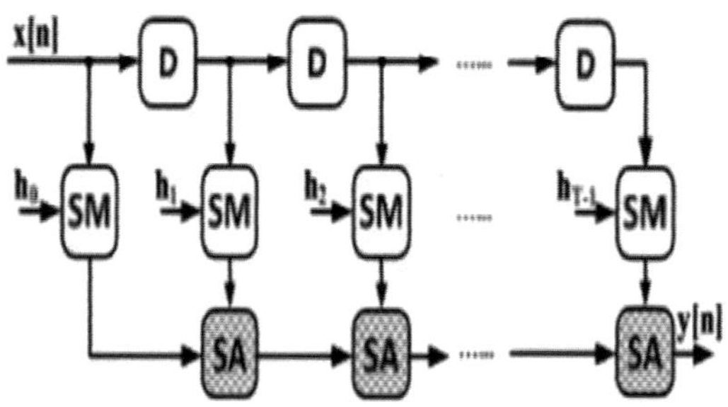

Figure: 7 - Stochastic FIR filter

First of all, it is a non-scaled adder. Although OR gate can be used as an approximated non-scaled adder, in some applications the OR-gate based adder suffers from huge approximation errors as well as the limitation for positive only addition. Thereby, causing the OR-gate based adder has very severe accuracy loss if we consider the negative inputs. Secondly, the proposed stochastic adder has very high accuracy in terms of signal to noise ratio (SNR). Especially, for the adder chain case. Figure 8 shows two-line stochastic multiplier. By using two-line based non-scaled stochastic adder and multiplier, the hardware architecture of the high accuracy stochastic FIR filter was developed. The overall architecture was chosen as either direct-form or lattice format. Due to this proposed stochastic FIR achieves both low hardware cost and high computation accuracy and is suitable for practical systems.

M. Ganasekaran et al (2014) elaborated the design of low delay high compact FIR filter using reduced Wallace multiplier [12]. In this proposed system, the adder was modified by reduced carry save adder in the last stage of multiplier. So that, the performance of the over-all FIR filter structure is improved in terms of area and delay. The operation of carry save adder is, it computes the sum in parallel without waiting for its carry. The carry is computed in the next stage of adder structure. So, no adder performs the addition of 'm' numbers faster than normal or regular addition. And it has a full adder like ripple carry adder. But, the carry output from each bit is taken out to form second result vector rather than being directed to the next most significant bit.

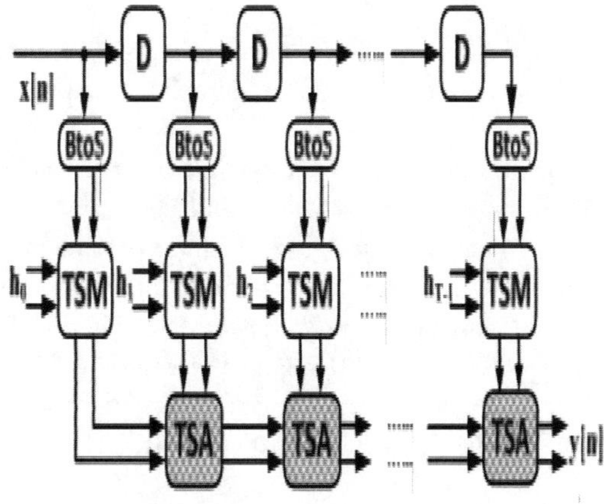

Figure: 8 - Two-Line Stochastic Multiplier

3. CONCLUSION

The elaborated design of high compact FIR filter, and its different structures are described. The computational blocks are described, Such as Adders and Multipliers. The performance of the FIR filter structure are analysed in terms of area and delay.

REFERENCES

[1] Deepika, Nidhi Goel, "Design of FIR Filter Using Reconfigurable MAC Unit", 2016, 3rd International Conference on Signal Processing and Integrated Networks.

[2] Deepika, Nidhi Goel, "RTL Design of Reconfigurable Multiplier", IEEE International conference on Soft computing Techniques and Implementations, pp 24-28, 2015.

[3] Evangelos Kyritsis, Kaimal Pekmestzi, "Hardware Efficient Fast FIR Filter Based on Karatsuba Algorithm", 2016 5th International Conference on Modern Circuits and Systems Technologies (MOCAST).

[4] www.synopsys.com/dw/doc.php/doc/dwf/ datasheets/dw02 mult.pdf.

[5] Ravi.T "10-Nanometer carbon nano tube field effect transistor based high celerity transposed polyphase decimation filter" Materials Today: Proceedings, Volume 3, Issue 6, 2016, pp: 1799–1807

[6] Thamizharasan.V, Parthipan.V, "An Efficient VLSI Architecture for FIR Filter Using Computational Sharing Multiplier", International Journal of Computer Applications (0975 – 8887), Volume 54– No.14, September 2012.

[7] W. Jeong, K. Roy, and C. Koh, "High-performance low-power carry-select adder using dual transition skewed logic," in Proc. ESSCIRC, 2001,pp. 172–175.

[8] Ravi.T "Design and performance analysis of ultra low power RISC processor using hybrid drowsy logic in CMOS technologies" International Journal of Applied Engineering Research (IJAER) Volume 10, Number 2 (2015) pp. 4287-4296

[9] Mayan J.A, Arifa S, Pavithra R,"Semantic based multi lexical ranking technique for an effective search in protected cloud,2016 International Conference on Control, Instrumentation, Communication and Computational Technologies (ICCICCT), Kumaracoil, 2016, pp. 570-576..

[10] Ravi.T, Kannan.V "Modeling and performance analysis of ballistic carbon nanotube field effect transistor (CNTFET)" International Conference on Recent Advances in Space Technology Services & Climate change – RSTS&CC-2010, November – 2010, pp. 327-331

[11] G.Savitha and S.Karthikeyan,"Architecture for diverse modulation techniques in SDR", concurrency computation practice and experience,John wiley & sons, Ltd, February 2019,PP: 1-10.

[12] B. Brown and H. Card, "Stochastic neural computation I: computational elements, "IEEE Trans. Comput., vol.50,no.9,pp.891-905,Sept.2001.

[13]. R.Ramya and T.Ravi, "Design of cache memory mapping techniques for low power processor" ARPN Journal of Engineering and Applied Sciences, VOL. 10, NO. 11, JUNE 2015, pp:4783-4788.

IOP Conf. Series: Materials Science and Engineering **590** (2019) 012043 doi:10.1088/1757-899X/590/1/012043

[14]. Micheal Priyanka.M and Ravi.T, "Survey on role of memristor in electronics "2015 International Conference on Control Instrumentation Communication and Computational Technologies, ICCICCT 2015, 18-19 Dec 2015, Pp:738-744. .

[15] M.Gnanasekaran, M.Manikandan, "Low delay-High compact FIR filter using Reduced Wallace Multiplier", IEEE 2014,IEEE Conference Number - 33344,July 8, 2014, Coimbatore, India.

[16] Brij Bhushan Ojha, S.Karthikeyan ,"Design of an interconnection network using VLSI photonics: Issues and challenges " International Journal of Pure and Applied Mathematics, Volume 118 No. 17 2018, PP: 211-224

[17] S. Vimalsree and S. Karthikeyan, "A Low Power Multibit Flip Flop Merging Technique Using WSN Nodes", ARPN, vol. no:11, No.15, August 2016, Page no: 9358-9363.

[18] Balaji and S.Karthikeyan, "A survey on moving object tracking system using Image Processing", 11th international conference on Intelligent systems and control, IEEE explorer conference, vol. , Jan 2017, PP: 11-16.

[19] R. M. Joany, E. Logashanmugam, M.Ooha, "Algorithmic Approach on Video Compression", International Journal Control Theory Applications, 9(7), 2016, pp. 339-343,ISSN : 0974-5572 , International Science Press.

[20] R. M. Joany and E. Logashanmugam, "Evaluation of Efficient Method to Control the Functionality of DMD", IEEE International Conference on Communication and Signal Processing-ICCSP'15, IEEE Xplore Digital Library,DOI: 10.1109/ICCSP.2015.7322857, pp-0015 – 0019,2-4, April, 2015, Print ISBN:978-1-4799-8080-2.

[21] Anna Devi, E., Martin Leo Manickam, J., "A light weight hop count based link failure detection protocol in wireless sensor networks", International Journal of Pure and Applied Mathematics, Volume 116, No. 10, PP: 425-433, 2017, ISSN: 1314-3395.

[22] Anna Devi, E., Martin Leo Manickam, J., "A novel technique for prolonging the connectivity of wireless sensor networks", International Review on Computers and Software, Vol 10, No 2 (2015), eISSN: 2533-1728.

IOP Conf. Series: Materials Science and Engineering **590** (2019) 012044 doi:10.1088/1757-899X/590/1/012044

Implementation of Li-Fi Technology in Classrooms

Shubhendu Apoorv[1]
Department of Electronics and Communication Engineering,
Sathyabama Institute of Science and Technology,
Chennai, India.
[1]shubhenduapoorv172@gmail.com

Sudharshan Kumar Bhowmick[2]
Department of Electronics and Communication Engineering,
Sathyabama Institute of Science and Technology,
Chennai, India.
[2]sudharshan.bhomick@gmail.com

E.Annadevi[3]
Assistant professor, Dept of ECE,
Sathyabama Institute of Science and Technology,
Chennai, India
[3]annadevimit@gmail.com

Abstract-In this paper the implementation of Li-Fi technology in classrooms has been investigated. It is well known that the visual information is better perceived by the brain as compared with the textual information. Using visuals to explain a concept helps student get a better perception about the topic. To ensure the best of the visual experience, there is a big need of an efficient way of communication between the devices used by students and the teacher.The traditional methods like Wi-Fi or wired connections are not much efficient due to the speed limitation and other factors. Light fidelity, also known as Li-Fi is a type of data transmission technology which uses light rays to transmit data. By implementation of this technology, high volume of data can be transmitted at extremely high speed as it gives higher bandwidth. It also ensures efficiency, security, as well as the mobility of the device. Hence, we are looking to implement this technology to enhance the learning experience of the students. Every student gets a visual input on what, when, where and how anything happens and the concepts are understood easily and thoroughly. Also, it is cheap and safe to the human body.

Keywords:Li-Fi (Light Fidelity), Spectrum Crunch, LED (Light Emitting Diode), Orthogonal Frequency Division Multiplexing (OFDM).

1. INTRODUCTION

This system will encourage both visually challenged person and illiterate person to perform their work In the current time, the wireless technology has become an integral part of our life. We are constantly connected to some form of wireless instrument. Internet plays a very important role in data exchange in our day to day life. With the improving technology, the demand of high speed data is constantly increasing. Now the radio frequency based networks are not sufficient for the bandwidth intensive application. Hence it often leads to a situation called as spectrum crunch. So we need to move on to a better method of transmission of information. With the development of Light Fidelity or Li-Fi, we can meet this high demand of extensive amount of data transfer at efficient rates. This term was first coined by professor Harald Hass at 2011 TED

IOP Conf. Series: Materials Science and Engineering **590** (2019) 012044 doi:10.1088/1757-899X/590/1/012044

global talks. The light fidelity system uses visible spectrum of light to transmit the data. It has been observed that the Li-Fi provides an efficient data transfer with high security.

The basic principle in this Li-Fi system in that the data will be sent through an LED and will be received by the photodiodes which convert the light signals into electrical signals. The data will be transmitted in the binary form (0, 1) by blinking of the LEDs. Due to its advantages Li-Fi finds its application in many fields like vehicle to vehicle communication, vehicle to infrastructure communication, data transmission in a place with high security and even in common places like hospitals and schools, etc.

In this paper, a design to implement the Li-Fi technology in the classroom has been done to enhance the learning experience. It is observed that the visual information is more likely to stay in our mind rather than reading it from a book. The crux points of the topic can be easily explained, zoomed in to show the relevant visuals, stopped and annotated when and where it needs to be emphasized along with appealing animations, colours, and sounds .Thus ,the teachers gain complete attention and interest of every child in the class. Every student gets a crystal clear idea of the topic explained.

In our current wired technology, all the devices are connected either by wires like optic fibre, twisted cables or a Wi-Fi connection. But there are many disadvantages of using LAN as it is wired; hence it requires proper installation procedures such as drilling holes properly, fitting sockets and redecorating. It gets worn and torn if it is kept in open space. Maintenance of wires is quite difficult and most important of all, it clutters the workspace. These methods require complex circuits for transmitting and receiving data. Also, it can cause a tripping hazard. Installation needs to be done by skilled technicians, and even a virus can spread more easily.

In case of Wi-Fi, the RF communication system this harms human body because of its constant radioactive radiation. These communication systems cannot give a high data transfer rate. Li-Fi does not harm our human body because the radiations are not harmful as it mainly consists of light waves, which are not even of that high intensity which would hurt the human eye.

2. THEORY

The Li-Fi system consists of key components like LED array, LED driver circuit, and a receiver. The LED array acts as a light source. It can be dimmed up and down at an extremely high speed without being visible to the human eye. The data is received from the server to the LED driver. Then the data is sent to the LED by the LED driver, which uses Signal Processing Technology to convert the message signal into binary codes that can be further transmitted. The LED blinks and transmits data which is embedded in its beam at rapid speed. The transmitted light wave is received by a photo detector. The rapid change in lights intensity of LED bulb is converted into electrical signal by the photo detector. The signal is converted into binary data by decoding and then amplified. This binary data is generally a web video or audio application that runs on internet .Li-Fi uses orthogonal frequency division multiplexing modulation method to optimize data rates, multiple access and Energy Efficiency.

The RF communication system can provide frequency up to 300 GHz whereas Li-Fi can provide a frequency of 790 Terahertz. Li-Fi can give up to 224 gigabytes per second which is quite higher than the highest speed achieved by RF communication systems yet.

3. DESIGN

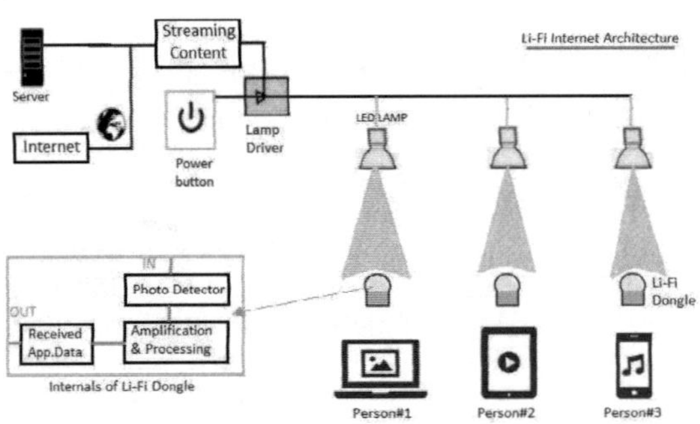

Fig.1. General Architecture of a Li-Fi system.

The Infrastructure needed to build a smart class contains main components like a server, transmission source, receiving element, device to interface with common devices and interactive board for lecture. The server is used as a database of smart class and to access contents from internet. Transmitter source is used to transfer data. It contains two main components, namely LED driver circuit and the LED grid or array. The LED array contains high brightness LED that can be controlled by the LED driver circuit. These LED grids will be installed at the ceiling so that it can provide light at every point of the class. The receivers will be installed on every table. Accessing by the general devices can be done by using wires between receiver and the devices or attaching a separate compatible device to the main device (example Li-Fi dongle) to ensure continuous reception of data to the receivers,. In our design, LED grid is installed with a slight inclination to expand the field of light signals. A projector can also be installed so that the teachers can demonstrate thing to the class. Since, the student get instant access to the notes, they do not need to waste time by copying the notes. The internet can be accessed by a computer and a projector in case the teacher wants to show something from internet or from the database. The pictorial representation of the proposed model is shown in figure 3.

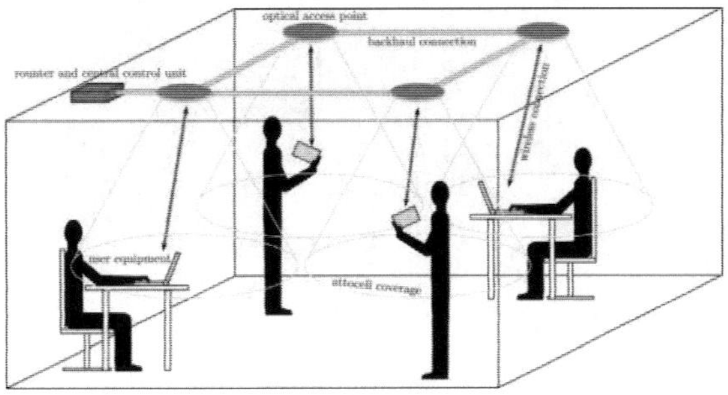

Fig.2. General Model of implementation of Li-Fi in a classroom.

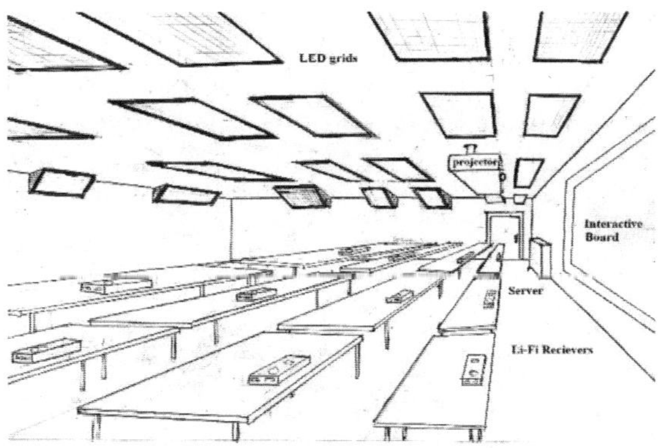

Fig.3. Proposed design of a Li-Fi enabled classroom.

4. WORKING

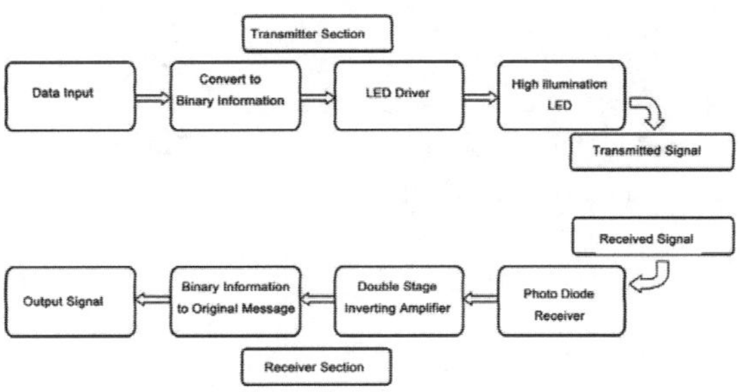

Fig.4.Block diagram of working of the Li-Fi system

LED array will be installed on the ceiling of the class. The information will be fed by the server to LED driver circuit which will transmit data. The LED will send the data by blinking at a very high speed and thus, binary data in form of 0s and 1s will be sent to the receiver. The LED driver circuit controls the working of the LED thus controlling the data flow. The receivers installed at each table will receive the data and convert it into electronic signals. The electronic signals will be sent to the general devices from the receiver using a data cable. In case of the Li-Fi dongle, it can directly receive the data and feed to the device the user is using like phones, tablets or computers, etc. Since, the led matrix is slightly inclined to increase the field of light signals, so even if there is an obstruction in the way for one LED the other LED can compensate for the loss of signal. The projector displays data over the smart board. When the projector is being used the intensity of light can be decreased so that the communication will still happen and the projected data can be seen easily. The smart board can also detect what is being written on the board and that can be transferred to the students thus saving time for students.

ADVANTAGES

- It will enhance the learning experience for the students by interactive learning.
- The data transfer can be done in much faster way compared to the traditional wired networks or Wi-Fi.
- It can handle heavy network traffic.
- The soft copy of the data can be shared with students easily.
- There will not be any harmful effects from the signals as they constitute of medium intensity light waves unlike the radio waves.
- The design ensures the signal to be uniformly spread in every direction as it will help to ensure continuous flow of data at every point.
- The devices can be easily installed without much need of external structure. Hence it is less labour intensive and also cost effective.
- It is ideal for the students who are working in groups. Every student in class gets the same interactive experience.
- It provides a wireless means of communication hence no need of installing extra wires that clutters the workplace.
- The mobility of the devices is increased since it can be accessed all over the room.

DISADVANTAGES

- External sources of light like sun and other bulbs can create interference in the communication.
- If the sources of light are even slightly blocked, the transmission and reception cannot happen properly since it is a line of sight communication.
- It is only a one way connection as the receivers cannot send back the data.
- The reception of data to general devices can be done by Li-Fi dongle. If, the interfacing to the common devices is done by wires, there can be a limitation in the volume of data being transferred.

5. PRACTICAL EXAMPLE

Since, this technology is quite new, there are not many real life examples present today. The world's first Li-Fi enabled classroom has recently been implemented in Hegel-Gymnasium,Stuggart,Germany. This installation is a trail by the Fraunhofer Heinrich Hertz Institute and is funded by the City of Stuggart.In Britain , a school in Kent has also implemented this technology.

Fig.5 .Li-Fi installed classroom in Hegel-Gymnasium, Stuggart, Germany.

6. CONCLUSION

As the technology is expanding, the students have to be educated well to understand these things. Some topics can not be explained by the conventional system of teaching so we need an interactive place so that the students can have a clear insight of the knowledge. Evolving technologies like 3D printing and nanotechnology can be easily understood visually. Traditional teaching methods cannot give a deeper insight about it. So, digital technology is much needed to make students understand well. An efficient data transmission technology can easily accomplish a difficult task of transmitting heavy data. There is a continuous development happening in communication method in many areas, so to keep it up with the time we also need to implement these new methods in education field as well. Also, traditional method of teaching required very skilled teachers so that the student can have better knowledge. Which visual technology teachers with a little less knowledge can also teach students efficiently? Hence, it does not give much work load to the staff as well. We hope that the design proposed in this paper will be implemented and will work efficiently.

REFERENCES

[1] Kirti Tanwar, Stuti Gupta, Smart Class Using Li-Fi Technology, The International Journal Of Engineering And Science (IJES) || Volume || 3 || Issue || 7 || Pages || 16-18 || 2014 || ISSN (e): 2319 – 1813 ISSN (p): 2319 – 1805www
[2] Ekta,Ranjeet Kaur,"Light Fidelity(LIFI)- A Comprehensive study, IJCSMC,Vol 3,Issue4,April 2014,pg-475-481.
[3] Yash Khare,Vijay Prakash Tiwari,Akshay B. Patil,Prof. Kumkum Bala,LIFI Technology – Implementations and Applications,International Reasearch Journal of Engineering Technology (IRJET),vol 03,issue 04,Apr-2016,e-ISSN-2395-0056,p-ISSN: 2395-0072
[4] S.Tarannum,"Data transmission through Smart Light Communication Technology" International Journal of Technical Research and Applications (IJTRA),Vol 4no. 2,pp.136-142,2016.
[5] S.Rajan,A. Reena,"Visible light communication" International Journal of computer Trends and Technology.(IJCTT),Volume-4 Issue-4,Apr 2013.
[6] Alao O.D., Joshua J.V.,Franklyn A.S.,Konolafe O."Light fidelity(Li-Fi):An emerging Technology for The Future",IOSR Journal Of Mobile Computing & Application,Volume-3,Issue-3,May-Jun-2016,PP-18-28,e-Issn:239-0050,P-ISSN: 2394-0042.
[7] Arya V,Priya P.,Reshma Omanakuttan,Shilby Baby,"LiFi:The Future Technology in Wireless Communication",International Journal of Advanced Reasearch inElectrical and Electronics And Instrumentation EngineeringVol 4,Issue 4,April 2015,pISSN-2320-3765,eISSN:2278-8875.

[8] Akash Gupta,Parul Garg,Nikhil Sharma,"Hybrid LiFi-Wi-Fi indoorBroadcasting System",IEEE explorer,Feb-2015,eISSN-2166-9589.
[9] R.Karthika,S.Balakrishnan,"Wireless Communication using Lifi Technology",SSRG International Journal of Electronics and Communication Engineering,Volume2,Issue 3,March .
[10] Satyajit Banarjee,Mayank Gupta,Vaibhav Sharma,Alpanna Sahu,"LI-FI –An Implementation to future Wireless Communication",SSRG International Journal of Electronics and Communication Engineering,Volume 4,Issue 3,-March -2017,Issn-2348-8549.
[11] https://www.slideshare.net/shwrvppt/li-fi-tch
[12] https://www.techopedia.com/definition/29483/spectrum-crunch
[13] https://www.bartleby.com/essay/Li-Fi-Technology-A-Visible-Light-Communication-PKC829RAVJ.
[14] http://luxreview.com/article/2017/11/world-s-first-li-fi-classroom-opens-in-germany.
[15] http://luxreview.com/article/2013/10/britain-s-first-li-fi-classroom-logs-on
[16] G.Savitha and S.Karthikeyan,"Architecture for diverse modulation techniques in SDR", concurrency computation practice and experience,John wiley & sons, Ltd, February 2019,PP: 1-10.
[17] S.Karthikeyan, S.Jayashri, "Energy Utilization Strategies using Novel Approach in Wireless Sensor Networks", International Conference on Software Engineering and Mobile Application Modelling and Development, Dec 2012, pp: 1-5.

IOP Conf. Series: Materials Science and Engineering **590** (2019) 012045 doi:10.1088/1757-899X/590/1/012045

Optimization of Thin film for the design of H-Alpha Filter

L.Jegan Antony Marcilin
Research Scholar,
Sathyabama Institute of Science and Technology,
Chennai – India.
jegan25@hotmail.com

N.M.Nandhitha
Professor & Dean,
Sathyabama Institute of Science and Technology,
Chennai – India.
nandhi_n_m@yahoo.co.in

R.Kural Mathi
Final Year UG student,
Sathyabama Institute of Science and Technology,
Chennai – India.
kuralmathirajendran@gmail.com

Abstract: Electrical power plays an important part and it is a critical contributing element for a most of the developments. Numerous power plants are accessible for the production of electrical energy and the most significant power sources are coal, solar power, wind, atomic, gas, Biomass, and so on. Power is also produced from plasma which is an ionized gas. It is said that while producing power from plasma because of some unprecedented reason the plasma begins to deteriorate which leads to the emission of enormous amount of gases including H-alpha which is of wave length 656.3 nm and it is necessary to design a filter to detect the H-alpha discharges. In this paper, a filter design for the H-alpha emission is discussed.

1. Introduction

Aditya Tokomak is device at INSTITUTE FOR PLASMA RESEARCH in GUJARAT in INDIA utilized for the generation of power. It is being worked for over a decade. It was authorized in 1989. It has a major radius of 0.75 meters and a minor radius of 0.25 meters. It has a greatest field quality 1.2 testla created by 20 toroidal field coils divided symmetrically the toroidal way. It has two power supplies, a capacitor bank Power supplies, a capacitor bank and the APPS (ADITYA pulse power supply).

Fig:1: Tokomak

The plasma parameters amid capacitor bank releases are : Ip ~30kA,shot length ~25ms,Central electron temperature ~100ev and center plasma thickness ~1019 m-3and the common parameters of APPS operation is ~100 kA plasma current,~100 ms span, Central electron temperature ~300ev and ~3×1019 m-3 center plasma density. Various diagnostics utilized as a part of ADITYA incorporate electric and attractive probes, microwave interferometry, Thomas Scattering and charge trade spectroscopy. By utilizing controlled Nuclear fusion reaction, electric current is generated. The plasma is made at high temperature and heavy magnetic field is delivered to perform nuclear fusion reaction. Also the chamber must be kept up at its working temperature. When the temperature isn't kept up and furthermore because of some unusual reasons at the external boundary of the plasma begins to crumble which prompts colossal measure of arrival of gases like helium, carbon, neon, H-alpha, H-beta, carbon emissions, nitrogen, etc., Therefore it is important to locate a symptomatic method to recognize the disturbances so it can be utilized to mollify the discharges in future.

2. Literature Survey

T.FURUKAWA, K. TAKIZAWA, D. KUWAHARA and S. SHINOHARA (2017) in their paper 'Electrodeless plasma acceleration system using rotating magnetic field method' have proposed an alternate procedure for quickening the plasma. They have utilized a radio recurrence receiving wire for producing the plasma and they are quickened by RMF reception apparatus which has two sets of coils looking inverse to each other. The RMF parameters like RMF current and furthermore the present stage contrast can be varied. As a result, the warm pushes increments with the expansion in RMF current. Results demonstrates that the age of plasma and speeding up to a higher rate in this strategy.

M.B.CHOWDHURI, J.GHOSH, R.MANCHANDA (2014) in their paper Measurement of spatial and transient conduct of H-alpha emanation from Aditya tokomak utilizing a symptomatic in view of a photomultiplier tube cluster detailed the spatial and fleeting conduct of h-alpha emissions. By utilizing a photomultiplier tube exhibit based spectroscopic, collimated light is gathered from the plasma and the examinations are conveyed out. It is watched that plasma initiated in high field side in common releases of Aditya. The plasma development is coordinated with invalid field area assessed through reenactment.

K.KITTUI, M. M MWAMBURI and F. GAITHO (2014) have revealed the optical properties of Titanium dioxide in their paper' Optical Properties of Tio2 Based Multilayer Thin–films Application to Optical Filters'. They have utilized a high refractive list material tio2 and a low refractive file material sio2 .A five alternating stack of these materials acts as a bandpass with notch filters. It is said that as the thickness of the individual layer increases the number of band passes also increases. Transmittance and Reflectance are measured with spectrophotometer which is equipped with sphere coated with barium sulphate. For 375nm the transmittance was 70%and 80% was the transmittance for 617nm,2500 nm. Reflectance was 34.9% , 9.8%, 30.3% and 98% at

325 nm,375 nm,617 nm and 2500 nm respectively.80% was the highest transmittance and 9.8% was the lowest reflectance realized at 2500nm for this filter.

LEYSON G.P and GONG H. (2014) their paper 'Design of Narrow Band pass filters using thin films' have planned a narrow band pass channel utilizing thin films and furthermore investigated their optical exhibitions. They have talked about a straightforward and fast strategy to ascertain transmittance and reflectance that is done in Abeles method. For their plan, titanium dioxide and silicon dioxide is utilized. The transmittance bend for it a sharp bend and that won't bolster for high transmission. In order to overcome that, a coupling Fabry-perot channels is used in the arrangement. Now the bend is to a greater degree a rectangular shape and this channel serves useful for high transmittance signals

SANTANUBANERJEE, J.GOSH, R. MANCHANDA (2010) in their paper on 'Observations of H-alpha emission profiles' in Aditya tokomak have described the H-alpha emissions from the Aditya tokomak. From the hydrogen Balmer alpha (656.28 nm) emissions are recorded for a huge plasma discharges using a Czerny-Turner spectrometer. It has eight simultaneous vertically collimated line of sights. The line of sights can be moved along the major radius to detect the emissions from the major-radial directions. Considerable h-alpha emissions are observed in the bulk plasma. Also, a second peak in h-alpha radial profile was also observed.

ASIM KUMAR CHATTOPADHYAY and ADITYA TEAM (2008) in their paper' Instability analysis in Aditya tokomak discharges with the help 0f soft x-ray' told about their examination at Instability Analysis in Aditya Tokomak Discharges with the assistance of delicate x-beam have endeavored to dissect the foundations for the disruptions. It is said that Sawtooth motions and other major disruptions are much of the time saw in Aditya tokomak. To know the reasons for these emanations they utilized Soft x-beam tomography. It is really finished with the assistance of a solitary exhibit of finders and it investigations the mode structure of interior disruptions. As a result, the presence of resonance that makes the signal non-sinusoidal are watched and furthermore said that it could undoubtedly couple in reverberation with the mode motions .Finally, it gives the idea that modes and resonance could be the conceivable reason for these disturbances.

DONGHYUN KIM and KIERON BURKE (2003) in their paper' Design of a grating based thin-film filter for broadband spectropolarimetry' proposed a straight forward system for the plan of integrated single-chip grating based thin-film filter. RCWA simulation shows that if the individually designed thin-film spectral and polarization filter are structurally adjusted, then the desired integration is achieved without significant performance degradation.

3. Proposed Method

In Aditya Tokomak, in order to identify the H-alpha emissions a filter is designed. It is said that the wavelength of the H-alpha emission is 656.3nm which falls in the visible light region. So, it is necessary to design an optical filter. Optical filters transmit light of selective wavelength rejecting all other wavelength. When one or more thin layers of optical materials are coated on the substrate, it acts as an optical filter .Zinc sulphide, Magnesium fluoride, Titanium dioxide, Silicon dioxide , Indium tin oxide, Aluminium doped zinc oxide, Zinc selenide are the materials that are mostly used for optical coating. The working process for the design of an optical filter is described below.

Among so many materials available for optical coating, the suitable material is determined. Refractive index of the material plays major role in it. Both material of high and low refractive index should be used in order to prevent complete reflection or refraction of the light incident on it. For the H-alpha filter design, Titanium dioxide (Tio2) and silicon dioxide (sio2) are used. Titanium dioxide is basically a tasteless and Odourless white powder of high refractive index and it has a wide band-gap. This material exhibits high transmittance in the visible region (380nm-760nm).Its optical properties and colour makes it suitable for optical coating and as opacifiers.Tio2 can be manufactured by different techniques like sputtering, sol-gel, pulse laser deposition, electrode position, etc. This material is cost effective for large area deposition and also pairs up well with other materials compared with other materials available. Silicon dioxide (Sio2) is the most abundant oxide on earth's crust. It is an amorphous and smooth material of lower refractive index. It is also durable and also it is insoluble nature plays a major role as it will not get dissolved in nuclear reactor and also these properties makes it suitable for optical coatings.

IOP Conf. Series: Materials Science and Engineering **590** (2019) 012045 doi:10.1088/1757-899X/590/1/012045

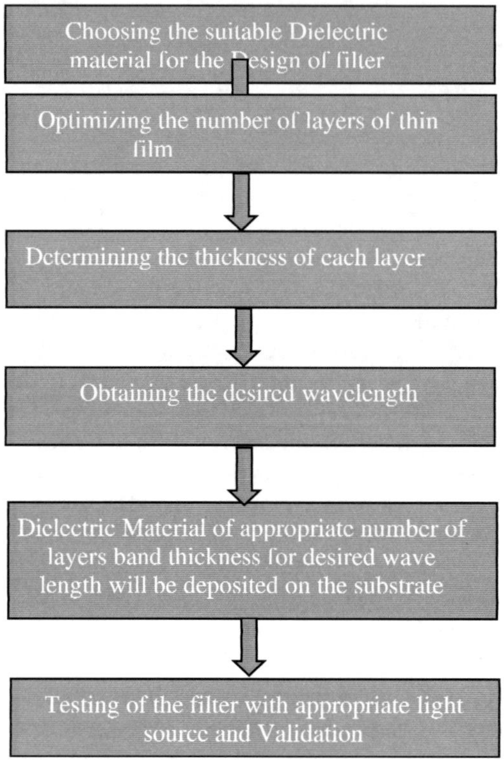

Figure 2. WORK DESCRIPTION

3.1 Substrate

The materials will be coated on the BK7 glass. It is a high quality optical glass with high homogeneity, low bubble and inclusion rate. It also shows an excellent transmittance in visible and near IR region. Due to their optical properties they are used in lenses, prisms and also most suitable for mirror and filter coatings.

3.2 Number of Layers and their Thickness

The process started with a two layer combination of the materials and many layers of combinations are worked out. For each combination, the thickness of the materials are varied. MATLAB tool plays a very important role for this process. The mathematical formulas for transmittance and reflectance are also framed and worked on Matlab and the transmittance for it is traced. Morethan number of outputs are obtained for this process .

4. Results and Discussions

The outputs obtained for different layers of combinations for different thickness are given below

tl=279,th=561

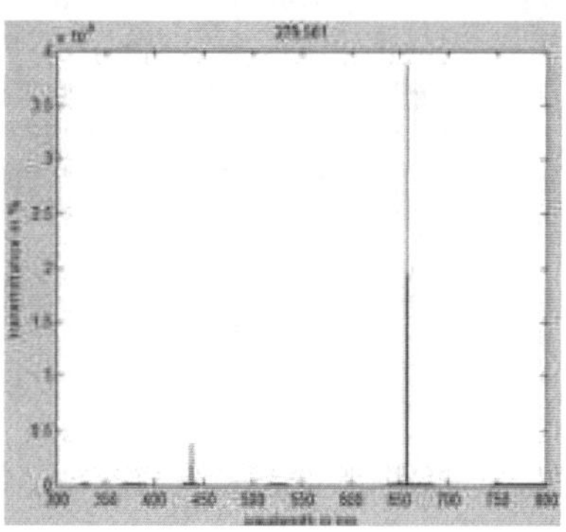

tl=552,th=561

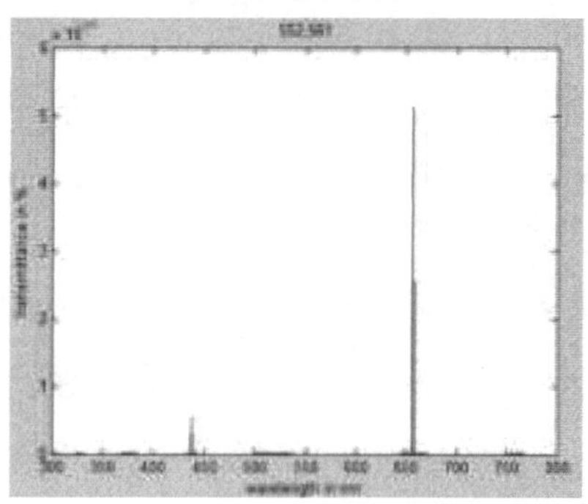

IOP Conf. Series: Materials Science and Engineering **590** (2019) 012045 doi:10.1088/1757-899X/590/1/012045

tl=277,th=561

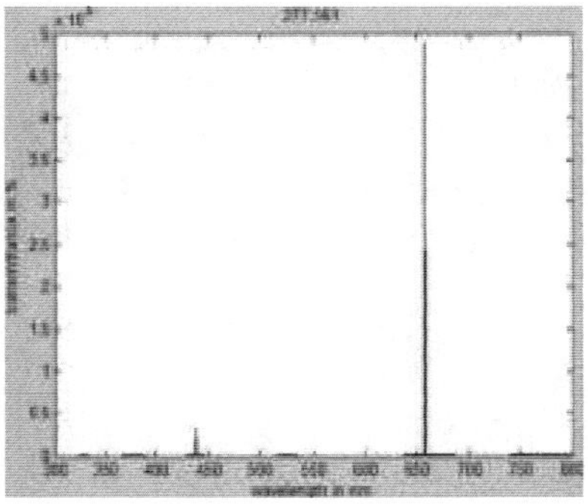

tl=399,th=561

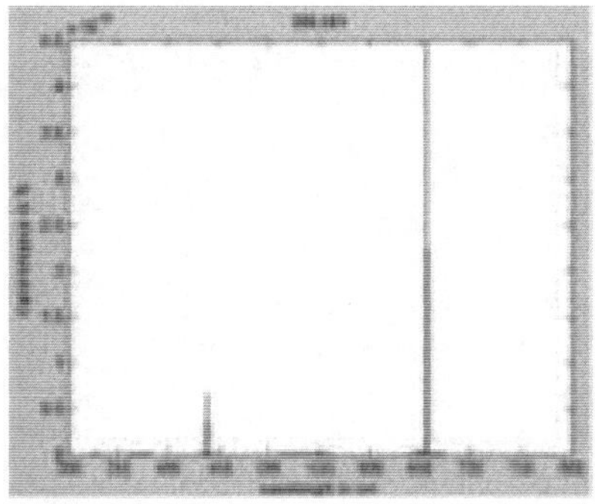

tl=283,th=561

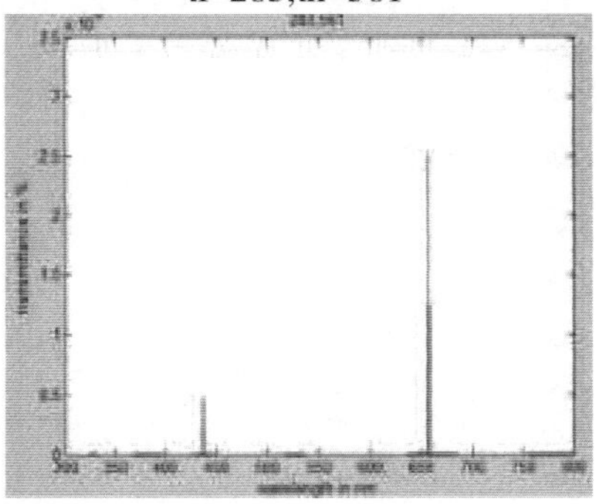

tl=569,th=561

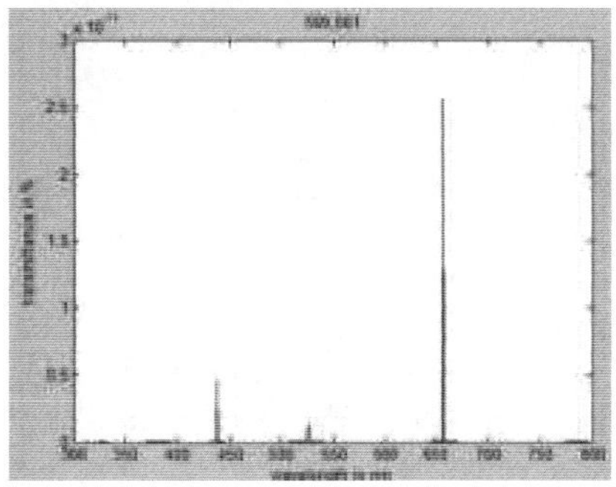

tl=274,th=561

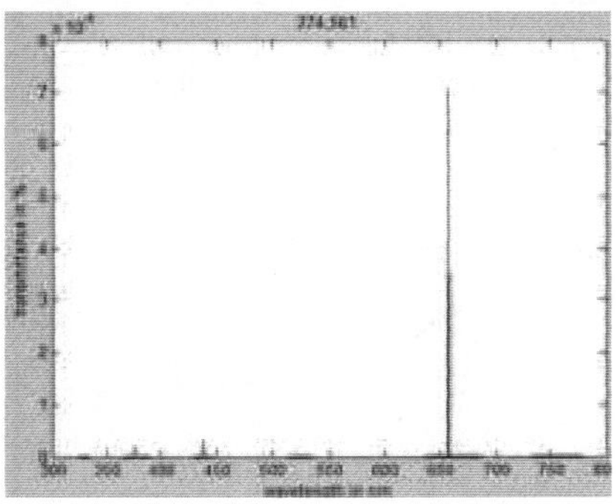

tl=397,th=561

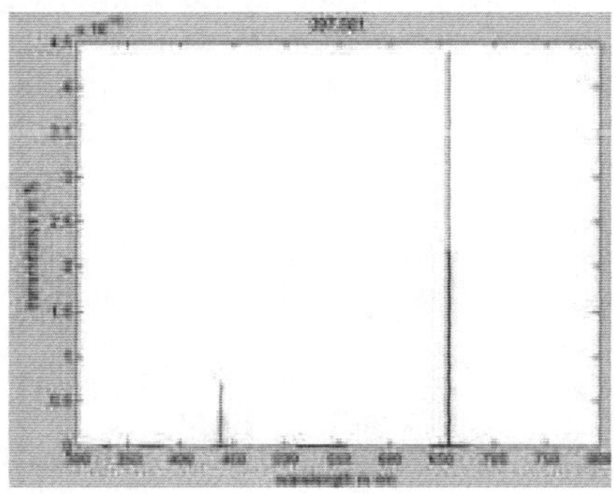

tl=201,th=561

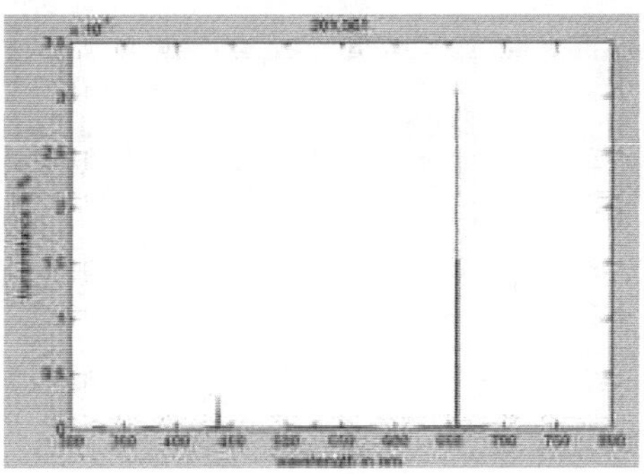

tl=1000,th=561

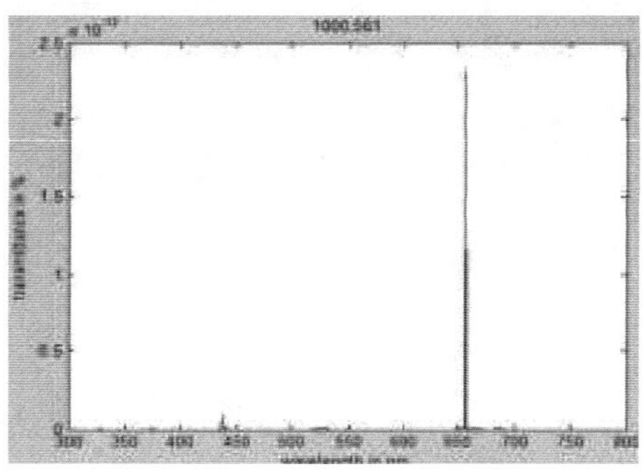

IOP Conf. Series: Materials Science and Engineering **590** (2019) 012045 doi:10.1088/1757-899X/590/1/012045

tl=245,th=561

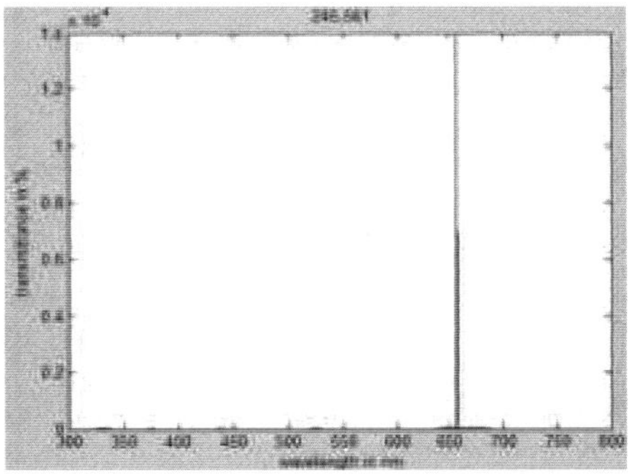

tl=250,th=279

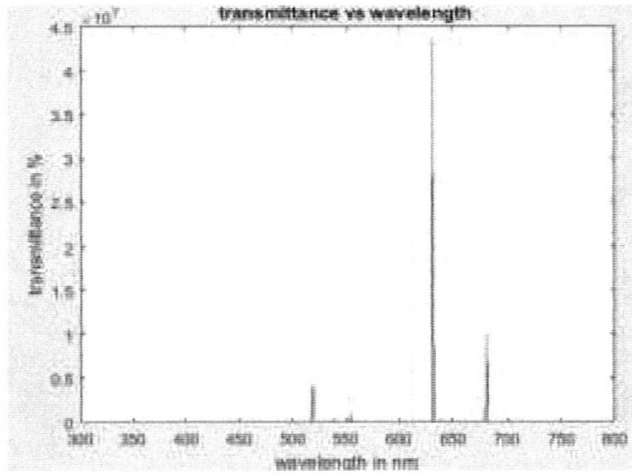

Figure 3. Results of various tl and th

5. Conclusion

In this paper, the working of the tokomak and the problems during the operation and also the emissions from the tokomak are also discussed. It also clearly says the need for an optical filter. For the filter design, Tio2 of high refractive index and sio2 of low refractive index are optimized because of their optical properties. From the MATLAB tool, number of layers and thickness of each layer is determined using Transmittance and reflectance coefficient. Many number of outputs of desired wavelength are obtained for various layer combinations and thickness. The suitable output will be used for the design of this optical filter for H-alpha.

References

[1] Asim kumar Chattopadhyay and Aditya Team (2008), 'Instability analysis in Aditya tokomak Discharges with the help of soft x-ray'.
[2] Beggs.D.M, M. A. Kaliteevski, S. Brand, R. A. Abram, 'Optimization of an optical filter with a Square-shaped pass band based on coupled micro activities'.
[3] Chowdhuri M.B, J. Ghosh, R.Manchanda (2014)'Measurement of spatial and temporal behavior of H-alpha emission from Aditya tokomak using a diagnostic based on a photomultiplier tube arrays'.
[4] Daniela M.Topasna, Gregory A. Topasna, 'Numerical modelling of thin film optical filters.
[5] Donghyun Kim and Kieron Burke (2003),' Design of a grating based thin-film filter for broadband spectropolarimetry'.
[6] Furukawa .F, K. Takizawa , D. Kuwaharaands.Shinohara, (2017)' Electrodeless plasma acceleration system using rotating magnetic field method'.
[7] Kitui.M, M.M Mwamburi, F.Gaitho,(2014)'Optical properties of Tio2 based multilayer thin-films: Application to optical filters, International Journal of thin films science and technology.
[8] Leyson G.P, Gong.H, (2014) 'Design of narrow bandpass filters using thin films'.
[9] Santanubanerjee, J.Gosh, R. Manchanda (2010) 'Observations of H-alpha emission profiles in Aditya tokomak'.
[10] Tibulaec.S, R. Magnusson,' Narrow-line width band pass filters with diffractive thin-film layers'.

Simulation and Design of Beta Filter for Tokamak Diagnostics

N.L.Deepika[1]
Final Year UG student, Dept of ECE,
Sathyabama Institute of Science and Technology
OMR Road,Chennai, India
P.Manasa[1]
Final Year UG student, Dept of ECE,
Sathyabama Institute of Science and Technology
OMR Road,Chennai, India
L.Jegan Antony Marcilin[2]
Assistant Professor, Dept of ECE,
Sathyabama Institute of Science and Technology
OMR Road,Chennai, India
jegan25@hotmail.com
N.M.Nandhitha[3]
Professor&Dean, Dept of ECE,
Sathyabama Institute of Science and Technology
OMR Road,Chennai, India
nandhi_n_m@yahoo.co.in

Abstract- Simulations of H-beta filter based on thin films is proposed and demonstrated in this paper. The beta filter is optical thin film pass band filter of having the wavelength of 486.13nm.H-beta filter is used to reduce the transmittance of H-beta wave. This H-beta filter is used in Aditya Tokamak which is a plasma reactor. when the plasma ring in the plasma breaks and releases ionized gases. H-beta is one of them. In order to reduce the transmittance of the H-beta wave this filter is used in the reactor. The H-beta filter consists of thin films on the substrate. The thin films are the substance having the optical properties .The thin film consists of low and high refractive index. The simulation of the H-beta filter program was done by using matlab.

Keywords:Transfer matrix, Thin film, Optical pass band filter, Aditya Tokamak, Plasma reactor, Plasma ring Refractive index, Transmittance, Reflectance

1. INTRODUCTION

H-beta filter is used in the Aditya Tokamak which is at IPR-Indian plasma research. IPR is autonomous hysics research institute located in India. The institute is involved in research in aspects of plasma science including basics plasma physics research on magnetically confined hot plasmas and plasma technologies for industrial applications. It is a large and leading plasma physics organization in India. Aditya Tokamak is a medium sized Tokamak installed for plasma research in India. Tokamak is a device that uses a powerful magnetic field to confine plasma in the shape of torus. The Tokamak is one of the several types of magnetic confinement device being developed to contain the hot plasma needed for producing controlled thermonuclear fusion power. Aditya Tokamak has a major radius of 0.75 meters and minor radius of 0.25 meters.The maximum field strength is 1.2 tesla produced by 20 toroidal field coils spaced symmetrically in the toroidal direction.

Usually fusion reactors fuse two lighter atomic nuclei to form a heavier nucleus releasing energy. Devices designed to harness this energy are known as fusion reactors. The fusion reaction normally takes place in a plasma of deuterium and tritium heated to millions of degrees. Plasma is a state of matter resembling an ionized gas. It is one of the four fundamental states of matter solid, liquid, gas, plasma. Through partially ionized plasma also exist. This ionization makes plasma effected by electromagnetic fields. Plasma and ionized gases have unique properties and display behavior unlike those of other states, and the transition between them is mostly a matter of nomenclature and subject to interpretation. In Aditya Tokamak inside the plasma reactor a constant temperature should be maintained. If the temperature changes the plasma reaction will get distracted, then high intensity radiation will released. This radiation consists of h-alpha, h-beta, carbon-III and so many other ionized gases which are very harmful to the environment and living bodies.

Fig:1 Aditya Tokamak

H-beta filter is used to reduce the transmittance of H-beta rays released form the reactor. H-beta is having the wavelength of 486.13 nm. H-beta occurred in the series of spectral emission lines of hydrogen atom. In the hydrogen spectrum when the electrons from the higher levels down to the energy level with principal quantum number n=2 H-beta will occur H-beta filter is a optical thin film filter. It consists of one substrate and two thin films one is high refractive index and the other is low refractive index. By using transfer matrix we are simulating and optimizing this filter in matlab.

2. LITERATURE SURVEY

NUMERICAL MODELING OF THIN FILM OPTICAL FILTERS: Authors-Daniela M. Topasna and Gregory A .Topasna

The transfer matrix is used to represent the film and characterize its performance. A homogenous isotropic film with index of refraction n1 deposited on a substrate with index of refraction ns and placed in a medium with n0. The incident beam undergoes an external reflection at the first interface and the transmitted beam undergoes an internal reflection and transmission at the second interface.

WIDELY TUNABLE SINGLE PASS BAND MICROWAVE PHOTONIC FILTER BASED ON DFB-SOA-ASSISTED OPTICAL RECOVERY: Authors-Ye Deng, Ming Li, Jian Tang, Shuqian Sun, Ningbo Huang, Ninghua Zhu. Oct 2015

Recovered the suppressed optical carrier from a pass band filter by the wavelength selective amplification via a DFB-SOA(distributed feedback semiconductor optical amplifier). Tuning range from 5 to 35 GHZ.The 3db band width and the out of band suppression ratio are measured to be 4GHZ and 20db respectively.

SYSTEM PERFORMANCE OPTIMIZATION OF FREQUENCY SWEPT PUMP BASED RECTANGULAR BRILLOUIN OPTICAL FILTER: Authors -M. Shi, L.Yi, W. Wei, G. Pu, W.Hu. Feb 2017

IOP Conf. Series: Materials Science and Engineering **590** (2019) 012046 doi:10.1088/1757-899X/590/1/012046

The frequency sweep pump is the key parameter to achieve acceptable signal performance in the frequency swept pump based SBS gain filter. Signal is fed into filter with period ragging from 0.1 to 100us in terms of BER. The results show that the optimal frequency sweep pump period is always 1/120 of signal propagation time.

INVESTIGATION OF ALL OPTICAL TUNABLE NOTCH FILTER BASED ON SUM FREQUENCY GENERATION IN PPLN WAVEGUIDE: Authors- Zhefeng Hu, Min Hou, Haiyang Liu, Fushenchen.Oct 2015

Optical tunable notch filter based on a sum frequency generation process in a periodically poled lithium niobate (PPLN) wave guide. Maximum signal suppression at central wavelength by varying the power of input pump light. Maximum will be 34.4db at 1545nm.

3. EXISTING MODEL

Optical filters are devises that selectively transmit light of different wavelengths. Usually implemented as plane glass or plastic devices in the optical path which are either dyed in the bulk or have interference coatings used in photography, optical instruments, color stage lighting. There are two types of optical filters absorptive and dichroic filters. Dichroic filters are thin film filters, can be made by coating a glass substrate with a series of optical coatings using interference principle. Their layers form a sequential series of desired wavelength. Other wavelengths destructively cancel or reflect as the peaks and troughs of the wave overlap. These are expensive. There are long pass optical filters, band pass optical filters, short pass optical filters.

4. PROPOSED METHOD

H-beta filter is an optical band pass thin film filter. In this there are two thin film materials one is having high refractive index and the other is having low refractive index.Titanium nitride having refractive index 1.37 is low refractive index film and silicon nitride having refractive index 2.35 is used as high refractive index materials in H-beta filter. Glass which is having refractive index 1 is using as substrate in this filter. The transmittance depends on the interface of light or H-beta rays on these materials.

Transfer matrix is used to represent the film and characterize its performances

$$t = \frac{2\gamma_0}{\gamma_0 m_{11} + \gamma_0 \gamma_s m_{12} + m_{21} + \gamma_s m_{22}}$$

$$r = \frac{\gamma_0 m_{11} + \gamma_0 \gamma_s m_{12} - m_{21} - \gamma_s m_{22}}{\gamma_0 m_{11} + \gamma_0 \gamma_s m_{12} + m_{21} + \gamma_s m_{22}}. \quad (1)$$

In this formulae t and r represents the coefficients of transmittance and reflectance. Where, Transmittance(T) =|t*t| Reflectance (R) = |r*r|. The simulations are done by using this in matlab.

5. RESULT

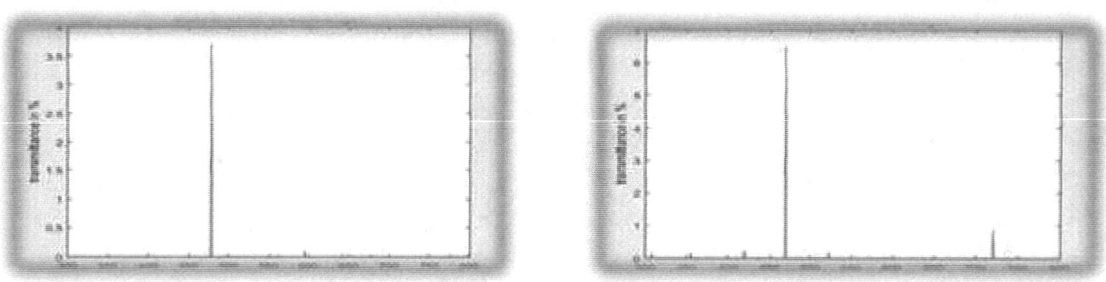

Figure 3. Result analysis

IOP Conf. Series: Materials Science and Engineering **590** (2019) 012046 doi:10.1088/1757-899X/590/1/012046

6. CONCLUSION

Titanium nitride and silicon nitride are used as low and high refractive index thin film materials in H-beta optical thin film band pass filter. After optimizing so many combinations of these using transfer matrix in matlab, getting the transmittance value about 5.3*10^5 at 486nm wavelength which is the desired H-beta wavelength.

REFERENCES

[1] Investigation of all optical tunable notch filter based on sum frequency generation in a PPLN waveguide-Zhefeng Hu, Min Hou, Haiyang Liu, Fushenchen. Dec 2015.

[2] Introduction to modern optics- Fowles, Grant R.

[3] Introduction to optics –Pedrotti, Frank L,S.J. and pedrotti, Leno s.

[4] Numerical modeling of thin film optical filters –Daniela M. Topasna and Gregory A .Topasna.

[5] System performance optimization of frequency-swept pump- based rectangular brillouin optical filter-M. Shi, L.Yi, W . Wei, G. Pu, W .Hu. Feb 2017.

[6] Thin film physics – Heavens,O.S

[7] Pravin A and Srinivasan S (2012), "An efficient programming rule extraction and detection of violations in software source code using neural networks," 2012 Fourth International Conference on Advanced Computing (ICoAC), pp. 1-4.

IOP Conf. Series: Materials Science and Engineering **590** (2019) 012047 doi:10.1088/1757-899X/590/1/012047

DESIGN AND IMPLEMENTATION OF C III FILTER FOR MEASURING DISRUPTION IN ADITYA TOKAMAK

L. Jegan Antony Marcilin[1]
Assistant Professor, Department of ECE,
Sathyabama Institute of Science and Technology,
Jeppiaar Nagar, Rajiv Gandhi Salai,,Chennai-India
jegan25@hotmail.com
N.M.Nandhitha[2]
Professor & Dean, Department of ECE,
Sathyabama Institute of Science and Technology,
Jeppiaar Nagar, Rajiv Gandhi Salai, Chennai - India
nandhi_n_m@yahoo.co.in
Gayathri Sam [3,]
Final Year UG student, Department of ECE,
Sathyabama Institute of Science and Technology
Jeppiaar Nagar, Rajiv Gandhi Salai,,Chennai-India
vinisam05@gmail.com
M.Gladies Jaya Chitra [3]
Final Year UG student, Department of ECE,
Sathyabama Institute of Science and Technology
Jeppiaar Nagar, Rajiv Gandhi Salai,,Chennai-India
mgladiesthangaraj@gmail.com

Abstract- The main objective of this project is to filter carbon gas among many other gases emitted from the plasma in Aditya Tokamak. The aim is to measure the disruptions released by plasma due to some unpredictable conditions. C3 filter detects carbon gases from the various gases emitted from the plasma in Aditya Tokomak. The filter is to be designed with 5-6 layers of the material chosen. Then the signal segmentation for transmittance vs. wavelength is to be achieved using MATLAB software.
Keywords: Aditya Tokamak, C III Filter, Matlab.

1. INTRODUCTION

In thermonuclear fusion research using magnetic confinement, tokamak is the most promising candidate to demonstrate fusion as achievable energy source. A tokamak (“**TOroidalnaya KAmera MAgnitnaya Katushka** “or **"Toroidal Vessel with Magnetic Coils”**) is a toroidal device which uses a strong toroidal magnetic field, to confine high temperature plasma within the torus for a sufficiently long time. Final goal of the tokamak research is to reach fusion of deuterium and tritium nuclei for production of electricity. In diagnostic signals of Aditya tokomak certain wavelength 464.7 nm is highly significant. Hence it is necessary to analyse the signals throughout the operation of the Tokamak . Range of wavelengths for which a filter is highly reflecting or transmitting can be altered by changing the characteristics of the thin film components. The transmittance is affected by the choice of the dielectric material, thickness of the dielectric layers, number of these layers, angle of incidence light on the filters. The main objective of

IOP Conf. Series: Materials Science and Engineering **590** (2019) 012047 doi:10.1088/1757-899X/590/1/012047

this project is to filter carbon gas among many other gases emitted from the plasma in Aditya Tokomak . The aim of this project is to measure the disruptions released by plasma due to some unpredictable conditions .C3 filter detects carbon gases from the gases emitted from the plasma in Aditya Tokomak. The filter is to be designed with 5-6 layers of the material chosen. Various types of filters are available but we are going to use optical filter since it has lots of advantages over other filters. In optical filters absorptive filters and interference filters are most commonly used for the selective transmission light of certain wavelength and helpful in removing harmful UV rays. Its properties describing the changes of refractive index with respect to the changes in the wave length. To prevents mounting difficulties to ensure the optical system performance. Mechanical properties like elasticity for withstanding the highest pressure, density for heaviness, hardness for scratching resistance. Applications like helpful in fluorescence microscopy, spectroscopy, clinical chemistry and imaging. ADITYA TOKAMAK means a Russian word which means confining the plasma in the torus shape. 'Tokamak' means the Magnetic confinement device needed for producing controlled thermo nuclear fusion power. 'Aditya' means Medium sized tokomak installed at the research centre for plasma research. The tokamak is in the torus shape which is a revolving circle in 3D axis containing polloid and toroid.

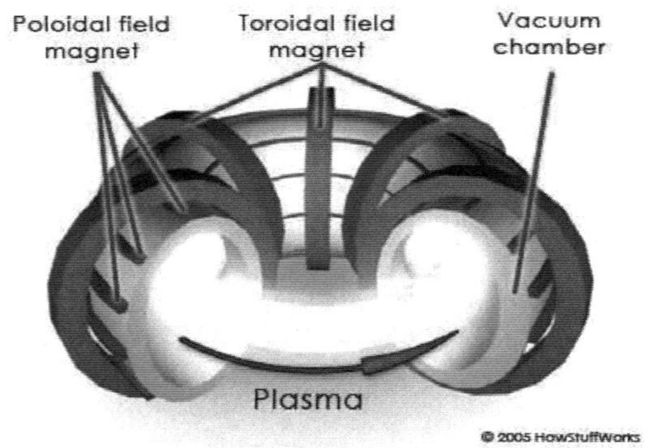

Fig1. Diagram representing tokomak internal view.

To contain the hot plasma needed for producing controlled thermonuclear fusion power (i.e.) electricity for generating heat. Deuterium and tritium has been used which has to be heated up to 1000 degree. There is a vacuum placed inside the torus which contains the hot plasma. The hot plasma is maintained in a certain temperature range to avoid disruptions. The horizontal magnetic windings is known as "poloidal field magnet" and the vertical windings is known as "toroidal field magnet". This magnetic confinement device contains hot plasma to provide thermo nuclear fusion power. Its radius ranges from 0.25m-0.75m.

2. EXISTING SYSTEM

The two dielectric material with high refractive index (ZnS) and low refractive index (MgF_2) selection is made depending on their property and uses. MgF_2 for highly pure, well suited for optical coating, used for many years in anti-reflection and multi layer coatings. It is insoluble and hard if deposited on hot substrate. MgF2 evaporates cleanly and completely, without out gassing or spitting and leaves a uniform coating on the substrate. Resist to thermal and mechanical shocks. ZnS for its insoluble in water and well suited for optical coating, substrate BK7 glass. They optimized magnesium fluoride by thermal evaporation method. The number of layers of optical coating of dielectric material onto the substrate is optimized to its efficient values, 6 layers are used as dielectric stack. The thickness of each layers in terms of nm optical coating of dielectric material are predicted from the MATLAB tool using mathematical equations. In heuristic method: H alpha (HLHHLH) (th=279, tl=250), H beta (LHLHLL) (th=347,tl=300), carbon III (LHLLHH) (th=164,tl=107). In deterministic method: (HLHHLH) H alpha (th=561), H beta (th=100), carbon III (th=95).

Thus the results are analyzed with the comparison of wavelength in nm Vs transmittance coefficient.

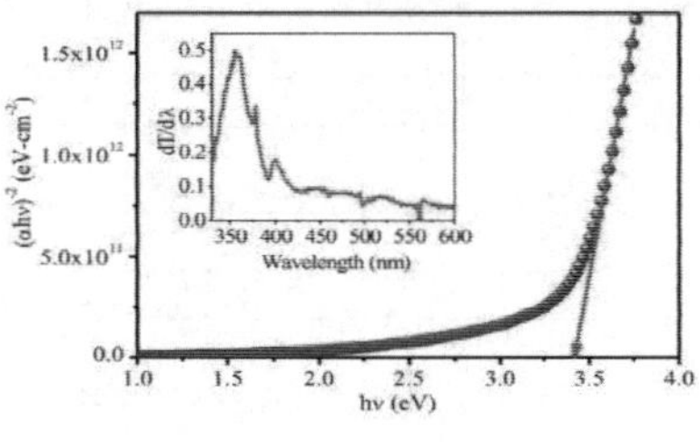

Fig2. Transmittance vs wavelength

3. PROPOSED SYSTEM

Thermo nuclear fusion power provides heat to generate electricity. Fusion is the process of fusing two small nuclei together to form a heaviest nuclei. This heaviest nuclei releases some energy. In nuclear fusion, there are two types of reactors stable enough to conduct fusion: magnetic confinement reactors and inertial confinement reactors. The former method of fusion seeks to lengthen the time that ions spend close together in order to fuse them together. Inertial confinement reactors, unlike magnetic confinement reactors, use laser fusion and ion beam fusion in order to conduct fusion. However, with magnetic confinement reactors you avoid the problem of having to find a material that can withstand the high temperatures of nuclear fusion reactions. The heating current is induced by the changing magnetic fields in central induction coils and exceeds a million amperes. Magnetic fusion devices keep the hot plasma out of contact with the walls of its container by keeping it moving in circular or helical paths by means of the magnetic force on charged particles and by a centripetal force acting on the moving particles. The most serious problem caused by magnetic islands is the major disruption, it is the disturbance caused in any event or process. The vacuum consists of hot plasma inside it. Plasma is sometimes the fourth state of matter, sometimes referred to as ionised gas. Disruption is sudden and complete loss of plasma confinement and a collapse of the plasma current, which result in large electromagnetic and pressure forces in the surrounding structure. These generally occur at the sufficiently high plasma density or high plasma current. Due to some unpredictable conditions, disruptions will occur. This disruption leads to the emission of various gases like H alpha, H beta, Carbon, etc., One mechanism for the disruption is nonlinear coupling of two islands which generate a large stochastic magnetic region within the plasma volume destroying the plasma confinement. Magnetic islands, when uncoupled or only loosely coupled to each other, grow on a time scale of milliseconds. This growth rate is too slow to explain the sudden nature of the hard disruption which happens on a microsecond time scale. If, however, two magnetic islands of different felicity were to grow and overlap, The resulting region engulfed by the islands may explosively grow into a stochastic magnetic region of poor confinement. If the initial islands were large enough, a large percentage of the plasma confinement would be destroyed there by terminating the discharge. Thus we are going to detect CIII gases which get transmitted in the wavelength of 464.7nm.

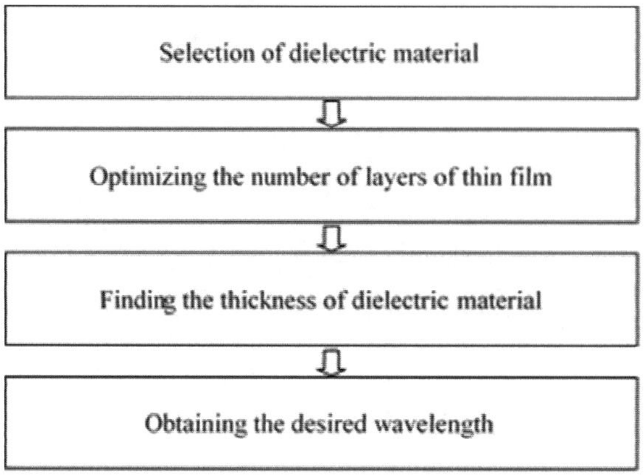

Figure 3 Flow diagram

4. HARDWARE ARCHITECTURE

4.1 Titanium dioxide(TiO_2):
Its refractive index n=2.4335,extinction coefficient K=0.0001,Appearance-White solid,Odour-Odourless,Melting point-1843°C(116 K), Boiling point-2972°C(245K),Solubility –insoluble,Refractive index-2.488,Density-3.78 g/cm^3
Natural occurring mineral (i.e) earth mineral. Helps in thickening, whitening, and lubricating process. It helps from harmful UVA and UVB radiations. It is considered to have no risk to skin sensitivity.

4.2 Silicon dioxide(SiO_2):
Its refractive index-n=1.4432, extinction coefficient K=0.00002 , Appearance-White solid , Odour-Odourless, Melting point-1713°C(1986 K), Boilingpoint-2950°C (3220K) ,Solubility-insoluble,Refractiveindex-1.4585,Density-2.2 g/cm^3
Earth mineral containing quartz used to make glasses which helps in providing layers of substrate. No risk to skin sensitivity.

4.3 Zinc Sulphide(ZnS) :
Its refractive index : n=2.2719 ,wavelength : 1.53846 um ,Appearance-White solid ,Melting point-1850°C(120 K),Boiling point-2230°C(245K), Solubility –negligible, Refractive index-2.3677 , Density-4.090 g/cm^3 It is an organic compound formed by chemical vapour deposition. It is usually product from waste material. By product from ammonia synthesis using CH_4 usage of ZnS thus causes irritation, itching skin. Thus it is poisonous and hazardous.

5. LITERATURE CITED

Bin Alexander et al (2017) analyzed the acceleration of rotating plasma flows in crossed magnetic fields. A solenoid and permanent magnets were used. The change of axial magnetic flux induces an azimuthal electric field that rotates electron and ion flows in different direction, which produces induction acceleration. Using the linearity of the Maxwell equations, the total electric and magnetic fields of the system were presented. The interaction of azimuthal electron and ion flows with the radial component via the Lorentz force leads to the axial acceleration of electrons and ions in the same axial direction. In an axial direction, the twirling plasma flow in a cylindrical vortex are accelerated which resulting In an energy/momentum transfer in the axial direction. This will help in the creation of plasma thrusters [1].

IOP Conf. Series: Materials Science and Engineering **590** (2019) 012047 doi:10.1088/1757-899X/590/1/012047

Scott Robertson et al (2017) modelled the acceleration of plasma from a source region of high magnetic field to a region of lower field. The model combined the electron and ion equations of motion and includes a source term to replenish plasma in the source region, the acceleration from the ambipolar electric field, the acceleration from the magnetic mirror force. The gyro fluid equation for axial motion to finding the acceleration of plasma from a source region in strong magnetic field into region of lower field was obtained. The model is applicable to plasma experiments with helicon sources and plasma processing configurations [2].

Geraldo Roberson et al (2017) constructed an area preserving and integrable map to represent magnetic surfaces with triangularity in single null divertor tokamak. The magnetic surfaces obtained by the map can assume different asymmetric geometries. The shape of cross section of these magnetic surfaces played an important role on plasma stability and confinement in fusion reactors includes trajectory integration method and Equilibrium model. This map can be used to investigate the dependence of stability and deposition patterns in terms of different equilibrium parameters. The application of this methodology permits to obtain the escape patterns of magnetic field lines on the divertor plates [3].

Qin Hang et al (2017) analyzed the experimental advanced Superconducting Tokamaks. A good knowledge of the total plasma inductance was benefitted to map the discharging of vertical magnetic field by Maxwell's equation. The relationship of vertical magnetic field and line averaged density was derived during flat-top phase for given plasma current. An increase in magnetic strength of vertical field will allow high temperature, high density, high beta and high bootstrap current fraction to be achieved. High confinement mode and long-pulse steady-state operation was an important goal of current magnetic fusion energy research [4].

M.Afsharmanesh et al (2017) simulated the study of the factors affecting the collisional power dissipation in helicon plasma. Plasma density and magnetic field strength were modified to maximize the collisional power absorption in the helicon plasma. Helicon plasma sources were designed to allow helicon wave propagation. Simulations were done for two common types of RF antennas namely; half helix and Nagoya type-3.The simulations were performed with the CST Microwave studio code. The method was reliable. CST code had the capabilities that calculate the power absorption for different antenna types. The power deposition was significantly asymmetric when the ratio was more than a determined value. Asymmetric power absorption was favorable for the efficient ion beam extraction from the helicon plasma [5].

Dingee Yang et al (2017) conducted an experiment to improve the controlling effect of arc plasma and to understand the arc characteristics and the effect of magnetic field on the plasma jets. The Axial Magnetic Field (AMF) is the field of contacts were conducted in a Helmholtz coil. It is used to generate the different external imposed AMF's. It may lead to transfer of arc column when the composite AMF in inter electrodes to different directions at a certain moment. Appearance of cathode plasma jets was analyzed by three methods. They are appearance of cathode plasma jets, inclination of cathode plasma jets and its mechanism. It can be concluded that the appearance of individual cathode plasma jets can be manipulated by the variation of AMF. It was used to study the interaction between the arc plasma and electrodes. By introducing the external imposed AMF, the composite magnetic field vector direction changed [6].

Denis Eremin et al (2017) developed a model for resonant surface wave excitation in a large CCP reactor. Low pressure plasmas produced in Capacitive Coupled Plasma (CCP) reactors operated were used in large amount in the plasma processing industry. By using the resonant excitation of surface modes that a CCP reactor can support, they explained it. It showed how such modes are triggered and how they were affected the radial plasma density. This paper discussed some aspects of the physics of large scale CCP reactors driven at very high RF frequencies, which might affect uniformity of the plasma density. It was showed that in highly collisional plasmas, the non uniformities caused by the modes disappear due to the large mode damping [7].

6. RESULT AND DISCUSSION

Thus we executed for various layers starting from 2 layer combinations to 6 layers result is much closer to carbon wavelength. We got exact peak for 6 layer combination. All the existing systems proposed using maximum of 2 layers but we are using maximum of 6 layers in order to increase the efficiency. Using of TiO2, SiO2 is cost efficient and hazardless. These are the output closer to our work.

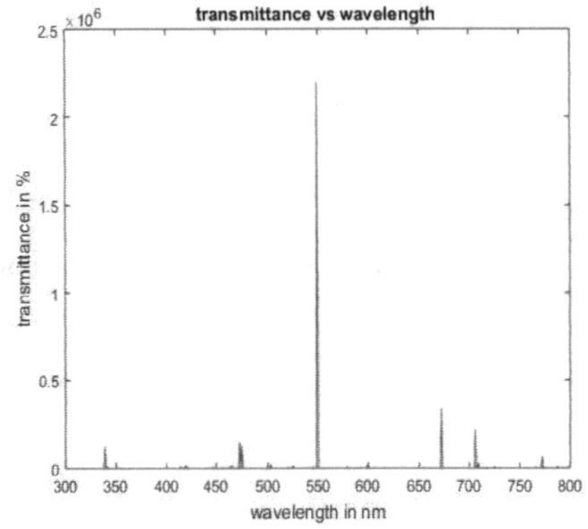

Fig4. Closest output

Thus we got the exact output i.e. CARBON III(464.7nm) for 6 layer combination using material like SiO2,TiO2 with thickness of dielectric material of about tl is 164 and th is 107 with refractive index of substrate is 1.5240, the refractive index of thin film material SiO2 is 2.4533 and refractive index of thin film TiO2 is 1.3809 with dielectric stack (HLHHLL) FWHM of 2nm

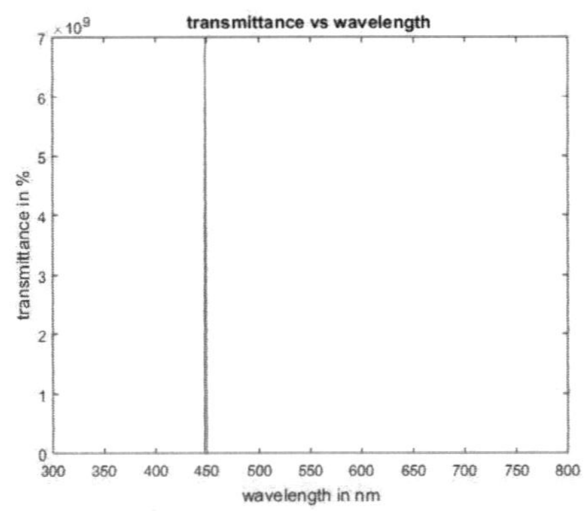

Fig5. Carbon III output characteristics

7. CONCLUSION

In Aditya Tokomak for generation of electric power and various kind of disruption occurring in the reactor due to unpredictable reasons is the Major challenges, in this design identifying a suitable thin film, determining the thickness of thin film, number of layers. The thickness of each layers in terms of nm optical coating of dielectric materials are predicted from the mat lab tool using mathematical equations. The results are analysed with the comparison of wavelength in nm Vs. transmittance coefficient.

REFERENCES

[1] BinAlexander R. Karimov and Paul A. Murad, "Acceleration of rotating plasma flows in crossed magnetic fields," IEEE transactions on plasma science, vol.45, no.7, July 2017.

[2] Scott Robertson, "Gyrofluid model of plasma expansion in a diverging magnetic field," IEEE transactions on plasma science, vol. 45. no. 10, oct 2017.

[3] Geraldo Roberson,Marisa Roberto,Tiago Kroetz, "Shaping Diverted Plasmas with Symplectic maps," IEEE transactions on plasma science, 2017.

IOP Conf. Series: Materials Science and Engineering **590** (2019) 012047 doi:10.1088/1757-899X/590/1/012047

[4] Qin Hang and Ge Li, "Discharge Analysis of East H-Mode for Designing Compressed Plasma," IEEE transactions on plasma science, 2017.
[5] M.Afsharmanesh and M.Habibi, "A Simulation study of the Factors Affecting the Collisional Power Dissipation in a Helicon Plasma," IEEE transactions on plasma science, 2017.
[6] Dingee Yang,Shenli Jia,Weidong Qi, "Influence of axial magnetic field on cathode plasma jets in high-current vaccum arc," IEEE transactions on plasma science, 2017.
[7] Denis Eremin, "Modelling of Resonant Surface Wave Excitation in a Large CCP Reactor," IEEE transactions on plasma science, 2017.
[8] L.Pigatto, M.Baruzzo, P.Bettini, "Control system optimization techniques for Real-time application in fusion plasmas: the RFX- mod experience," IEEE transactions on nuclear science, DOI.10.1109/TNS.2017.
[9] Tautvydas Maceina, Paolo Bettini, Gabriele Manduchi, "Fast and efficient algorithms for computational electromagnetics on GPU architecture," IEEE transactions on nuclear science, 2017.
[10] Gasser F. Abdelal and Adrain Murphy, "A Multiphysics Simulation Approach for efficient modeling of lightning strike tests on Aircrafts structures," IEEE transactions on plasma science, 2017. 55
[11] Bin Ma, Zhengxiang Huang, Qiangqiang Xiao, "Effect of External magnetic field loaded at the initial period of inertial stretching stage on the stability of shaped charge jet," IEEE transactions on plasma science, 2017.
[12] Lijun Wang, Kang Qin, Lilan Hu, Xiao Zhang, "Numerical simulation of vacuum arc behavior considering action of adjacent phases in vacuum circuit breakers," IEEE transactions on plasma science, vol.45, no.5, may 2017.
[13] Ram K. Singh, Subodh Kumar, Ram P. Sharma, "High-Power Terahertz Wave Generation by Optical Rectification of a Gaussian Laser Pulse Propagating in Magneto plasma," IEEE transactions on plasma science, 2017.
[14] Igor O. Girka and Manfred Thumm, "Excitation of Electromagnetic Waves Above the Upper-Hybrid Frequency by Internal Gyrating Electron Beam in a Coaxial Waveguide," IEEE transactions on plasma science, 2017.
[15] Manabu Tanaka and Yoshihiro Okuno, "Performance of a SeedFree Disk Magneto hydrodynamic Generator With Self-Excited Joule Heating in the Nozzle," IEEE transactions on plasma science, 2017.
[16] A.Mesyats and Igor V.Uimanov, "Semiempirical Model of the Microcrater Formation in the Cathode Spot of a vauum Arc," IEEE transactions on plasma science, 2017.
[17] P. C. de Vries, M. F. Johnson, I. Segui, and JET EFDA Contributors, Nucl. Fusion 49, 055011 (2009).

IOP Conf. Series: Materials Science and Engineering **590** (2019) 012048 doi:10.1088/1757-899X/590/1/012048

A Robotic Platform to Identify Gas Pipe Leakage Using IOT

L.K. Joshila Grace*[1], K.SaiTeja[2],J.V. Sai Kishan Reddy[3]

[1]Professor, [2,3]Student, Computer Science and Engineering

Sathyabama Institute of Science and Technology,
Chennai, India
[1]joshilagracejebin@gmail.com

Abstract. Gas pipes play a very vital role in cities as well as industries thus directly related to the growing economy of our country. Gas leakages in our present day world lead to many fire accidents and pose a greater threat. To place sensors at each section of the pipe is difficult as well as time taking and costly. So, we are proposing an innovative robot that clings to the outer surface of a gas pipe and moves along the pipe to detect gas leakages. The robot keeps on moving along the pipe surface and will monitor any gas leakage, when it detects it will send the transmit location using an interface GPS sensor to the IOT login system, we use a software called UBIDOTS to display the gas leakage position and it also sends alert messages over IOT. At last, we design a robot which is automated and has an insect like structure that moves along the gas pipe and it detects the gas leakages at a very low cost.

Keywords— Internet of Things, Gas detection Sensors, Smart Alerting Techniques, Prediction, Data Analytics.

1. Introduction

Internet of Things aims towards making life simpler by automating every small task around us. The Internet of things (IOT) is between frameworks organization of physical contraptions, vehicles (in like manner suggested as "related devices" and "splendid devices"), structures, and distinctive things—embedded with equipment, programming, sensors, actuators, and framework arrange that engage these articles to assemble and exchange data. Safety has always been an important criteria while designing homes, buildings, industries as well as cities. The increased concentration of certain gases in the atmosphere can prove to be extremely dangerous. These gases might be flammable at certain temperatures as well as in some specific humidity conditions, after exceeding the specified concentration limits or even a supplementary factor in the air pollution of an area leads to problems such as smog and reduced visibility which can in turn cause severe accidents and also have adverse effect on the health of people.

In order to achieve this, we are implementing a robotic Platform for gas pipe leakage detection in industries. The heart of the system is Raspberry Pi which uses Raspbian Os which functions like a mini CPU. Various gas sensors such as Co2, Methane-MQ2, H2S and GPS Sensor are

IOP Conf. Series: Materials Science and Engineering **590** (2019) 012048 doi:10.1088/1757-899X/590/1/012048

used. We attach the Raspberry Pi equipment to a Motor Driver which supplies additional voltage than the inbuilt voltage which helps in the better functioning of Dc 0Motors. Raspberry Pi also uses cloud technology to display alert messages and to store the values for future reference. This system not only alerts us through audio but it also gives us the video coverage, giving us the exact location of leakage.

2. Literature Survey

Dedy Rahman Wijeya & Riyanarto Sarno in their paper discussed about the concept of Mobile Electronic Nose (MoLen) which is very promising for Sensing as a Service (S2aaS) applications. In Internet of Things (IoT) era, MoLen can be implemented in many real-world applications such as food quality assessment, medical applications, gas leakage detection, home or automotive safety system, environment monitoring, etc.

Akihiro Shibata & Masami Konishi in their research analysed and mitigated the industry, for example, oil refinery industry, there may happen different sorts of security issues for pipelines matured after its developments. To acknowledge preventive support of pipelines, there are substantial requirements for the determination innovation of gas spillage. In this examination, gas spillage sounds created from the splits of channels are broke down and attempted to be utilized for location of the gas spillage. Sound information, for investigation are produced and gathered in the plant where foundation commotion is'nt insignificant.

Pushpendu Kar, in his research paper used three different models which have been developed to detect the flaws in a gas pipeline network, which is one of the significant weaknesses towards the sheltered gas appropriation among the end clients. Along these lines, it is huge to find the flaws in a pipeline framework. In this work, a novel sensor based blame confinement conspires, FLoGPN is proposed for recognition and restriction of flaws in a gas conveyance arrange. Here hole in gas pipeline organize is considered as blame.

According to José A. Ramos, another way to deal with gas spillage identification in high weight gaseous petrol transportation systems . The pipeline is demonstrated as a Linear Parameter Varying (LPV) System driven by the source hub mass stream with the gas stock variety in the pipe (line pack variety, relative to the weight variety) as the planning parameter. The mass stream at the off take hub is taken as the framework yield.

Gas Detectors have been in the market for a very long time and have been vastly used. They have broad assortment of employments and can be found in mechanical plants, refineries pharmaceutical gathering, paper crush processing plants, flying machine and ship building facilities, wastewater treatment workplaces, vehicles, indoor air quality testing and homes. We can detect gas pipe leakages through various gas detectors and they are characterized on the basis of the gas which they detect it also involves the technology behind the making of the sensor , even the components also reflect. Gas Detectors are classified into two types. They are Fixed and portable.

2.1. GSM Technique

This designed project proposes an economical, automatic LPG booking, spillage discovery and constant gas observing framework. In this particular frame, LPG Gas Spillage is detected with the help of sensor and this data is sent to the assigned client via SMS and also alarms the client with the help of a GSM module. Then again this framework persistently screens the level of LPG exhibit in the chamber with the assistance of load sensor. The Gas client can supplant the old chamber with new one in time and naturally book the barrel utilizing a GSM module. This model uses conventional method for GSM and which is somewhat intense for uneducated individuals to get itSome text.

2.2. GSM with no secure block

Presently a day's each one need an office which decreases their endeavours, time and give a way to do their work effortlessly. For cooking sustenance we as a whole utilize LPG gas. It is generally utilized as local fuel, modern fuel and car fuel. In INDIA numerous wholesalers utilize IVRS, SMS or ONLINE reserving for LPG which are time taking procedures in quick running life. It is badly designed for both Educated and additionally uneducated individuals because of absence of time and absence of information separately and Safety is likewise a critical concern. As we as a whole realize that numerous mishaps occur because of gas spillages. So to stay away from these challenges we are building up a paper. We plan a paper by considering wellbeing issues and furthermore give an Easy method to LPG booking. In the paper MQ-6 gas sensor is utilized to detect the spillage gas. Once the spillage is distinguished, it will close the controller and send the SMS to the client through GSM.
This model is very new but it has some drawbacks like heat-up of motor (easily) maybe in the case of an accident, and even more to make use of the GSM model.

2.3. Amprobe GSD600 Gas Leak Detector

This is a portable gas detector for detecting gases such as methane and butane. It has a stainless steel probe. The probe allows the user to get into the hard- to-reach places. It was designed for detecting gas within closed piping system and it has an audible alarm [4].

IOP Conf. Series: Materials Science and Engineering **590** (2019) 012048 doi:10.1088/1757-899X/590/1/012048

2.4. Analox Sensor Technology

Ensuring Campus Safety in lab environment, canteens or any other place where there is a chance of gas leakage. To detect the gas leakage several devices have been used such as O2NE+, A50 etc. These are provided by ANALOX Sensor Technology many other equipments as such are being manufactured by several other manufacturers all over the world. Other devices like First Alert CO615,Kidde KN-COPP-B-LPM,KN-COEG-3 Nighthawk,PNG2000A

Natural Gas Detector Pen,Safe-T-Alert 30 Series and many more, to name only a few are available in the market for gas detection purpose.

These are a few of the mostly used devices and all of them have a very high precision. Their price range from 30$ to 180-200. These devices are mostly battery operated and alarms) and visual (digital or LED) techniques. Our IoT based model for gas leakage detection is taking this traditional approach a notch up by adding the feature of immediately intimating the concerned authorities and also updating the sensor readings regularly to the cloud.

2.5. Handheld EGD01

This Handheld EGD01 conveys high-affectability, and is effortlessly movable for identifying a wide assortment of burnable gases, including methane, propane, and butane. It is utilized by building investigators [4]. As the name of the item recommends, it is a convenient gadget and thus battery worked. It has both sound and light alert.

2.6. LM35 Temperature Sensor

The LM35 can be associated successfully correspondingly as the other joined circuit temperature sensors. It can be adhered or built up to a surface and its temperature will be around 0.01§C of the surface temperature. This presumes the enveloping air temperature is about the same as the surface temperature, if the air temperature were fundamentally higher or lower than the surface temperature, the genuine temperature of the LM35 would be at a centre temperature between the surface temperature and the air temperature.

3. Gas Leakage Detectors

Types of equipments are used for various purposes in our day to day life and most of them have the capability of emitting some kind of gases or some compounds in the air while in use. It is very important to keep a check on the concentration levels of the gases and other compounds as some of them, after exceeding the safe concentration level, are flammable under the room temperature and humidity condition.

Blends of scattered ignitable materials, (for example, vaporous or vaporized energizes, a few cleans) and air will consume just if the fuel fixation exists in very much characterized lower and upper limits decided tentatively, alluded to as combustibility points of confinement or hazardous cut off points. Ignition can run in savagery from deflagration, through explosion, to blast [10].

A customary gas recognition framework checks for the fixation levels and cautions individuals about the spillage through sound and visual alerts. The IOT based model of the customary gas spillage recognition framework does caution individuals by sound caution as well as endeavours to alarm the concerned work force through a call and instant message on their telephone so that even if no one is show in the territory of spillage, they are made mindful of the circumstance.

Alongside this the model additionally sends the points of interest containing sensor readings recorded when the caution got impelled which could be utilized by an individual or an association to settle on choice about what kind of consideration is required in the territory of gas spillage.

Another most significant feature of this prototype is that it can record the sensor readings and maintain a database of these readings of concentrations of gases at different timestamps. This data could be utilized for carrying out analytics on it. The analysis of the sensor readings can help in understanding the usual conditions of the area, under what conditions do the readings of the sensor usually go up and when does it actually start getting dangerous and needs attention. This will increase the precision of the system, reducing the false alarms and hence actually becoming a very reliable system within a span of a few days. Along with all these features, this system also comes with a power cut off option.

In case the concentration of flammable gases is about to reach their lower explosive limit (LEL) the system trips the main switch which causes a complete power cut off of the building or office or her ever the system is installed. This will serve two purpose that is one, it will prevent any disaster due to electricity or power and two, in case the leakage is being caused by any such device turning off will prevent any further leakage, this purpose will be served by the power cut off feature. Description about the components used and the preparation steps involved are explained in detail.

3.1. Components

The main components used in the making of this prototype are as follows:

IOP Conf. Series: Materials Science and Engineering **590** (2019) 012048 doi:10.1088/1757-899X/590/1/012048

3.1.1. RasberryPI :

The Raspberry Pi is a movement of minimal single-board PCs made in the United Kingdom by the Raspberry Pi Foundation to propel the teaching of fundamental programming designing in schools and in making countries. The main model wound up significantly more unmistakable than anticipated, offering outside its target advertise for uses, for instance, mechanical self-rule. It excludes peripherals, (for instance, consoles, mice and cases).

3.1.2 CO2 Sensor:

A carbon dioxide sensor or CO_2 sensor is an instrument utilized for estimating carbon-di-oxide gas. The most normal standards for CO_2 sensors are infrared gas sensors (NDIR) and synthetic gas sensors. Measuring carbon dioxide is critical in observing indoor air quality, the capacity of the lungs as a capnograph gadget, and numerous mechanical procedures.

3.1.3. RASPBIAN OS:

Raspbian Stretch and Raspbian Jessie. Since 2015 it has been formally given by the Raspberry Pi Foundation as the fundamental working structure for the gathering of Raspberry Pi single-board PCs. Raspbian was made by Mike Thompson and Peter Green as a self-governing endeavour. The basic frame was done in June 2012. The working system is still under powerful headway. Raspbian is significantly upgraded for the Raspberry Pi line's low-execution ARMCPUs.

3.1.4 Methane Sensor:

Methane gas sensor recognizes the gathering of methane gas and sends its examining as a straightforward voltage. MQ-4 gas sensor has high affectability to Methane, furthermore to Propane and Butane. The sensor could be used to recognize different ignitable gas, especially Methane, it is low in cost and fitting for different applications.

3.1.5 Hydrogen Sulphate Sensor(H2s Sensor):

Hydrogen sulphate sensor detects the concentration of hydrogen sulphide gas. It is used to accurately monitor the areas where this gas can be very hazardous such as iron smelters, breweries, food processing plants. It has an abundant importance in the nuclear industry because nuclear reactors use a lot of uranium which in turn gives a lot of H_2S gas as a wastage into the atmosphere. The farming industry also uses H_2S as a agricultural disinfectant which is highly flammable and toxic in nature, It can be detected using the H_2S sensor.

IOP Conf. Series: Materials Science and Engineering **590** (2019) 012048 doi:10.1088/1757-899X/590/1/012048

3.2. Block Diagram

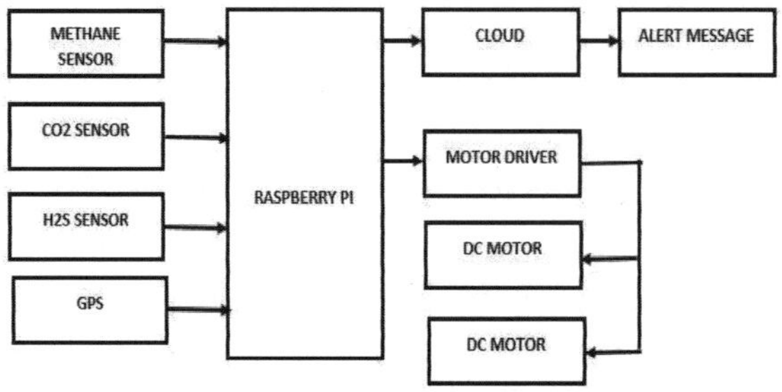

Figure 1 Block Diagram

Processing Steps:

1. The MQ-2 gas sensor of the MQ-x family detects the gases. In order to stabilize the sensor it is important to pre heat it.
2. When the MQ-2 sensor is exposed to the gases it will detect and then it could send the response to the Raspberry Pi.
3. The concentration of other gases such as CO2,H2S can also be detected by CO2 and H2S sensors respectively.
4. Every value recorded by the sensors is checked against a threshold value. This threshold value is decided based upon the nature of the gases and the behaviour of the sensors.
5. We also attach a GPS Sensor to the Raspberry Pi which can detect the exact location of the leakage.
6. The various gas concentrations when recorded by the sensors are sent to the cloud storage for analysing the conditions of the place.

where the leakage took place at the same time alert messages are sent to the registered devices.

7. An email about the focus level of gases is sent to the concerned expert as readings all the time.
8. The sensor readings that are uploaded to the cloud can be used for Data Analytics. The average reading of 24 hours can be calculated, every day since the day of deployment.
9. We also use a motor driver for additional voltage to run the dc engines which are associated with the Raspberry Pi.

4. Conclusion

Hence, from the above discussion we can conclude that the paper (Gas pipe leakage detection using IoT) is absolutely ethical for application by the users who work in a gaseous industry . It not only helps in making the work easier but also plays a major role in the security avoidance of accidents to the industry and helps in leading a easy life. The implementation of interface between GSM module and Gas detection sensor is executed zestfully. The interfacing is properly done without a little collision among GSM and the sensor. The design and realization of interface between GSM modem and GSM network for data transmission is done successfully. Thus, the robot is monitored time to time with the help of the sensors and also a SMS is sent to the registered user whenever gases are present around the pipe. The designed system provides a real time and low cost solution to the gas leakage problem and prevents many accidents.

5. Future Work

We can further extend this project by adding some more features which can make it more efficient and security oriented. The camcorder can also be used to track all the activities of the unknown intruders. An emergency Alert can also be made to police and fire brigade as well. The complexity of the algorithm of the system can be increased by introducing number of sensors to make the system more efficient.

Analytics could be performed on the sensor readings. The readings from sensors could be used for forming predictions of situations where there can be a mishap. Instead of straightaway alarming when the concentrations have gone high, algorithms could be worked upon which could determine such situations prior to their occurrence.

A dedicated mobile application could be made for the system. The features of the application would be: Getting the details of the concentration levels of the gas pipe within a tap of a button. Since it is a safety device it is important for it to be perfectly calibrated and maintained at all times. The app can make sure to send reminders about getting the system checked every once in a while. The user can add or remove the recipients who will receive the information of leakage whenever they require.

References

[1] Akanksha Singh, Arijit Pal, Bijay Rai ."GSM Based Home Automation, Safety and Security System Using Android Mobile Phone (IJERT) ISSN: 2278-0181, Vol. 4 Issue 05, May-2015

[2] Vaibhav Sharma, Chirag Fatnani , Pranjal katara,Vishnu Shankar. "advanced low- cost security system using sensors, arduino and gsm communication module".

[3] Gas Detector on Wikipedia., 2016, [online] Available: https://en.wikipedia.org/wiki/Gas_detector.

[4] Introduction to Gas Detection System on Draeger website., 2016,Available:https://www.draeger.com/Library/Content/intro duction_gds_fl_9046421_en_1.pdf.

[5] Best Gas Detectors on ezvid website., 2016, [online] Available: https://wiki.ezvid.com!best-gas-detectors.

[6] Analox Sensor Technology, 2016, [online] Available: https://www.analoxsensortechnology.com/downloads/142315 26008722_Lab_brochure15.2.pdf.

[7] Feng Zhang, Min Liu, Zhuo Zhou, Weiming Shen, "An IoT Based Online Monitoring System for Continuous Steel Casting", IEEE Internet of Things Journal, 2016, [online] Available: 10.1109/JIOT.2016.2600630.

[8] Industrial Internet of Things and Communications at the Edge by Tony Paine CEO Kepware Technologies, 2016.

[9] Somova. Andrey, Alexander Baranovb, Denis Spirjakinb, "A wireless sensor-actuator system for hazardous gases detection and control" in a: CREATE-NET Via alla Cascata 56/D 38123 Trento TN Italy-Russian State Technological University Orshanskaya 3 121552, Moscow, Russia:published in Science Direct, Sensors and Actuators, 2014.

IOP Conf. Series: Materials Science and Engineering **590** (2019) 012049 doi:10.1088/1757-899X/590/1/012049

An Advanced Attendance Marking system using facial Recognition

Jarugula.Vamsikrishna[1]
Final Year UG student-Dept of ECE,
Sathyabama Institute of Science and Technology,
Chennai – India.
vamsikrishna0777@gmail.com

Kollipara.Anudeep[1]
Final Year UG student-Dept of ECE,
Sathyabama Institute of Science and Technology,
Chennai – India.
kolliparaanudeep1234@gmail.com

L.Jegan Antony Marcilin[2]
Assistant Professor, Department of ECE
Sathyabama Institute of Science and Technology,
Chennai – India.
jegan25@hotmail.com

V. Balamurugan[3]
Assistant Professor, Department of ECE
Sathyabama Institute of Science and Technology,
Chennai – India.
vbbala@gmail.com.

Abstract. With advancement in computing and telecommunications technologies, digital images and videos are playing main role in the current information era. Human face is an important bio-metric object in image and video database of surveillance systems. Detecting and location of human faces and facial features in an image sequence is an important task in dynamic environments, such as videos where noise conditions, illuminations, locations of subjects and pose can vary significantly from frame to frame. An automated attendance system is used for marking human face recognition in real time for college to mark the attendance for their staff and students. Smart automated attendance marking using real time face recognition is a real world solution which comes with day to day activities of handling employees. Here multiple user faces are detected and recognized with the data base trained multiple texture based notability.

1. Introduction

The attendance maintaining system is difficult process if it is done manually. The smart and automated attendance system for managing the attendance can be implemented using the various ways of biometrics. Face recognition is one of them. By using this system, the issue of fake attendance and proxies can be solved. In the previous face recognition based attendance system, there were some disadvantages like intensity of light problem and head pose problem. In this project, Dominant Rotated Local Binary Pattern is used.

The major steps in this system are detecting the faces and recognizing them. After this, the comparison of detected faces will crosscheck with the database faces. This smart system will be an effective way to maintain the attendance and records of students. The attendance details will be sent through E-mail. If the

IOP Conf. Series: Materials Science and Engineering **590** (2019) 012049 doi:10.1088/1757-899X/590/1/012049

person is entering the class it will cross check with the database and if their record is present in the database then their name will be send through e-mail. If a person is leaving the room their photo along with their name will be sent to the concern people.

2. Pattern

In this paper the pattern used is dominant rotated local binary pattern (DRLBP). By using "vision.Cascade Object Detector" keyword only the face of the person is recognized. This recognition can be designed as required by the programmer like shape and color of the recognition box.

The recognized face will be converted into pixels. A 3*3 matrix formed from the pixels will be considered and the middle most (5th) element will be taken for reference. The other elements of the matrix will be compared with reference. If the value is greater than the reference value, it will be assigned to Zero. If the value is less than the reference value, it will be assigned to One. Now the newly obtained matrix will comprise of only binary values (0 &1) and it is compared with the database faces.

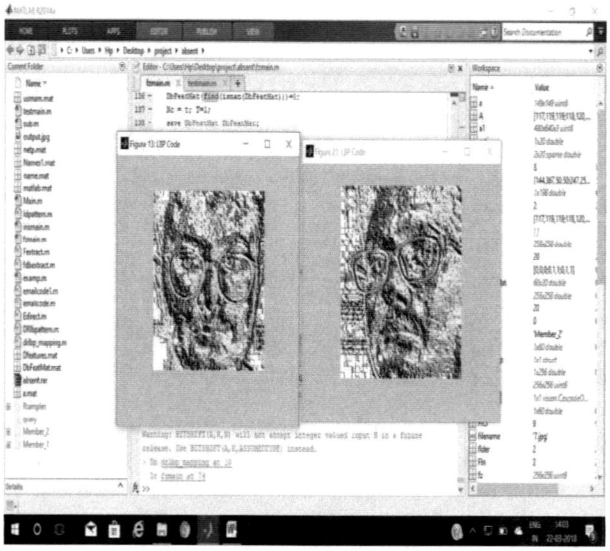

Fig.1 LBP Code image

3. Description

The system contains two modules mainly

1) Data base creation.
2) Attendance Marking Through Face recognition

3.1 Database Creation

In this module, the system is designed to store the faces and their facial features of the persons. The Flow Diagram of the database creation is described below in detail.

IOP Conf. Series: Materials Science and Engineering **590** (2019) 012049 doi:10.1088/1757-899X/590/1/012049

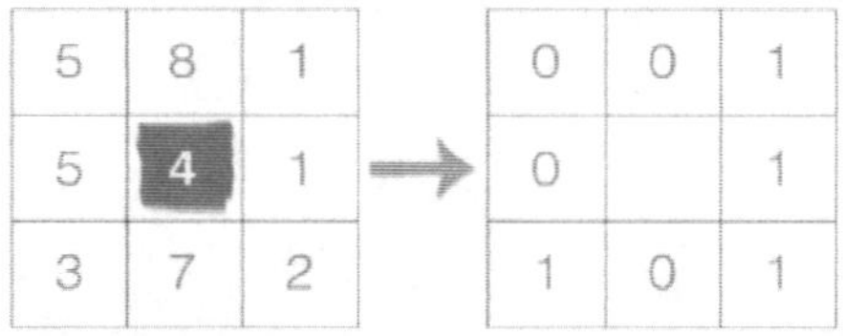

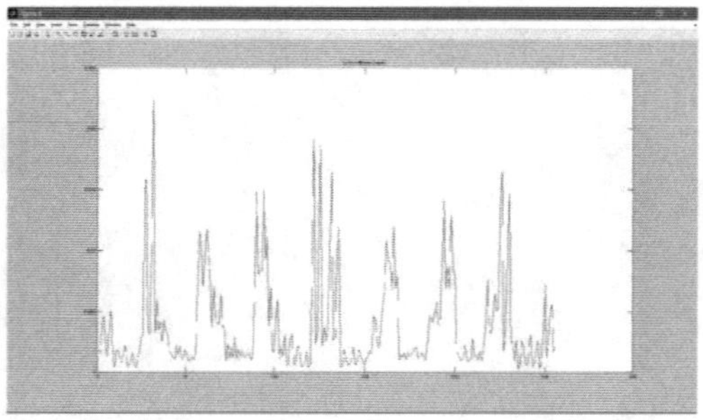

Fig.2 DRLBP Histogram

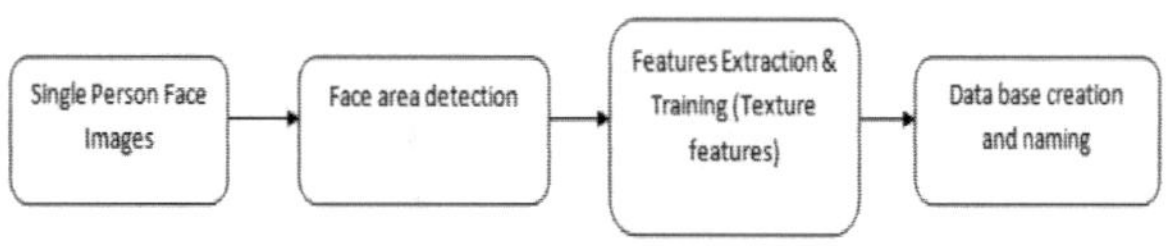

Fig.3 Flow diagram of database creation

3.1.1 Single person Face Image

The Images of the persons will be taken individually using digital cameras and will be stored in the server/data storage devices. The data of persons will be stored in a specified folder and their names also will be stored in the folder in order to mark attendance. If the server is having more images of a particular person then we can get accurate results.

3.1.2 Face Area Detection

Image is processed to extract face regions from input image which has normalized intensity. Here MATLAB vision tool box will be utilized to detect the face from the input and it uses DRLBP algorithm. In the MATLAB vision tool box the face boundary vector [X_{min}, Y_{min}, Width and height] are cropped to get the desired region for further analysis.

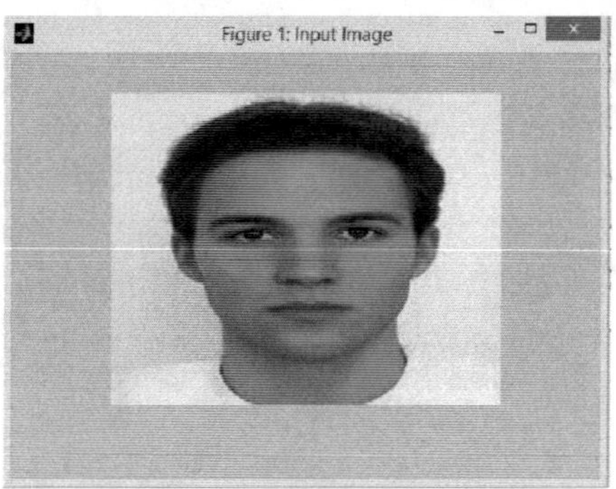

Fig.4 Single person

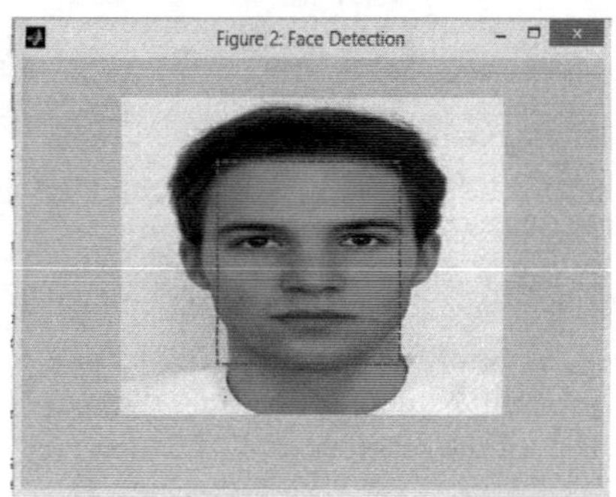

Fig.5 Face Area Detection

IOP Conf. Series: Materials Science and Engineering **590** (2019) 012049 doi:10.1088/1757-899X/590/1/012049

3.1.3 Features Extraction and Training

It will be useful when image sizes are large and a reduced feature representation is required to quickly complete tasks such as image matching and retrieval. Facial land features like nose, eyes and lips will be helpful for the person recognition.

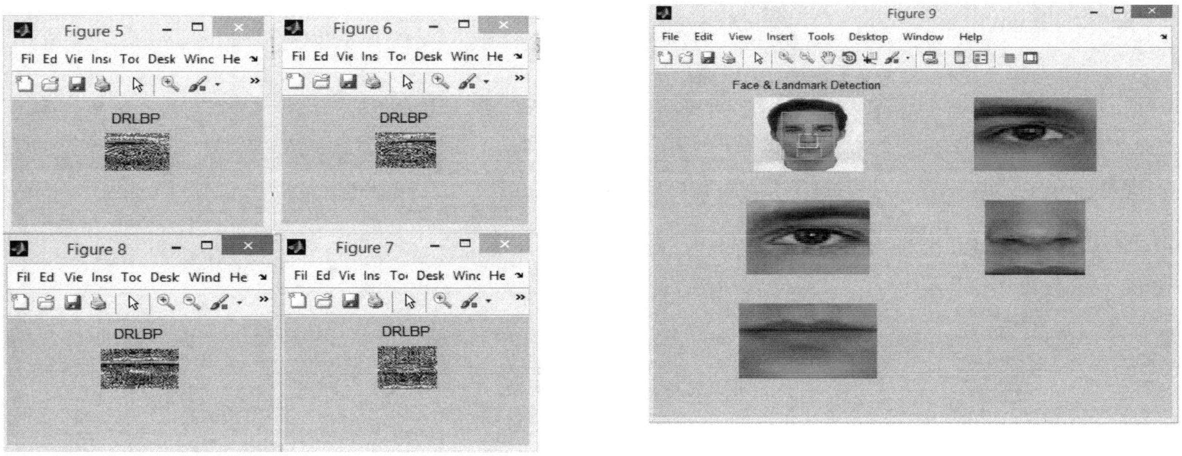

Fig.6.Features Extraction

3.1.4 Data base creation

The images taken from the digital cameras will be stored in a folder and will be assigned a name to the particular person and this process should be continued until all the persons data is stored.

Fig.7 Database Created

3.2 Attendance Marking through Face recognition

In this Module, Real Time Face will be detected and pre-processed, compared with database and attendance will be marked and sent through E-mail and records will be maintained in the email server which will be helpful for the marking of attendance.

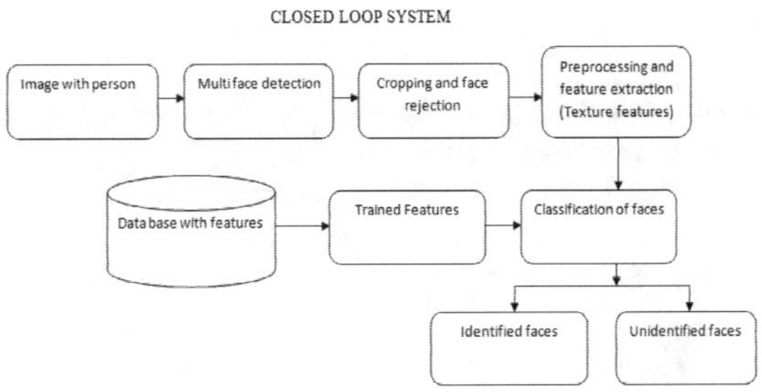

Fig.8 Facial recognition and Attendance Marking

3.2.1 Image with person & Multi face Detection

When a person is entering/leaving Real time frames will be taken and it will check each and every frame taken and faces are detected by using vision tool in MATLAB. If two or more persons are encountered in a particular frame, then Each and every person faces also will be detected.

Fig.9 Multi Face Detection

3.2.2 Cropping and Face Rejection

The person image is cropped by using vision tool and it will be cross check with the database. If the person face in that particular frame is not clear/distorted then it will be automatically rejected otherwise it will go to further process.

3.2.3 Classification of Faces

When a person is entering/leaving, the person is detected in real time then the features will be extracted from the detected face and it will cross check with the data stored in the database if it is matched then it will mark the attendance either present or absent through E-mail.

4. Result

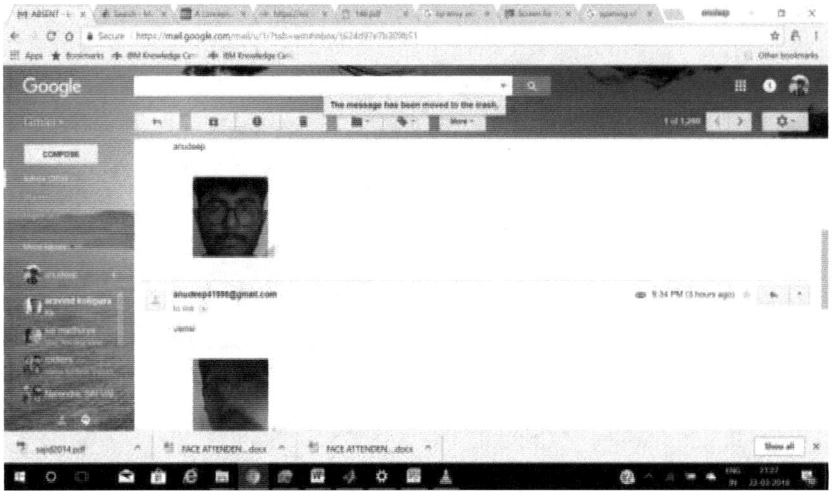

Attendance will be automatically marked and it will send to E-mail shown in the above Fig. Day to Day Student Attendance Database can be maintained and it is secured.

References

[1] Dr.Ramaprasad.P, Rekha.E ,"An Efficient Automated Attendance Management System Based On Eigen Face Recognition",7th International Conference on Cloud Computing, Data sciences and Engineering-Conference,2017.

[2] Jiachen Chen, W.Kenneth Jenkins, "Facial Recognition with Principal Component Analysis and Machine Learning Methods", International Journal of Scientific Research in Computer Science and Engineering,2017.

[3] Ahmed Aldhahab, Wasfy B.Mikhael, "A facial recognition method based on DMW Transformed Partitioned Images ."2017.

[4] Jin Liu,Yue Chen "Face Recognition Based on Multi-Direction Local Binary Pattern" 3rd IEEE International Conference on Computer and communications-2017.

[5] Khem Puthea,Rudy Hartantoand Risanuri Hidayat," A Review Paper on Attendance Marking System Based on Face Recognition", 2nd International Conference on Information Technology,Information Systems and Electrical Enginnering-2017.

[6] C.B.Yuvaraj,M.Srikanth,V.Santoshkumar,Y.V.SrinivasaMurthy and Shashidhar G.Koolagudi,"An Approach to Maintain Attendance Using Image Processing Techniques",10th International Conference on Contemporary Computing -2017.

[7] Refik Samet,Mohammed Tanriverdi,"Face Recognition-Based Mobile Automatic Classroom Attendance Management System", International conference on Cyberworlds -2017.

[8] Visar Shehu, Agni Dika, "Using Real Time Computer Vision Algorithms in Automatic Jiachen Attendance System ," International Conference on Information Technology Interfaces ,June 2010.

[9] Robert C.Schultz , Robert W.Ives , "Biometric Data Acquisition using MATLAB GUIs," 35th ASEE/IEEE Frontiersin Educational Confernce,2005.

[10] LI-Ying Lang, Wei-Wei Gu, "A Robustness and real time face detection algorithm in complex background." 2009.

IOP Conf. Series: Materials Science and Engineering **590** (2019) 012050 doi:10.1088/1757-899X/590/1/012050

INCREMENTAL FREQUENT MINING HUMAN ACTIVITY PATTERNS FOR HEALTH CARE APPLICATIONS

[1] G. Vignesh Krishna Sai, [2] Pavan Sai Kumar, [3] Dr. A. Viji Amutha Mary

[1,2] U.G Student, Dept of CSE, Sathyabama Institute of Science and Technology, Chennai.

[3] Associate Professor, Dept of CSE, Sathyabama Institute of Science and Technology, Chennai-119.

[1]vigneshghanta@gmail.com, [2]pavansaitanneru@gmail.com, [3]vijiamumar@gmail.com

Abstract: Nowadays, there is a consistently expanding relocation of individuals to urban territories. Human services administrations is a Standout amongst the most difficult perspectives that is extraordinarily influenced by the immense deluge of individuals to downtown areas. In such change a great many homes are being with brilliant gadgets which create huge volume s of indexical information that can be dissected to help savvy city administrations. In this paper, we propose a model that uses savvy home huge information as a methods for learning and finding human movement designs for wellbeing applicati ons.for this we utilize visit design mining ,bunch investigation and forecast to gauge and dissect vitality utilization changes the tenants behaviour. Since, individuals propensities are generally disti nguished by ordinary schedules, demonstrates in dividuals' troubles in taking tend to themselves, for example, not planning nourishment or not utilizing shower/bath.our places of business the need to break down transient vitality utilization at mach ine level, which is straightforwardly identified with human exercises. This exploration utilizes the UK Domestic Appliance Level Electricity dataset(UK-Dale) time arrangement information of energy uti-lization gathered from 2012-2015 with time determ ination of six seconds for five houses with 109 app aratus from SouthernEngland.The information from shrewd meters are mined in the quantum/information cut of 24hrs, and the outcomes are kept up cro-sswise over progressive mining exercises.the after effect of distinguishing human movement desig-ns from machine use are displayed in points of interest.

Keywords:. Big data, smart homes, smart urban areas, human service applications, conduct examinations, visit pattern, cluster investigation, incremental information mining, association rules, prediction.

1.Introduction

Studies demonstrate that by year 2050, 66 percent of total populus are living in urban communities. The interest for the He althcare assets are significantly experiencing this huge inflow of people to town fo-cuses. This new demographic alteration places stupendous burden on urban areas to reexamine the standard methodo-logies of giving human services administrations to homes. In reacting to new needs and difficulties, urban communities are by hold tremendous computerized change in a short to support the urban groups, and supply more advantageous environment. In such a change, a few homes are been furnished with great gadgets that produce gigantic volumes of fine grained and reality learning which will be broken down to help social insurance administrations. Headway of enormous infor-mation preparing innovations which offer implies that of procedure huge amount of learning for amount of line insites, will help USA in understanding people mo-ve their life.since individuals propensities are essentially known by regular schedules.

IOP Conf. Series: Materials Science and Engineering **590** (2019) 012050 doi:10.1088/1757-899X/590/1/012050

2. Existing System

Nowadays, there is a ceaselessly extending movement of people to urban domains. Social protec-tion organizations is a champion among the testing perspectives that is remarkably impacted by the gigantic of people to downtown regions. Along these lines urban groups far and wide are placing strongly in automated change with a true objective to give more useful natural frameworks to people.

3. Proposed System

We propose the utilization of successive example mining, bunch examination and forecast to gauge and break down vitality use changes started by te-nants conduct. Since individuals propensities are for the most part distinguished by ordinary sched-ules, finding these schedules enables us to percei ve abnormal exercises that may shows individuals challenges in taking watch over themselves, our street numbers to examine worldly vitality utilization designs at the Appliance level, which is specifically identify with human exercises.

4. Architectural Diagram

A system model that explains how data is provided to the cloud platform, processed in PU and the data is encrypted to the undiagnosed patient is depicted in fig.1.

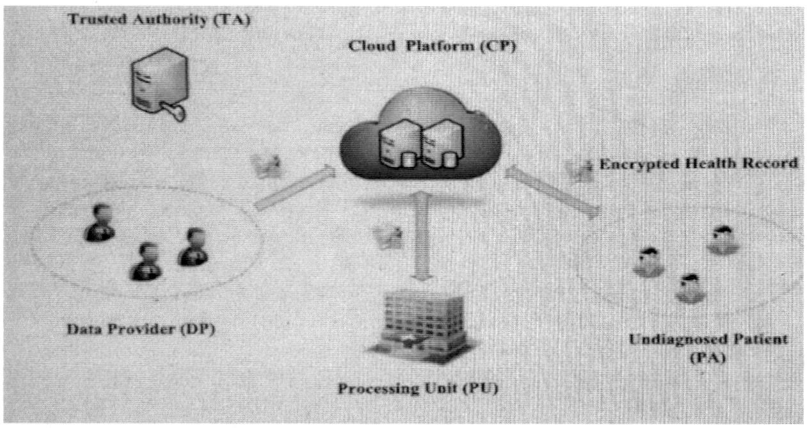

Figure 1. System model

5. Proposed Algorithm - "Incremental Frequent Pattern Mining"

As said prior, the objective is to find human action designs from reliant meters information. Our aim is to distinguish the examples of this exercises so a medicinal service applications, that screens patients with intellectual debilitation, can send opportune alarm to human services suppliers in such process, all apparatus that are registered dynamic amid the 30 minute time interim are incorporated into source database for visit design information mining.

6. Conclusion

In this paper, we exhibited a model for perceiving human movement designs from low determination shrewd meters information. individuals propensiti es and conduct take after an example that could be utilized as a part of wellbeing applications to track the prosperity of people living alone are those with self-restricting Conditions. The major ity of these exercises can be scholarly application liance to apparatus and machine to time affiliations. We introduced incremental incessant mining and anticipate particle display in view of bayesian sys-tem. In our present Work, through tests, we found that 24 hour duration Was ideal for information mi-ning, yet we constructed model to Operate on any Quantum of time. From the investigation comes about we have exhibited the relevance of proposed model to effectively recognize numerous machine utilization make short and long haul forecast at high exactness.

7. Future Enhancement

For future work, we are intending to refine the model and present appropriated taking in of huge information mining from numerous houses in a continuous way. This will push wellbeing applications to instantly take activities, for example sending alarm to patients or care providers. Furthur more,we are intending to manufacture consequently delineate missions to potential activities. This implies we can productively prepare the framework and increment the precision of recognising human exercises.

REFERENCES

[1] C.Chalmers, W.Hurst, M.Mackay and P.Fergus, "Smart meter profiling for health applications" published in Neural Networks(IJCNN) in 2015 .

[2] M.S.Hossain, "A patient's state recognition system for healthcare using speech and facial expression" published in journal of medical systems in december 2016.

[3] M.S.Hossain and G.Muhammad "Cloud-assisted industrial internet of things(IOT)-enabled framework for health monitoring" published in the international journal of computer and telecommunications networking in june 2016.

[4] Albert Mayan J , Kumar Akash Deep , Mukesh Kumar , Livingston Alvin , Siva Prasad Reddy , "Number plate recognition using template comparison for various fonts in MATLAB ", IEEE International Conference on Computational Intelligence and Computing Research, ICCIC 2016 ,PP: 1-6 , DOI: 10.1109/ICCIC.2016.7919542

[5] P.Pouladzadeh, P.Kuhad, S.V.Peddi, A.Yassine and S.Shirmohammadi "Mobile cloud based food calorie measurement" published in multimedia and expo workshops(ICMEW) in july 2014.

[6]Alam, Roy, Petruska and Zemp. "Smart-energy group anomaly based behavioural abnormality detection" published in wireless health (WH) in 2016.

[7]Viji Mary, Justin Samuel, "A novel technique to secure the acute myocardial infarcta images by the enhancement of privacy level", Biomedical Research 2018; 29 (1): 15-21.

IOP Conf. Series: Materials Science and Engineering **590** (2019) 012051 doi:10.1088/1757-899X/590/1/012051

TO EFFECTIVELY CONTROL THE WORKLOAD BY HIERARCHICAL APPROACH

S.L.Jany Shabu
Department of Computer Science & Engineering
Sathyabama Institute of Science and Technology ,Chennai, India
janyshabu@gmail.com

B.Bhaskarreddy
Department of Computer Science & Engineering
Sathyabama Institute of Science and Technology
Chennai, India

A.Sreekanth
Department of Computer Science & Engineering
Sathyabama Institute of Science and Technology
Chennai, India

Abstract: The productive administration of geologically dispersed server farms has turned into an imperative issue not just for huge organizations that claim a few destinations, yet additionally because of the rising of between Cloud foundations that permit heterogeneous server farms to coordinate. These conditions open phenomenal roads for the help of a colossal measure of workload, yet they require the meaning of novel calculations and methodology for their administration, where adaptability is a need. The intricacy determines by the span of the framework and by the requirement for achieving a few and some of the time clashing objectives, among which: stack adjusting among various destinations, anticipation of dangers, workload union, and decrease of costs, expended vitality and carbon outflows. In this paper a various leveled approach is displayed, which protects the self-rule of single server farms and in the meantime takes into consideration a coordinated administration of heterogeneous platforms. The system is deliberately nonexclusive yet can be customized to the particular prerequisites of single situations. Exhibitions are examined for a particular Cloud framework made out of four server farms.

Keywords: Cloud computing, Hierarchical, Distributed Data Center, Workload, Energy Saving.

1. INTRODUCTION

Present day distributed computing frameworks work in another and dynamic world, described by constant changes in the earth and in the framework and execution necessities that must be fulfilled. Nonstop changes happen all of a sudden and in an erratic way, which are outside the control of the cloud supplier. Along these lines, propelled arrangements should be produced that deal with the cloud framework in a progressively versatile manner, while persistently giving administration and execution ensures: (I) lessen costs, (ii) enhance levels of execution, and (iii) upgrade accessibility and constancy. Inside this structure, it should first be noticed that, everywhere benefit focuses, the quantity of servers are developing altogether and the unpredictability of the system foundation is additionally expanding. This prompts a gigantic spike in power utilization: IT investigators foresee that before the

IOP Conf. Series: Materials Science and Engineering **590** (2019) 012051 doi:10.1088/1757-899X/590/1/012051

finish of 2012, up to 40% of the financial plans of cloud benefit focuses will be given to vitality costs. Vitality effectiveness is in this way one of the principle central focuses on which asset administration ought to be concerned. Moreover, suppliers need to conform to Service Level Agreement (SLA) gets that decide the incomes picked up and punishments brought about based on the level of execution accomplished. Nature of Service (QoS) ensures must be fulfilled regardless of workload changes, which could traverse a few requests of extent inside a similar business day .

In addition, if end-client information and applications are moved to the cloud, foundations should be as dependable as telephone frameworks : Although 99.9% up-time is promoted, a few hours of framework blackouts have been as of late experienced by some cloud showcase pioneers (e.g., the Amazon EC2 2011 Easter blackout and the Microsoft Azure blackout on 29th February 2012. As of now, Infrastructure as a Service (IaaS) and Platform as a Service (PaaS) suppliers incorporate into SLA contracts just accessibility, while execution are disregarded. If there should arise an occurrence of accessibility infringement (with current figures, notwithstanding for huge suppliers, being around 96%, much lower than the qualities expressed in their agreements), clients are discounted with credits to utilize the cloud framework for nothing. The idea of virtualization, an empowering innovation that permits sharing of the same physical machine by various end-client applications with execution ensures, is principal to creating suitable strategies for the administration of current cloud frameworks. From an innovative point of view, the union of various client workloads on the same physical machine diminishes costs, yet this additionally converts into higher use of the physical assets. Consequently, unexpected load vacillations or equipment disappointments can have a more noteworthy effect among various applications, making a cost productive, trustworthy cloud frameworks with QoS assurances of vital significance so as to acknowledge expansive appropriation of cloud frameworks.

Various self-overseeing workload forecasts. The greater part of these methodologies have actualized brought together arrangements: A focal controller deals with all assets involving the framework, normally following up on a hourly premise. This time-scale is regularly received [8] in light of the fact that it speaks to a bargain in the exchange off between the recurrence of choices and the overhead of these choices. For instance, self-overseeing structures may choose to quiets down or shut down a server, or to move virtual machines (VMs) starting with one server then onto the next by abusing the VM live-relocation component. Notwithstanding, these activities expend a lot of vitality and can't be performed too every now and again. As already noted, cloud frameworks are constantly developing as far as size and many-sided quality: Today, cloud benefit focuses incorporate up to 10,000 servers , and every server has a few VMs. In this specific situation, brought together arrangements are liable to basic plan impediments , including an absence of versatility and costly checking correspondence costs, and can't give viable control inside a one hour time skyline. In this paper we take the viewpoint of a PaaS supplier that deals with the value-based administration uses of its clients to fulfill reaction time and accessibility ensures and limit vitality costs in substantial cloud benefit focuses. We propose a dispersed progressive system in light of a blended whole number non-straight streamlining of asset administration over various time-scales. At the base of the order, a Central Manager (CM) parcels the workload (applications) and assets (physical servers) over a 24-hour skyline to acquire bunches with homogeneous workload profiles one day ahead of time. For each group, an Application Manager (AM) decides the ideal asset portion in an incorporated way over a subsystem of the cloud stage on a hourly premise.

All the more unequivocally, AMs can choose the arrangement of utilizations executed by every server of the group (i.e., application position), the demand volumes at different servers inside the bunch (i.e., stack adjusting), and the limit dedicated to the execution of every application at every server (i.e., limit designation). AMs can likewise choose to switch servers into a low-control rest state contingent upon the group stack (i.e., server changing) or to

IOP Conf. Series: Materials Science and Engineering **590** (2019) 012051 doi:10.1088/1757-899X/590/1/012051

decrease the operational recurrence of servers (i.e., recurrence scaling). Moreover, since stack adjusting, limit distribution, and recurrence scaling require a little measure of time to be performed, they are executed by AMs each few (5-15) minutes. The periodicity of the situation choices is researched inside the trial investigation in §VI. Broad examinations over a wide assortment of setups exhibit the viability of our approach, where our answers are near those accomplished by a concentrated strategy that requires requests of greatness more opportunity to take care of the worldwide asset portion issue in view of an entire perspective of the cloud stage. Our answer additionally renders net-income upgrades of around 25-30% over different heuristics from the writing that are as of now conveyed by IaaS suppliers . In addition, our calculations are appropriate for parallel usage that accomplish almost direct accelerate. At last, we have likewise examined the shared impacts and interrelationships among the various time-scales (24h/1h/5-15 minutes) by shifting the attributes of the forecast mistakes and indentifying the best fine-grained time-scale for run-time control.

2. RELATED ARTICLES

Bernardetta Addis, DaniloArdagna, Barbara Panicucci, Mark S. Squillante, Fellow, IEEE, and Li Zhathey [4] "A Hierarchical Approach for the Resource Management of Very Large Cloud Platforms", in this paper they have proposed in perspective of considering only the largeness of the moving toward undertaking to disseminating the advantage in light of this they may have different clashes and for portion of asset to virtual machines they have just think about the limit designation, recurrence, scaling as a result of this effectiveness of asset allotment has been decreased.

D. Ardagna, B. Panicucci, M. Trubian, and L. Zhang, "Vitality Aware Autonomic Resource Allocation in Multitier Virtualized Environments", in this paper they proposed in view of forecast of approaching task.They for the most part centered around different timescales [3] they have look at undertakings in light of exchange amongst recurrence and overhead of these choice .Self-overseeing systems [11], [3] they choose to themselves whether virtual machines from the server can quiets down or shut down a server, and furthermore VM movements starting with one then onto the next for these activities they required huge vitality so they won‟t performed so as often as possible.

T. Nowicki, M.S. Squillante, and C.W. Wu, "Essentials of Dynamic Decentralized Optimization in Autonomic Computing Systems", in this paper, they have a point of view of a PaaS supplier oversees finish the exchange administration of the client undertaking to fulfill reaction time and accessibility and to lessen vitality costs in extensive cloud benefit focuses. They propose an appropriated various leveled structure [17] in view of a blended whole number nonlinear streamlining of asset administration over different timescales. At the best progressive system, a focal chief (CM) allotments of the approaching workloads in view of some obliges and assets (physical servers) over a 24-hour skyline to make bunches with homogeneous workload profiles one day ahead of time. So in each group, an application director (AM) assigns the assets in incorporated way in hourly premise to the subsystems. AMs need to choose the which application need to executed by every server of the bunch putting application to the subsystems they check for the asked for volume of assets and accessibility of the asset in the group need to check the limit required to the execution application at every server.AMs can likewise choose whether server can change to a low-control rest state in view of the bunch stack or to diminish the operational recurrence of servers.

A current report [21] they proposed various leveled control arrangements, stick this they for the most part centered around bunch level control engineering that play out different server control controllers inside a virtualized server group [21]. The higher layer decides limit distribution of VM in server and furthermore VM relocation inside a bunch, while the inward controllers need to check the power level of individual servers. Notwithstanding, application and VM position inside each administration focus group is thought to be given before execution. For the fulfillment of the undertaking is predominantly in light of the coordination among different power controllers acting at rack walled in areas and at the server level is finished.

Vitality Aware Autonomic Resource Allocation in Multitier Virtualized Environments [3] in this have been proposed to accomplish fine-grained dynamic asset provisioning strategy. They proposed booking the undertakings in light of Xen"s credit scheduler. They can progressively control the utilization of asset like circle I/O transfer speed among the allotted virtual machines. Through these techniques for VMs can distribute assets are utilized for dynamic parcel or else reassembled of assignments to address the issues of the clients. These are the extraordinary fate of the VMs in cloud which isn't conceivable to apply in most Grid frameworks [12], [13], [14].

Numerous arrangements have been proposed for the self-administration of cloud benefit focuses, each trying to meet application necessities while controlling the hidden foundation. Five fundamental issue zones have been considered in framework portion strategy plan: 1) application/VM situation, 2) affirmation control, 3) limit assignment, 4) stack adjusting, and 5) vitality utilization. While every region has regularly been tended to independently, it is important that these issue arrangements are firmly related. For instance, the demand volume decided for a given application at a given server relies upon the limit apportioned to that application on the server. An essential commitment of this paper is to coordinate each of the five issue territories inside a binding together structure, giving exceptionally effective and vigorous arrangements at various time-scales. Three primary methodologies have been created: (I) control theoretic criticism circle strategies, (ii) versatile machine learning methodologies, and (iii) utility-based streamlining procedures. A fundamental favorable position of a control theoretic criticism circle is framework soundness ensures. Upon workload changes, these strategies can likewise precisely demonstrate transient conduct and alter framework arrangements inside a brief period, which can be settled at configuration time. A past report actualized a constrained lookahead controller that decides servers in the dynamic express, their working recurrence, and the assignment of VMs to physical servers. Nonetheless, this execution considers the VM situation and limit assignment issues independently, and versatility of the proposed approach isn't considered.

A Dynamic Optimization Algorithm for Task Scheduling in Cloud Computing With Resource Utilization , Ram Kumar Sharma, Nagesh Sharma they proposed basically on booking endeavors and effectively assigning the advantages for the virtual machines The objective of this paper is to the best use on client and server side getting to the cloud condition.

3. HIERARCHICAL APPROACH FOR THE RESOURCE MANAGEMENT

In this strategy by allotting the asset in zenith to base way, when ask for is stopping by distributed the asset for the approaching undertaking is going to apportioning in light of the heaviness of the assignment, spending plan, asset vector and given to appropriate VMs in view of each bunch in the application director they contrast and load adjusting, limit designation, recurrence scaling, vitality productivity, benefit separation for the effective asset portion of virtual machines in the server. The PaaS supplier need to help the numerous value-based administrations to execute , every exchange with various client with their distinctive application. The facilitated administrations can be heterogeneous concerning asset requests, workload profile, organize transmission capacity and QoS prerequisites. Administrations with various property of workload profiles are isolated into free demand classes, where they set serves a set R of demand classes. Fig. 1 design with various leveled approach. The framework incorporates a set S of heterogeneous servers, each running a virtual machine screen (VMM, (for

IOP Conf. Series: Materials Science and Engineering **590** (2019) 012051 doi:10.1088/1757-899X/590/1/012051

example, VM Ware or Xen, IBM POWER Hypervisor), the figuring assets are topped and saved for the execution of individual VMs. The physical assets like CPU, hard circles, transmission capacity of a server are apportioned among each in view of weight of the undertaking, spending plan, asset vector to make groups of virtual machines.

Resource Allocation:

- Server ON/OFF
- Applications arrangement
- Frequency Scaling (DVFS)
- Load adjusting
- Capacity assignment

So now any class can bolster any kind of use in the numerous levels. Each VM is facilitated in a solitary application level, where additionally numerous VMs are in a similar application level can be keep running in parallel on various hosts. Under less work stack conditions server move to standby mode or to close down fundamentally to lessen the vitality cost. These servers move back to dynamic state amid the pinnacle hours.

Acknowledge that every server has a solitary CPU supporting dynamic voltage and recurrence scaling (DVFS) by picking the two its supply voltage and working recurrence from a restricted arrangement of qualities, noticing that this single-CPU bolsters that is without loss of diagram in substantial rush hour gridlock and can be effectively casual. By utilizing DVFS isn't turned out to be overhead to framework yet to hibernate and reestablishing a server both require time and vitality. In this [13], they proposed embrace full framework control models and accept that the power utilization of a server relies upon its working recurrence/voltage and the present CPU use. Our asset administration system depends on a various leveled engineering [17]. At the most elevated amount of the chain of command, a CM follows up on (i.e., at regular intervals) and is in charge of apportioning the classes and servers into groups (long haul issue). The protest is to shape bunches C, in view of same work stack profile it diminishes the server exchanging better grained timescales. Besides, by signify by Rc the arrangement of demand classes relegated to group C and by Sc the arrangement of physical servers in bunch C at time interim t. At a lower level of the chain of command, AMs midway oversee singular bunches (i.e., on a hourly premise). Each AM can choose (medium-term issue):

1. Setting the applications inside the group,
2. Dealing with the VMs capacities with respect to running on every server.
3. Load adjusting bolsters in same application level for different VMs.
4. Exchanging servers into dynamic or low-control rest states, and
5. Expanding/diminishing the working recurrence task of the CPU of a server.

It doesn't deal with a lot of information, and prompts handle greater intricacy objectives. There can be an issue in the meaning of procedures in light of a separated administration of various classes of VMs, both in the task stage and in the movement stage In existing framework, the calculation utilized are web based planning calculation in view of lyapunov enhancement.

DISADVANTAGES OF EXISITNG SYSTEM

- It isn't effective
- It does not deal with a lot of information.

- It prompts handle greater unpredictability objectives.

4. PROPOSED SYSTEM

In the proposed framework, a straightforward approach is utilized for workload administration and vitality cost sparing. Here we are utilizing Eco multicloud calculation for workload and movement. And furthermore workload moving strategy is utilized to make the workload administration more effective.

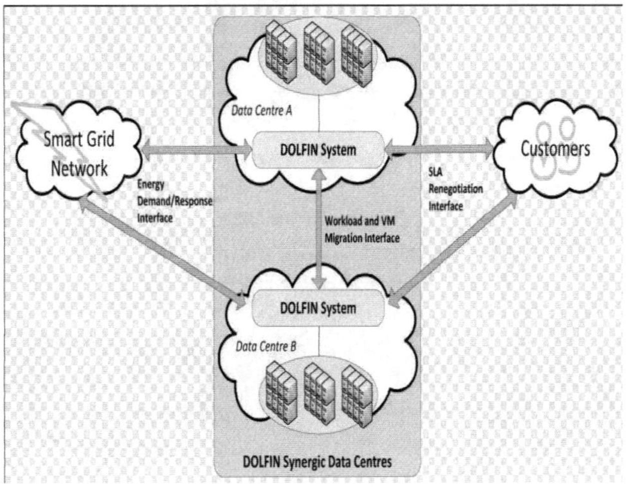

Fig1 Overview of Proposed System

ADVANTAGES OF PROPOSED SYSTEM

- The proposed plot is requires ostensible measure of computational
- cost and correspondence cost contrast with other related plans.
- It is productive.

Selecting the cloud service providers.

We need to recognize the capacity of every framework and their undertaking. So we ought to keep up the database of all cloud specialist co-ops. Cloud innovation associates a system of virtualized PCs that are powerfully provisioned as processing assets, in view of arranged understandings between specialist organizations and clients.

Task assignment

Each sort of process in a distribution center commonly requires specialists who have a particular arrangement of qualities that compare with the errand to be performed.

Maintaining work load

Workload administration is the procedure of powerful workload dissemination which is made to empower the representatives to accomplish ideal execution and efficiency levels. An adjusted appropriation of workload helps the administration of an association to improve the efficiency of their current workforce.

5. CONCLUSIONS

This paper has introduced EcoMultiCloud, a various leveled approach that means to enhance the workload administration of a multi-site server farm. The related engineering contains two layers, the upper layer for the task/relocation of workload among remote destinations, and the lower layer that allots Virtual Machines to physical has inside each nearby site. The approach is adaptable and can be used to accomplish and adjust diverse objectives, among which decreases of costs/expended vitality/carbon outflows, stack adjusting, and so forth. The paper has concentrated on the examination of a four-site framework for the situation that the objectives to be accomplished are the diminishment of carbon emanations and the heap adjusting among server farms. Execution investigation has demonstrated that the various leveled approach accomplishes about an indistinguishable quantitative outcomes from a reference unified arrangement, however offers better functionalities as far as adaptability – it can be adjusted to the particular objectives determined by the administration – and self-sufficiency of single information centers. And as they are free to adopt any internal algorithm for workload management.

REFERENCES

[1] C. Adam and R. Stadler, "Service Middleware for Self-Managing Large-Scale Systems," IEEE Trans. Network and Service Management, vol. 4, no. 3, pp. 50-64, Dec. 2007.

[2] M. Armbrust, A. Fox, R. Griffith, A.D. Joseph, R.H. Katz, A. Konwinski, G. Lee, D.A. Patterson, I.S.A. Rabkin, and M. Zaharia,"Above the Clouds: A Berkeley View of Cloud Computing," http://www.eecs.berkeley.edu/Pubs/TechRpts/2009/EECS- 2009-28.pdf, 2013.

[3] D. Ardagna, B. Panicucci, M. Trubian, and L. Zhang, "," IEEE Trans. Services Computing, vol. 5, no. 1, pp. 2-19, Jan.-Mar. 2012.

[4] Bernardetta Addis, DaniloArdagna, Barbara Panicucci, Mark S. Squillante, fellow, ieee, and li zhang ,"A Hierarchical Approach for the Resource Management of Very Large Cloud Platforms", ieee transactions on dependable and secure computing, vol. 10, no. 5, september/october 2013

[5] J. Cao, K. Hwang, K. Li, and A.Y. Zomaya, "Optimal Multiserver Configuration for Profit Maximization in Cloud Computing," IEEE Trans. Parallel Distributed Systems, preprint, no. 99, 2012.

[6] M. Cardosa, M. R. Korupolu, and A. Singh, "Shares and utilities basedpower consolidation in virtualized server environments," in Proceedingsof the 11th IFIP/IEEE Integrated Network Management (IM 2009), LongIsland, NY, USA, June 2009.

[7] M.D. Dikaiakos, D. Katsaros, P. Mehra, G. Pallis, and A. Vakali, "Cloud Computing: Distributed Internet Computing for IT and Scientific Research," IEEE Internet Computing, vol. 13, no. 5, pp. 10- 13, Sept./Oct. 2009.

[8]Gartner,"2012CloudComputingPlanningGuide,"http:// gartner.com/portal/server.pt?open=512&objID=249&mode=2&PageID=864059&resId=1837017&ref=Browse, 2013 .

[9] P. Graubner, M. Schmidt, and B. Freisleben, "Energy-efficient virtualmachine consolidation," IT Professional, vol. 15, no. 2, pp. 28–34, 2013.

[10] Josephine Brinda,S.L.Jany Shabu,. A Trustworthy Ewom in Social Networks. International Conference on Information Communication and Embedded Systems.

[11] D. Kusic, J.O. Kephart, N. Kandasamy, and G. Jiang, "Power and Performance Management of Virtualized Computing Environments via Lookahead Control," Proc. Int"l Conf. Autonomic Computing (ICAC), 2008.
[12] F. Longo, R. Ghosh, V.K. Naik, and K.S. Trivedi, "A Scalable Availability Model for Infrastructure-as-a-Service Cloud," Proc. IEEE/IFIP 41st Int"l Conf. Dependable Systems and Networks (DSN), 2011.
[13] G.S.G.E.D. Lazowska, J. Zahorjan, and K.C. Sevcik, Quantitative System Performance, Computer System Analysis Using Queueing Network Models. Prentice-Hall, 1984.
[14] H. Li and S. Venugopal, "Using Reinforcement Learning for Controlling an Elastic Web Application HostingPlatform," Proc. ACM Eighth Int"l Conf. Autonomic Computing (ICAC), 2011.
[15] Mayan J.A, Arifa S, Pavithra R,"Semantic based multi lexical ranking technique for an effective search in protected cloud,2016 International Conference on Control, Instrumentation, Communication and Computational Technologies (ICCICCT), Kumaracoil, 2016, pp. 570-576.
[16] Microsoft, "Windows Azure," http://msdn.microsoft.com/ en us/library/windowsazure/dd163896, 2013.
[17] T. Nowicki, M.S. Squillante, and C.W. Wu, "Fundamentals of Dynamic Decentralized Optimization in Autonomic Computing Systems," Self-Star Properties in Complex Information Systems, pp. 204-218, Springer-Verlag, 2005.

[18] R. Raghavendra, P. Ranganathan, V. Talwar, Z. Wang, and X. Zhu, "No „Power" Struggles: Coordinated Multi Level Power Management for the Data Center,"SIGARCH Computer Architecture News, vol. 36, no. 1, pp. 48-59, 2008
[19] S. Rivoire, P. Ranganathan, and C. Kozyrakis, "A Comparison of High-Level Full-System Power Models," Proc. Conf. Power Aware Computing and Systems (HotPower), 2008.
[20] Sheng Di, Cho-Li Wang," Dynamic Optimization OfMultiattribute Resource Allocation In Self-Organizing Clouds" IEEE Transactions On Parallel And Distributed Systems.
[21] F. Wuhib, R. Stadler, and M. Spreitzer, "A Gossip Protocol for Dynamic Resource Management in Large Cloud Environments," IEEE Trans. Network and Service Management, vol. 9, no. 2, pp. 213- 225, June 2012.

DESIGN AND FABRICATION OF MICRO STRIP ANTENNA FOR CUBESAT APPLICATIONS

Tharun.K[1]
BE, Electronics and Communication Engineering,
Sathyabama Institute of Science and Technology,
Chennai
jaitharunchinna@gmail.com

Vivekanand.V[2]
BE, Electronics and Communication Engineering,
Sathyabama Institute of Science and Technology,
Chennai
viveks4567@gmail.com

T. Ravi[3]
Associate Professor,
School of Electrical and Electronics Engineering,
Sathyabama Institute of Science and Technology,
Chennai
ravivlsi123@gmail.com

M.Sugadev[4]
Associate Professor,
School of Electrical and Electronics Engineering,
Sathyabama Institute of Science and Technology,
Chennai

Abstract- In this work we have designed an antenna for 450 MHz in UHF band using rectangular micro-strip patch antenna. It is used for most complex multi band characteristics. Micro-strip antenna has two elementary systems for cubesat. Micro-strip antenna consists of conducting patch on a ground plane separated by di-electric substrate radiation from ground plane has different dielectric configurations. Two antennas are including slot and micro-strip patch antenna. Slot antennas can be placed around solar cells and the slots are narrow enough to be fit into the space between solar cells. Launching the large conventional satellite takes more time and expensive to design the satellite. Cubesats has been proposed to serve the purpose in satellite application cubesat has no more than 1.33kg/unit which reduces the size of satellite and make it less expensive. The proposed antenna at 450Mhz in UHF band and we have achieved 4.2771 of average of gain with parameters of 60*60*1.6mm.

***Keywords**:* Cubesat, UHF band, Ultrawideband, Gain, Directivity, Return loss, VSWR.

1. INTRODUCTION

A cubesat is a type where it is used in space research is made up of multiples 10*10*10 cm cubic units. It have a mass of no more than 1.33 kilograms per unit and often it is used in commercial of electronics and structures. Primary it is in space station and secondary it is a payloads on a launch vehicle. It covers the UHF band for amateur satellite communications 400-470MHz.It is compatible with endurosat solar panels. Gain >0 dBi.

IOP Conf. Series: Materials Science and Engineering **590** (2019) 012052 doi:10.1088/1757-899X/590/1/012052

Maximum radio frequency output power is 3.5watts. Recently they are being designed by many universities globally to demonstrate the technology as well as involve more researchers into this study. The link which is placed side by side laterllay and the communication happens simultaneously. NASA two satellite nodes demonstrated space craft network operation using two or more space crafts in an autonomously configured space network which allows crosslink communications and scientific measurements. The main challenging factor in design is the need of nadir pointing attitude for earth observation it can be made possible using high gain directive antennas and hence employee lower transmitting power to achieve good results and less transmitting power.

Jordipuig-suari has been developed the cubesat where it is been working in LEO and it is small in size and less in cost to make a satellite and it will be communicate between the link earth and cubesat in antenna. In this mission to deploy the picosatellite has resulted in the development launcher system that was hopelessly complicated and it can only to be work for most of the time. Bob twigs known that the projects are going slow he has been sought the DARPA funding the result and redesign the mechanism of the satellite[3].

QB50 it is proposed in international network of 50 cubesats for multipoint the measurements in the lower thermosphere (90-350)kms and re-entry research. QB50 where founded by the European commission as a part of FP7. Where 2U cubesats been in the size of 10*10*20 cm are been developed in 1U [4].QB50 has been launched in the 2012. Insight mission is upcoming satellite to be launched in the May of 2018 cubesat to be landed in mars to provide relay communication from insight to earth during entry and landing[7]. This is the first mission where cubesat technology going to deep in space where MarCO is mars cube one each one is a six unit cubesat 14.4 inches by 9.5 inches by 4.6 inches.

Ms. Sayalidixit[2015]: Analysed micro strip antenna with good antenna parameters has been designed and fabricated at two different resonant frequencies(8.12Ghz and 9.42Ghz) and the return loss -17db and -44db thus a good antenna parameters and wide bandwidth at resonant frequencies is well suitable for satellite applications. VSWR is 1.26db and 1.2db.

Ambresh.P.A: A technique in which slits are introduced on the radiating edges of rectangular patch antenna and using high dielectric constant which reduces the size of the antenna. Thetwo resonant frequency 2.03Ghzand 3.79Ghz. VSWR is 1.850 and 1.108. Return loss is -15db and -11.27db.A high dielectric constant makes the antenna to work at low frequencies and slits plays a vital role not only in obtaining good Antenna parameters. Gain is 2.00db and 1.90db for two frequency.

Vaishalikamboj[2016]: A single layered coaxial fed micro strip antenna with polygon, rectangular and square slots are placed for miniaturization and for obtaining good radiation properties[6]. The resonant frequency is 3.16Ghz and return loss is -26.78db. Slots at effective positions makes antenna to radiate effectively and decreases the size. Gain is 4.44db and VSWR is 1.69.

Constantine G.kakoyiannis/2008:A tapered peripheral slit micro strip antenna has designed with miniature size and but in capability for circular polarization.Gain is 0.7db .Dual feed technique in antenna produces circular polarization which is most essential characteristic of cubesat.Dielectric loading have a detrimental effect on band width.Directivity is 4.9db[9].

Kamal sarabandi [2006]: proposed has compact wideband UHF patch antenna on a reactive impedance substrate. Antenna designed at three frequencies covering the 410-480Mhz. UHF antenna is operating frequency range 420-450 Mhz gain is 5.2 is to desired for terrestrial communication. Frequency 480Mhz gain is 5.9. Return loss is -10db[8].

Antenna types are printed circuit board (or) whip antenna, raised-strip antenna, conventional whip antenna, helix antenna, patch antenna.

IOP Conf. Series: Materials Science and Engineering **590** (2019) 012052 doi:10.1088/1757-899X/590/1/012052

2. PROPOSED MICROSTRIP ANTENNA

Microstrip antenna is also known as printed antennas is to be fabricate the printed circuit board.it is mostly used in microwave frequencies. Microstrip antennas where it is individual consists of patch of metal foil and been in different shapes on the printed circuit board in the metal foil of ground plane on the other side of the board. Where it is been designed for narrowband communication at UHF and microwave ranges.

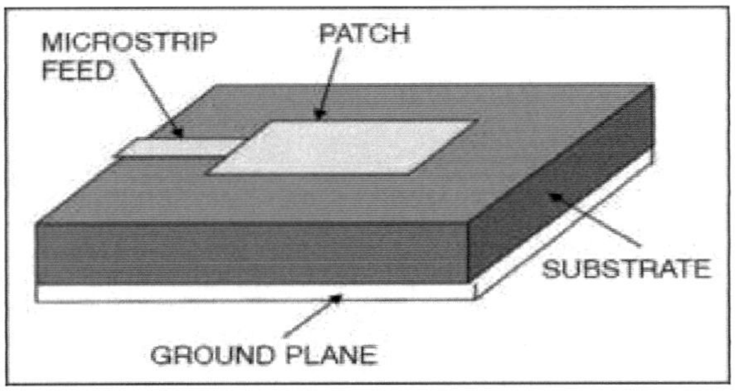

Fig:1 basic design of microstrip antenna

The advantage of microstrip antenna is light weight, small in size, not expensive, easy to design and manufacture. For low frequency size is larger at 450Mhz 60*60*1.66mm.

Proposed UHF Microstrip antenna:

To design the microstrip antenna using cubesat application. Done with the resonant frequency, gain, return loss, directivity. Our antenna is designed in 3 bands UHF, s-band, SHF.

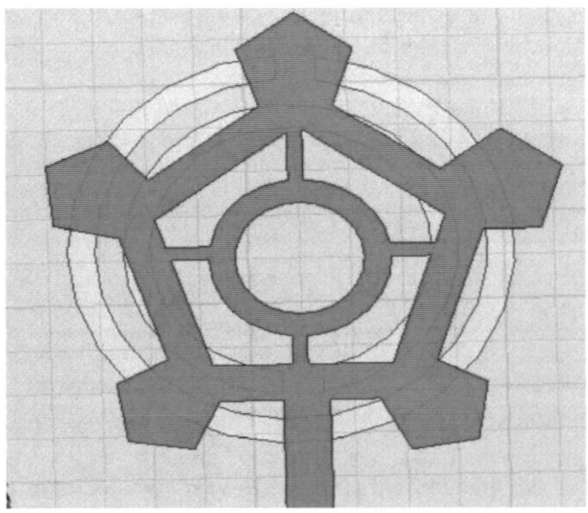

Fig 2: without circular ring microstrip patch antenna

Fig 2.1: with circular ring microstrip patch antenna

Antenna design at 450Mhz has been finally designed with the hexagonal shapes with the material of using fr4. Fr4 is easy to design and fabricate the antenna parameters. Working in mobile communications, military uses, weather purposes. Where it is acting in different bType equation here.ands in Ultra high frequency(UHF) 30-500Mhz, S-band parameter 2-4Ghz, Super high frequency(SHF) 3-300Ghz. Gain, directivity, return loss, VSWR has good in antenna design. Cubesat acting in low frequencies due to high dielectric resonant where it will be maintain in LEO.

Length of the patch :20mm
Width of the patch :21.5mm
Length of the substrate :38mm
Width of the substrate :38mm
Height of the substrate :1.6mm
Feed type : microstrip line feed
Feed position :17.5 ,0 ,0

- **Patch Width**
- $W = \frac{c}{2f}\sqrt{\frac{2}{\epsilon r+1}}$
- w= width of the patch
- C= velocity of light (3mm)
- f = Resonant frequency
- Dielectric constant of substrate
- **Patch Length**
- L=$L_{eff} - 2\Delta L$
- L= length of the patch
- L_{eff} = Effective length, and it is given by

$$L_{eff} = \frac{c}{2f\sqrt{\epsilon_{reff}}}$$

$$\Delta L = 0.412h\frac{(\epsilon_{reff} + 0.3)(\frac{w}{h} + 0.264)}{(\epsilon_{reff} - 0.258)(\frac{w}{h} + 0.8)}$$

- ϵ_{reff} =Effective dielectric constant
- **Substrate Length and Width**
- L_g=L+6h
- $Wg_v = W + 6h$
- $h = \frac{0.0606\lambda}{\sqrt{\varepsilon r}}$

3. RESULTS AND DISCUSSION

The antenna is stimulated in HFSS(high frequency simulated system)software.

A. S11 Return loss:

Return loss is the loss of power in the signal returned/reflected by the discontinuity in a transmission line or optical fiber. It is usually expressed as a ratio in decibles(db).

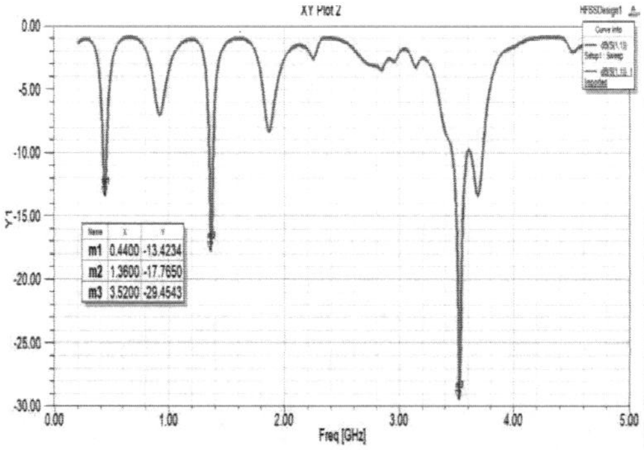

Fig 3: Return loss at 450Mhz is -13.4234.

B. VSWR(voltage standing wave ratio):

The parameter VSWR is a measure that numerically describes how well the antenna is impendence matched to the radio or transmission line it is connected to.VSWR stands for voltage standing wave ratio, and is also referred to as standing wave ratio.

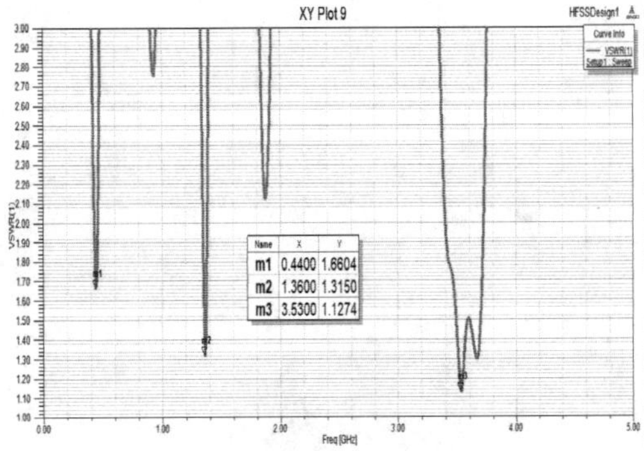

Fig 3.1: VSWR at 450Mhz is 1.66

a. Gain:

An antenna power gain or simply gain is a key performance number which combines the antenna directivity and electrical efficiency. In a transmitting antenna, the gain describes how well the antenna converts input power into radio waves headed in a specified direction Gain is represented in terms of db.

GAIN

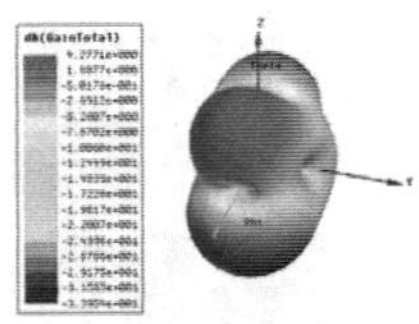

Fig 3.2: Gain at 450Mhz is 4.2771e+000

b. Directivity:

Directivity is a fundamental antenna parameter. It is a measure of how directional antenna radiation pattern is. An antenna that radiates equally in all directions would have effectively zero directionally, and the directivity of this type of antenna would be 1

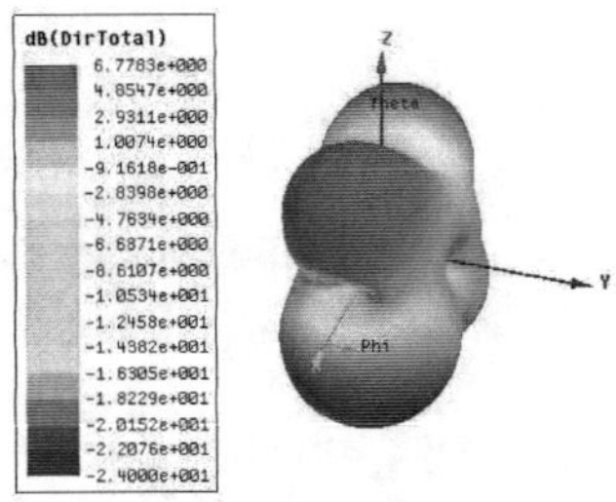

Fig3.3: Directivity at 450Mhz is 6.7783

IOP Conf. Series: Materials Science and Engineering **590** (2019) 012052 doi:10.1088/1757-899X/590/1/012052

c. Electric field:

The electric field[E], flows from the positive charge to negative charge placed on the elements by voltage applied to the antenna.

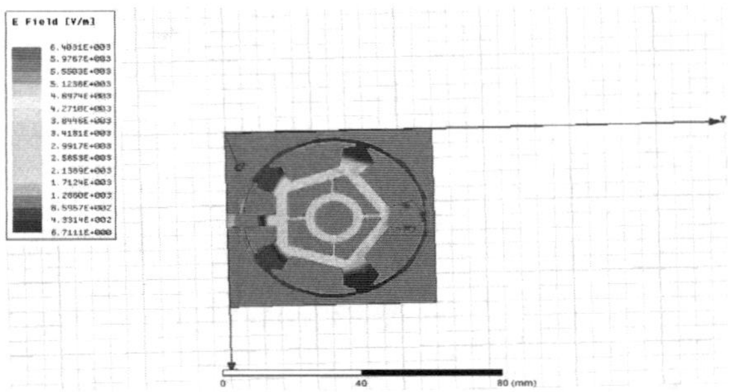

Fig 3.4: Electric field 6.4031

f. Magnetic field:

The magnetic field[H] is a vector quantity has a magnitude and direction around a wire of moving charge.

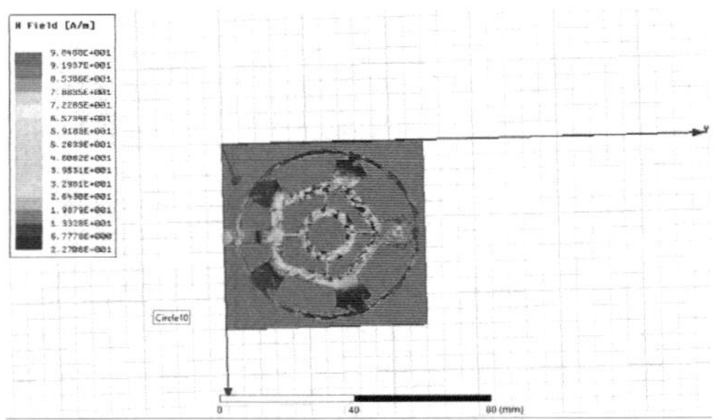

Fig3.5: Magnetic field 9.048

g. 2D pattern:

Two dimensional or planar arrays with uniform spacing and with phased weights. The extension of two dimensional arrays is straight forward.

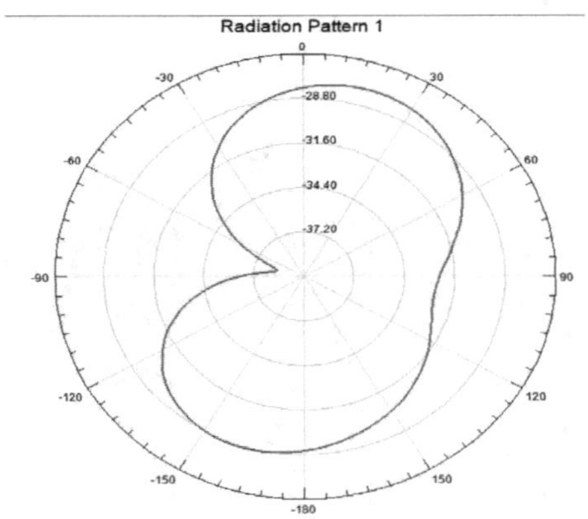

Fig3.6:from the radiation this is acting at 90 degress in bidirectional.

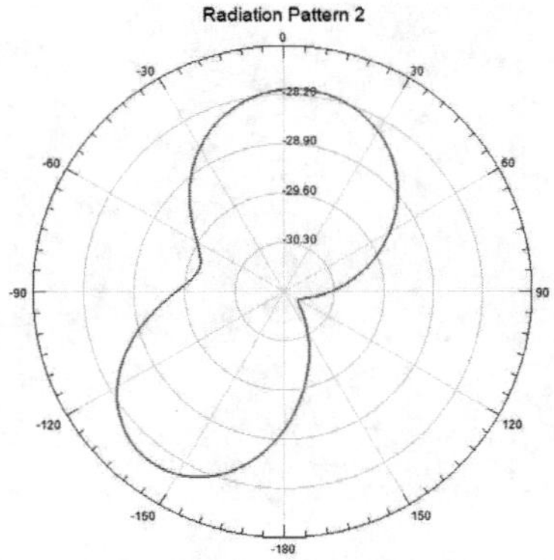

Fig 3.6.1: From the radiation pattern it is acting at 0 degress is going in bidirectional.

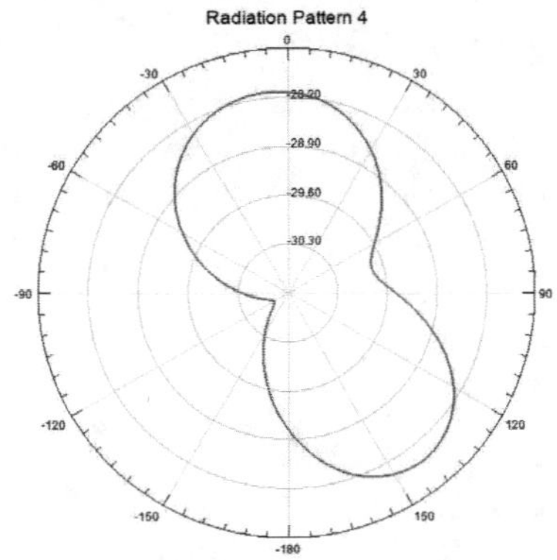

Fig 3.6.2:This pattern is acting at 180 degrees and moving in bidirectional.

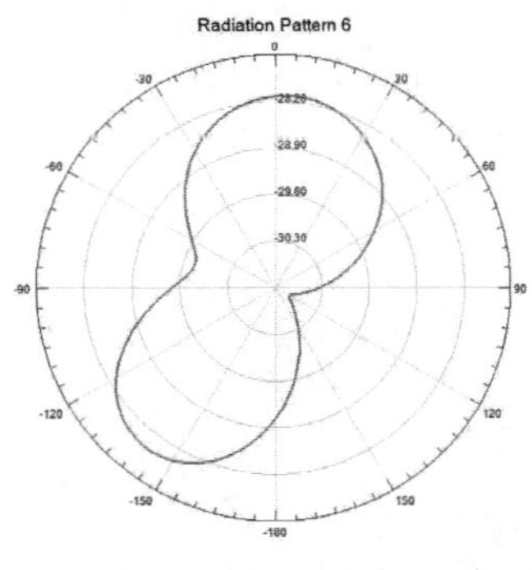

Fig3.6.3:Radiation pattern is moving in bidirectional and is acting at 360 degress.

Fig 3.7:Fabrication of antenna design(450Mhz) has been done this design the process is working in gain, directivity, VSWR, return loss. Return loss is increasing in when frequencies are increased.

4. CONCLUSION

Microstrip antenna has been done with cubesat application using the UHF band. Antenna design (450Mhz) working in 3 bands UHF, S-band, SHF band. This antenna is very compact in size. It is very useful in monoethic application and also in mobile satellite radar, biomedical. The resonant frequency is 450Mhz and the return loss -13.4234. omni-directional radiation characteristics. VSWR is 1.66 with this frequency the antenna of the patch has been settled in this frequency. Directivity 6.7783db and the Gain is 4.2771db.

REFERENCES

[1] S. Gao, K. Clark, M. Unwin, J. Zackrisson, W. A. Shiroma, J. M. Akagi, K. Maynard, P. Garner, L. Boccia, G. Amendola, G. Massa, C. Underwood, M. Brenchley, M. Pointer, and M. N. Sweeting, "Antennas for modern small satellites," *IEEE Antennas Propag. Mag.*, vol. 51, no. 4, pp. 40–56, 2009.

[2] M. A. Swartwout. (2016, Aug. 31). CubeSat database.[Online].Available: https://sites.google.com/a/slu.edu/swartwout/home/cubesat-database

[3] Ravi.T and Kannan.V "Effect of N-type cntfet on double edge triggered D flip-flop based PISO shift register" IEEE Proceedings of the International Conference On Emerging Trends in Science Engineering and Technology: Recent Advancements on Science and Engineering Innovation, INCOSET 2012, Dec-2012, pp. 344-349.

[4] R.Ramya and T.Ravi, "Design of cache memory mapping techniques for low power processor" ARPN Journal of Engineering and Applied Sciences, VOL. 10, NO. 11, JUNE 2015, pp:4783-4788.

[5] State of the Art of Small Spacecraft Technology [Online]. Available: https:// sst-soa.arc.nasa.gov/03-power

[6] Ravi.T, Kannan.V "Modeling and performance analysis of ballistic carbon nanotube field effect transistor (CNTFET)" International Conference on Recent Advances in Space Technology Services & Climate change – RSTS&CC-2010, November – 2010, pp. 327-331.

[7] P. A. Ambresh, P. M. Hadalgi and P. V.Hunagund, Gulbarga-India. "Compact Omni-Directional Patch Antenna For S-Band Frequency Spectra "International Journal of Advances in Engineering & Technology, Sept 2011. ISSN:2231-1963

[8] Micheal Priyanka.M and Ravi.T, "Survey on role of memristor in electronics "2015 International Conference on Control Instrumentation Communication and Computational Technologies, ICCICCT 2015, 18-19 Dec 2015, Pp:738-744.

[9] Brij Bhushan Ojha, S.Karthikeyan ,"Design of an interconnection network using VLSI photonics: Issues and challenges " International Journal of Pure and Applied Mathematics, Volume 118 No. 17 2018, PP: 211-224

[10] S. Vimalsree and S. Karthikeyan, "A Low Power Multibit Flip Flop Merging Technique Using WSN Nodes", ARPN, vol. no:11, No.15, August 2016, Page no: 9358-9363.

PERFORMANCE ANALYSIS OF BELIEF PROPAGATION POLAR CODE DECODER

KARTHICK C [1]
Assistant professor, Dept of ECE,
Sathyabama Institute of Science and Technology
OMR Road,Chennai, India
karthivlsi13@gmail.com

VINO T [2]
Assistant professor, Dept of ECE,
Sathyabama Institute of Science and Technology
OMR Road,Chennai, India
vinodhevan@gmail.com

BABJI T [3]
Student, Dept of ECE,
Sathyabama Institute of Science and Technology
OMR Road,Chennai, India

NAVEEN P [4]
Student, Dept of ECE,
Sathyabama Institute of Science and Technology
OMR Road,Chennai, India

Abstract- Attributable to their ability accomplishing execution and low encoding and interpreting intricacy, polar codes have gotten noteworthy consideration as of late. Successive cancellation decoding (SCD) and belief propagation decoding (BPD) are two mainstream approaches for disentangling polar codes. SCD, in spite of having less computational intricacy when contrasted and BPD, experiences long inertness because of the serial idea of the SC calculation. BPD, then again, is parallel in nature and is more alluring for low- dormancy applications. Nonetheless, because of the iterative idea of BPD, the required inertness and vitality dispersal increment straightly with the quantity of emphasis. In this paper, we propose a novel plan in light of sub-factor diagram solidifying to lessen the normal number addition the normal number of cycles required by BPD, which specifically converts into bring down inactivity and vitality dissemination. Besides, the equipment design for the proposed plot is created and contrasted and the best in class BPD executions for (1024,512) polar codes.

Keywords: Li-Fi transmission, Transmitter chip, receiver chip, Android application, prolific cable.

1. INTRODUCTION

Distinctive limit moving toward codes, for example, Turbo codes [1] and low-thickness equality check (LDPC)codes [2], have been planned and utilized for quite a while in applications, for example, remote correspondence and information stockpiling, to accomplish high information rates. As of late, the principal provable limit accomplishing codes called polar codes, were developed by Arikan [3]. Being the principal group of codes known to accomplish the channel limit with unequivocal development, polar codes have pulled in a great deal of consideration since their innovation. Polar codes have been demonstrated to accomplish the limit with respect to twofold info symmetric memory less channels [3] and additionally discrete and constant memory less channels [4]. In addition, an express development strategy for polar codes has been given, and it is demonstrated that they can be proficiently encoded and decoded with many-sided quality O(n log n), where n is the code length. Various translating strategies have been proposed for polar codes [5] [2], and among these, successive cancellation decoding (SCD) and conviction engendering interpreting (BPD) are the two most famous techniques. . Because of the serial idea of the calculation, SCD experiences long idleness, despite the fact that it requires less calculation as contrasted and BPD. A few strategies have been proposed to lessen the idleness of SC decoders to accomplish a high throughput [6]. In addition, list translating and stack disentangling, which depend on SCD, have been proposed to enhance the mistake rectifying execution for polar codes with short code lengths [1]. Then again, polar BP decoders [5] have the natural favorable position of parallel preparing. Along these lines, contrasted and their SC partners, polar BP decoders are more appealing for low-idleness applications. Be that as it may, because of their iterative nature, the required inertness and vitality dissemination of BP decoders increment directly with the quantity of cycles. The necessity of countless outcomes in high calculation many-sided quality, and subsequently makes BPD less algorithm its SC partner. To diminish the region cost, as of late a memory effective BP decoder was proposed by Sha et al. [2] in light of consolidating two neighboring phases of a BPD factor chart. To diminish the calculation many-sided quality, another deciphering technique, called delicate cancelation (SCAN) disentangling, was proposed in [2]. By limiting the delicate data proliferation plan for the unraveling procedure, the computational intricacy of SCAN is much lower than that of BPD. Be that as it may, not quite the same as BPD, the SCAN task is serial in nature, prompting an any longer disentangling dormancy.

2. METHODOLOGIES

In this paper, we go for the plan of a low- dormancy polar codes decoder, and thus, we focus on the parallel BPD execution. To address the issues of the vast number of emphases and high calculation multifaceted nature inalienable in BP decoders, Yuan and Parhi [9] proposed a G-network based early ceasing plan (named G-lattice BPD, for future reference), which depends on the idea that iterative decoders ordinarily unite before achieving a settled most extreme number of cycles. The G-framework based halting measure is utilized to stop the calculation if union has been come to. It is accounted for in [1] that the vitality utilization is lessened by33% when contrasted and BPD without a halting model (named benchmark BPD, for future reference), and a coded throughput of 9.1 Gb/s (the data throughput of 4.55 Gb/s) is accomplished at 3.5 dB. Another early halting technique in view of the understanding of three sequential hard choices was presented by Park et al. in [7],[9]. They proposed a twofold section design (named twofold segment BPD, for future reference) for BPD of a (1024, 512) polar code and accomplished a coded throughput of4.68 Gb/s. The early end conspire empowers the proposed design to be more vitality proficient and decreases the dormancy for the deciphering activity.

To additionally lessen the calculation many- sided quality of BP decoders, in our past work [5], we proposed a technique in light of the joining of the subfactor-charts, which is come to at a considerably prior stage. Obtaining a thought from SCD, a portion of the subfactor-diagrams are checked amid every cycle, and in the event that they have united, they are a standout amongst the most prevalent subjects in data hypothesis and have pulled in a ton of consideration. A few deciphering techniques are accessible for disentangling polar codes [6], SCD and its variations and BPD are two famous strategies. SC decoders experience the ill effects of long inertness because of the serial idea of the SC calculation. In any case, the SC calculation requires less calculation when contrasted with BPD. In light of this property, a few high-throughput ease SC decoders were accounted for in [7]. Another preferred standpoint of the SC calculation is its capacity to accomplish great blunder redressing execution for long code lengths For short code length, in light of the SCD, the rundown deciphering or stack translating technique likewise accomplish great mistake rectifying execution [1] [5]. Then again, polar BP decoders [6] have the inherent favorable position of parallel preparing. Along these lines, contrasted and their SC partners, polar BP decoders are more appealing for low-idleness applications. For iterative decoders, (for example, polar BP decoders), the required inertness and vitality dispersal increment straightly with the quantity of cycles. Notwithstanding, the requirement for countless influences BP decoders to experience the ill effects of high calculation many-sided quality, and thus polar BP decoders are still not as appealing as their SC partners. To this end, another unraveling strategy, called delicate cancelation (SCAN) disentangling, is proposed in [2],[3]. By limiting the delicate data proliferation plan for the disentangling procedure, the computational unpredictability of SCAN is much lower than that of BPD. Be that as it may, unique in relation to BPD, the SCAN task is serial in nature, prompting an any longer translating inertness. Subsequently, going for the low-idleness polar codes decoder, we focus on the BPD in this work. separation width of arbitrarily created standard diagrams. Luczak and McDiarmid[1]-[9] additionally think about the base separation width of diagrams produced by a conveyance unique in relation to our own. Besides, our examination is of arbitrary bipartite charts, instead of irregular standard diagrams. Also, our outcome makes just powerless presumptions on the hub degree circulation, without requiring a degree-normality supposition, rather than past work.

3. EXISITNG SYSTEM

Various interpreting techniques have been proposed for polar codes, and among these, progressive cancelation unravelling (SCD) and conviction engendering translating (BPD) are the two most prevalent strategies. Because of the serial idea of the calculation, SCD experiences long idleness, in spite of the fact that it require less calculation as contrasted and BPD. A few techniques have been proposed to decrease the idleness of SC decoders to accomplish a high throughput. Also, list disentangling and stack translating, which depend on SCD[10], have been proposed to enhance the mistake redressing execution for polar codes with short code lengths. Then again, polar BP decoders have the natural preferred standpoint of parallel handling. Along these lines we should get number of iterations contrasted and their SC partners, polar BP decoders are more alluring for low-idleness applications. In any case, because of their iterative nature, the required idleness and vitality dispersal of BP decoders increment straight with the quantity of emphases.

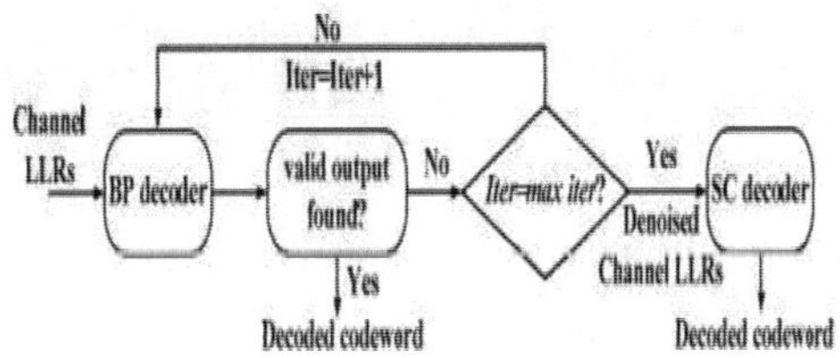

Fig.1. Block Diagram Of successive cancellation and belief propagation decoder

The necessity of an extensive number of emphases brings about high calculation many-sided quality, and consequently makes BPD less alluring than its SC partner[10]. To decrease the region cost, as of late, a memory productive BP decoder was proposed by Sha et al. in light of consolidating two neighbouring phases of a BPD factor chart[10]. To decrease the calculation multifaceted nature, another translating technique, called delicate cancelation (SCAN) deciphering, was proposed. altogether less memory for putting away log-probability proportions (LLRs) as contrasted and the SCAN decoder, and by wiping out [10]

4. PROPOSED SYSTEM

Polar Codes and Belief Propagation Decoding: Polar codes are direct square codes in view of the wonder of channel polarization, in which singular channels are recursively consolidated and part, to such an extent that their common data inclines toward either 1 or 0. At the end of the day, some of these channels turn out to be totally clamor free, while the others turn out to be totally loud[9]. Moreover, the portion of silent channels inclines toward the limit of the basic twofold symmetric channels. CSFG Freezing Concept: The proposed BP disentangling plan depends on a CSFG solidifying idea to accomplish bring down multifaceted nature. At a specific cycle t, if a CSFG at arrange j can accurately unravel its relating constituent code, it is solidified and no message passing or refreshing inside the CSFG will be required in the resulting emphases. The subtle elements of how to check whether a CSFG can be solidified will be displayed later. At the point when the unravelling achieves a specific stage, its CSFGs are checked for solidifying.

In the event that the CSFG can't accurately unravel its constituent code, at that point it can't be solidified and the message passing and refreshing are executed by the PEs at that stage. From that point forward, we move to the following stage and check the meeting of the relating CSFGs. When we move to the following stage, the quantity of CSFGs is multiplied. This solidifying checking system proceeds from stage to arrange until the finish of the BPD. Fig demonstrates the SCD planning tree and the factor diagram of then=8 polar code. We can see that the best CSFGs at the distinctive stages relate to the initial few subtrees that came about because of the profundity first traversal of the SCD planning tree. CSFG Freezing Criterion: MLD depends on thorough pursuit, and consequently has a colossal computational unpredictability. To lessen the unpredictability, a novel checking measure is proposed to effectively discover the MLD consequence of the constituent code.

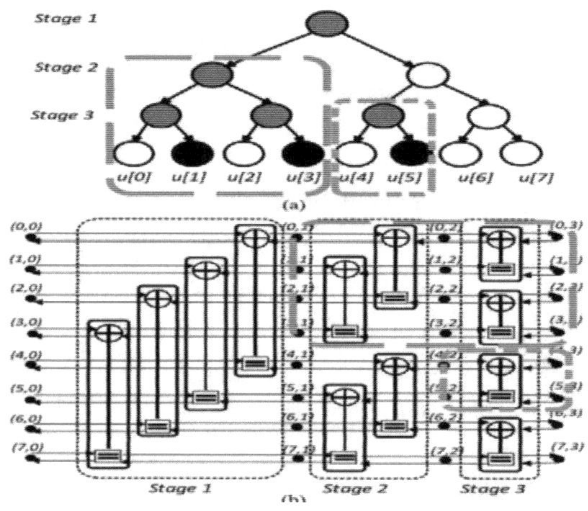

Fig. 2. Correspondence between SCD scheduling tree and BPD factor graph. (a) Subtrees in SCD graph (bottom)

5. WORKING PRINCIPLE

Polar codes have gotten much consideration from data scholars on account of their ability accomplishing property fig 3. Progressive cancelation (SC) [1] and conviction proliferation (BP) [2] calculations are the two primary translating approaches for polar codes. Be that as it may, to date the useful uses of polar codes are blocked by their long disentangling inertness and constrained blunder rectifying execution issues. Earlier examinations have proposed answers for beat these issues at both calculation and VLSI execution levels. In [3-5], SC list (SCL) calculation and its variations were proposed to enhance the execution of polar codes. In the meantime, [6-10] introduced a few ways to deal with lessen the deciphering idleness of SC and SCL decoders. Moreover, upgrades in deciphering execution of BP calculation were accounted for in [1-7]. In any case, even with the utilization of the above endeavours, polar codes have not in any case met the prerequisites of short inactivity and enhanced disentangling execution with short basic way delay. In view of our current advance on early ceasing criteria for a BP decoder, in this paper we propose another BP-SC half and half translating plan. Unique in relation to a L-estimate SCL decoder, the proposed half breed decoder just comprises of one pre-processing BP decoder and one SC decoder.

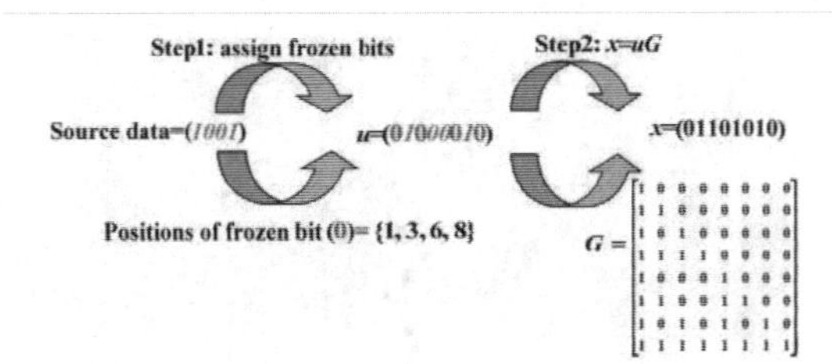

Fig.3. Polar Codes

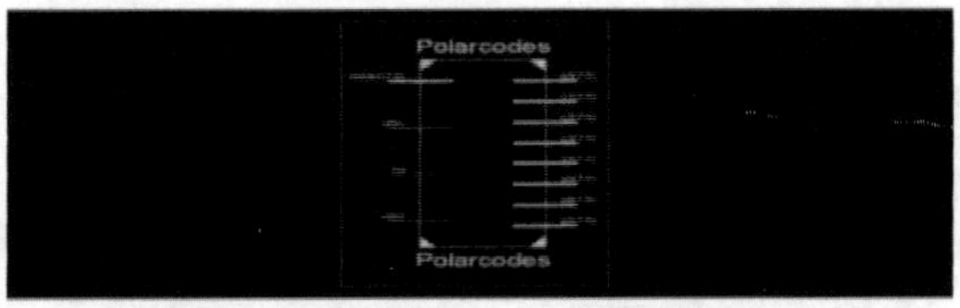

Fig. 4. Simulation Diagram of Polar Codes

With the assistance of pre- processing BP decoder, the normal unravelling inertness of whole interpreting is altogether diminished. Then again, the de noised channel data yield from the pre- processing BP decoder can likewise enhance the mistake revising execution of the SC part decoder Therefore, the whole half breed decoder accomplishes a substantially shorter interpreting idleness with enhanced unravelling execution. For (1024, 512) polar codes, fig 2 simulation diagram comes about demonstrate that the proposed approach prompts no less than 0.2dB pick up contrasted with a traditional BP decoder with a similar most extreme number of cycles for the whole SNR locale. What's more, the proposed decoder likewise accomplishes 0.2dB increase over the BP decoder with a similar most pessimistic scenario inertness in the high SNR locale. Also, contrasted with a customary SC decoder, the proposed approach prompts a 0.2dB pick up in the medium SNR district with substantially less normal disentangling dormancy. Besides, we likewise build up the low-many-sided quality bound together equipment design of the half and half decoder, which can execute SC and BP calculations on a similar equipment.

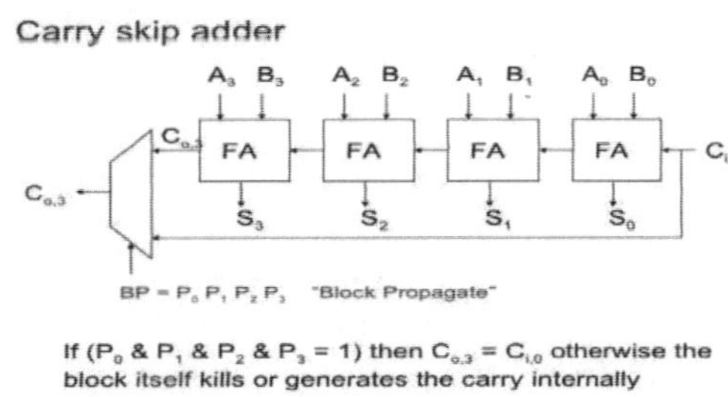

Fig. 5. Carry Skip Adder

Carry skip Adder is a viper usage that enhances the postponement of a swell convey with little exertion contrasted with different adders This will skirt the convey to conquer the deferral and zone and power. Besides, the equipment design for the proposed plot is created and contrasted and the best in class BPD executions for (1024, 512) polar codes. An interpreting throughput of 13.9 Gb/s is accomplished alongside a 60% 73% change in vitality lessening and two times increment in equipment proficiency when contrasted and the current BPD usage. The proposed engineering of this paper investigation the rationale size, territory and power utilization utilizing.

6. RESULT

By using Verilog Xilinx software we simulated our proposed one by taking clock pulse as high and time period as 100us that we check in the simulated output

IOP Conf. Series: Materials Science and Engineering **590** (2019) 012053 doi:10.1088/1757-899X/590/1/012053

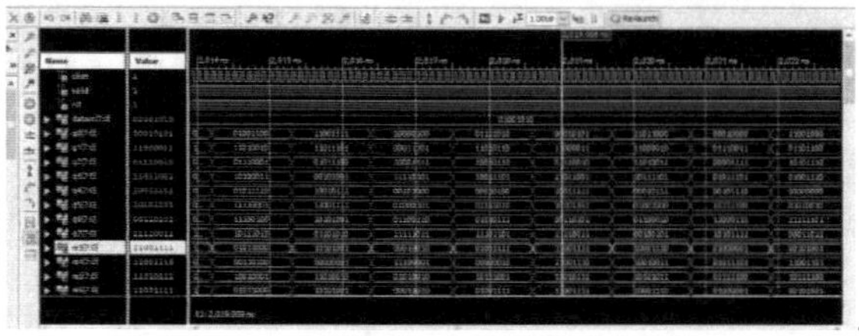

Fig. 6. Output Waveform

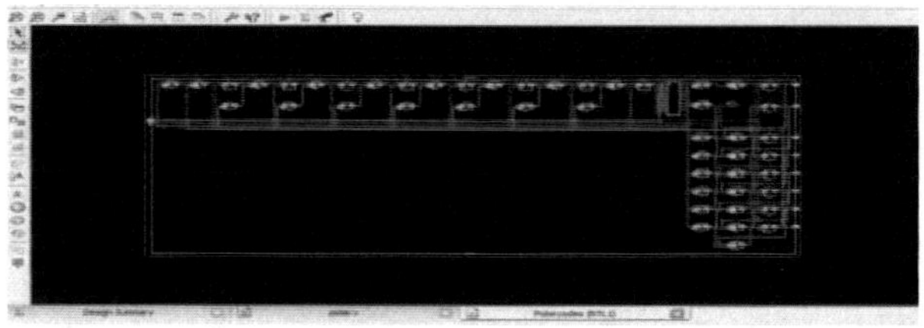

Fig. 7. Block diagram of Polar Codes

Table 1.Performance Comparison

Analysis	**Existing method**	**Proposed method**
Power consumption	0.0020watts	0.0018watts
Time delay	7.523ns	6.378ns
Frequency range	190.63MHz	156.783MHz
Memory usage	411MB	352MB

7. CONCLUSION

This paper introduces a half and half unravelling plan for polar codes. With the utilization of link of BP and SC decoders the half and half decoder can at the same time accomplish great blunder amending and equipment execution. This paper explores the execution of stochastic SC polar code decoder. Different potential methodologies that can enhance translating execution are Analyzed and examined. It is demonstrated that the stochastic SC decoder accomplishes comparable blunder remedying execution to its deterministic partner. This work makes ready for future VLSI plan of stochastic polar decoders

REFERENCES

[1] C. Berrou, A. Glavieux, and P. Thitimajshima, " Near Shannon limiterror – Correcting coding and decoding: Turbo codes" in Proc IEEEInt. Conf. Commun.(ICC), vol. 2. May 1993, pp. 1064, 1070.

[2] R. G. Gallager, Low-Density Parity-Check Codes". Cambridge, MA,USA: MIT Press, 1963.

[3] E. Arikan," Channel polarization : A method for constructing capacity achieving codes for symmetric binary input memoryless channels". IEEEInt. Theory, vol. 55, no. 7, , pp. 3051 3073, Jul. 2009.

[4] E. Sa,So'glu E.Telata, , and E. Arikan, " Polarization for arbitrary discrete memoryless channels," in Proc . IEEE Inf . Theory Workshop (ITW),Oct. 2009, pp. 144 148.

[5] A. Alamdar- Yazdi and F.R. Kschischang, "A simplified successive cancellation decoder for Polar Codes," IEEE Commun . Lett., vol. 15, no 12, pp 1378 1380, Dec. 2011.

[6] C. Leroux, I. Tal, A. Vardy, and W. J. Gross, Hardware architecturesfor successive cancellation decoding of Polar codes," in Proc. IEEE Int.Conf. Acoust., Speech, Signal Process.(ICASSP), May, 2011,pp. 1665 1668.

[7] C. Leroux, A. J. Raymond, G. Sarkis, and W. J. Gross, " A Semi Parallel Scccessive- Cancellation decoder for polar codes," IEEE Trans., vol 61, no 2, pp. 289 299, Jan. 2013.

[8] Y. Fan and C. Y. Tsui, " An efficient partial – sum network architecture for semi-parallel polar codes decoder implementation ." IEEE Trans SignalProcess., vol. 62, no. 12, pp. 3165 3179, Jun 2014.

[9] G. Sarkis and W.J. Gross, "Increasing the throughput of polar decoders," IEEE Commun. Lett., vol. 17, no. 4, pp. 725 728, Apr. 2013.

[10] Karthikeyan S., Karthick C., Shibu Prasath S.V. "Human tracking system for victims trapped from collapsed building" ARPN Journal of Engineering and Applied Sciences, 2016.

[11] Subudhi, J., Karthick C, "Implementation of vedic divider on RSA cryptosystem". ICIIECS 2015 - 2015 IEEE International Conference on Innovations in Information, Embedded and Communication Systems,. 2015.

[12] Kumaresan, R., Senthilkumar, S., Karthick, C. "Biometric high secure and cost effective finger vein authentication system for ATM" ARPN Journal of Engineering and Applied Sciences, 2015.

IOP Conf. Series: Materials Science and Engineering **590** (2019) 012054 doi:10.1088/1757-899X/590/1/012054

IOT Based Omni Directional Robot Control by Using ARM-Series

K.N.HITESH [1]
Department of Electronics and Communication Engineering
Sathyabama Institute of science and Technology
Chennai, Tamilnadu, India
hitesh96.knh@gmail.com

JUTURU MANOJ KUMAR REDDY [2]
Department of Electronics and Communication Engineering
Sathyabama Institute of science and Technology
Chennai, Tamilnadu, India
j.manjreddy@icloud.com

ILAYARAJAA KT [3]
Assistant professor
Department of Electronics and Communication Engineering
Sathyabama Institute of science and Technology
Chennai, Tamilnadu, India

JOANY RM [4]
Assistant professor
Department of Electronics and Communication Engineering
Sathyabama Institute of science and Technology
Chennai, Tamilnadu, India

VIJAYAKUMAR V [5]
Assistant professor
Department of Electronics and Communication Engineering
Sathyabama Institute of science and Technology
Chennai, Tamilnadu, India

Abstract— On this research, controlling the omni directional mobile robot is divided by two controls, high level control which covers positioning and path tracking, also the low-level control which covers omnidirectional kinematic, speed control, heading lock and communication. This also covers thread management, priority fixation and stack memory fixation. Program is divided into several main threads, i.e., IMU, speed controller, communication and motion thread (kinematic). Stack memory configuration for the above threads can be allotted as IMU thread with 1024 bytes of stack memory, speed control with 512 bytes, motion management with 512 bytes, communication with 128 bytes. IMU execution time lapses for 5.6uS, speed control elapses for 19.7uS, whereas motion elapses for 11uS and communication elapses for 1.2mS. Robot's low-level control which includes robot movement kinematic (e.g., heading lock, speed control etc) is required to support the predefined system and enhance the robot sensor's response in navigation.

Keywords: IMU, speed control, communication, kinematic

1. INTRODUCTION

In this paper, fabrication of omni-directional robot can be done to control the robot by using the low-level control which includes kinematic like heading lock, speed control etc., support in the robot navigation. High-level control comprises path tracking as well as positioning. Both low-level and high-level programming are commonly done sequentially which will not be ideal for mobile robot. Hence RTOS is needed for robot task management due to its ability to perform priority-based multitasking. The low-power embedded memory and an ARM Cortex-M0 core that operate at 30 MHz were fabricated in combination with a 60-nm c-axis aligned crystalline indium– gallium–zinc oxide FET and a 65-nm Si CMOS. The embedded memory adopted a structure in which oxide semiconductor-based 1T1C cells are stacked on Si sense amplifiers. This memory achieved a standby power of 3 nW while retaining data and an active power of 11.7 μW/MHz by making each bitline as short as each sense amplifier. The M0 core adopted the flip-flop in which an oxide semiconductor- based 3T1C cell is stacked on the Si scan flip-flop cell without area overhead and achieved a standby power of 6 nW while retaining data. The combination of the embedded memory and the M0 core provided high-performance, low-power Internet of Things devices operating with a broad range of active standby power ratios [1].16.07pJ/cycle 31MHz ARM Cortex M0 core in 40nm CMOS. The system was designed using differential transmission gates in an extended standard cell flow, taking into account variability, speed, energy and scalability. Extensive measurements over a range of 25 dies show it achieves sub-20pJ/cycle operation in a 330-500mV 10-48MHz range and is fully functional down to 190mV. Compared to state-of-the-art, a 40x speed and 4.8x EDP improvement is reported at the MEP. With low variation (σ/μ) on the clock frequency (3.5% at the MEP) and energy consumption (18.2% at the MEP), it combines the low variability, high speed and low energy of full custom work with the ease and design time of standard cell design [2].This research aims to create a RTOS-based mobile robot so that there will be better and much easier performances of the program multitasking like distance calculation, path tracking etc.

2. LOW-LEVEL CONTROL

A. IMU thread

Initial thread used is the IMU thread, which performs tasks and process the IMU data by using the MPU-6050 microcontroller. IMU stack memory. IMU thread is used to measure the angular rate, robot body's specific force and sometimes magnetic field around the robot by using the accelerometers and gyros which are inbuild within the IMU [6].

B. Speed control

Speed control thread has the second highest priority in real time. The stack memory allotted for speed control is 512 bytes. It will process the rotary encoder data of the three motors and uses them as input of PID controller. It takes 25ms to get the rotary encoder data. The set point value for all three motors will be given by motion thread [8].

C. Motion thread

Motion thread accesses the IMU thread and also the kinematic of the robot. It takes the third highest priority in real time. The stack memory allocated for motion thread is 512 bytes.

D. Communication thread

It is purely used for communicating with IMU thread to send IMU data and speed control thread to get the motor speed data for monitoring in PC. This takes the fourth priority in real time and the stack memory allocated to communication thread is 128 bytes.

3. DESCRIPTION

The robot mannerisms controlled by the IMU, speed, motion andcommunication threads are operated using the arm cortex M4STM32F407VGT6 microcontroller. The instructions to the microcontroller can be given by using the mobile phone by interfacing each other with Bluetooth technology [11].This robot will have three wheels which rotate with the help of motors within it. Each wheel will have ultrasonic sensors which is solely used for obstacle detection.

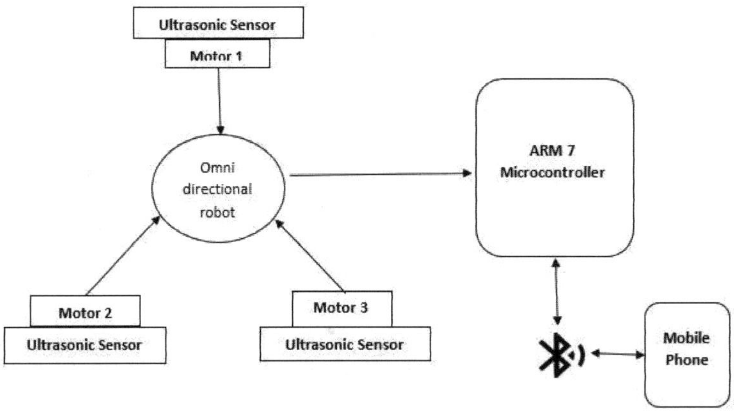

Fig 1 Block diagram of omni directional robot using ARM 7

4. HARDWARE DESCRIPTION

- ARM CORTEX M4
- MPU 6050
- Ultrasonic sensors
- Bluetooth
- Motors and Wheels
- Gears and Battery

A. ARM controller

The controller in this project is from ARM family and indetail the chip used is LPC2148. The program designed using the Keil software is dumped into ARM controller LPC 2148. The entire control of the omni directional robot is done by using this controller. It has 8 to 40 Kb of on-chip static RAMand 32 to 512 Kb of memory for flash program. It has power saving mode which includes Idle and power-down. Processor is moved from power saving by using external interrupts,USB, Brown-Out Detect and Real-Time Clock. CPU operating voltage is in the range of 3.0V to 3.6V with 5V tolerance

B. MPU 6050

The MPU 6050 is used to calculate angular rate. It has an accelerometer and gyroscope. The MPU-6050 devices combine a 3 wheeled gyroscope and a 3 wheeled accelerometer on the same silicon die which processes complex 6-axis Motion fusion algorithms. MPU-6050 allows the devices to gather a full set of sensor data without intervention from the system processor.

C. Ultrasonic sensors

It has three ultrasonic sensors, each having four pins (Vcc, TRIG, ECHO, GND). An Ultrasonic sensor is a device that can measure the distance to an object by using sound waves. It measures distance by sending the sound waves at fixed frequency and waiting for that sound waves to reflect back. By recording the elapsed time between the sound wave being generated and the sound wave bouncing back, it is possible to calculate the distance between the sonar sensor and the object. We send the sound waves using the TRIG pin and observed the distance where it turns out to be echoed. The ECHO pin is used to calculate the distance from which the waves are being reflected.The formula used to calculate the distance can be inferred as, Distance = (pulse start time + pulse end time) * 17150 .The same is repeated for all other sensors in all directions

D. Bluetooth

In our project , Bluetooth is used to establish the connection between the microcontroller and the mobile phone and control the robot or send instructions to the robot. An application called Arduino Bluetooth is used to send the control to the robot using the Bluetooth connection

E. Motors and Wheels

In this project, four motors are used out which three are 60rpm DC geared motor and one is the 3.5RPM DC geared motor. The three 60rpm motors are fixed to the three wheels which will used to rotate and move in different directions, these three motors are uniform in direction which means that they all move in the same direction at a time.The other 3.5rpm motor will be placed at the centre of the robot which proved the mechanical strength for the dinion gears to rotate and transmit energy from battery to the 60rpm DC geared motors.Three 3*4 motor wheels are used for the movement of the robot fixed with the 60rpm motors each.

F. Gears and Battery

The power supply to the robot will be provided by the 12V1.3amp battery. Seven dinion gears will be mounted on the top of the robot where the middle most gear is connected to3.5RPM motor. These gears will be connected in a triangular shape in coordination with each other and finally will be connected to wheels via 60rpm motors. The purpose of these gears is to transfer power from the battery to the wheels and this is done by using the motors as intermediate.

5. SOFTWARE DESCRIPTION

- Keil
- Flash magic

IOP Conf. Series: Materials Science and Engineering **590** (2019) 012054 doi:10.1088/1757-899X/590/1/012054

- OrCAD

A. Kiel

Kiel is used for developing IDE (Integrated development environment) in this project. The IDE provides the development of software to be written to the microcontroller. A single program can be used for all types of functionalities. Modifying, compiling, debugging etc., is done at a time using Kiel software.

B. Flash Magic

Flash magic is used to dump the code created by the Kiel into the microcontroller chip.

C. OrCAD

Orcad is used design a PCB board. The OrCAD PCB design supports a full suite of photo tooling and bare-board fabrication, test outputs can be generated and bare-board test in a variety of formats can be done.

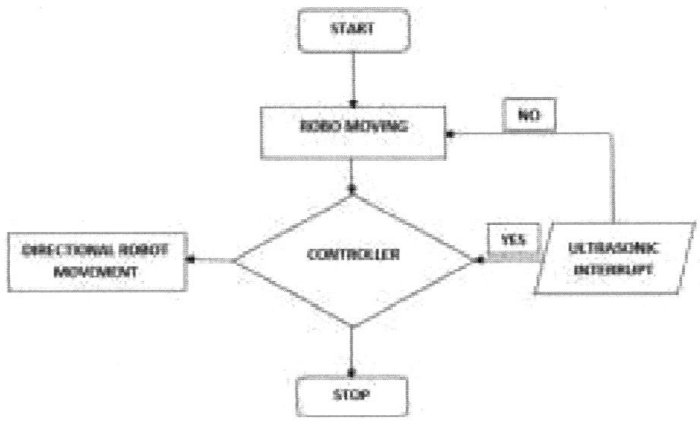

Fig 2 Flow Chart of omni Directional Robot

6. RESULT

In this project we finally made the navigating robot which automatically diverts and takes the shortest path to its destination by detecting the obstacles.

Fig 3 Hardware connection

Here we observe the shortest distances calculated by the ultrasonic sensor in all the directions. The code for short distance is compiled in python code. The output of the sensor can be observed from fig 6.2

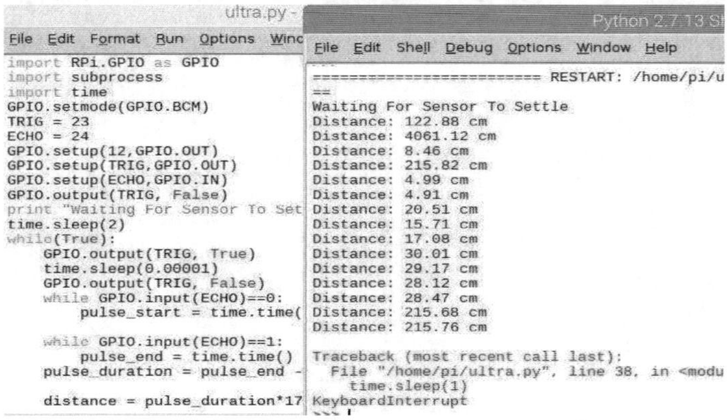

Fig 4 Simulation Output

REFERENCES

[1] Architecture Design of Low Level Control Omni Directional Robot with RTOS-RTX Arm Cortex-M4‖, Muhammad Ismi Musyafani, Fernando Ardilla, Mohammad Mobed Bachtiar Computer Engineering ElectronicEngineering Polytechnic Institute Of Surabay, Indonesia.

[2] Akhilesh Murikipudi, V. Prakash, T. Vigneswaran. ‒PerformanceAnalysis of Real Time Operating System with General Purpose Operating System for Mobile Robotic System‖. School of Electronics Engineering, VIT University, Chennai–600127, India.

[3] Wong Chia Yeak, Muhammad Fauzi Zakaria, Hazwaj Mhd Poad. ‒Real Time Operating System for Mobile Robot Using PICos18‖. Department of Mechatronics and Robotics Engineering, Departement of Computer Engineering, Universiti Tun Hussein Onn Malaysia.

[4] Paratama, Derry. "Sistem Penyeimbang Beban Asimetri Pada *Balancing Robot* Dengan Kendali *PID"*. Program Studi Teknik Komputer, Politeknik Elektronika Negeri Surabaya, Surabaya. 2015.

[5] Mayan, J.A., Priya, K.L.," Novel approach to reuse unused test cases in a GUI based application", IEEE International Conference on Circuit, Power and Computing Technologies, ICCPCT 2015,pp.1-7

[6] R.Ramya and T.Ravi, "Design of cache memory mapping techniques for low power processor" ARPN Journal of Engineering and Applied Sciences, VOL. 10, NO. 11, JUNE 2015, pp:4783-4788.

[7] Huang, Irene. ‒Introduction to Keil IDE and RTX‖. Electrical and Computer Engineering, University of Waterloo. https://ece.uwaterloo.ca /~yqhuang/labs/ece254/document.html. 2012.

[8] Ravi.T, Kannan.V "Modeling and performance analysis of ballistic carbon nanotube field effect transistor (CNTFET)" International Conference on Recent Advances in Space Technology Services & Climate change – RSTS&CC-2010, November – 2010, pp. 327-331.

[9] EE8205: Embedded Computer System, ‒Multitasking ARM- Applications with uVision and RTX‖. Electrical and Computer Engineering, Ryerson University. https://www.sase.com.ar/2012/files/2012/09/RLarmSteps.pdf

[10] Rhomadhon Syuhuri Akbar, Muhammad. ‒Sistem Navigasi Robot OmniDirectional 3 Roda Menggunakan Umpan Balik sensor Rotary Encoder dan Gyro‖. Program Studi Teknik Elektronika, Politeknik Elektronika Negeri Surabaya, Surabaya. 2013.

[11] Samuel Lazar.M and Ravi.T, "Li-Fi design for high speed data transmission" ARPN Journal of Engineering and Applied Sciences, ARPN Journal of Engineering and Applied Sciences, Vol. 10, No. 14, August 2015, pp: 5999-6003.

IOP Conf. Series: Materials Science and Engineering **590** (2019) 012055 doi:10.1088/1757-899X/590/1/012055

Implementation of Internet Connected Industrial Robot Using Hog Algorithm based Object Recognition

AnuPriya George[1]
Research Scholar, ECE Department,
NICHE, TN, India.
anuabrain@gmail.com

Dr.X.Felix Joseph[2]
Associate Professor,
Bule Hora University.

Abstract -The main focus is on object searching and recognition which is a popular area which increases tremendously in industries and research labs. The proposed work is based searching of an object which is done with the help of computer vision techniques. But in real time object searching will have many difficulties when the recognizing of an object is been done. To overcome this problem there is an improved technique called HOG Algorithm is been used in the real time situations. As the result the performance of the object searchingin the present experimental scenario is evaluated through the integrated hardware, which is synchronized with a camera and embedded system based wheeled robot.

Keywords:Cortex-m4, camera, Wi-Fi, PC, Object recognition, Histogram of oriented gradients algorithm, Local binary pattern features, Support vector machine classifier.

1. INTRODUCTION

The object management in the industrial environment is very important because the environment present in the industry is very tidy. The works that fully depend on human interactions like searching, packing, carrying, dispatching etc. Repetitive type based activities are important for object management to do tedious work. Avoiding human interaction between these sort of works will reduce more errors in the management[28].

This paper describes the Traffic light mapping,localization, and state detection for autonomous vehicles [1]. The paper involves the Traffic light recognition using image processing compared to learning processes [2]. This paper says the Real time visual traffic lights recognition based on spot light detection and adaptive traffic lights templates [3]. This paper illustrates the Robust and Real-Time Traffic Lights Recognition in Complex Urban Environments [4]. This paper determines the Traffic Light Mapping and Detection [5]. This paper explains the Traffic light detection with colour and edgeinformation [6]. The Recognition and Tracking of Traffic Lights Based on Colour Segmentation and CAMSHIFT for Intelligent Vehicles is done in [7]. This paper involves the Traffic Lights Detection in Adverse Conditions using Colour [8]. The paper says theSuspended TrafficLights Detection and Distance Estimation Using Colour Features [9]. Detection of Traffic Lights for Vision Based Car Navigation System is discussed in [10]. Traffic light detection during day and night conditions by a camera is described in [11]. Bayesianfiltering for location estimation is done in [12].

2 IMAGE PROCESSING USING MATLAB SOFTWARE:

2.1 Object searching system:

The automatic object searching and recognition has been done in the proposed system. To reach this goal the implementation of computer vision techniques is been done. By using the improved version of HOG algorithm, the real time object recognition has been implemented. For a real time robust object recognition scale invariant features should be considered. Integrating local feature matching techniques to HOG algorithm makes more efficient. The local feature approach can easily incorporate novel feature types because extra features contribute to robustness when they provide correct matches, but otherwise do little harm other than their cost of computation. Therefore, future systems are likely to combine many feature types.

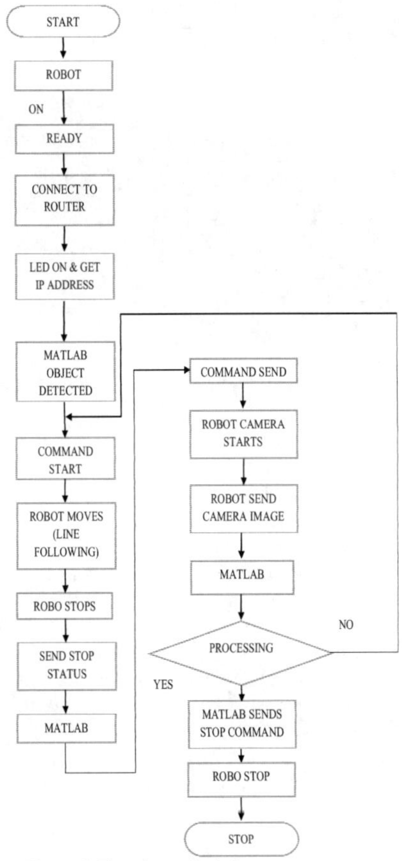

Figure 1: Flowchart of existing system

A MATLAB image processing with all the robot and camera controls will be developed. A database of all the images of the objects are created and preserved. This database is interfaced with MATLAB. The images of objects were assigned with different object ID nos. These images can be accessed using that ID nos. when the system operator wants to get one object, then he inputs the corresponding object ID no. Then the MATLAB process matches the product ID with its image in database and it sends a start command to the robot unit to gather a new image which is moving on a pre-planned path is shown in figure1[15].

The MATLAB process running on the PC collects each image and compares with the image extracted from the database using Real Time HOG algorithm,Until a match is found. If there is any match, the MATLAB process sends a stop command to the robot, which will end the surveillance process. If not, it issues a new start command to the robot, extending the surveillance.

IOP Conf. Series: Materials Science and Engineering **590** (2019) 012055 doi:10.1088/1757-899X/590/1/012055

3. HARDWARE INTEGRATION

3.1. Line following sensor:
3.1.1. IR Transmitter and Receiver:

Figure: 2 LM 358 based IR sensor for line following

A photodiode is a type of photo detector capable of converting light into either current or voltage, depending upon the mode of operationis shown in figure2. Many diodes designed for use specifically as a photodiode will also use a PIN junction rather than the typical PN junction. Most photodiodes is similar to a light emitting diode[14]. They will have two leads, or wires, coming from the bottom. The shorter end of the two is the cathode, while the longer end is the anode. See below for a schematic drawing of the anode and cathode side. Current will pass from the anode to the cathode, basically following the arrow. The best frequency for the job is between 30 and 60 kHz, the most used is around 36kHz.So, remote controls use the 36 kHz (or around) to transmit information. Infra Red light emitted by IR Diodes is pulsated at 36 thousand times per second, when transmitting logic level "1" and silence for "0". To generate a 36 kHz pulsating infrared is quite easy, more difficult is to receive and identify this frequency. It has an output pin that goes high (+5V) when there is a pulsating 36kHz infrared in front of it, and zero volts when there is not this radiation[16].

A photodiode is a PN junction or PIN structure. When a photon of sufficient energy strikes the diode, it excites an electron thereby creating a mobile electron and a positively charged electron hole. If the absorption occurs in the junction's depletion region, or one diffusion length away from it, these carriers are swept from the junction by the built-in field of the depletion region. Thus holes move toward the anode, and electrons toward the cathode, and a photocurrent is produced[17],[18],[]19].

The IR light that was emitted from another led of the exact same type! This is an electrical property of Light Emitting Diodes (LEDs) which is the fact that a led produces a voltage difference across its leads when it is subjected to light. As if it was a photo-cell, but with much lower output current. In other words, the voltage generated by the led's can't be - in any way - used to generate electrical power from light, It can barely be detected[20].

As the name implies, the sensor is always ON, meaning that the IR led is constantly emitting light. This design of the circuit is suitable for counting objects, or counting revolutions of a rotating object, that may be of the order of 15,000 rpm or much more. However this design is more power consuming and is not optimized for high ranges[21]. In this design, range can be from 1 to 10 cm, depending on the ambient light conditions.

The sender is composed of an IR LED (D2) in series with a 470 Ohm resistor, yielding a forward current of 7.5mA[23].

The receiver part is more complicated, the 2 resistors R5 and R6 form a voltage divider which provides 2.5V at the anode of the IR LED (here, this led will be used as a sensor). When IR light falls on the LED (D1), the voltage drop increases, the cathode's voltage of D1 may go as low as 1.4V or more, depending on the light intensity. The output will be High when IR light is detected, which is the purpose of the receiver.

IOP Conf. Series: Materials Science and Engineering **590** (2019) 012055 doi:10.1088/1757-899X/590/1/012055

Using infrared signals, controller is able to localize the position of devices. With the relevant upgrade, microcontrolleralso allows two-way data exchange between two devices. The transmitter and the receiver face each other. The transmitter and receiver have to be positioned so that they are opposite of and within sight of one another. There is a window for the optical signals at the front (IR receiver/transmitter, status LEDs).

The transmitter should be mounted in a protected environment. The receiver is to be mounted in some place. Transmitter mounted at any place. Transmitter and receiver must be parallel in their axes and point towards one another.

3.2 Four Wheel Robot Setup:

3.2.1. L293D Motor driver:

The typical motor driver used to drive the motor in any directions is L293D. The simultaneous rotation of two DC motors is controlled by a 16pin IC called L293D. Here inside the L293D a Dual H-bridge is used to control two DC motors[27].

The microprocessor or controller is used to transmit the relative signals to motor through the L293D. There will be two voltage pins used; one of these is used to draw current for the working of the L293D and the second one is the applied voltage to the motor. The supply voltage does not change due to this.

The motor driver is used as a type of current amplifier; the function of the motor driver is to operate in higher-current signal which drives a motor with low –current control signal.

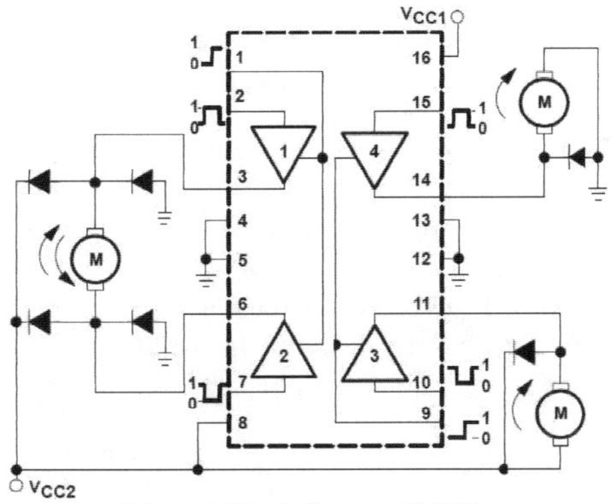

Figure 3: Circuit diagram of L293D

The L293D is an integrated monolithic circuit in a 15-lead Multiwatt and PowerSO20 packages. It is a high voltage, high current dual full-bridge driver designed to accept standard TTL logic levels and drive inductive loads such as relays, solenoids, DC and stepping motors.

Drivers are not used only for motors. They are used for any device that usually draws more than 50-100 mA. Maximum current of microcontroller output (typically 10-20mA) is not enough to drive motor coil. Connecting motor directly to microcontroller will damage microcontroller output transistor.

An H bridge is an electronic circuit that enables a voltage to be applied across a load in opposite direction. These circuits are often used in robotics and other applications to allow DC motors to run forwards or backwards.

3.2.2 Operation of four wheel Robot:

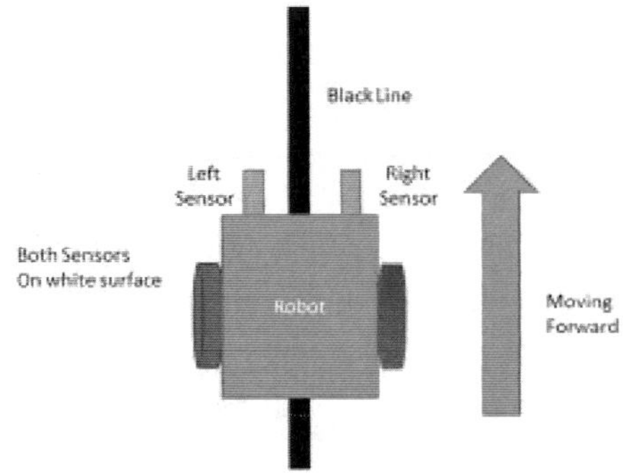

Figure: 4 Four wheel robot

The above shown figure shows the four wheel set up of the Robot. Here the robot is been operated with the help of L293D motor driver as the circuit of motor driver works with the H-Bridge circuit and pulse generated from the controller.

4. RESULT

Object classification is an important task in many computer vision applications, including surveillance, automotive safety, and image retrieval. For example, in an automotive safety application, you may need to classify nearby objects as pedestrians or vehicles. Regardless of the type of object being classified, the basic procedure for creating an object classifier is:

- Acquire a labeled data set with images of the desired object.
- Partition the data set into a training set and a test set.
- Train the classifier using features extracted from the training set.
- Test the classifier using features extracted from the test set.

Here OV2640 is used to capture the image and after the image is captured image will be send to STM32f 476zi an embedded 32 bit ARM microcontroller and that image will be spitted into several parts and it will be send to memory card from the memory the image will be send to CC 3200 through UART serial communication Via Wi-Fi and LAN the image will be send and received to process it in MAT Lab image processing. By this way the object recognition is done and robot will move according to the command send from the MAT LAB Image processing.

Figure: 5 Output image of vision based object detection

When the MAT Lab sends the START command the device searches the object until the object is been identified after the object is identified MAT Lab sends the stop command. Then, again the procedure will be repeated until the object searching is completed.

IOP Conf. Series: Materials Science and Engineering **590** (2019) 012055 doi:10.1088/1757-899X/590/1/012055

5. CONCLUSION

In the proposed system the image based object recognising using MAT Lab and IOT based image transfer and receiving techniques is been done. Here two types of 32 bit arm controllers are effectively used. First in one controller the camera module is been connected with the help of camera, image is been captured and send to the STM32F426zi and through the UART to CC3200 WI-FI based controller through which the image will send to the MAT Lab to do the image comparison with object database which is already stored in our MAT Lab. This image capture and transfer mechanism is done in a real time environment with the help of a four wheel robot and the robot will be moved through black line following which is done by IR sensor. By this way the proposed system is been effectively utilised.

REFERENCE

[1] J. Levinson, J. Askeland, J. Dolson and S. Thrun, Traffic light mapping,localization, and state detection for autonomous vehicles, In proceedingsof ICRA 2011.
[2] R de Charette, F Nashashibi. Traffic light recognition using imageprocessing compared to learning processes. In proceedings of IROS 2009,pp 333-338.
[3] R de Charette, F Nashashibi. Real time visual traffic lights recognitionbased on spot light detection and adaptive traffic lights templates. Inproceedings of IV 2009, pp 358-363.
[4] C. Wang, T. Jin, M. Yang, B. Wang. Robust and Real-Time TrafficLights Recognition in Complex Urban Environments. International Journalof Computational Intelligence Systems, Vol. 4, No. 6, Dec 2011, pp1383-1390.
[5] N. Fairfield, C. Urmson. Traffic Light Mapping and Detection. Inproceedings of ICRA 2011, pp 5421 – 5426.
[6] M. Omachi, S. Omachi. Traffic light detection with color and edgeinformation. In proceedings of ICCSIT 2009. pp 284 - 287
[7] J. Gong, Y. Jiang, G. Xiong,C. Guan, G. Tao and Huiyan Chen, TheRecognition and Tracking of Traffic Lights Based on Color Segmentationand CAMSHIFT for Intelligent Vehicles. In proceedings of IV, 2010.
[8] G.Siogkas, E.Skodras and E.Dermatas. Traffic Lights Detection inAdverse Conditions using Color, Symmetry and Spatiotemporal Information.In proceedings of VISAPP 2012.
[9] M. Diaz-Cabrera, P. Cerri and J. Sanchez-Medina, Suspended TrafficLights Detection and Distance Estimation Using Colour Features, Inproceedings of ITS'2012, pp. 1315-1320.
[10] T. Hyun,J. In-Hak,C Seong-lk. Detection of Traffic Lights for Vision BasedCar Navigation System, In proceedings of PSIVT 2006.
[11] C. Yu, C. Huang, Y. Lang. Traffic light detection during day and nightconditions by a camera. In proceedings of ICSP 2010, pp. 821-824. 10thInternational Conference on Signal Processing.
[12] D. Fox, J. Hightower, L. Liao, D. Schulz, and G. Borriello, "Bayesianfiltering for location estimation," IEEE Pervasive Computing, vol. 2,pp. 24–33, 2003.
[13] G. Bradski and A. Kaehler, Learning OpenCV, 1st ed. O'ReillyMedia, Inc., 9 2008, p. 112.
[14] J. F. S. Hill and S. M. Kelley, Computer Graphics Using OpenGL (3rdEdition). Upper Saddle River, NJ, USA: Prentice-Hall, Inc., 2006
[15]G. Karudaiyar, S.Karthikeyan and B. Sainath, "Encryption and Decryption Scheme by Using Finite State Machine", BIOSCIENCES BIOTECHNOLOGY RESEARCH ASIA, December 2014 Vol. 11(3), 1867-1872.
[16]S.Karthikeyan, S.Jayashri, "Energy Utilization Strategies using Novel Approach in Wireless Sensor Networks", International Conference on Software Engineering and Mobile Application Modelling and Development, Dec 2012, pp: 1-5.
[17]S.Karthikeyan, S.Jayashri, "Energy Efficient System for Heterogeneous Wireless sensor Networks", European Journal of Scientific Research, Volume 72,No 4 march 2012, PP:599-607. ISSN: 1450-216X.
[18]S.Karthikeyan, S.Shanmugapriya and S.Jayashri, "Sub-head Transmission of Heterogeneous Data by Cloned Agent to Android Mobile",Research Journal of Applied Sciences, Engineering and Technology, 8(1):pp.24-34, 2014
[19] S.Pavithra, S.Karthikeyan, V.J.K.KishorSonti, S.Jayashri, "Competent Realization of Co-operative spectrum sensing in cognitive radio systems", International journal of Engineering systems modeling and simulation, Inder science publishers, Vol. 7, No. 2,pp 103-110 2015.(FREE JOURNAL)
[20]K. Bharathi,S.Karthikeyan, "A Novel Implementation of Image Segmentation for Extracting Abnormal Images in Medical Image Applications", Indian Journal of Science and Technology, Vol. 8(S8), 333–340, April 2015.

[21]Meenakshi. R, Pooja Suppaiah , Siva kumar .V.G, Karthikeyan.S,"COST EFFECTIVE REAL TIME HOME ENERGY MONITORING AND MANAGEMENT SYSTEM" Journal of Advanced Research in Dynamical and Control Systems, 2017, vol. No:17 , PP: 1103-1111

[22]Aashik JPM , Ajith M ,Mr.V.G.Sivakumar,Mr.S. Karthickeyan, " AN INTERNET OF THINGS APPROACH FOR MOTION DETECTION AND CONTROLLING HOME APPLIANCES USING RASPBERRYPI WITHOUT CLOUD SERVER",Journal of Advanced Research in Dynamical and Control Systems, 2017, vol. No:17 , PP: 1112-1123

[23]T.V.N.Surendra Gopi, Illuru Phani Kumar, Dr. S.Karthikeyan , Dr. V.G.Sivakumar "A Novel Method Of Dynamic Mobile Data Collection Scheme By Fuzzy Logic", Journal of Advanced Research in Dynamical and Control Systems, 2017, vol. No:17, PP: 1136-1149

[24]Kumar, M.K., Surya, J., Kumar, V.G.S., Karthikeyan, S."Online prediction of driver distraction based on brain waves activity patterns", International Journal of Pure and Applied Mathematics, Volume 117 No. 21 2017, PP: 139-146

[25]Larisha Rayen, B., Monica, B., Karthikeyan, S., Sivakumar, V.G.," Multi-band circularly polarized antenna for 2.4/5.3/5.8 ghz WLAN and 3.5 ghz WI-MAX applications", Journal of Advanced Research in Dynamical and Control Systems, 2017,vol. No:16, PP: 1311-1322

[26]Karthikeyan, S., Venkata Nagi Reddy, M., Mazeed Ahamed, M., Sivakumar, V.G., "A novel approach to cost-aware energy based routing protocol for wireless sensor networks", International Journal of Pure and Applied Mathematics, Volume 117 No. 21 2017, PP: 107-112

[27]Manikandan, R., Karthikeyan, S., Balaji, S.R., Vasudevan, V., Shalini, S. " An internal startup circuit for pacemakers using body temperature", International Journal of Pure and Applied Mathematics, Volume 117 No. 21, december 2017, PP:155-162.

PERFORMANCE ENHANCEMENT OF RECONFIGURABLE MICRO STRIP PATCH ANTENNAE USING METAMATERIALS

G.Jegan 1 [1]
Assistant professor, Dept of ECE,
Sathyabama Institute of Science and Technology
OMR Road,Chennai, India
jjsfss@gmail.com

M.Florence Silvia [2]
Assistant professor, Dept of ECE,
Jeppiaar SRR Engineering College
OMR Road,Chennai, India
silviajegan@gmail.com

A.Vimala Juliet [3]
Professor & Head, Dept of EIE,
SRM University,
Kattatankullathur,Chennai, India

Abstract— Micro strip reception apparatuses are broadly utilized in wireless applications as a result of its simple design, compactness and economical. These small scale strip fix receiving antennas are made utilizing materials like FR4, Rogers RT duroid. These common materials are being supplanted with a helpful material named as Meta materials. These materials are unnatural materials having non-indigenous properties which are controlled with basic properties for the acquirement of updated properties. This whole procedure work with heterogeneous kinds of reconfigurable small scale strip fix receiving antenna utilizing Meta materials. Thus the exhibitions of smaller scale strip reception apparatuses are superior to whatever other conventional radio wires which likewise fill in as an inconceivable use for receiving antenna geeks and architects. The dynamic tuning of such receiving antenna parameters is commonly accomplished by changing the switch conditions. A solitary model can be utilized to help various capacities at different recurrence groups. Objective is to plan a reconfigurable micro strip patch antenna which works at four unique frequencies in X-band and Ku band. The reproduced reception apparatus emanates at five unique frequencies of 5.48 GHz, 9.88 GHz, 11.32 GHz, 12.68 GHz and 15.08 GHz and has an arrival loss of - 24.6103 dB, - 15.095dB, - 24.6103 dB, -30.0870 and - 30.8473 dB individually. And afterward it is manufactured utilizing FR4 as substrate and afterward it is observed to be in great match with the reproduced comes about.

***Keywords**:* Antennae, Micro strip patch antenna, DGS, Meta Material, HFSS.

1. INTRODUCTION

In the world moving towards being wireless, antennas play a vital role in converting an electrical quantity into electromagnetic wave that radiate and propagate in space from a transmitting antenna, and at the same time on the receiving end some of the transmitted power that the receiving antenna intercepts is converted into an electric potential that is proportional to the transmission. Such a requirement of energy conversion is done by the antenna that has improved a great deal over the years at the end of the transmitter and at the receiver or in a transceiver.

* *Author for Correspondence*
E-mail address: jjsfss@gmail.com

IOP Conf. Series: Materials Science and Engineering **590** (2019) 012056 doi:10.1088/1757-899X/590/1/012056

The microstrip patch antenna is a type of narrow band radio antenna which is designed on the microstrip line basis. The increased usage of microstrip antenna in various industries as it can be printed directly onto a circuit board. Due to its planar configuration and ease of integration with microstrip technology, the microstrip antenna proves to be the compact structure and can be widely used in modern wireless transmitters and receivers.

2. ANTENNA DESIGN

The proposed design is etched on standard FR4 substrate that has relative permittivity 4.4 and loss tangent 0.02. The thickness of the substrate is taken as 1.6 mm and metamaterial structure is used on the ground plane in order to increase performance characteristics of the proposed antenna [9]. In this design defected ground structure is used in the bottom layer whose length is equal to the length of the feed line which is used in the top layer with the structurally modified radiating patch layer as shown in the figure 1.1. In the ground layer, Split Ring Resonator as a metamaterial is used along with DGS [1]. Inverted T shaped slots help to achieve a higher level of frequency with reduced dimension. The addition of capacitive loading patch segments provides impedance matching.

The inclusion of the inverted T slot and small rectangular capacitance slots modifies the structure of the smooth radiating patch layer causes an impedance mismatch. This can be resolved by the attachment of Split Ring Resonator on the top of the ground structure and keeping the remaining portion as it is according to the calculation [2]. Therefore the narrow band output with appreciable gain and directivity is obtained in the designed resonance frequency as well as additional resonance frequencies are also produced by the parasitic capacitance effects due to structural modifications at the top and ground layers [3].

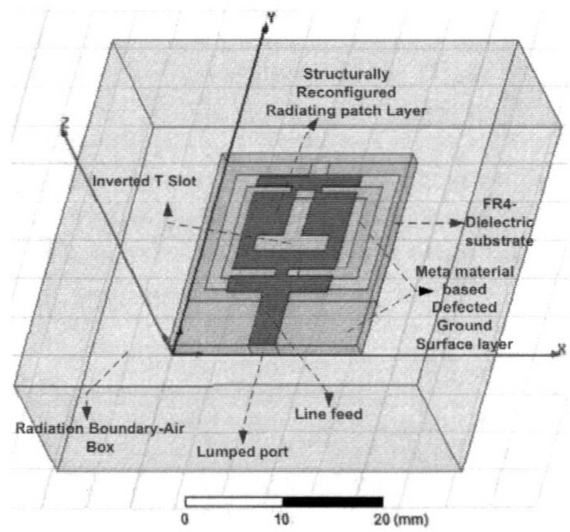

Figure 1.1:- HFSS 3D design structure of proposed antenna

3. ANTENNA DESIGN STRUCTURE AND WORKING PRINCIPLE

The antenna structure designed and considered in this case is a slotted structure with a detective ground and the effect if the dipole structure and the defective ground are studied to observe its property. From the figure 1.2 the ground plane has the split ring resonator that exhibits isolated sections of the ground coupled by the inductive nature of the substrate. This approach makes it possible to have a resonant circuit that is inductively coupled through the radiating patch [4].The radiating plane is made up of a patch of 15.21*11.3 mm and has an inverted T shaped slot etched into it. The 3 mm feeder line is connected to it and has the ground plane unaltered below the feeder line. The rest of the ground structure is the slotted split ring that completes the base. T patch is built using FR-4 Epoxy board that is double sided with the radiating patch etched on the top surface and the slotted ground below.

IOP Conf. Series: Materials Science and Engineering **590** (2019) 012056 doi:10.1088/1757-899X/590/1/012056

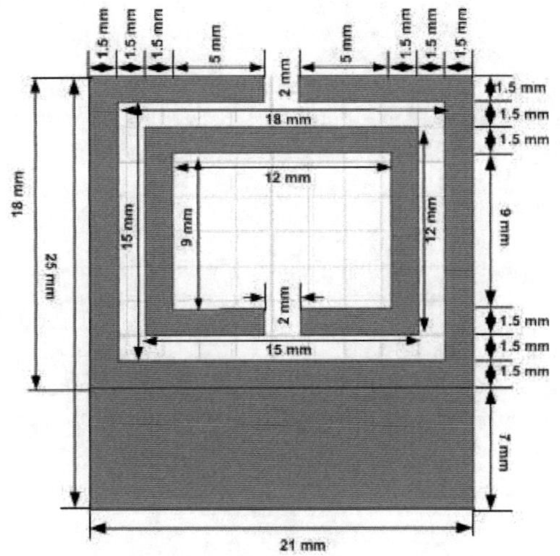

Figure 1.2:- Defected Ground layer (Meta material based) of proposed antenna

4. RESULTS AND DISCUSSIONS

In order to analyse the performance of an antenna, the important aspects that always be considered in the design antenna such as VSWR, Return loss, Gain and directivity of the antenna. Thus, all the desired simulation result parameter of the patch antenna as mention above can be implemented by adjusting the antenna ground plane length; antenna feed length, feed width, substrate material, thickness [10].

Return Loss

The return loss is the figure-of-merit of an antenna. It can be observed as the proportion of radio waves incoming at antenna input that are reprobated as the ratio against the radio waves that are accepted. It is commonly specified in decibels (dB).The Figure 1.3 shows, the return loss graph of the proposed antenna modelled with the operating frequency of 5.5GHz.The observed return loss is -21.3192dB with the centre frequency of 5.48GHz, marked as 'm1'. This indicates that approximately 96.25% of power is transmitted by the antenna and remaining 3.75% of power is reflected. Thus, the bandwidth calculated from the return loss graph at -10dB level is 5.35-5.60GHz, which signifies 250MHz. Similarly, the same antenna has additional peak resonant frequencies at 9.88GHz, marked as 'm2'with the maximum return loss of -15.095dB.

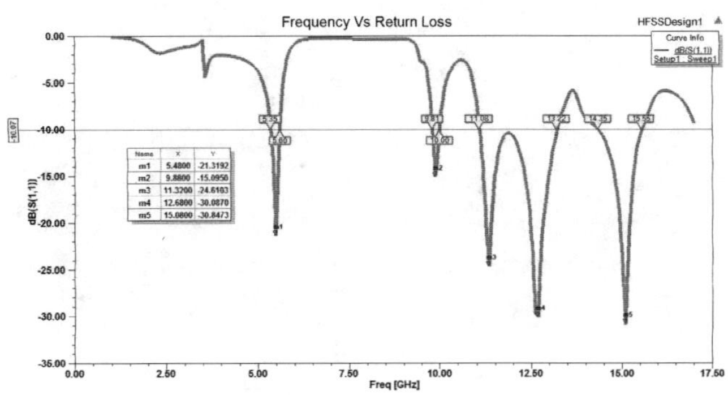

Figure 1.3:- Return loss versus Frequency

IOP Conf. Series: Materials Science and Engineering **590** (2019) 012056 doi:10.1088/1757-899X/590/1/012056

Both 11.32GHz and 12.68GHz, considered as same band between markers 'm3' and 'm4', have below -10dB return loss of -24.6103dB and -30.0870dB. Finally, 15.08GHz is marked as'm5'with the maximum return loss of -30.8473dBrespectively. The -10dB bandwidth calculated for the above bands are190MHz (9.81-10.00GHz), 2.14GHz (11.08-13.22 GHz) and 1.2GHz (14.35-15.55GHz). Thus the proposed structurally modified antenna provides a wide range of bandwidths for covering many wireless applications [11].

VSWR

VSWR is a measure of how an antenna is matched well to the cable impedance. It describes how much power is reflected back or transferred into the cable, with its value acting as a function of reflection coefficient [5]. A perfectly matched antenna will have a VSWR of 1:1.From the figure 1.4, the obtained VSWR values are 1.188, 1.4269, 1.1250, 1.0646 and 1.059 at the corresponding resonant frequencies of 5.48GHz, 9.88GHz, 11.32GHz, 12.68GHz and 15.08GHz respectively. Thus the antenna impedance is well matched with cable impedance and it is qualified of achieving effective transmission and reception, accomplishing 96.25% efficiency at above frequencies.

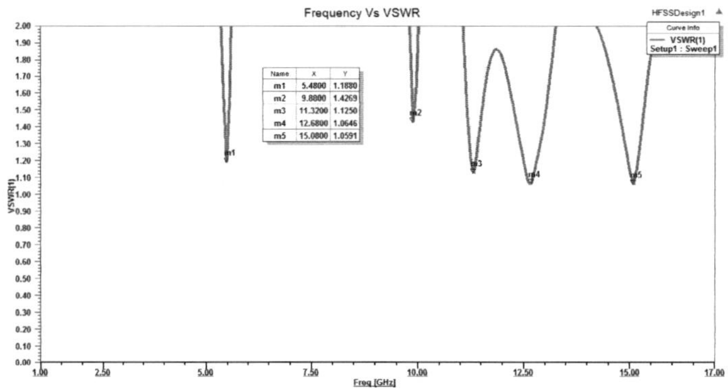

Figure 1.4:- VSWR versus frequency

Gain and Directivity:

The overall efficiency of an antenna can be calculated from total gain and total directivity through the imbibed Figure 1.5. The proposed antenna, yields an outcome of peak recognized total gain and total directivity at 3.0489dB and 3.1674dB respectively. Efficiency calculated as follows

$$\text{Overall Efficiency } \dot{\eta} = \frac{\text{TotalGain}}{TotalDirectivity} \text{x } 100 = \frac{3.0489}{3.1674} \text{ x } 100 = 96.25\% \quad (1)$$

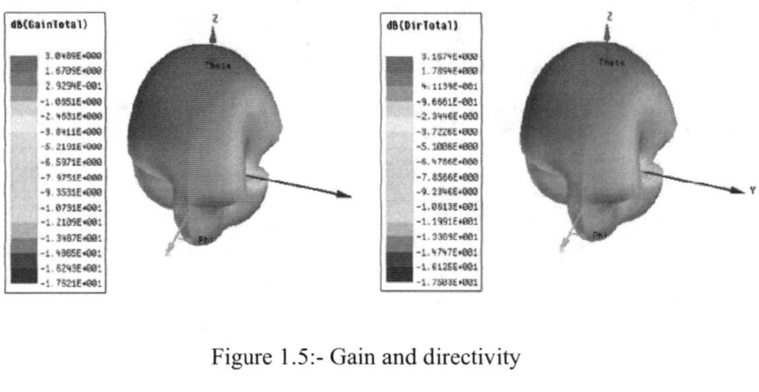

Figure 1.5:- Gain and directivity

5. OPTIMIZATION APPROACH PROCEDURE

Table 1. Optimization approach

Property	Optimization-1	Optimization-2	Optimization-3	Optimization-4
Design	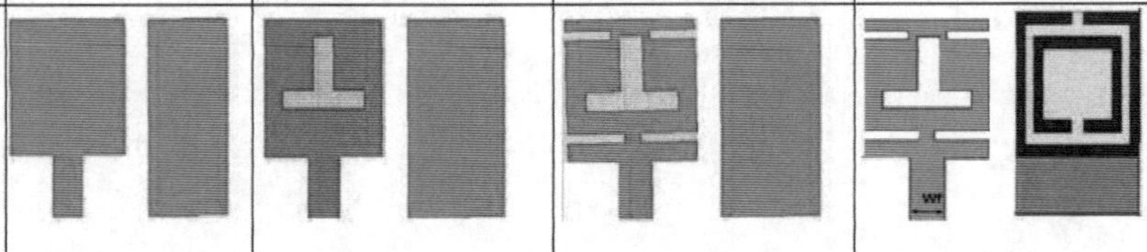 Patch Ground	Patch Ground	Patch Ground	Patch Ground
Return loss s(1,1)	-5.59dB	-11.7dB	-18.7dB	-28.5dB
Gain	2.5dB	2.59dB	3.5dB	3.5dB
Directivity	3.45dB	2.76dB	3.82dB	3.75dB
Efficiency	72.4%	93.76%	90.9%	95.73%

From the optimization approach procedure it is observed that with the use of notch on the radiating layer and metamaterial structure on the ground layer the performance characteristics of the antenna have been increased with good efficiency, gain and minimum return loss[6].

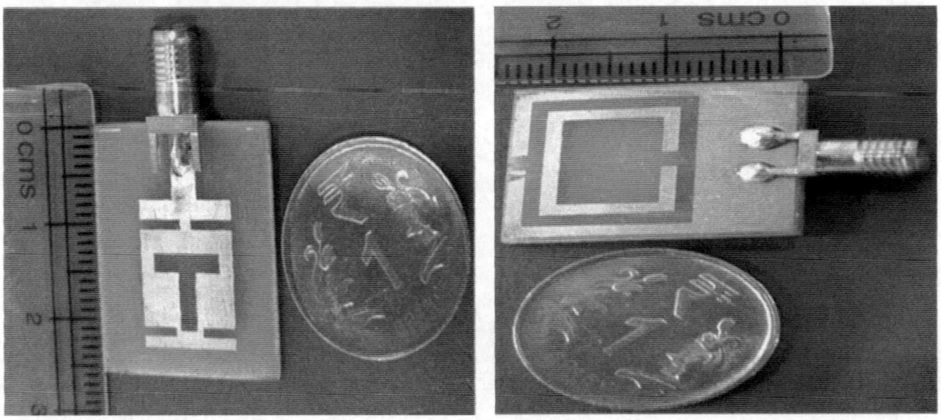

Fig1.6:- Fabricated antenna **a**. Top layer, **b**. Ground plane

From the figure 1.6 the fabricated antenna has two parallel dipoles separated by a patch that has an inverted T shaped slot that is connected to the slotted grounded plane through the substrate. This structure is etched on a double sided fiber board that is porous to act as a material that has its dielectric property un-uniformly loaded to help match the impedance of the antenna [7].

IOP Conf. Series: Materials Science and Engineering **590** (2019) 012056 doi:10.1088/1757-899X/590/1/012056

6. MEASURED RESULT

The proposed fabricated antenna is tested using the network analyzer and the measured result is shown in figure 1.7. For the measured result it is observed that peak resonance is obtained at 5.46 GHz, 10.11 GHz, 11.8 GHz and 14GHz with the return loss value of -35.34 dB,-22.31 dB,-15.17 dB and -10.10 dB.

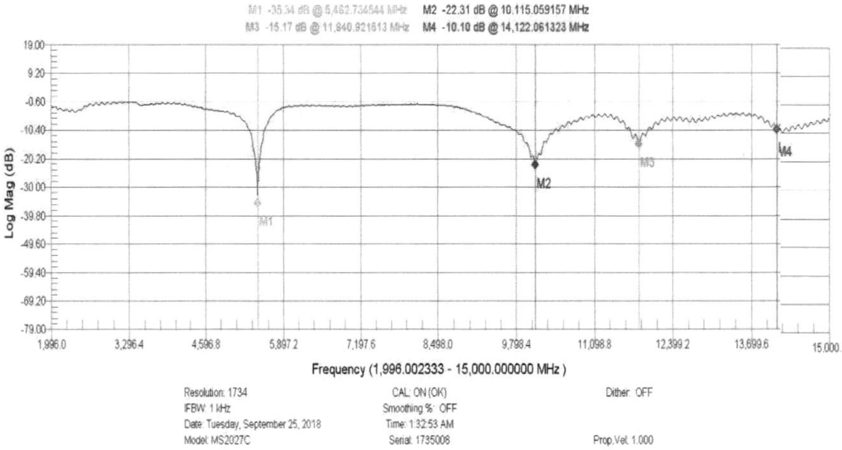

Fig 1.7:- Measured result of fabricated antenna

Table 2:-Return Loss Comparison

Simulated		Measured	
Observed (Center) Resonance frequency (GHz)	Return loss (dB)	Observed (center) Resonance frequency (GHz)	Return loss (dB)
5.48	-24.6103	5.46	-35.34
9.88	-15.0950	10.11	-22.31
11.32	-24.6103	11.8	-15.17
12.68	-30.0870	14.00	-10.10
15.08	-30.8473	-	-

7. CONCLUSION

The Impedance mismatching is resolved by the structural modification carried out in both radiating patch and ground layer. Switching loss totally removed (highly minimized) and very sharp resonance frequency with narrow band is achieved with the design. Thus the proposed structurally modified antenna provides a wide range of bandwidths for covering many wireless applications such as Wi-Fi, Bluetooth, Wi-Max, WLAN and the antenna impedance is well matched with cable impedance and it is qualified of achieving effective transmission and reception, accomplishing 96.25% efficiency at above frequencies.

REFERENCES

[1] Somarith, S., Kang, H. and Lim, S. (2014), "Frequency reconfigurable and miniaturized substrate integrated waveguide inter digital capacitor (SIW-IDC) antenna", IEEE Transactions on Antennas and Propagation, Vol.62, No.3, pp.1039-1045.

IOP Conf. Series: Materials Science and Engineering **590** (2019) 012056 doi:10.1088/1757-899X/590/1/012056

[2] Nugroho, B.S., Zulkifli, F.Y. and Rahardjo, E.T. (2014), "Lossy-Transmission-Line Analysis Of Frequency Reconfigurable Rectangular-Ring Microstrip Antenna", International Journal of Microwave Science and Technology.

[3] Row, J. S. and Tsai, C. W. (2016), "Pattern Reconfigurable Antenna Array With Circular Polarization", IEEE Transactions on Antennas and Propagation, Vol. 64, No.4, pp. 1525-1530.

[4] Huang, C., Pan, W., Ma, X. and Luo, X. (2016), "A Frequency Reconfigurable Directive Antenna with Wideband Low-RCS Property", IEEE Transactions on Antennas and Propagation, Vol.64, No.3, pp.1173-1178.

[5] Baracco, J.M., Ratajczak, P., Brachat, P. and Toso, G. (2015), "A dual frequency Ka-band printed Fresnel reflector for ground terminal applications", IEEE Transactions on Antennas and Propagation, Vol. 63, No. 10, pp.4352-4366.

[6] Jegan G, Vimala Juliet A. and Florence Silvia M. (2016), "Design of Circularly Polarized Frequency Reconfigurable Wearable Microstrip Patch Antenna for Wireless Applications", Research Journal of Pharmaceutical, Biological and Chemical Sciences, Vol. 7(6), ISSN: 0975-8585, RJPBCS Publishing Limited, Indexed in Scopus.

[7] Obeidat, K. A., Raines, B. D., Rojas, R. G. andStrojny, B. T. (2010), "Design of frequency reconfigurable antennas using the theory of network characteristic modes", IEEE Transactions on Antennas and Propagation, Vol.58, No.10, pp. 3106-3113.

[8] G. Jegan, A. Vimala Juliet, and R. Himanshu Singhvi, "A Novel Design Approach Of Reconfigurable Patch Antenna For Wireless Applications", Applied Mechanics and Materials Vols. 336-338 (2013) pp 1935-1938,© (2013)TransTechPublications,Switzerland,doi:10.4028/www.scientific.net/AMM.336-338.1935.

[9] M. Si. Moussa and A. B. Kouki, "Design of a Rectangular Micro strip Patch Antenna at 1 GHz." 978-1-4577-0972-2/11/$26.00 ©2011 IEEE.

[10] Mohit Anand, "Applications of Metamaterial In Antenna Engineering", International Journal of Technical Research and Applications e-ISSN: 2320- 8163, www.ijtra.com Volume 2, Issue 1 (Jan-Feb 2014), PP. 49-52.

[11] G. Jegan, A.Vimala Juliet , M. Florence Silvia and G. B. V. Sai Krishna, "Novel Design and Performance Analysis of Rectangular Frequency Reconfigurable Micro Strip Patch Antenna using Line Feed technique for Wireless Applications", Indian Journal of Science and Technology, Vol 9(29), DOI: 10.17485/ijst/2016/v9i29/98582, August 2016.

Digital Video Steganography Technology For Security Application

E. Anna Devi[1]
Research Scholar, Department of ECE,
Sathyabama Institute of Science and Technology, Chennai-600119.

R. M. Joany[2]
Research Scholar, Department of ECE,
Sathyabama Institute of Science and Technology, Chennai-600119.
mariajoany@gmail.com

S. Yogalakshmi[3]
Research Scholar, Department of ECE,
Sathyabama Institute of Science and Technology, Chennai-600119.

L. Magthelin Therase[4]
Research Scholar, Department of ECE,
Sathyabama Institute of Science and Technology, Chennai-600119.

Abstract - The paper proposes a technique on the digital video steganography. Various techniques are employed in order to achieve a higher secure data transmission these includes the cryptography, watermarking and steganography the steganography finds its application in the audio-video synchronization,copyright control and T.V.broadcasting specially in military areas the paper analyses the various improvements made in the security f the data transmission as well as the various methodologies adapted in achieving the same.the major difference between the steganography and the cryptgraphy is the latter has the cipher text which is coded and then transmitted while the former is embedded within a image or in a video frame so as to achieve the intended use.

Keywords - Encryption, Cryptography, LSB, steganography adaptive lsb, watermarking, AES.

1. INTRODUCTION

With the improvement and with the advancement in technology and events security plays a major role. Where the privacy concern of an individual has increased. Thus it becomes mandatory for an individual to protect his privacy security and this paper proposes a methodology where the images are hidden or embedded with in images or with in the video frame which finds a large application in security fields and military applications as well as the proposed technique makes use of RC7. Encryption which is far more better the conventional method also the image embedded this paper gives an insight about the various advantages as well as the disadvantages that are applicable in the industry, using the MATLAB the application is processed.

Steganography is classified in to two types as 1. Linguistic steganography 2. Technical steganography also watermarking is one such technique which is used in the data transmission which achieves a higher secrecy the watermarking can be classified into 2 types they are 1. Robust 2. fragile the block diagram of the steganography both in the embedding as well as the extraction side is as shown below in the fig.

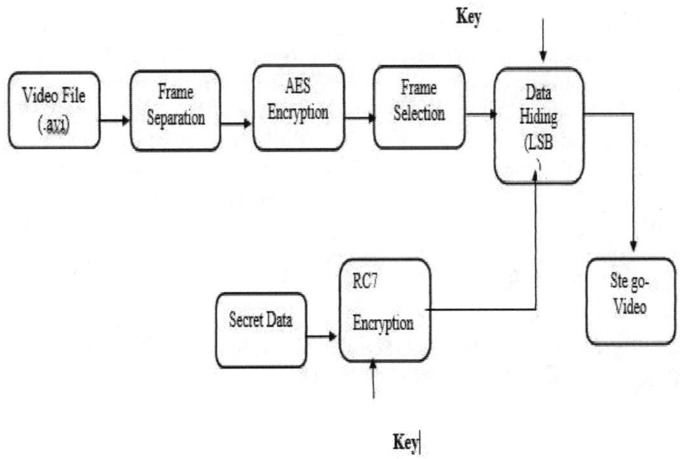

Fig 1. Embedding

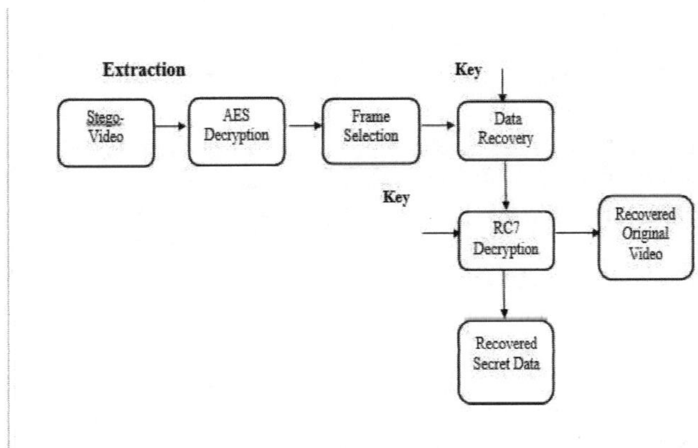

Fig.2.Extraction

2. Theoritical Framework

2.1 DOUBLE CODING MECHANISM FOR ROBUST AUDIO DATA HIDING IN VIDEOS

The author of this paper describes about the double coding mechanism by which two codes are used such that it becomes complex to track down the original information content this is performed by pseudo codes and morse codes the performance of the system is measured in terms of mean square error(MSE) And peak signal noise ratio(PSNR) initially the transformation between the domains are made (i.e) from transform domain to spatial domain. The spatial domain as a disadvantage of more payload the mathematical tools used in transformation of the domain fourier transform discrete cosine transform and the discrete wavelet transform .the conclusion of this paper better methods can be adopted and the loopholes as well the paper as a better MSE and PSNR further more write after the embedding of data the resultant image is compressed and even on compressing of the resultant image the method prof to be effective

2.2 AN IMPROVED IMAGE STEGANOGRAPHY TECHNIQUE BASED ON MSB USING BIT DIFFERENCING

On conventional steganography the data is embedded with in the video or image frame the LSB as know vitel information so that data to be hidden is concealed within the LSB and then it is transmitted this paper overcomes the conventional steganography by using the MSB . the different formats of data that can be processed are image takes and audio the theratical frame work used in this MSB technique the bite 5 shows the secret bit based on the difference in bit 5 and 6 of thr cover image three important parameters in the steganography is discussed in this paper which are 1.payload capacity 2.robustness 3.security where the payload capacity plays an important role in spatial domain the difference between in cryptography steganography is also discussed where as in a cryptography cipher text is used but in a steganography the data is emdedded and hidden to the rest of them other than the intentent recipient the conclusion of this paper is that MSB based data hiding technique is more efficient and provides a better PSNR ratio over the conventional way

2.3 MULTICLASS DETECT OF CURRENT STEGANOGRAPHIC METHODS FOR JPEG FORMAT BASED RE-STEGNOGRAPHY

The goals of JPEG Image steganography detection are not only distinguished cover images (the normal images) and stego images, but also determine the steganography algorithms of the stego images, and extract the embedded information, of which the precondition is to determine the image steganography algorithm. The changes of storage size from re-steganography to image is much smaller than that from steganography to image. Specifically, the storage size of JPEG image, with the first JPhide steganography, is reduced about 50% than before; but with the second JPhide steganography to the image, the change of the storage size was relatively small, almost no change. an effective steganalysis multi-class classification algorithm in this paper, which outperforms the state-of-the-arts in detecting the modern steganographic methods plan to further improve the detection reliability of the proposed steganalysis algorithm.

Firstly, expanding the application range of the algorithm, we want the proposed algorithm outperform in other steganographic methods: MB1, MB2, SSIS and H4PGP. Secondly, we want the algorithm to further simplify and run with more high efficiency.

2.4 THE VIDEO HIDING TECHNIQUE BASED ON DWT AND GENETIC ALGORITHM WITH BOOLEAN FUNCTIONS

The author of the paper makes use of the discrete wavelet transform(DWT) wavelet is the short lived pulse unlike a sinusoidal signal. The wavelets can be classified in to many types based on that amplitude and the shape of the waveform sum of them are by orthogonal 2.2,daubauchies so this paper deals with the DWT as a tool in data hiding technique .DWT is useful in digital globe the activity of wavelet transform is analyised to the attacks the data hiding technique they are by preventing from the hackers or attackers.from this paper the embedding image Is randomized and is moved towards the higher LSB layers the conclusion of this paper is that gives us DWT and the generic algorithm for hiding data effectively and efficiency.

2.5 A NEW WATERMARKING APPROACH FOR REPLAY ATTACK DETECTION IN LQG SYSTEMS

They are major three techniques in data hiding one such technique is watermarking the author of this paper deals with the watermarking is a process of hiding data embedding in a noise tolerant signal such as an audio video or image data the digital information is hidden in a carrier signal both the steganography and digital watermarking employee's steganography techniques to embedded data in a noise signal. The watermarking technique is combined along with the steganography and it enhance the PSNR and provides better solution for the noise attack further. this paper gives an idea about the classification of steganography which is of two types linguistic and technical for the more watermarking is divided into two types robust and fragile various parameter like PSNR and MSE and embedding capacity for evaluated and the results are discussed in the paper

2.6 CONTENT-ADAPTIVE PENTARY STEGANOGRAPHYUSING THE MULTIVARIATE GENERALIZED GAUSSIAN COVER MODEL

The vast majority of steganographic schemes for digital images stored in the raster format limit the amplitude of embedding changes to the smallest possible value. In this paper,the author investigate the possibility to further improve the empirical security by allowing the embedding changes in highly textured areas to have a larger amplitude and thus embedding there a larger payload. The approach is entirely model driven in the sense that the probabilities with which the cover pixels should be changed by a certain amount are derived from the cover model to minimize the power of an optimal statistical test. The embedding consists of two steps. First, the sender estimates the cover model parameters, the pixel variances, when modeling the pixels as a sequence of independent but not identically distributed generalized Gaussian random variables. Then, the embedding change in the probabilities for change in each pixel by 1 or 2, can be transformed to costs for practical embedding using syndrome-trellis codes, and they are computed by solving a pair of non-linear algebraic equations

3. CONCLUSION AND FUTURE WORK

Hence this paper analyses the various complexities and the problems that are associated with the Digital Video Steganography technology for security applications along with the remedies that are solved in the consecutive papers the future work of this can be extended to design an hardware that is capable of making video steganography which is intended for the use by a general consumer.

REFERENCE

[1] Shaikh Shakeela, P. Arulmozhivarman, Rohit Chudiwal, Samadrita Pal Double Coding Mechanism for Robust Audio Data Hiding in Videos Recent Trends in Electronics, Information & Communication Technology (RTEICT), IEEE International Conference on 20-21 May 2016.

[2] Ammad Ul Islam, Faiza Khalid ,Mohsin Shah,An improved image steganography technique based on MSB using bit differencing Innovative Computing Technology (INTECH), 2016 Sixth International Conference on 24-26 Aug. 2016.

[3] G. Sowmya; K. Meghana; B. Subbarayudu; R. Ch. A. Naidu; G. KeerthiWangqiang Niu The video hiding technique based on DWT and genetic algorithm with boolean functions 2016 2nd International Conference on Next Generation Computing Technologies (NGCT) Year: 2016 ,Pages: 843 – 849.

[4] Steganography over video file by hiding video in another video file, random byte hiding and LSB technique Rachna Patel; Mukesh Patel 2014 IEEE International Conference on Computational Intelligence and Computing Research Year: 2014 Pages: 1 – 5.

[5] Datta, Biswajita, Souptik Tat, and Samir Kumar Bandyopadhyay. "Robust high capacity audio steganography using modulo operator."International Conference on Computer, Communication, Control and Information Technology (C3IT), IEEE, (2015, December 21-24), Himachal Pradesh, India.

[6] Guohua Wu, Xiaodong Zhou, 2008. "A Fast Audio Digital WatermarkMethodBasedonCounter-propagation Neural Networks", InternationalConference on Computer Science and Software Engineering, pp. 583-586 Hartung, F. and Girod, B. 1998.

[7] Watermarking of uncompressed and compressed video, SignalProcessing, Special Issue on Copyright Protection and Access Control for Multimedia Services, 66 (3): 283-301. Ko-Chin Chang., Chien-Ping Chang., Ping S. Huang., and TeMing Tu,:

[8] A Novel Image Steganographic Method Using Tri-way Pixel-Value Differencing, Journal of Multimedia, VOL. 3, NO. 2, JUNE 2008 Lee, Y. K. and Chen, L. H. 2000.

[9] High capacity image steganographic model, IEE Proceedings on Vision, Image and Signal Processing, Vol. 147, No.3, pp. 288-294. Maher El' Arbi et al. 2006. "Video Watermarking Based On Neural Networks", ICME, pp. 1577-1580.

[10] S.Karthikeyan, S.Jayashri, "Energy Utilization Strategies using Novel Approach in Wireless Sensor Networks", International Conference on Software Engineering and Mobile Application Modelling and Development, Dec 2012, pp: 1-5.

[11] S.Karthikeyan, S.Jayashri, "Energy Efficient System for Heterogeneous Wireless sensor Networks", European Journal of Scientific Research, Volume 72,No 4 march 2012, PP:599-607. ISSN: 1450-216X.

[12] S.Karthikeyan, S.Shanmugapriya and S.Jayashri, "Sub-head Transmission of Heterogeneous Data by Cloned Agent to Android Mobile",Research Journal of Applied Sciences, Engineering and Technology, 8(1):pp.24-34, 2014

[13] S.Pavithra, S.Karthikeyan, V.J.K.KishorSonti, S.Jayashri, "Competent Realization of Co-operative spectrum sensing in cognitive radio systems", International journal of Engineering systems modeling and simulation, Inder science publishers, Vol. 7, No. 2,pp 103-110 2015.

Cut Node Recovery Protocol for Wireless Sensor Network

E. Anna Devi[1]
Assistant professor, Deptof ECE,
Sathyabama Institute of Science and Technology
OMR Road,Chennai, India
annadevimit@gmail.com

J. Martin Leo Manickam[2]
Professor, Deptof ECE,
St.Joseph's College of Engineering
OMR Road,Chennai, India
josephmartin74@yahoo.co.in

Abstract- In wireless sensor network (WSN), energy efficiency and network lifetime are the important attributes to be improved for enhancing the performance of the network. Due to the failure or cut of a node, performance of the network is affected. So cut node detection and recovery are presented in this paper. Initially, sensor nodes in the search area are clustered using the node weight of the sensor nodes. The node with maximum weight is selected as cluster head (CH) and this CH performs as a monitor node which monitors all non-CH members in a cluster. Backup nodes (BNs) are selected in each cluster using Particle Swarm Optimization (PSO) algorithm. Cut node in the network is detected based on its hop-count and link cost value. This cut node is recovered by the selected backup nodes and mobile nodes (MNs) in the network. Simulation results show that the performance of the proposed approach outperforms that of the existing approach in terms of overhead, energy efficiency, network lifetime and BN selection time.

Keywords: *Wireless sensor network, Node weight, Backup Node (BN), Particle Swarm Optimization (PSO), Cut node recovery.*

1. INTRODUCTION

WSNs are these days conveyed for a substantial assortment of uses are as yet picking up enthusiasm for a few fields [1-4]. The network comprises of battery-fueled wireless gadgets, with sensing and correspondence capacities, that gather information about the earth or around an occasion of intrigue (e.g. light, temperature, movement). The gathered information is then handed-off toward a "sink" in a multi-hop way [5]. Other than the requirements as far as preparing force and capacity limit, the most critical issue in WSNs is the constrained battery term of sensors and the resulting restricted network lifetime. As a result, WSN arrangements should adapt to the issue of sensor disconnection and multiple passing of hubs after some time.

In WSNs, connectivity and coverage of the network are critical all through the entire network task time for meeting the coveted application prerequisites. For example, if WSNs need to occasionally transmit the sensed information effectively to the sink node, at that point the network ought to be associated constantly. In like manner, for observing all spots of a district precisely, the network ought to give alluring coverage amid whole lifetime of the network. In nearness of successive disappointments, giving wanted nature of administration (QoS) is all the more difficult in remotely sent WSNs, where manual substitution of failed node isn't conceivable. Along these lines, keeping in mind the end goal to give self-recuperating capacity against such disappointments, mobile.Sensor nodes are broadly considered as a choice. In spite of the fact that, the utilization of mobile nodes in sensor networks expands the cost of the network, however is extremely helpful for connectivity rebuilding, supplanting failed nodes and dynamic adjustment of the network. Mobile nodes are significantly more adaptable than static sensor nodes as they can be sent in any situation and can adapt to quick topological changes. Rather, a blend of mobile and static sensor nodes has for the most part been favored.

Likewise, the utilization of mobile sensor nodes is very clear from some ongoing plans proposed for taking care of disappointments and topology administration where such nodes are broadly utilized [2-4]. Along these lines, to recover the failure node or cut node in the network, Backup Node (BN) based cut node recovery is exhibited.

2. PROBLEM STATEMENT

Krishna Pal Sharma and TeekParval Sharma [6] have presented failure recovery based on zone. In this approach, authors have selected a zone monitor using the approach based on consensus and agreement. The process of zone monitor was to monitor the activities of failure recovery. Based on the backup nodes and mobile nodes, failure nodes in the network have been recovered. The authors have used mathematical model to select the backup nodes from the available sensor nodes. Because of this selection method, backup node selection time has been increased and also energy efficiency of the network has been decreased. To overcome these issues, an optimization algorithm is presented in this paper for backup node selection.
Contributions of this paper are described below:

- Based on the parameters residual energy and average distance of the sensor nodes, cluster head (CH) is selected and cluster formation is done by this selected CH
- Backup Nodes (BNs) are selected using the Particle Swarm Optimization (PSO) algorithm in each cluster
- After establishing a path between source and destination using hop-count and link cost value, the source node detects whether the path has cut node using the percentage of difference
- Once a cut node is detected, the CH activates the BN if it is closer to the cut node to replace the cut node, otherwise nearby mobile node (MN) is activated replace the cut node

3. RELATED WORKS

In this section, some previous literatures are survived. These research works focused on the failure node or cut node recovery for improving the performance of the network. Krishna Pal Sharma and TeekParval Sharma [9] have presented failure node recovery method based on zone. Authors have selected a zone monitor using the consensus and agreement based scheme. The activities of failure node recovery were monitored by this zone monitor. This monitor node detected the suspicious node by receiving the heartbeat messages from the sensor nodes. The failure node was recovered using the backup nodes those were overlapped with the sensing range of cut node.Abdollahzadeh, Sanay and NimaJafariNavimipour. [2]have developed a failure detection and recovery method. In this method, authors have generated agent packet from the sink node and also they have formed a query path to the failure node using the agent. Based on these agent packets, the receiving node has detected the failure node. This failure node has been recovered using the Least-Disruptive Topology Repair.Mehajabin, Nusrat[3] have proposed an innovative solution for the issue of the connectivity due to the failure nodes. Authors have presented two steps to replace the failed nodes. Redundant nodes were identified using the selected cluster heads in the first step. Then the connectivity was restored in the second step. Cerulli, Raffaele, et al [4] have enhanced the network lifetime of WSN by presenting fault node recovery algorithm. This algorithm was performed based on the grade diffusion algorithm together with the genetic algorithm. Using these combined algorithms, authors have detected failure nodes.

4. BACKUP NODE SELECTION BASED CUT RECOVERY

4.1 Overview

Initially, cluster head (CH) is selected based on the node weights. Node weight is calculated using the parameters average distance and residual energy of the node. Then the selected CH selects the rest of the sensor nodes those are in its communication range and forms a cluster. CH of each cluster plays as a monitor node which monitors all cluster members in the cluster. In each cluster backup nodes (BNs) are selected by CH using particle swarm optimization algorithm. These selected BNs are in sleep mode initially. Before data transmission, the source node identifies the cut node based on the hop-count. This cut node is recovered using the selected

IOP Conf. Series: Materials Science and Engineering **590** (2019) 012058 doi:10.1088/1757-899X/590/1/012058

BN. If a cut node is detected, the CH turns on the BN which is in on hop distance to the cut node. The BN has sensing coverage of cut node so it replaces the cut node. If there is no BN in one hop distance to the cut node, a mobile node (MN) is turned-on active by the CH. Then MN moves to the location of cut node and replaces the function of cut node. Figure 1 shows the block diagram of this proposed approach.

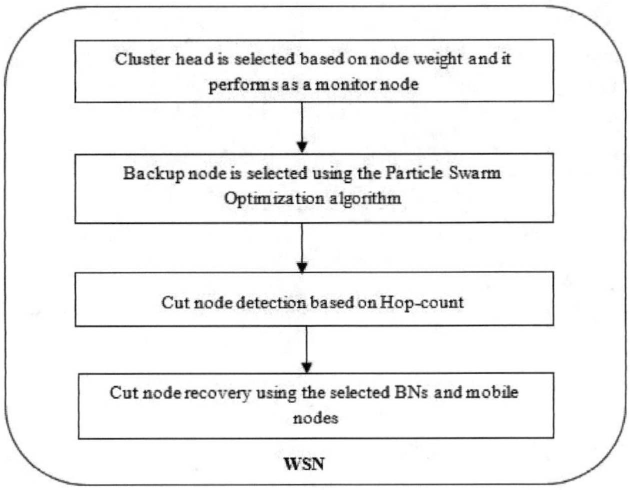

Figure1: Block diagram of the proposed system

4.2 Cluster formation

Sensor nodes in the network are grouped as number of clusters by selecting the cluster head. This cluster head is selected based on the node weight. Also it performs as a monitor node by monitoring the activeness of cluster members in an each cluster. Weight of each node is calculated based on the parameters such as average distance and residual energy.

Average distance: It is defined as the average distance between each sensor node (SNi) to the base station (BS) in the network. Average distance of sensor node from the base station should be minimal.

$$o_1(i) = \left(\frac{1}{\frac{1}{m} d(SN_i, BS)} \right) \quad (1)$$

Where, m – denotes amount of sensor nodes in the sensing area

$\frac{1}{m} d(CH_i, BS)$- defines the average distance between a base station and SNi

Residual energy:

It is defined as the difference between the initial energy and consumed energy of the sensor node (SNi).

$$o_2(i) = E_{initial\ i} - E_{cons\ i} \quad (2)$$

Where, $E_{initial\,i}$ and $E_{cons\,i}$ represent the initial energy and consumed energy of SNi.

Using equations (1) and (2), node weight (NW) is calculated as follows

$$NW(i) = o_1(i) \times \mu + o_2(i) \times (1-\mu), \quad 0 < \mu < 1 \quad (3)$$

Where, μ is the weight factor varying in the range of [0, 1].

Node with maximum value of node weight is selected as optimal cluster head. Rest of the sensor nodes those are in the communication range of CH join to the corresponding CH. Finally, clusters are formed. The selected CH collects information about the all non-CH members in a cluster.

4.3 Backup node selection

To recover a node from the cut, backup nodes are selected in each cluster by corresponding CH using particle swarm optimization algorithm. Before the selection of backup nodes, k-coverage [2x] areas are formed for set of sampling points which denoted as qm.

As shown in figure 2, qm points are considered in each cluster to cover sensor nodes based on their sensing range (rs). Euclidian distance (e) is calculated between each sensor node ni and the point qm. This calculated distance is compared with the sensing range of the sensor node. If distance of k-non-CH members in each cluster is less than the sensing ranges from point qm, then k-coverage areas are formed for point qm i.e.,

$$d(n_i, q_m) < r_s, \quad i = 1,2,\ldots,N, \quad m = 1,2,\ldots M \quad (4)$$

Where, N denotes the number of sensor nodes or non-CH members, M denotes the number of points in a cluster. Sensing ability of each sensor node is calculated based on its distance to the point qm i.e.,

$$S(n_i, q_m) = \frac{\alpha}{[e(n_i, q_m)]^{\beta}} \quad (5)$$

Where, the positive constants α and β represent the sensor technology dependent parameters.

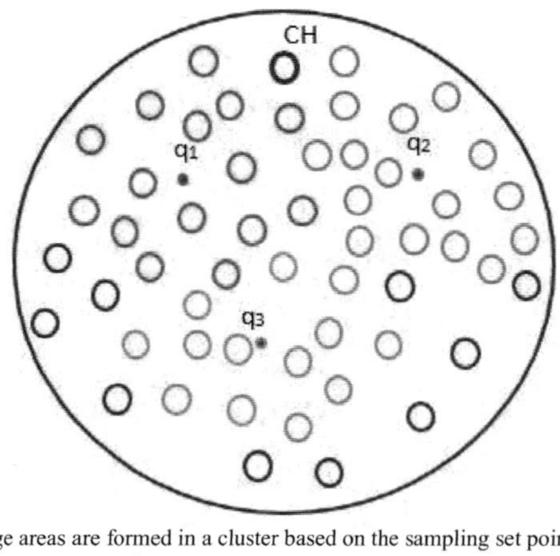

Figure 2: k-coverage areas are formed in a cluster based on the sampling set points (q1, q2 and q3)

After forming the k-coverage areas with respect to the sampling points, backup node (BN) is selected for each area in a cluster. These BNs are selected using the Particle Swarm Optimization (PSO) algorithm. Particle swarm optimization (PSO) is a nature inspired swarm intelligence based algorithm, modelled after observing the choreography of a flock of birds, i.e., how they can explore and exploit the multi-dimensional search space for food and shelter. Selection BNs using this algorithm is described as following,

Initialization: Particles or solutions of this approach are initialized. Backup nodes which are to be selected are the solutions. These solutions or backup nodes are initialized with its positions as follows,

$$P_i = \left[Y_{i,1}(t), Y_{i,2}(t), \ldots\ldots, Y_{i,D}(t)\right] \quad (6)$$

Where $Y_{i,d}(t)$ denotes the position of the ith particle in the dth dimension. This is also represented as

$$Y_{i,d}(t) = \left(x_{i,d}(t), y_{i,d}(t)\right), \quad 1 \le i \le N_P, \ 1 \le d \le D \quad (7)$$

Fitness: Fitness of each solution is calculated based on the sensing coverage of the sensor nodes. The closest sensor node to the sampling points or the node with maximum sensing ability is selected as the optimal node or BN. This fitness function is calculated using equation (8),

$$Fitness = Max\left(S\left(n_i, q_m\right)\right) \quad (8)$$

Updating: Each particle or solution is initialized with a random position and velocity to move in the search space. In each generation, particles find their own best values also known as personal best value (Pbesti) and also find the global best value (Gbest). Particle is updated using position (X) and velocity (V) equations until obtain the global best solution. Velocity is calculated using the Pbesti and Gbest. New position of the particle is calculated using the velocity. Velocity and position of the particle are calculated as follows,

$$V_{i,d}(t+1) = V_{i,d}(t) * w + a_1 * \gamma_1 * \left(YPbest_{i,d} - Y_{i,d}\right) + a_2 * \gamma_1 * \left(YGbest - Y_{i,d}\right) \quad (9)$$

$$Y_{i,d}(t+1) = Y_{i,d}(t) + V_{i,d}(t+1) \quad (10)$$

Where, a_1 and a_2 represent the acceleration coefficients varying in the range [0, 2], γ_1 and γ_2 represent the random values varying in the range [0, 1], w denotes the inertia weight varying in the range [0, 1].

After the particle is updated to the new position, fitness value of the particle Pi is compared with the fitness of previous Pbesti. If the fitness (Pi) is greater than the fitness (Pbesti), then the particle Pi is updated as the Pbesti. Similarly, if the fitness (Pi) is greater than the fitness (Gbest), then the particle Pi is updated as the Gbest.

Termination: The above phases are continued until find the optimal solution or BNs. Once the optimal solution is obtained the algorithm will be terminated.

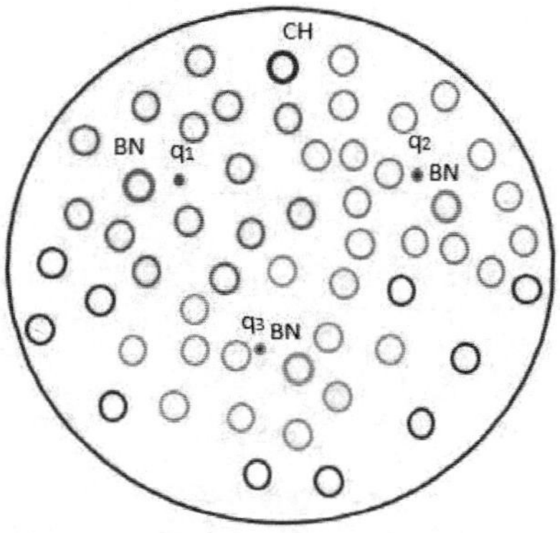

Figure 3: BNs are selected in each k-coverage area

Once the BNs are selected, these nodes are in SLEEP mode initially i.e., the BN doesn't sense data or transmit the data but it can listen to the remote medium to detect a packet. In sleep node, the BNs consume low power. These BNs are played as the alternative nodes to replace the cut node in the network. Figure 3 shows the BNs selection.

4.4 Cut node detection

Prior to the packet transmission, a source node transmits the information packet Route Request (RREQ) message to the destination by means of the assigned next hop nodes. After acknowledgment the RREQ message, the end transmits the Route Reply (RREP) message to the source node. At that point the source node passes on the packet by means of the assigned ensuing hop node. In this transmission, packet and neighbour table of every node are refreshed with the information of neighbour nodes like residual energy, hop tally, size of queue, link quality and link cost until achieve the destination.

As a result of the quantity of transmissions, couple of nodes may lose its energy level that is recognizes as cut nodes. Again the source node diffuses the RREQ to the destination, if the cut node available in the course, the source node doesn't acquire the RREP message inside the round trip time. So the source node retransmits the parcel and each hop node in the course checks link cost estimations of neighbour nodes. To identify the cut node in the course, contrast inside the normal link cost esteems and the most extreme link cost estimation of the neighbour nodes is outlined and the level of distinction is related with the threshold esteem. On the off chance that the threshold esteem (40%) is more prominent than the percentage of difference, at that point the node is to be viewed as that the node is to be cut. The percentage of difference is intended as

$$diff(\%) = \frac{C_{High} - C_{Avg}}{C_{Avg}} \times 100\% \quad (11)$$

Where, C_{High} and C_{Avg} are signified as maximum link cost value of a neighbour node with minimum hop count and average link cost value of all neighbour nodes except the node with minimum hop count correspondingly. This link cost value is calculated as

$$C_{ij} = \frac{E_{residual_j}}{E_{initial_j}} \times cc_E + \frac{Q_{free_j}}{Q_{total_j}} \times cc_Q + LQ_{ij} \times cc_L \quad (12)$$

Where, cc_E, cc_B and cc_L are signified as the constant coefficients, $E_{residual_j}$ and $E_{initial_j}$ signify the residual and initial energy of node j correspondingly, Q_{free_j} and Q_{total_j} represent the free queue size and maximum queue size of node j correspondingly and LQ_{ij} signifies the link quality between two nodes i and j.

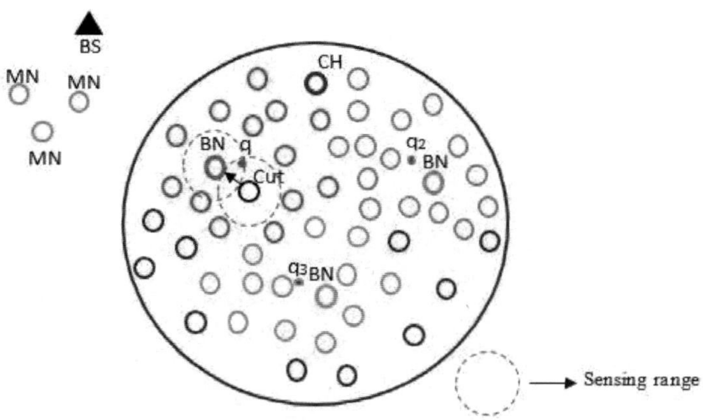

Figure4: Cut Node recovered by BN (34)

4.5 Cut node recovery

After detecting the cut node, the source node sends the message to the corresponding CH. This message includes the ID and status of the cut node. After receiving the message, the CH verifies the Neighbour table of the cut node and decides whether the BN is in one hop distance to the cut node. If the BN places in one hop distance to the cut node, then the CH instructs the BN that turn on the BN as ACTIVE node itself. The coverage of this BN is overlapped with the sensing range of other neighbour nodes in the area of k-coverage so that BN replaces the cut node to continue the data transmission as shown in figure 4. New BN is to be selected when the BN replacing the cut node. If the cut node doesn't has the BN in one hop distance, then the CH finds the nearby mobile node (MN) with the ACTIVE status and the MN moves to the received position of the cut node and replaces it.

5. RESULTS AND DISCUSSION

The proposed approach Backup Node selection for Cut node Recovery (BNCR) is implemented in the platform of Network Simulator (NS2). In this simulator, 500 sensor nodes are executed in the search space 1000m×1000m.

5.1.1 The performance metrics

Performance of this proposed approach is evaluated based on the following metrics. The performance metrics of the proposed Backup Node selection for Cut node Recovery (BNCR) is compared with that of the existing ZBFR [9]. The entire scenario is simulated in ns2 and various parameters are calculated.

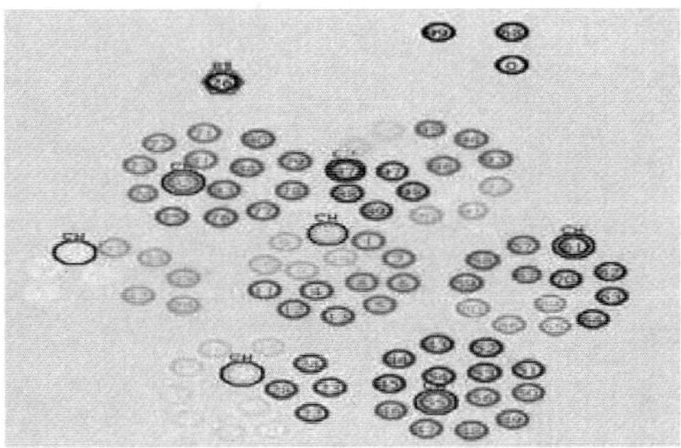

Figure 5: Simulation output for 100 nodes

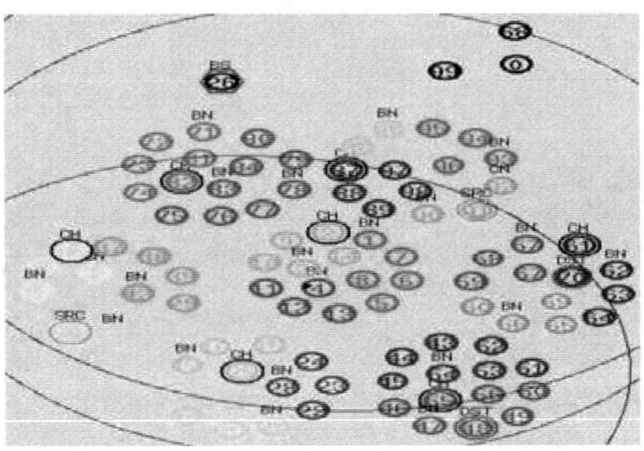

Figure 6: Cut node (40) is recovered by BN (34)

IOP Conf. Series: Materials Science and Engineering **590** (2019) 012058 doi:10.1088/1757-899X/590/1/012058

Energy consumption: It is defined as the amount of energy consumed by individual node during the transmission. Also it is referred as the difference between the current energy and initial energy of a node. Unit of this parameter is Joule (J).

$$Energy\ consumption = Initial\ energy - current\ energy$$

Network lifetime: It is referred as the time of network disconnection due to the failure of one or more sensor nodes. Unit of this parameter is seconds (s).

BN selection time: It is defined as the time to select the number of backup nodes within the period of simulation time.

Coverage ratio: It is defined as the ratio of the covered area to the total area.

Message overhead: It is defined as the number of messages exchanged during the process of BN selection and cut node detection.

Recovery time: The time at which a cut node is to be recovered using the BNs or MNs in the network.

Delivery ratio: It is the ratio of the number of packets received successfully and the total amount of packets transmitted.

$$Delivery\ ratio = \frac{Number\ of\ packets\ received}{Amount\ of\ packets\ transmitted}$$

Packet drop: It is the amount of packets dropped during the data transmission.

Packet delay: The delay of network describes how long the network takes to transmit a bit to the destination. Unit of this parameter is seconds (s).

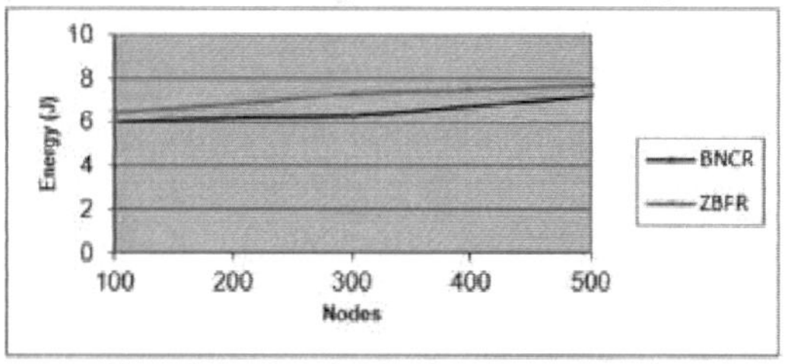

Figure 7: Number of nodes Vs Energy consumption

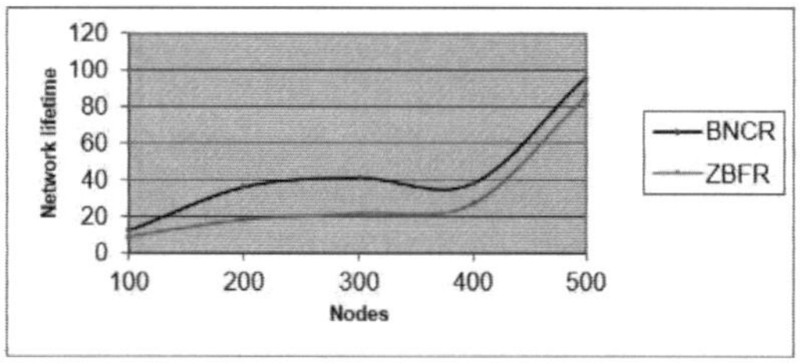

Figure.8: Number of nodes Vs Network lifetime

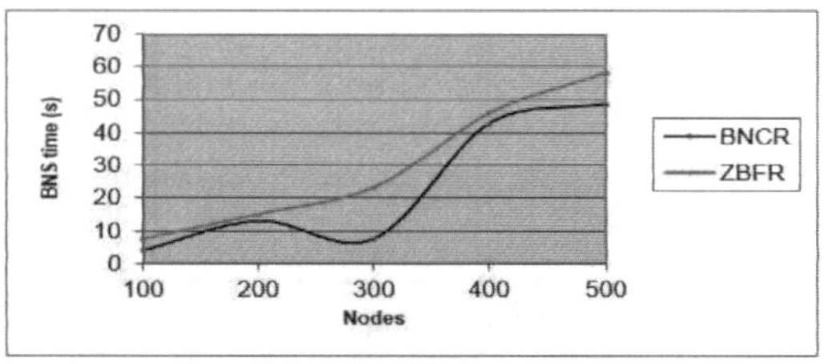

Figure 9: Number of nodes Vs BNS time

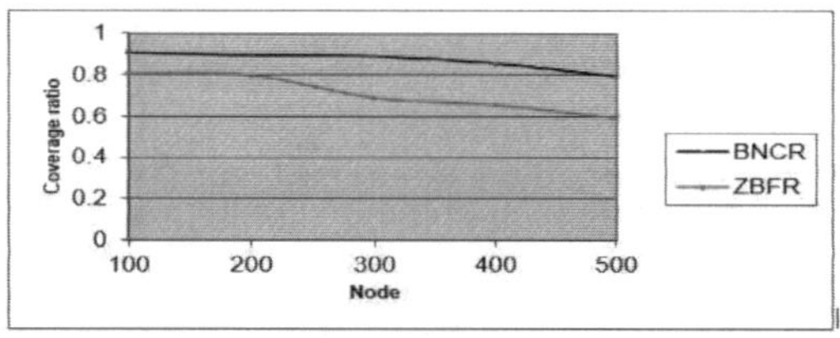

Figure 10: Number of nodes Vs Coverage ratio

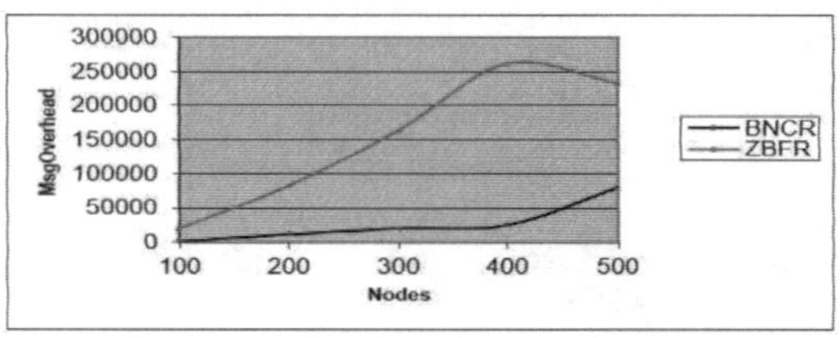

Figure 11: Number of nodes Vs message overhead

6. CONCLUSION

In this paper, Particle Swarm Optimization algorithm based backup node selection for cut node recovery has been presented and this proposed approach has been simulated in the Network Simulator.

IOP Conf. Series: Materials Science and Engineering **590** (2019) 012058 doi:10.1088/1757-899X/590/1/012058

Cluster formation has been done based on the node weight. In each cluster, backup nodes were selected using the proposed particle Swarm Optimization algorithm. Then the cut node in the network has been detected based on the hop-count and link cost value of each sensor node. This cut node has been recovered by the selected backup nodes if it was in one hop distance to the backup node, otherwise it has been recovered by mobile nodes in the network. Simulation results showed that the performance of the proposed approach outperformed that of the existing approach in terms of overhead, energy efficiency, network lifetime and BN selection time.

REFERENCES:

[1]Gherbi, Chirihane, ZiboudaAliouat, and Mohamed Benmohammed. "An Adaptive Clustering Approach To Dynamic Load Balancing And Energy Efficiency In Wireless Sensor Networks". Energy 114 (2016): 647-662. Web.

[2]Abdollahzadeh, Sanay and NimaJafariNavimipour. "Deployment Strategies In The Wireless Sensor Network: A Comprehensive Review". Computer Communications 91-92 (2016): 1-16. Web.

[3]Mehajabin, Nusrat et al. "Energy-Sustainable Relay Node Deployment In Wireless Sensor Networks". Computer Networks 104 (2016): 108-121. Web.

[4]Cerulli, Raffaele, Monica Gentili, and Andrea Raiconi. "Maximizing Lifetime And Handling Reliability In Wireless Sensor Networks". Networks 64.4 (2014): 321-338. Web.

[5]Ojha, Tamoghna, SudipMisra, and Narendra Singh Raghuwanshi. "Wireless Sensor Networks For Agriculture: The State-Of-The-Art In Practice And Future Challenges". Computers and Electronics in Agriculture 118 (2015): 66-84.

[6]]K. Sharma and T. Sharma, "ZBFR: zone based failure recovery in WSNs by utilizing mobility and coverage overlapping", Wireless Networks, vol. 23, no. 7, pp. 2263-2280, 2016.

[7]S.Karthikeyan, S.Jayashri, “Energy Utilization Strategies using Novel Approach in Wireless Sensor Networks”, International Conference on Software Engineering and Mobile Application Modelling and Development, Dec 2012, pp: 1-5.

[8]S.Karthikeyan, S.Jayashri, “Energy Efficient System for Heterogeneous Wireless sensor Networks”, European Journal of Scientific Research, Volume 72,No 4 march 2012, PP:599-607. ISSN: 1450-216X.

[9] G.Savitha and S.Karthikeyan,"Architecture for diverse modulation techniques in SDR", concurrency computation practice and experience,John wiley & sons, Ltd, February 2019,PP: 1-10.

IOP Conf. Series: Materials Science and Engineering **590** (2019) 012059 doi:10.1088/1757-899X/590/1/012059

A MILLIMETER WAVE BASED CIRCULAR SLOT LOADED MICROSTRIP PATCH ANTENNA FOR 5G COMMUNICATION

V.Karthikeyan[a], S. Vignesh[b], K. Ayyavar Reddy[b], G. Bhargavi Reddy[b], R. S. Aparna Sekhar[b]
Assistant Professor[a], UG Student[b]
[a,b]Department of Electronics and Communication Engineering
[a,b]Vel Tech Rangarajan Dr. Sagunthala R&D Institute of Science and Technology
Chennai, India.
vkarthikeyan652@gmail.com, vigneshmessi101@gmail.com, ayyavarreddy.ark@gmail.com, bhargavig516@gmail.com, aparnasekhar123@gmail.com

Abstract. This paper presents a microstrip patch antenna for 5G communication at the frequency range of 32GHz. The patch is loaded with multiple circular slots to improve the performance of the antenna. It consists of RT duroid substrate with low dielectric loss tangent 0.0009, dielectric constant 2.2 and thickness 0.5mm, PEC ground and PEC patch. Four various patch antenna with slot array based antenna has simulated on Ansoft HFSS (High Frequency Structural Simulator) Platform. The overall dimensions of designed antennas are 6mm x 6mm x 0.5mm. The performance of antenna is analysed by using simulated results of reflection co-efficient, VSWR, gain, bandwidth and directivity and antenna parameters are compared with four antennas. The proposed antennas of reflection co-efficient characteristics is below -10dB at the resonance frequency.

1. Introduction

The fifth generation is elevating the mobile network to not only interconnect people, but also interconnect machines and devices. It produces a high efficiency and next level of performance. It delivers ultra -low latency of 1ms, high peak rates and massive capacity due to availability of high bandwidth [1]. The 5G wireless signals is transmitted via huge number of small 5G cell which would be located or placed on a street lamp post and buildings. The use of large number of cells is necessary because 5G uses millimetre wave technology [3]. Millimeter wave band of frequency spectrum is between 30 GHz-300 GHz. It is also known as extremely high frequency (EHF) or very high frequency (VHF). It allows data rates up to 10Gbps [2, 4]. Millimetre wave has a range of 1mm to 10mm wavelength. RT/Duroid5880 is a microwave material which offers superior electrical and mechanical property. It has a highly depressed thermal coefficient of dielectric constant of -55°C to +150°C. RT/Duroid5880 substrate is mainly used for antenna design because it has low electrical loss, low moisture absorption and stable frequency [5, 9]. It is mainly used in applications such as airborne and ground based radar systems, millimetre wave application, and military radar systems. Micro strip patch antennas are becoming more important for mobile phone market. It can be printed directly on to a circuit board [6,8]. This type of antenna is low cost and easily fabricated and designed. They have simple construction, it consist of dielectric substrate, patch and ground. There are so many possible shapes like circular, rectangular and triangular. They have very large bandwidth, higher efficiency and better radiation. The circular slots loaded in the patch will improve the bandwidth and impedance hence enhances the radiation pattern and it also allows to compact the size of antenna significantly suit for the specific applications [7,10]. Here in this paper a comparison is made between the simple rectangular structure and proposed rectangular microstrip patch antenna loaded with various slots. Those various slots are 67 slots load antenna, 36 slots load antenna and 16 slots load antenna.

2. Design Equations of Microstrip Patch Antenna

It consists of patch which is usually a thin metallic strip placed on the dielectric substrate. Patch can be of any shape but here the shape of patch is rectangular. The design parameters of rectangular patch

antenna are length (L) and width (W) [12, 13]. These two parameters depend on dielectric constant, height of substrate and resonant frequency.
Width of patch (W) is given by,

$$W \leq \frac{c}{2f_r}\sqrt{\frac{2}{1+\varepsilon_r}} \quad (1)$$

Where c is the speed of light and f_r is the resonant frequency. In this design f_r is 32GHz.
The length of patch(L) is determined by ,

$$L = \frac{c}{2f_r\sqrt{\varepsilon_e}} - 2\Delta L \quad (2)$$

where ε_e is the effective dielectric constant

$$\varepsilon_e = \frac{1}{2}\left[\varepsilon_r + 1 + (\varepsilon_r - 1)\sqrt{\frac{1}{\left(1+\frac{12h}{W}\right)}}\right] \quad (3)$$

$$\Delta L = 0.412\frac{(\varepsilon_e + 0.3)(\frac{W}{h} + 0.264)}{(\varepsilon_e - 0.258)(\frac{W}{h} + 0.8)}h \quad (4)$$

From (1) & (2) we get W=3.32 mm and L=2.45 mm.
In this design the microstrip feed line width is 0.46mm and length is 1.57 to match the input impedance of patch element which should be close to 50Ω.

3. Antenna Designs

3.1 Rectangular patch antenna

The antenna consists of dielectric medium or substrate, patch or radiating element and ground. The structure of antenna is shown in the figure 1. This has a length and width of 2.75mm, 3.32mm respectively. Microstrip feed line length, width of 1.57mm and 0.46mm attached to the rectangular patch. The dielectric material in this design is used εr = 2.2 and substrate height=0.5mm.

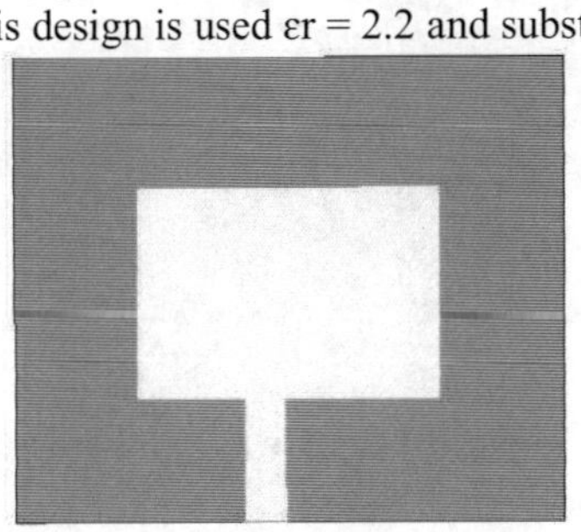

Figure 1. Structure of the Rectangular patch antenna

3.2 Rectangular microstrip patch antenna with 16 slots

The rectangular patch antenna with 16 slots is shown in the Figure 4. It is loaded with 16 circular slots. Circular slots are having a radius 0.05mm and a uniform spacing of 0.2mm is kept between them on the rectangular patch. Antenna bandwidth improves by inserting the slot and also the return loss, VSWR, gain increases compared to rectangular patch antenna.

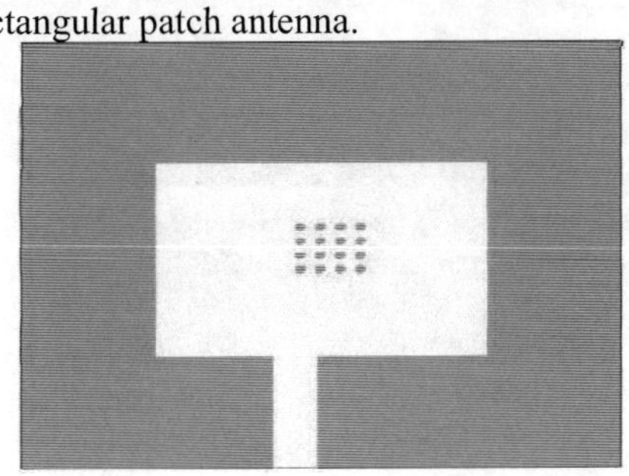

Figure 2. Structure of the rectangular patch antenna with 16 slots

3.3 Rectangular microstrip patch antenna with 36 circular slots

The structure of rectangular patch antenna with 36 circular slots is shown in figure 3. It is loaded with 36 circular slots of each having radius of 0.05mm and 0.2mm spacing between them on the rectangular patch antenna and Rectangular microstrip patch antenna with 36 circular slots has a improved return loss, VSWR and gain.

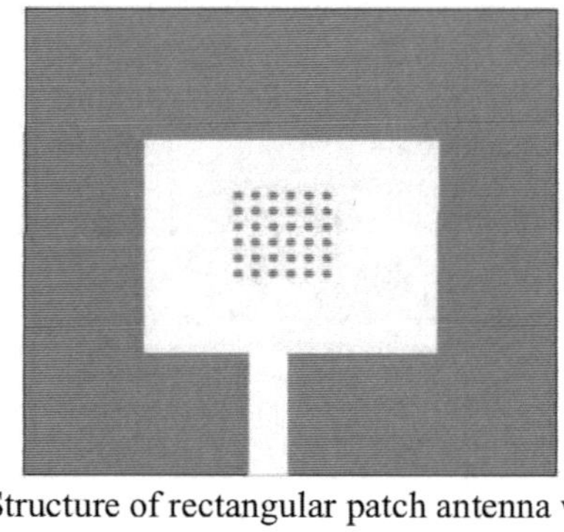

Figure 3. Structure of rectangular patch antenna with 36 slots

3.4 Rectangular Microstrip patch antenna with 67 circular slots

Top and bottom structure of proposed antenna is displayed in the figure 4. The patch length of 2.75mm and width of 3.32mm. It is loaded with 64 circular slots of radius 0.05mm. Ground is placed below the substrate with the dimension of 6mm x 6mm. It has presented in the figure 5.

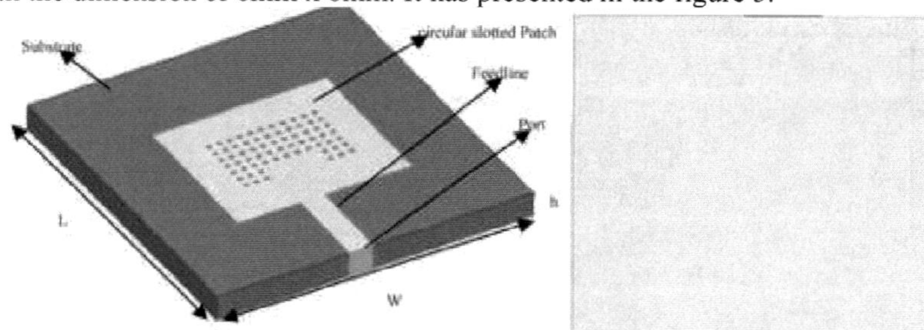

Figure 4. Top and bottom structure of the proposed antenna with 64 slots

4. Results and Discussion

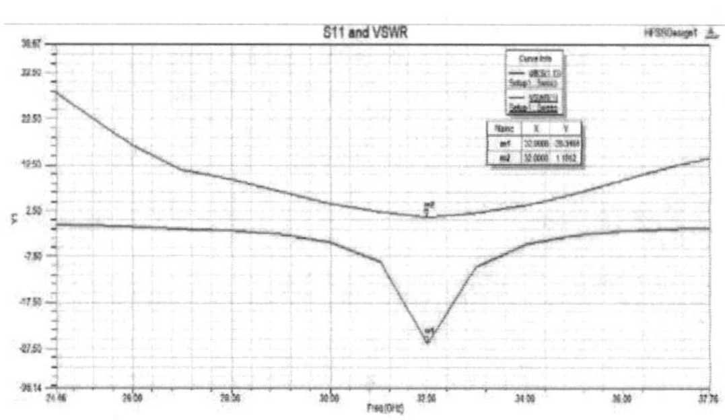

Figure 5. Return loss (s11) and VSWR characteristics of the simple patch antenna.

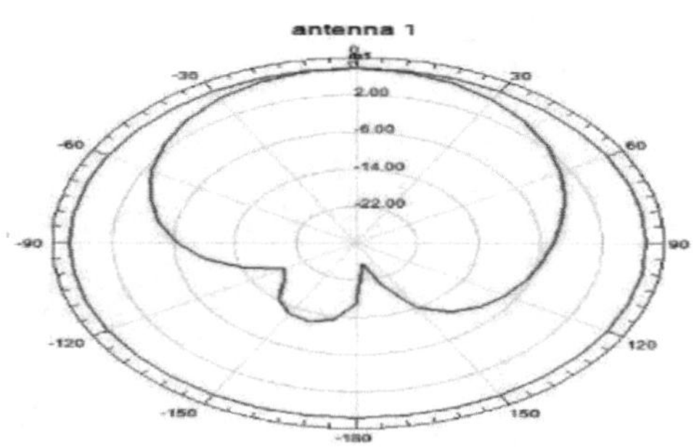

Figure 6. E-plane and H-plane of the simple patch antenna structure

The S11 parameter and VSWR for simple rectangular patch antenna is presented in the figure 5. The simulated results of simple patch antenna parameter -26.3464dB, 1.1012 and 7.6968 dB of reflection coefficient, VSWR and gain at the resonant frequency 32GHz. The E-plane and H-plane of simple patch antenna is presented in the figure 6.

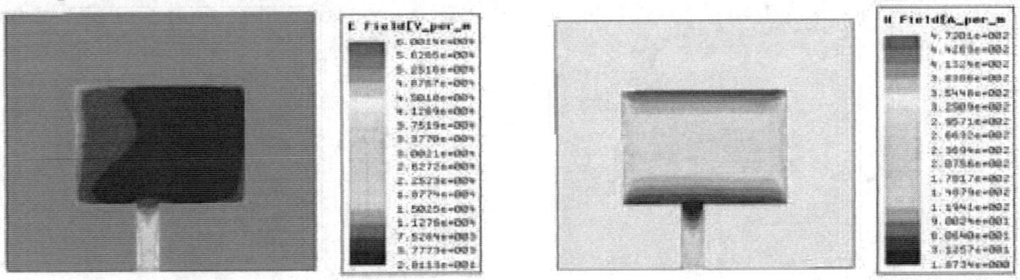

Figure 7. Electric and Magnetic field strengths of the simple patch antenna structure

Figure 8 shows that the maximum radiated electric field strength of the conventional rectangular patch antenna is 6.0014e+004 V/m. The maximum radiated magnetic field strength is that 4.7201e+002A/m and it has shown in the figure 7. Figure 8 shows the S11 parameter and VSWR for rectangular patch antenna with 16 slots. It has measured that the reflection coefficient is -26.5895dB and VSWR is 1.0983 at the resonant frequency 32GHz. E-plane and H-plane gains of the patch antenna with 16 slots are 7.6968 dB.

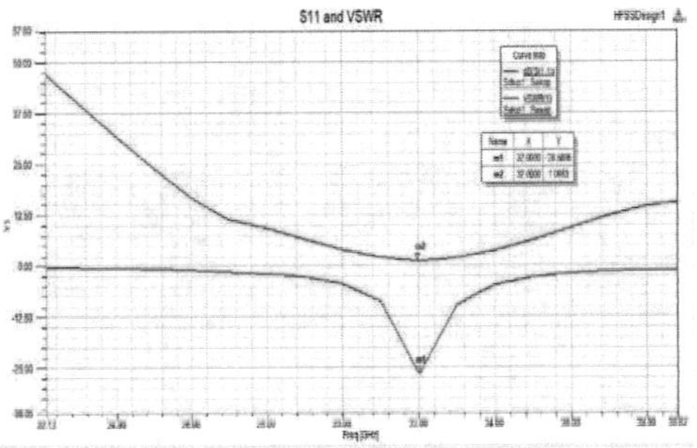

Figure 8. Return loss (s11) and VSWR characteristics of patch antenna with 16 slots.

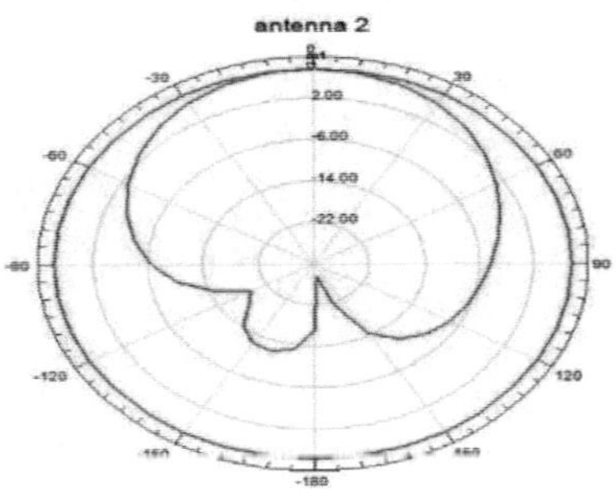

Figure 9. E-plane and H-plane of patch antenna with 16 slots.

The gain of E-plane and H-plane of patch antenna with 16 slots is shown in the figure 9. Figure10 shows that the maximum radiated Electric field strength is 5.6682e+004 V/m and maximum radiated magnetic field strength is 5.2674e+002 of rectangular patch antenna with 16 slot.

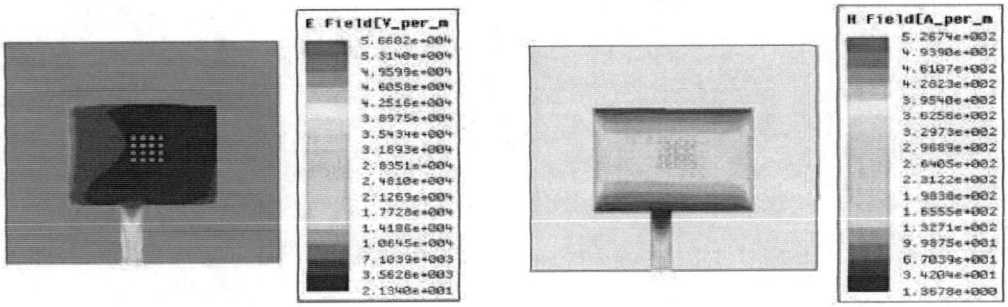

Figure 10. Electric and Magnetic field strengths of the patch antenna with 16 slots

Figure 11 shows that the S11 parameter and VSWR for rectangular patch antenna with 36 slots. It can be measured that the reflection coefficient is -27.5427dB and VSWR is 1.0876 at the resonant frequency 32GHz. E-plane and H-plane gains of rectangular patch antenna with 36 slots is 7.5494dB. The E-plane and H-plane of patch antenna with 36 slots is shown in the figure 12. Figure 13 shows that the maximum radiated Electric field strength of 7.2671e+004 V/m and the maximum radiated magnetic field strength of rectangular patch antenna with 36 slots is 5.1103e+002.

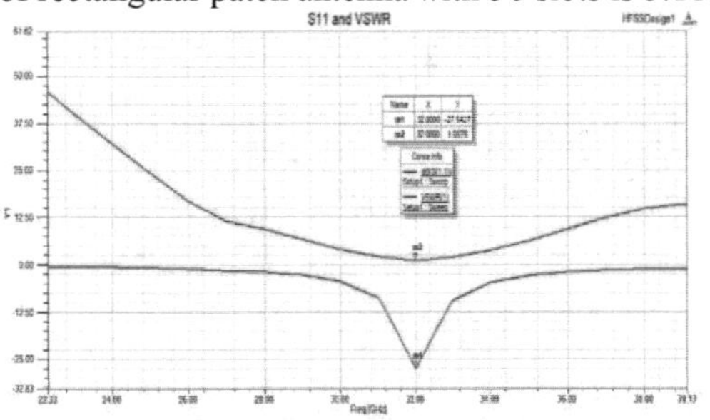

Figure 11. Return loss (s11) and VSWR characteristics of patch antenna with 36 slots.

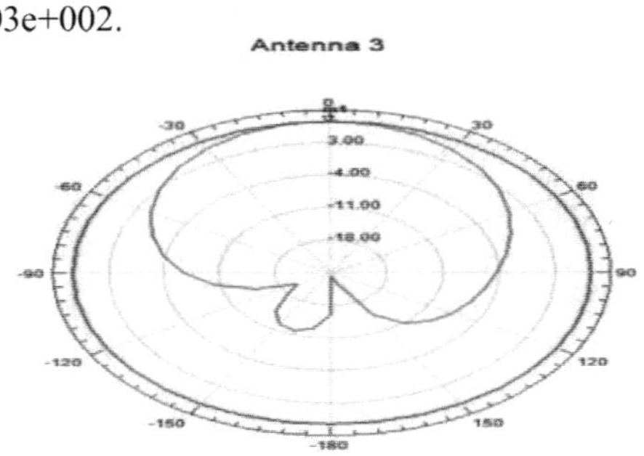

Figure 12. E-plane and H-plane of patch antenna with 36 slots.

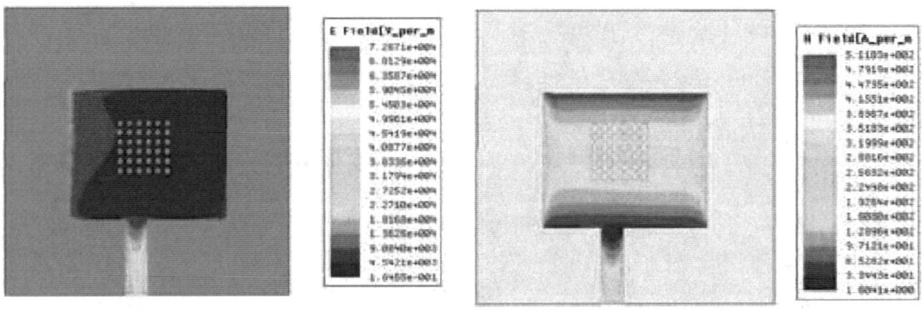

Figure13. Electric and Magnetic field strength of patch antenna with 36 slots

Figure 14 and figure 15 shows the S11 parameter and VSWR for proposed rectangular patch antenna with 67 slots. It can be measured that the reflection coefficient is -28.4701dB and VSWR is 1.0784 at the resonant frequency 32GHz. E-plane and H-plane of patch antenna with 64 slots as 7.6300dB. The E-plane and H-plane of patch antenna with 36 slots is shown in the figure 16.

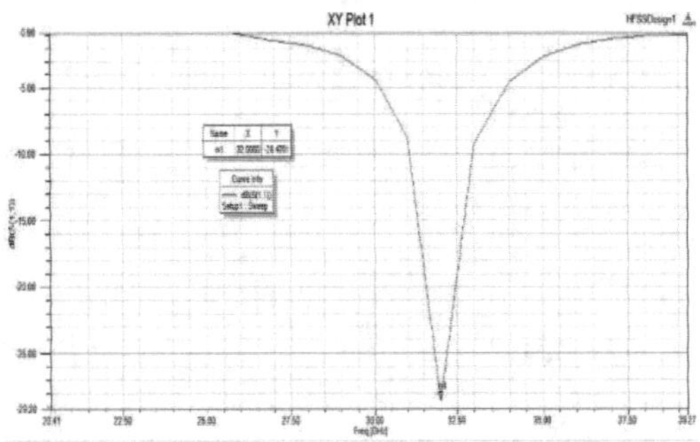

Figure 14. Return loss (s11) characteristics of patch antenna with 64 slots.

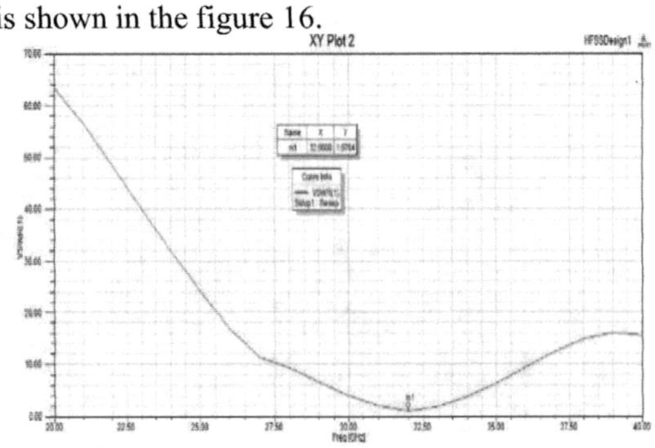

Figure 15. Return loss (s11) characteristics of patch antenna with 64 slots.

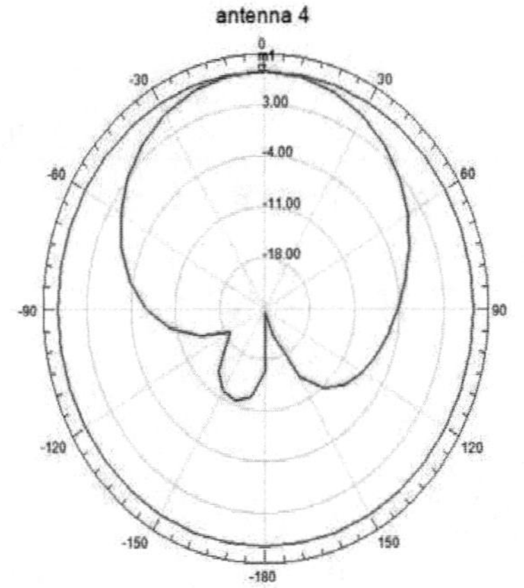

Figure 16. E-plane and H-plane of patch antenna with 64 slots.

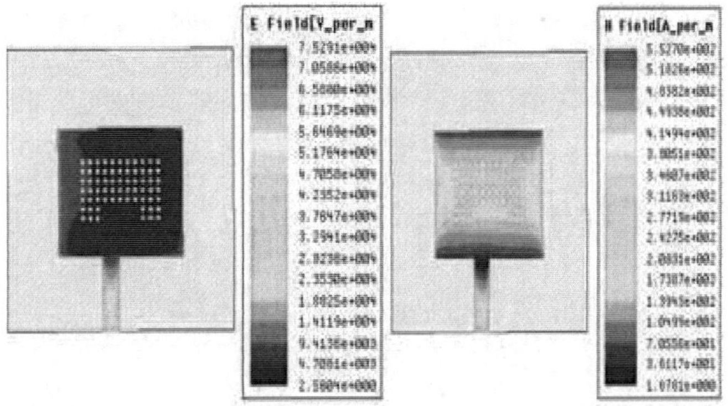

Figure 17. Electric and Magnetic field strength of patch antenna with 64 slots.

Figure 17 shows that the maximum radiated Electric field strength of 7.2671e+004 V/m and the maximum radiated magnetic field strength of 5.1103e+002 of proposed rectangular patch antenna with 64 slot. The rectangular patch antenna with 64 slots gives better return loss (S11 parameter), VSWR and gain compared with other proposed antennas such as antenna with 16 slots, 36 slots and 64 slots. The antenna parameters of all the four design have been compared in the table1. In the table 1, antenna 1, antenna 2, antenna 3 and antenna 4 represents the simple patch antenna, patch antenna with 16 slots, 36 slots and 64 slots respectively. Comparison of proposed antenna with other literature is mentioned in the table 2.

Table 1. Simulated results of designed Antenna parameters

Parameter	Antenna 1	Antenna 2	Antenna 3	Antenna 4
Return loss(dB)	-26.346	-26.589	27.512	-28.470
VSWR	1.1012	1.0983	1.0876	1.0784
Bandwidth (GHz)	1.591	1.898	1.899	2.22
Directivity	7.6755	7.6395	7.6252	7.6328
Incident power(W)	1	1	1	1
Accepted power(W)	0.9976	0.96153	0.86399	0.99927
Radiated power(W)	0.9025	0.95595	0.86380	0.99463
Radiation efficiency	1.0049	1.397	1.0004	1.7078
LHCP (V)	13.96	2.9138	13.037	2.2163
RHCP (V)	18.667	3.1681	11.423	3.2841

Table 2. Comparison of proposed antenna with other exiting antennas

S.No	Reference No.	Antenna Type	Dimension (mm$^{3)}$	Bandwidth (GHz)	S11 (dB)	VSWR	Gain (dB)	Directivity (dB)
1	[14]	U-Slotted antenna	46x56x2	0.051	-19.01	1.25	5.32	5.3
2	[15]	T-Slotted antenna	5x5x1.57	1.63	-38	1	6.34	3.2
3	[16]	Slotted rectangul ar antenna	12x8x6	0.012	-27	1.13	4.2	4.67
4	[17]	V-slotted antenna	60x90x3	0.028	-23	1.34	4.9	5.2
5	[18]	Triangula r antenna	32x46x2	0.0306	-22.56	0.87	6	7.3
6	[19]	S-shaped antenna	54x54x1	1.5	-10	1.005	3.1	6.78
7	[20]	Tapered slot antenna	7.4x6x1.4	0.74	-10.3	1	3	7.5
8	Proposed antenna	Circular slot	6x6x0.5	1.9	-28.47	1.078	7.63 0	7.632

5. Conclusion

A novel millimeter wave based microstrip patch antenna with circular slots array has been presented at the resonating frequency of 32 GHz. The resonance frequency range of simple patch antenna and antenna with 16, 36, 64 slots are 30.91- 32.501 GHz, 31.004-32.902 GHz, 31.1- 32.999 GHz, 30.8-33.02GHz. The proposed antennas of return loss characteristics is less than -26.346dB and VSWR is in the range of 1.0784 – 1.1012. The patch with 64 slots loaded antenna has given better performance than other designed antennas. The overall dimension and acceptable performance of the proposed antenna is better than other antenna designs. Therefore, the proposed millimeter wave based antennas design is a more potential candidate for future metal-rimmed mobile phone applications.

Reference

[1] Nikita M. Tarpara, Raju R. Rathwa and Dr. Nirali A. Kotak 2018 *Desigs of slotted microstrip patch antenna for 5G application* (International Research journal of engineering and technology) Volume 5 Issue 05.

[2] He Huang, Ying Liu and Shuxi Gong 2017 *A Dual-broadband, Dual-polarized Base Station Antenna with High Isolation for 2G/3G/4G Applications* (IEEE Antennas Wireless Propagation Letter) vol. 16, pp. 1111–1114.

[3] Yejun He, Zhengzheng Pan, Xudong Cheng, Yuan He, Jian Qiao, and Manos M. Tentzeris 2015 *A Novel Dual-Band, Dual-Polarized, Miniaturized and Low-Profile Base Station Antenna* (IEEE Transaction Antennas Propagation) vol. 63, no. 12, pp. 5399–5408.

[4] Huiqing Zhai, Lei Xi, Yiping Zang, and Long Li 2018 *A Low-Profile Dual-Polarized High-Isolation MIMO Antenna Arrays for Wideband Base-Station Applications* (IEEE Transaction Antennas Propagation) vol. 66, no. 1, pp. 191–202.

[5] K.-L. Wong and C.-Y. Tsai 2017 *Half-loop frame antenna for the LTE metal-casing tablet device* (IEEE Transaction Antennas Propagation) vol. 65, no. 1, pp. 71–81.

[6] A. Al-Dulaimi, S. Al-Rubaye, Q. Ni, and E. Sousa 2015 5G communications race: Pursuit of more capacity triggers LTE in unlicensed band (IEEE Vehicle Technology Magazine) vol. 10, no.

IOP Conf. Series: Materials Science and Engineering **590** (2019) 012059 doi:10.1088/1757-899X/590/1/012059

1, pp. 43–51.

[7] Y. Yan, Y.-L. Ban, G. Wu, and C.-Y.-D. Sim 2016 *Dual-loop antenna with band-stop matching circuit for WWAN/LTE full metal-rimmed smartphone application* (IET Microwave Antennas Propagation) vol. 10, no. 15, pp. 1715–1720.

[8] J. G. Andrews 2014 *What will 5G be?* (IEEE Journal of Selected Areas in Communication) vol. 32, no. 6, pp. 1065-1082.

[9] N. Valliappan, A. Lozano, R. W. Heath Jr. 2013 *Antenna Subset Modulation for Secure Millimeter-Wave Wireless Communication* (IEEE Transactions on Communications) vol. 61, no. 8, pp. 3231-3245.

[10] Ben-Dor, T. S. Rappaport, Y. Qiao and S. J. Lauffenburger 2011 *Millimeter-wave 60 GHz outdoor and vehicle AOA propagation measurements using a broadband channel sounder (* IEEE Global Communications Conference) pp. 1-6.

[11] C. Li, B. Ustundag, A. Kumar, M. Boenke, U. Kodak, and G. Rebeiz 2018 < *0.8 dB 46 dBm OIP3 Ka band SPDT for 5G communication* (IEEE 18th Top. Meeting Silicon Monolithic Integration Circuits RF Systems) pp. 1–3.

[12] M. Parlak and J. F. Buckwalter 2011 *A 2.9-dB noise figure, Q-band millimeter-wave CMOS SOI LNA* (IEEE Custom Integration Circuits Conference) pp. 1–4.

[13] Qinlong Li, S. W. Cheung, and Changfei Zhou 2017 A Low-Profile Dual-Polarized Patch Antenna With Stable Radiation Pattern Using Ground-Slot Groups and Metallic Ground Wall (IEEE Transaction Antennas Propagation)vol. 65,no. 10, pp. 5061–5068.

[14] WeiXing Liu, YinZeng Yin, WenLong Xu, and ShaoLi Zuo 2011 *Compact U Slot Antenna with Bandwidth Enhancement (*IEEE Antennas and Wireless Propagation Letters) VOL. 10.

[15] Yang, Y.-H., Guo, J.-L., Sun, B.-H., and Huang and Y.-H. 2009 *Dual-Band T Slot Helix Antenna for Global Positioning Satellite Applications* (IEEE Transactions on Antennas and Propagation) pp 5146–5152.

[16] Amit A. Deshmukh and Ray, K. P 2009 *Compact Broadband Slotted Rectangular Microstrip Antenna* (IEEE Antennas and Wireless Propagation Letters) pp 1410–1413.

[17] Tanawat Pantui, Chatchai Suppitaksakul,Paitoon Rakluea 2009 *Dual V shaped Slot microstirp Antenna for a bi- and uni-directional radiation pattern* (IEEE).

[18] Amit A. Deshmukh, Ankita R. Jain and K. P. Ray 2013 Analysis of Broadband Pair of slot cut Equilateral Triangular Microstrip Antenna (International Conference on Advances in Technology and Engineering).

[19] Guihong Li, Huiqing Zhai, Tong Li, Long Li, and Changhong Liang 2013 *CPW-Fed S-Shaped Slot Antenna for Broadband Circular Polarization* (IEEE Antennas and Wireless Propagation Letters) VOL.12.

[20] Fan Zhang, Fu-Shun Zhang, Gang Zhao, Chen Lin, and Yong-Chang Jiao 2011 *A Loaded Wideband Linearly Tapered Slot Antenna with Broad Beamwidth* (IEEE Antennas and Wireless Propagation Letters) VOL. 10.

IOP Conf. Series: Materials Science and Engineering **590** (2019) 012060 doi:10.1088/1757-899X/590/1/012060

A OCTA- RECTANGULAR SLOT ARRAY BASED PATCH ANTENNA FOR MULTIBAND APPLICATIONS

V.Karthikeyan[a], Muddineni Raveendra[b], Challa Hemasri[b], Male Praneetha Reddy[b], B.Harshitha[b]
Assistant Professor[a], UG Student[b]
[a,b]Department of Electronics and Communication Engineering
[a,b]Vel Tech Rangarajan Dr. Sagunthala R&D Institute of Science and Technology
Chennai, India.
vkarthikeyan652@gmail.com, muddineniravi962@gmail.com
challa.hemachowdary@gmail.com, praneethareddy799@gmail.com, harshithabandi77@gmail.com

Abstract. In this paper, a octa rectangular-slot array based patch antenna for multiband applications has been proposed. The operating frequency range of proposed antenna is 11-21.817 GHz in the X, Ku and K bands. This antenna designed with octa rectangular shaped slots on rectangular patch antenna. The FR4 (Flame Retardant)substrate material is used in this antenna, which has relative permittivity 3.66,relative permeability 1, loss tangent 0.0018 and thickness of 2mm.The design has been stimulated by using HFSS software. The performance of the antenna is analysed with the help of characteristics of return loss, VSWR, bandwidth, gain and directivity at each band. It gives resonance frequencies at 11.1GHz, 14.3GHz, 16.1GHz, 17.5GHz, 19.3 GHz and 20.3GHz.

1. Introduction

Antenna is one element of the integrated low profile wireless communication systems. Generally, patch has different outlines and sizes such as circular, elliptical, rectangular and triangular [1,9]. The feed length and width of patch were adjusted through HFSS simulations to match antenna impedance. Slot length and width are directly proportional to the low resonance. A microstrip patch antenna is the low profile antenna mounted over the ground plane and separated by dielectric substrate material [7]. Microstrip antenna also exhibits disadvantages that are low power capacity and narrow bandwidth. Multiband microstrip patch antenna maintains a consistent radiation pattern with stable gain. Multiband operations can be achieved by inserting a various rectangular shape slot elements [2,4]. By using electric and magnetic field behaviours, gain and directivity has been calculated .This work of octa rectangular slot array based patch antenna has to be established to fulfil the new levels of user proficiency and multi service capabilities. It has high oath and non-ionising properties [10]. Patch antennas are good choice due to easy design with low cost for multiband applications. It is recently used in radar systems and satellite communication [3].Hence developers are trying to develop this type of antennas. Slot array based antennas has become very popular in recent decades due to their planar profile which can be keen on the surfaces of user products, there is of manufacturization using printed circuit methods. Multiple reflectors security systems can be interchanged by multiband microstrip antenna, hence it has less intricate and easy restructuring of screening. The problem of path loss occurred due to atmospheric absorption can able to overcome by using slot array based antennas [6,8].

2. Antenna design

A hexa band octa rectangular slot array based patch antenna has been developed with good matching and radiation properties. This antenna is designed by making octa rectangular shaped slots on rectangular patch of the antenna. The length and width of patch is 16mm and 5mm respectively. The ground has been made with The complete size of the proposed antenna is 5mmx16mmx2mm.The FR4(Flame Retardant) substrate material is used in this antenna, to maintain the relative permittivity 3.66 and 1 as relative permeability as well as to maintain 0.0018 as loss tangent. The guided wavelength and the efficiency of the proposed antenna is enhanced for fair radiation [5, 11, 12]. The top view and bottom view of the proposed antenna has been shown in the figure 1.

IOP Conf. Series: Materials Science and Engineering **590** (2019) 012060 doi:10.1088/1757-899X/590/1/012060

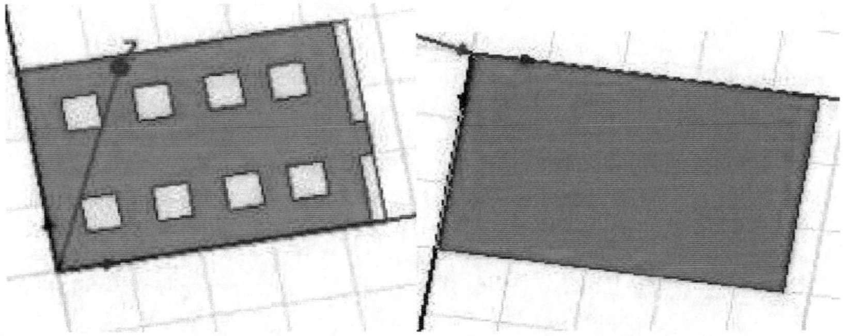

Figure 1. Top and bottom view of the proposed antenna

The following formulas are used to compute the performance of the antenna parameters such as reflection co-efficient, VSWR, directivity and gain. The lumped port has been assigned to the proposed antenna. The patch has assigned as PEC (Perfect Electric Conductor) to radiate the information signal. The total antenna design has kept inside the closed surface area to observe the radiations which are emitting from antenna. It has sown in figure 2.

Directivity is defined as

$$D = \frac{41253}{\theta_E \theta_H}$$

The reflection co-efficient (k) is defined as

$$|k| = \frac{Z0 - Zr}{Z0 + Zr}$$

Zo = Characteristic impedence of the feedline
Zr = Impedence of the antenna
VSWR (voltage standing ratio) is defined as

$$VSWR = \frac{1+|K|}{1-|K|}$$

The gain formulae is defined as

$$\text{Gain (dB)} = 10\log(\eta D)$$

where D=directivity

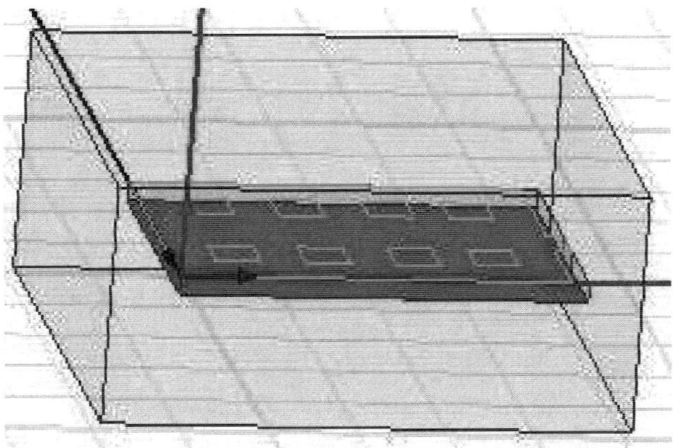

Figure 2. The structure of radiation box for designed antenna

IOP Conf. Series: Materials Science and Engineering **590** (2019) 012060 doi:10.1088/1757-899X/590/1/012060

3. Result and Discussion

The proposed antenna has been analysed by simulating antenna design using HFSS (high frequency structural simulation) software. The antenna impedance impeccably matched condition given a lower value of reflection coefficient (S11 <-10 dB) in the preferred. The reflection co-efficient of the antenna is achieved -36.9129dB, -23.6430dB, -29.7531dB, -37.3455dB,-31.0403dB and -22.5034dB at resonance frequency of 11.1GHz, 14.3GHz, 16.1GHz, 17.1GHz, 19.3GHz and 20.3GHz respectively. It means that 0.0204% power reflected back, remaining power 99.9796 is radiated at 11.1GHz by antenna, likewise radiated and reflected power is mentioned at remaining resonance frequency in the table 1. The reflection co-efficient and VSWR of the proposed antenna is displayed in the figure 3 and 4. The gain of proposed antenna at resonance frequencies are shown in the figure 5 and 6. Red colour shows that the E-plane radiation and blue colour shows the H-plane radiation in the figure 5 and 6. The antenna parameters of directivity, gain, VSWR and bandwidth are displayed in the table 1. The maximum electric field and magnetic field strengths are 5.088e+003 v/m and 9.3963e+001 v/m respectively those are presented in bellow figure 6. Table 2 is shown that the comparison of proposed antenna and other available antennas.

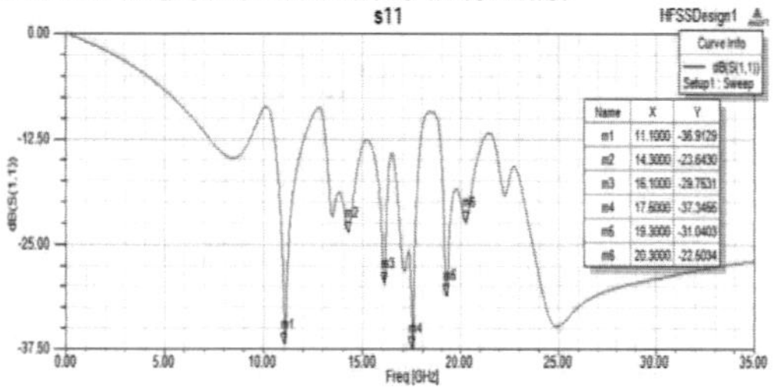

Figure 3. Return loss (s11) characteristics of the proposed antenna.

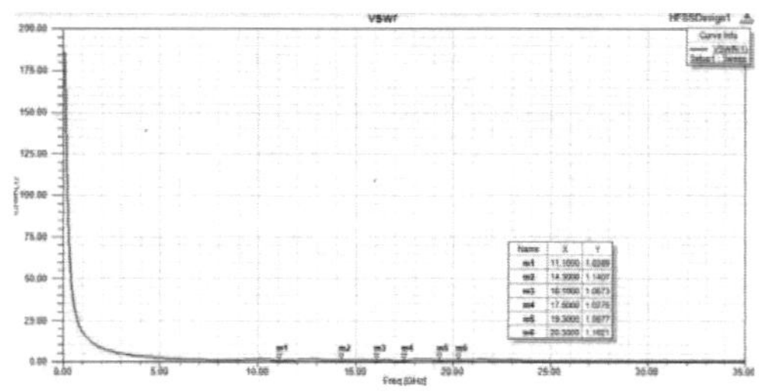

Figure 4. VSWR characteristics of the proposed antenna.

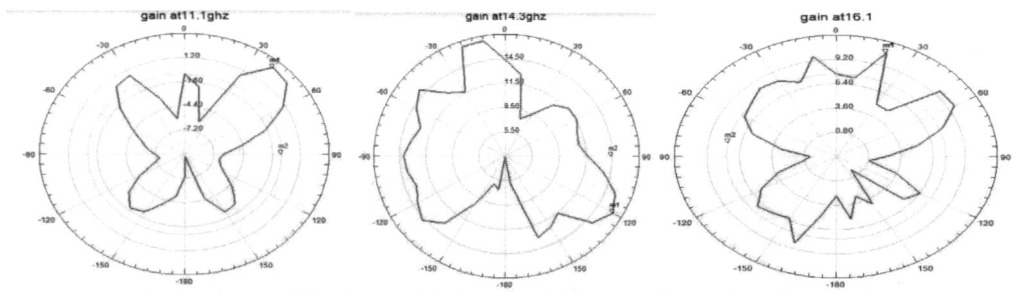

Figure 5. E- plane and H-plane at 11.GHz, 14.3GHz, 16.1GHz of the proposed antenna

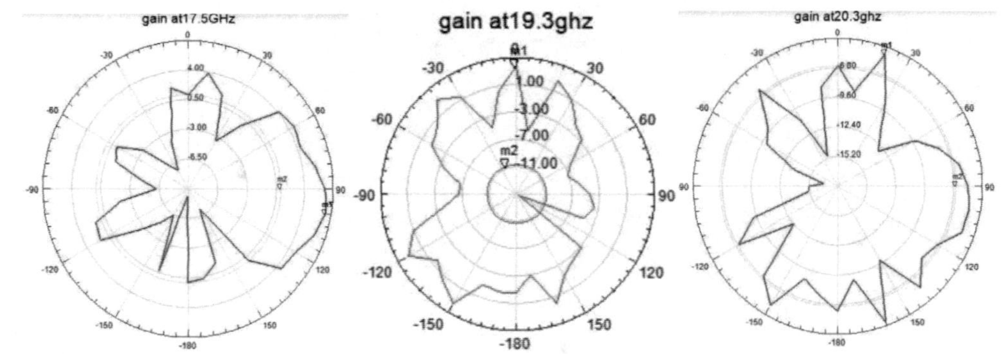

Figure 6. E- plane and H- plane at 17.5GHz, 19.3GHz, 20.3GHz of the proposed antenna

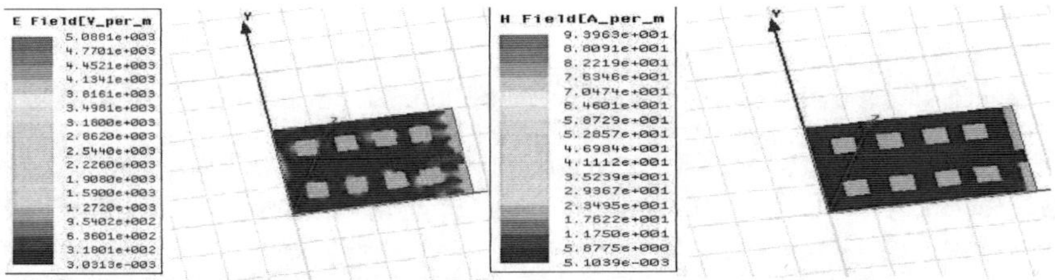

Figure 7. Electric and Magnetic field strengths of the proposed antenna

Table 1. Antenna parameters for multiple resonance frequencies

S.No	Frequency (GHz)	Bandwidth (MHz)	Return loss (dB)	Radiated Power (%)	Reflected Power (%)	VSWR	Gain (dB)		Directivity (dB)	
							E-plane	H-lane	E-plane	H-plane
1	11.1	2863	-36.912	99.9755	0.0245	1.0289	3.4575	-0.736	7.8909	4.0539
2	14.3	407	-23.643	99.568	0.432	1.1407	16.4279	14.398	5.6833	3.2565
3	16.1	303	-29.753	99.8944	0.1056	1.0673	10.7317	7.5152	6.6858	3.4693
4	17.5	189	-37.345	99.982	0.018	1.0275	6.8215	1.0976	7.7780	2.2329
5	19.3	650	-31.040	99.922	0.078	1.0577	3.7126	-10.68	6.8714	1.3247
6	20.3	2025	-22.503	99.439	0.561	1.1621	3.2807	0.9877	3.2807	0.9877

Table 2. Comparison of proposed antenna with other literature of multiband antenna

S.No	Reference	Antenna material	Antenna size (mm^3)	No. of bands	Operating frequencies	Gain (dB)
1	[13]	FR4	29x65x1.6	3	10.9/11.8/20.6	0.25/0.6/3.28
2	[14]	Rogers RT/ Duroid	49x60x0.5	3	11.5/20.4/15.8	-0.16/2.62/2.04
3	[15]	Rogers RT/ Duroid	25x22x2	3	12.5/13.5/18.8	1.98/3.15/2.68
4	[16]	FR4	50X50X1.6	3	14.4/15.2/15.8	1.9- 4.1
5	[17]	FR4	58.0x40.03x1.2	3	12.2/20.4/21.8	4.2/3.7/4.7
6	[18]	FR4	35x30x1.6	3	14.5/15.5/19.5	3.86/3.52/4.32
7	[19]	FR4	27x16x1.6	2	11.4/15.2	3.02/3.25
8	[20]	Rogers RT/ Duroid	60x45x0.5	3	11.4/14.2/19.8	0.1-2
9	[21]	FR4	28x30x1.6	4	11.6/12.5/15.8/ 19.5	2.9/2.4/3.1/1.8
10	Proposed	FR4	5x16x2	6	11.1/14.3/16.1/ 17.5/19.3/20.3	3.46/16.43/10.73/6. 82/3.71/3.28

4. Conclusion

A octa –rectangular slot array based patch antenna for multiple applications has been designed and it is resonating in the frequency range of 10.879-13.742GHz, 14.025-14.432GHz, 15.875-16.178GHz, 17.458-17.647GHz, 19.023-19.673GHz and 19.792-21.819GHz. The proposed antenna is achieved the reflection co-efficient less than -19.240dB for all resonating frequencies. Advantage of the antenna is high gain and directivity at satellite communication. The proposed antenna of gain, no. of resonating bands and the size of antenna is better that other exiting antennas. Furthermore, considering the simulated results of return loss, VSWR, gain, directivity is appropriate for radar applications, satellite communication, public safety applications and compact linear collider. Especially, this design is more suitable for space to earth communication.

Reference

[1] Cai Tong, Wang Guang-Ming, Zhang Xiao-Fei, Wang Ya-Wei, Zong Bin-Feng and He-XiuXu 2015 *Compact microstrip antenna with enhanced bandwidth by loading magnetoelectro-dielectric planar waveguided metamaterials* (IEEE Trans Antennas Propagation) pp 8–11.

[2] Lee Hakjune, Woo Duk-Jae and Nam Sangwoo 2015 *Compact and bandwidth-enhanced asymmetric coplanar waveguide (ACPW) antenna using CRLH-TL and modified ground plane* (IEEE Antennas Wirel Propag Letter) pp 10–13.

[3] Varamini Gohar, Keshtkar Asghar and Nasser-Moghadasi Mohammad 2018 *Miniaturization of microstrip loop antenna for wireless applications based on metamaterial metasurface (*AEU – Int J Electron Communication) pp 2–9.

[4] Yuan Hang-Ying, Shao-Bo Qu, Zhang Jie-Qiu, Wang Jia-Fu, Chen Hong-Ya, Zhou Hang and Zhuo Xu Zhang An-Xue 2016 *A metamaterial-inspired wideband high-gain fabry perotresonator microstrip patch antenna (*Microwave Opt Technol Letter) pp 1675–1678.

[5] Y. I. Abdulraheem Et Al. 2017 *Design Of Frequency Reconfigurable Multiband Compact Antenna Using Two PIN Diodes For WLAN/Wimax Applications* (IET Microwaves, Antennas & Propagation) Vol. 11, No. 8, Pp.1098-1105.

[6] S. Ahmed, F. A. Tahir and H. M. Cheema 2015 *A Flexible Low Cost Fractal-Slot Multiband Antenna for Wireless Applications* (European Conference on Antennas and Propagation) pp.72-78.

[7] Ali T, Khaleeq MM and Biradar RC 2016 *A multiband reconfigurable slot antenna for wireless applications* (AEU-International Journal of Electronics and Communications), pp 273–280.

[8] Ullah S, Ahmad S, Khan BA and Flint JA 2018 *A multi-band switchable antenna for Wi-Fi, 3G Advanced, WiMAX, and WLAN wireless applications* (International Journal of Microwave and Wireless Technologies) pp 1–7.

[9] Ullah S, Hayat S, Umar A, Ali U, Tahir FA and Flint JA. 2017 *Design, fabrication and measurement of triple band frequency reconfigurable antennas for portable wireless communications* (AEU-Int J Electron Communication); pp 236–242.

[10] Agrawal S 2017 *A wideband high gain dielectric resonator antenna for RF energy harvesting application* (AEU-International Journal Electron Communication) pp 24–31.

[11] Jing-Xian Liu and Wen-Yan Yin 2010 *A Compact Interdigital Capacitor-Inserted Multiband Antenna for Wireless Communication Applications* (IEEE Antennas and Wireless Propagation Letters) pp 567-575.

[12] Majid, H.A., Rahim and M.K.A., Dewan, R 2015 *Frequency reconfigurable square ring slot antenna* (IEEE Int. RF and Microwave Conf.) pp 678-684.

[13] Jwo-Shiun Sun, Han-Sheng Fang, Po-Yen Lin and Ching-Song Chuang 2015 *Triple-band MIMO Antenna for Mobile Wireless Applications* (IEEE Antennas Wireless Propagation Letter) pp 500-503.

[14] Rui Zhang, Hyung-Hoon Kim and Hyeondong Kim. *Triple-band ground radiation antenna for GPS, WiFi 2.4 and 5 GHz band application* (Electron Letter) pp 2082-2084.

IOP Conf. Series: Materials Science and Engineering **590** (2019) 012060 doi:10.1088/1757-899X/590/1/012060

[15] T. Ali and Biradar R.C 2017 *A compact multiband antenna using rectangular stub loaded with metamaterial for IEEE 802.11N and IEEE 802.16E* (Microwave Opt Technology Letter, pp 1000-1006.

[16] Lin, C. C., Yu, E. Z., and Huang, C. Y. 2012 *Dual-Band Rhombus Slot Antenna Fed By CPW For WLAN Applications* (IEEE Antennas And Wireless Propagation Letters) pp:362–364.

[17] Pan, Y., Ma, Y., Xiong, J., Hou, Z., Zeng, Y. 2015 *A compact antenna with frequency and pattern reconfigurable characteristics* (Microwave Optical Technology Letter) pp 49-57.

[18] Lin Dang, Zhen Ya Lei, Yong Jun Xie, Gao Li Ning and Jun Fan. 2010 *A compact microstrip slot triple-band antenna for wlan/wimax applications* (IEEE Antennas Wireless Propagation Letter) pp 1178-1181.

[19] T. Ali and Biradar. R. C 2017 *A compact hexagonal slot dual band frequency reconfigurable antenna for WLAN applications* (Microwave Optical Technology) pp 958-964.

[20] Tsai, L. C. 2014 *A Triple-Band Bow-Tie-Shaped CPW-Fed Slot Antenna For WLAN Applications* (Progress Electromagnetics Research) pp 167–171.

[21] Ali, T., Muzammil Khaleeq, M and Biradar, R.C. 2018 *A multiband reconfigurable slot antenna for wireless applications* (AEU-International journal of electronics and communications) pp 284-290.

IOP Conf. Series: Materials Science and Engineering **590** (2019) 012061 doi:10.1088/1757-899X/590/1/012061

Coplanar Waveguide fed MIMO RDR Antenna for ISM Band Frequencies

D.Dileepan[1], R.Sanmugasundaram[2], S.Natarajan[3]

[1,2,3]Department of Electronics and Communication Engineering,
Vel Tech Rangarajan Dr.Sagunthala R&D Institute of Science and Technology,
Chennai, India Type the author addresses here

[1]dileepand88@gmail.com, [2]rsanmu88@gmail.com, [3]drnatarajans@veltech.edu.in

Abstract. A Multiple-Input-Multiple-Output (MIMO) Rectangular Dielectric Resonator Antenna (RDRA) is presented in this paper for industrial scientific and medical applications with 2.45 GHz. The geomentry of proposed antenna is fabricated on low cost FR-4 substrate with dielectric constant of 4.4 with coplanar waveguide (CPW) inductive slot and its physical dimension is 69.3 X 73 X 1.6 mm3 (Length x Width x Height). In this work, RDRA uses two orthogonal modes with two different mechanisms and they are coplanar waveguide (CPW) and coaxial probe. The proposed MIMO RDR Antenna is expected to give various features such as isolation beyond 20 dBi range over the frequency of operation, positive gain, VSWR, Z-parameters display, 2-D current distribution using azimuthal and elevation patterns.

Index Terms— Rectangular Dielectric Resonator Antenna (RDRA), Multiple-Input–Multiple-Output (MIMO), Long Term Evolution (LTE).

1. Introduction

In human life wireless communications plays a prominent role which they can use anywhere at any time with high speed. Usage of internet became compulsory like downloading videos, software updating, uploading files and more usage of social media also. Our smartphone communications developed from 1G, 2G, 3G and 3.5G that is called as Long Term Evolution [1]. This gives perfect wide range of data services and also wireless communication services. The capacitance of the channel and data rates at high range for smart phones wireless networks had developed by Multiple Input-Multiple Output [2] –[3]. For representation of MIMO, micro strip antennas are viable for investigation on wireless communications.

It has pair of balanced inverted-L antennas builded on a rambled line, patch folding and having interlink capacitance strip linked to the patch folding [4] – [6]. Moreover for high frequencies, the efficiency of antenna decreases slowly and conductor loss becomes more. To solve this issue, a attractive feature is developed with ceramic material and applied on upper layer of the antenna radiating element [7]. These kind of antenna can be called as dielectric resonator antenna (DRA). The DRA's exibit many features that be utilized for proposal in wireless communication system because they are very small in size, light weight, conductive loss is small and easy to fabricate also having high radiation efficiency [8]. DRAs have feeding mechanisms and simplicity in shape and versatility [9] – [10]. The proposed antennas gives good isolation of above 20dB.

In this proposed work, an MIMO Rectangular Dielectric Resonator Antenna is designed to radiate at 2.45 GHz for ISM applications. To get more freedom in degree Rectangular shape is taken than considering a cylindrical or a spherical DRAs that could be utilized for determining impedance and antenna bandwidth. Techniques which are used to excite the given antenna: coplanar waveguides for port1 which is inductive slot and coaxial probe for port2. To make it to tuning at 2.45 GHz, we proposed to extend the length of port2 with a small metallic strip. The detailed analysis of simulation and scaled results and its performance are stated in detail.

2. Antenna Geometry

The structure has CPW fed and slots are introduced for inductive nature, which is fabricated on the FR4, this has the size of 63.3 × 70 × 1.6 mm3 (length×width×height). In this structure, the dimensions of rectangular dielectric resonator antenna is taken as a=37.8 and b=40 and having loss tangent and permittivity. Dielectric resonator is placed at the Mid-point of the FR-4 Substrate as depicted in Figure.1. Thus, this antenna is leaded to 49.99 impedance matching by imposing CPW of port1 & co-axial probe of port2 . In order to avoid the air gap, a coaxial probe has been used and it has been placed at the side of DRA. By achieved more than 20dB isolation. In direction to make the antenna radiate at 2.45 GHz DRA is coupled to the co-axial probe. The method ensure that the ground plane is connected to ground of port2.

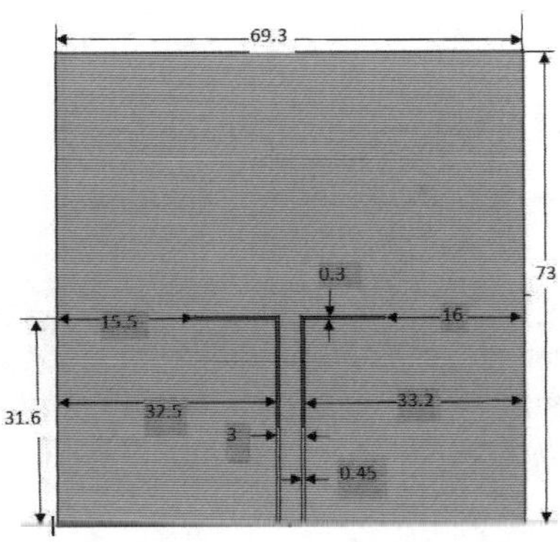

Figure 1(a) Top view.

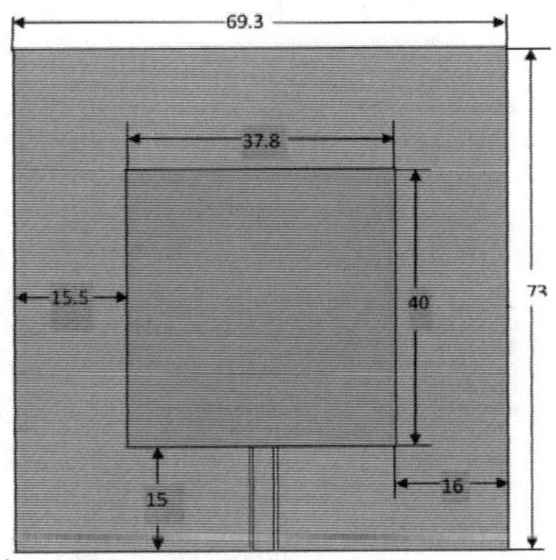

Figure 1(b) Bottom view.

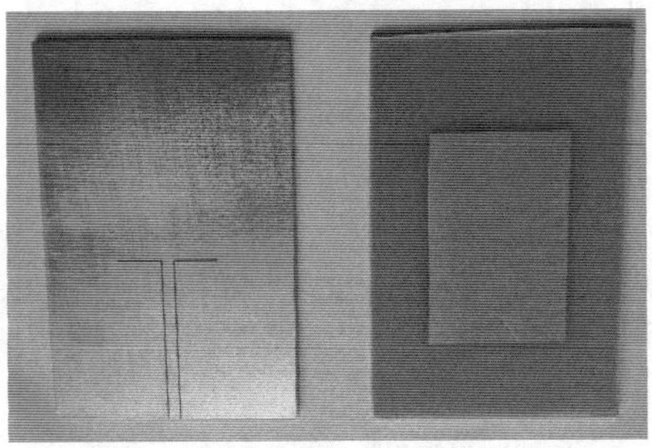

Figure 1(c) Fabricated Structure

To obtain the performance and parametric study of the dielectric resonator thickness, length of the strip and position of slot are vital.

3. Results and Discussion

To estimates the proposed antenna structure performances characteristics IE3D simulator is used to simulate first. By changing antenna parameters various investigation were carried out, to examine wide

IOP Conf. Series: Materials Science and Engineering **590** (2019) 012061 doi:10.1088/1757-899X/590/1/012061

bang coverage. The initial design of MIMO antenna is simulated and s-parameters value is shown in Figure.2. where S11 is -85dB at 2.45GHz. The antenna design is optimized with varying length, width and its characteristics were validated. After the antenna designing, it has been fabricated and tested in anechoic chamber. As per simulation results, in comparison with fabricated antenna fall within estimated band. The measured gain varies between 1 and 4 dB.

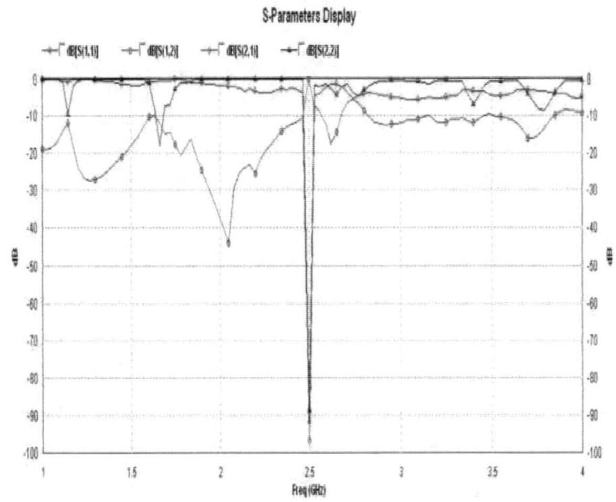

Figure 2 Return loss

Performance of patch antenna is good with respect to polarization, well in spite of the angles created in textile substrate. For θ=0 and φ=0 the propose antenna gain is 1dBi and it is shown in Figure.5 and Figure.6. The gain of the proposed structure is very high and good as the compatible materials used for the fabrication. At the frequency of 2.45 GHz, the surface field distribution & the radiation pattern of the structure is shown in Figure.4.

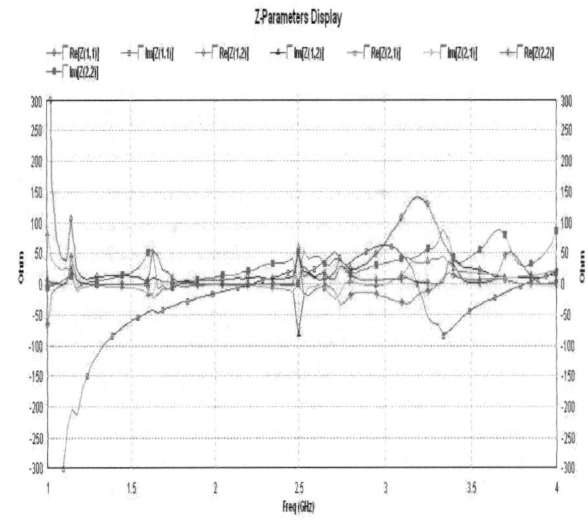

Figure 3 Z-Parameter

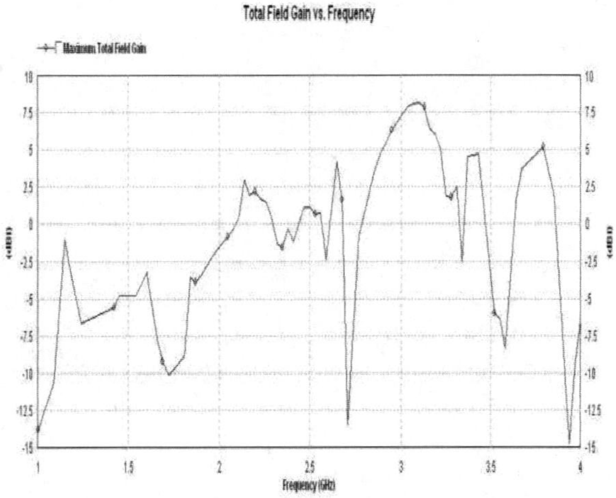

Figure 4 Gain vs frequency

Figure 3. shows Z-parameter of proposed structure has 50Ω impedance matching at 2.45GHz for square loop antenna. Figure 5 and 6 shows positive gain value for elevation and azimuth pattern. For wearable applications, the proposed structure is one of perfect solution for embedded in textile fabrics. Figure.7 shows current distribution of the structure were maximum at 2.45GHz.

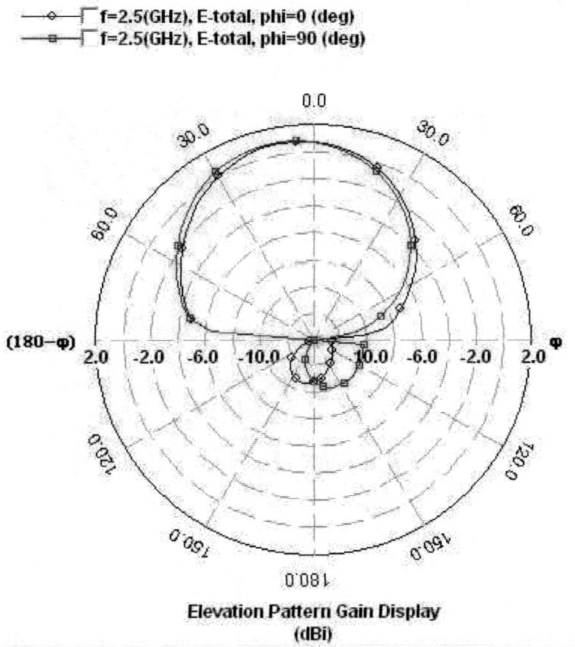

Figure 5 Elevation pattern

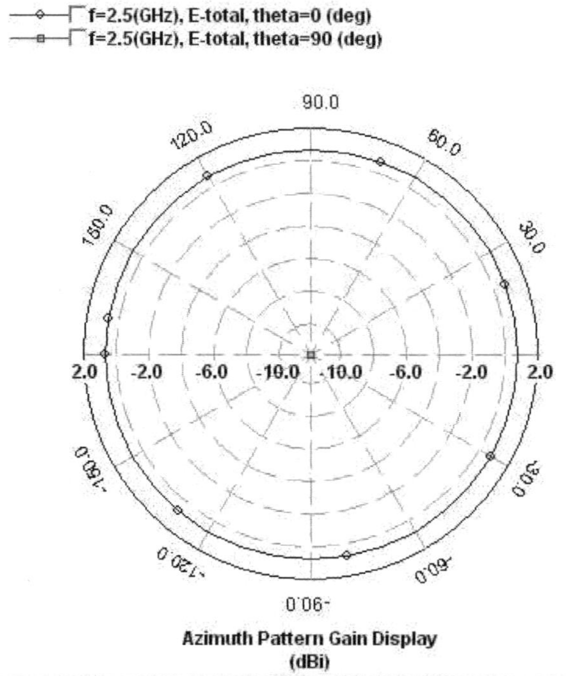

Figure 6 Azimuth pattern

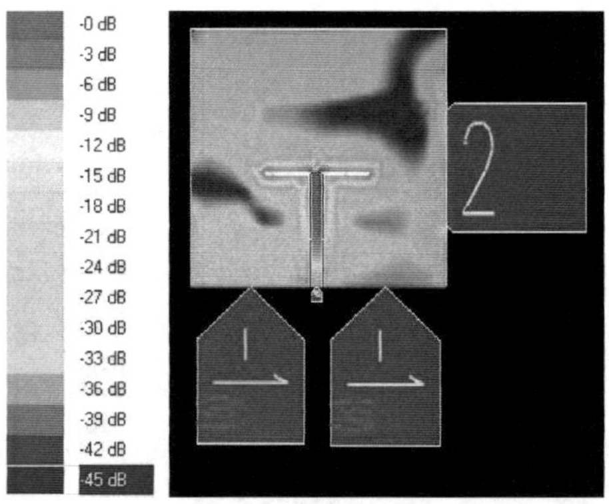

Figure 7 Current distribution

4. Conclusion

A MIMO RDRA is radiating at 2.45 GHz, with good S11, perfect impedance matching of 50Ω, VSWR at 1.2 and positive gain .The proposed structure show enough impression comparing results of measured and simulated. The antenna shows us uni-directional radiation pattern along broad side, through an operating frequency of gain for antennas. The given results shows that the antenna structure we proposed has good efficiency, gives agreeable contender for 4G applications. The investigation should be prolonged by decreasing the dielectric resonator height by improvising dielectric material with more thickness to satisfy utilization of smaller antennas in portable communication devices.

References

[1] L. Lizzi and A. Massa, "Dual-band printed fractal monopole antenna for LTE applications," *IEEE Antennas Wireless Propag. Lett.*, vol. 10, pp. 760–763, 2011.
[2] M. Mustaqim, K. Khan, and M. Usman, "LTE-Advanced: Requirements and technical challenges for 4G cellular network," *J. EmergingTrends Comput. Inf. Sci.*, vol. 3, no. 5, pp. 665–671, 2012.
[3] K. J. Babu, K. S. Ramakrishna and L. P. Reddy, "Reduction of mutual coupling in a MIMO array employing circular DRAs," *Int. J. Inf. Telecommun. Technol.*, vol. 4, no. 1, pp. 8–11, 2012.
[4] S. Ashok Kumar and T. Shanmuganantham, "Coplanar Waveguide Fed ISM Band Implantable Crossed Type Triangular Slot Antenna for Bio. Appl." *Int. J. of Mic & Wire. Tech.*, Vol.6(2), pp.167-172, 2014.
[5] Q. Rao and D. Wang, "A compact dual-port diversity antenna for longterm evolution handheld devices," *IEEE Trans. Veh. Technol.*, vol. 59,no. 3, pp. 1319–1329, Mar. 2010.
[6] Y. Gao, Z. Feng, and L. Zhang, "Compact CPW-fed dielectric resonator antenna with dual polarization," *IEEE Antennas Wireless Propag. Lett.*, vol. 10, pp. 544–547, 2011.
[7] M.Khalily, M.Rahim, A.Kishk & S.Danesh, "Wideband P-shaped dielectric resonator antenna," *Radio Engg.*, vol.22(1) pp. 281–285, 2013.
[8] W. A. M. Feed, R. K. Chaudhary, S. Member, and R. Kumar, "Wideband ring dielectric resonator antenna with annular-shaped microstrip feed,"*IEEE Antennas Wireless Propag. Lett.*, vol. 12, pp. 595–598, 2013.
[9] S. Ashok Kumar and T. Shanmuganantham, "Design and Development of Implantable CPW fed Monopole U Slot Antenna at 2.45 GHz ISM Band for Biomedical Applications" *MOTL, Wiley & Blackwel*l, US. Vol. 57, No. 7, pp1604–1608.July 2015.
[10] A. A. Kishk and W. Huang, "Size-reduction method for dielectric resonator antennas," *IEEE Antennas Propag.* Mag., vol. 53, no. 2, pp.26–38, Apr. 2011.

IOP Conf. Series: Materials Science and Engineering **590** (2019) 012062 doi:10.1088/1757-899X/590/1/012062

Detection and Content Retrieval of Object in an Image using YOLO

[1]Vinoth Kumar B, [2]Abirami S, [3]Bharathi Lakshmi R J, [4]Lohitha R, [5]Udhaya R B
[1]Associate Professor, [2,3,4,5]UG Scholars, Department of Information Technology, PSG College of Technology Coimbatore, TN, India.

[1]bvk.it@psgtech.ac.in
[2]abhishanmugam7@gmail.com
[3]bharathiravi0397@gmail.com
[4]lohithatk@gmail.com
[5]udhayaramesh1210@gmail.com

***Abstract** – It is easy for human beings to identify the object that is in an image. Even if the task is complex, human beings require only a minimal effort. Since computer vision is actually replicating human visual system, the same thing can be achieved in computers when they are trained with large amount of data, faster GPUs and many advanced algorithms. In general terms, Object detection can be defined as a technology that detects instances of object in images and videos by mimicking the human visual system functionalities. The motivation of the project is making the search process easier for the user i.e., if the object is very new for the user and he has no idea about it, he can upload a picture of that object and the algorithm will detect the object and gives a description about it. The objective of the project is to detect the object in an image, once the object is detected, the label i.e., the name of the detected object is searched in Wikipedia and few lines of description about that object is retrieved and printed. Also, the label is searched in google and the URL of the top pages with content related to the label are also displayed. The detection of object in an image is done using YOLO (You Only Look Once) algorithm with pre-trained weights. Previous methods for object detection, like R-CNN and its variations, used a pipeline to perform this task in multiple steps. This can take some time for execution, complex optimization may be involved because individual training of components is required. YOLO, does it all fastly with a single neural network. Hence, YOLO is preferred.*

***Keywords** – object detection, region proposals, optimization, YOLO, google search, description, Wikipedia, text to speech.*

I. INTRODUCTION

It is an easy task for a human being to identify the object that is in an image because of the faster and accurate neural and visual system. Even if a complex task is given, human beings can do it with minimal effort. Since computer vision is actually mimicking human visual system, the same thing can be achieved in computers by training them with large amount of data, faster GPUs and many advanced algorithms. In general terms, Object detection can be defined as a technology that detects instances of object in images and videos by mimicking the human visual system functionalities. The special feature that every object has on its own is used to classify the objects[6]. Example, when searching for circular objects, the objects at a specific distance from the center are searched. Likewise, when searching for square shaped objects, objects are checked for perpendicularity at corners and equality of side lengths. Similarly, for face detection applications standard features like eyes, lips, nose are considered and some other features like skin tone and distance between the eyes are also considered. Due to the circumstances there are some challenges faced during the detection of objects like:

- Lightning: the lightning conditions, weather conditions may vary during the entire course of the day.
- Positioning: the object in the image can be positioned in various aspects.
- Rotation: the object can be in various aspects in the image.

- Occlusion: some part of the object in the image may not be clearly visible.
- Scale: the size of the object may vary.

These are some challenges that should be taken into account while developing an object detection system.

II. ALGORITHMS FOR OBJECT DETECTION

There are several machine learning and deep learning algorithms for object detection. When machine learning approaches are used for detection, it is mandatory to define the features first. Deep learning approaches does not demand to specify the features instead they perform end-to-end detection. Machine learning methods use Support Vector Machine (SVM) and deep learning methods use CNN. R-CNN, fast R-CNN, faster R-CNN are some common algorithms for object detection.

A. R-CNN

R-CNN uses selective search. By this, bounding boxes are generated. Then, for each bounding box, image classification is done through CNN. Finally, each bounding box are refined using regression[7]. The problems with R-CNN are:

- It takes a huge amount of time to train the network as it requires classification of 2000 region proposals per image.
- It cannot be implemented real time due to the time constraints.
- Since the selective search algorithm is a fixed algorithm, no learning is happening at that stage.

B. Fast R-CNN

In fast R-CNN, the input image is fed into the CNN. Then, the region of proposals are identified and wrapped into squares[4]. Then, reshaping of regions is done using a RoI pooling layer. Then, a softmax layer is used to predict the class of the proposed region[3]. The problem with this fast R-CNN is:

- Performance degradation during testing.

C. Faster R-CNN

Both of the above mentioned algorithms uses selective search which is a slow and time-consuming process.

Faster R-CNN does not use selective search. It uses a separate network and by that it produces region proposals[5].

All of the previously explained algorithms use region based approach to detect the object in the image without looking at the complete image.

D. You Only Look Once (YOLO)

YOLO is also an object detection algorithm which uses only one convolutional network to predicts the bounding boxes and the class probabilities and thus YOLO differs from other region based algorithms[1].

III. WORKING OF YOLO

YOLO trains and tests on full images and directly optimizes detection performance. YOLO model has several benefits over other traditional methods of object detection like the following. First, YOLO is **extremely fast**. Since frame detection in YOLO is a regression problem there is no need of complex pipeline. We can simply run our neural network on any new image at test time to make predictions.

- Second, YOLO **sees the entire image** during training and testing unlike other sliding window algorithms which require multiple iterations to process a single image.
- Third, YOLO learns **generalizable object representations**. When trained on real time images and tested, YOLO outperforms top detection methods like DPM and R-CNN.

YOLO network uses features from the entire image to predict each bounding box. It also predicts all bounding boxes across all classes for an image simultaneously. This means our network reasons globally about

IOP Conf. Series: Materials Science and Engineering **590** (2019) 012062 doi:10.1088/1757-899X/590/1/012062

the full image and all the objects in the image. The YOLO design enables end-to-end training and real time speeds while maintaining high average precision[2].

Following are the steps how YOLO works,

- First it divides the input image into an S × S grid as shown in fig.1.

Fig. 1 Divide the image into S × S grid

- If the center of an object falls into a grid cell, that grid cell is responsible for detecting that object.
- Each grid cell predicts B bounding boxes and confidence scores for those boxes as shown in fig.2.
- These confidence scores reflect how confident the model is that the box contains an object. If no object exists in that cell, the confidence scores should be zero.

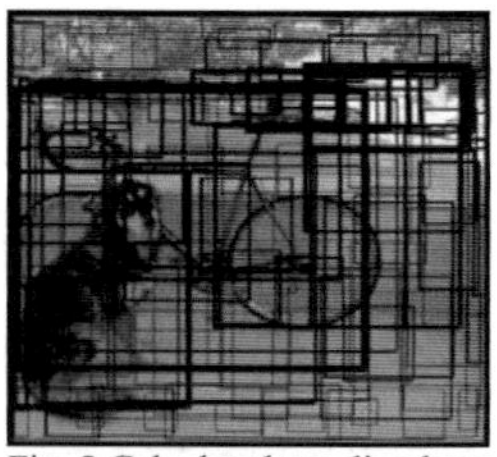

Fig. 2 Calculate bounding boxes and confidence score for each box.

- Each grid cell also predicts conditional class probabilities.
- These probabilities are conditioned on the grid cell containing an object. We only predict one set of class probabilities per grid cell, regardless of the number of boxes B.
- Finally, we multiply the conditional class probabilities as shown in fig.3 and the individual box confidence predictions which gives us class-specific confidence scores for each box as shown in fig.4.

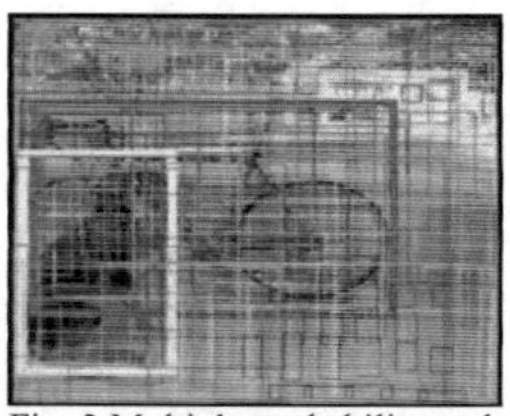

Fig. 3 Multiply probability and confidence scores.

Fig. 4 Final Output

IV. Web Scraping and Text to Speech Conversion

IOP Conf. Series: Materials Science and Engineering **590** (2019) 012062 doi:10.1088/1757-899X/590/1/012062

Web scraping is a technique that is used to retrieve the content from websites. It consists of two phases namely fetching the web page and later extracting the required content from it. Here two types of web scraping is done one is extracting the content from Wikipedia and other is top google search links for that label. The required modules are installed to system using pip.

A. Content Retrieval from Wikipedia

After detecting the object from image will use that labelled class to retrieve data from Wikipedia. It is a free encyclopedia in web. So by extracting data from Wikipedia helps the user to get a idea about what the object is and its uses. Wikipedia is a python library that will help to access and extract data from Wikipedia. In that module with a help of a predefined function Summary(), label(object name) and filter(no of lines from Wikipedia) are arguments for this function and returns a string that contains the extracted data.

B. URL Retrieval from Google

By using the label(object name) will extract top google URL's from google with the help of python module Googlesearch. By using pre defined function called Search() will extract the required URL's. In this function we can pass arguments like label(object name) , no of links need to be extracted etc. With these links they can refer more about the object other than Wikipedia content[9].

C. Text to Speech Conversion

This step will convert the label(object name) and Wikipedia content to voice so that everybody can understand better. The module used for text to speech conversion is pyttsx which is platform independent and it can convert in offline too.
But pyttsx is supported only in python 2.x versions so pyttsx3 module can used in both python 2.x and 3.x versions. Inorder to use pyttsx3 init() function need to be called to initialize the process and use a predefined method say() with argument text which needs to be converted to voice[8]. Finally use `runAndWait() to run the speech.`

V. PERFORMANCE ANALYSIS

To analyze the performance of YOLO, it compared with algorithms like R-CNN, fast R-CNN, faster R-CNN on various performance measures like time taken, accuracy and the frames per second. When analysis was done based on time taken by the algorithm to detect the objects as listed in table 1, it is found that R-CNN takes around 40 to 50 seconds, fast R-CNN takes 2 seconds, faster R-CNN takes 0.2 seconds, and YOLO takes just 0.02 seconds. From this analysis it can be inferred that, YOLO performs 10 times quicker that faster R-CNN, 100 times quicker than fast R-CNN and more than 1000 times quicker than R-CNN.

TABLE I
PERFORMANCE EVALUATION BASED ON TIME TAKEN

Algorithm	**Time taken** (in sec)
R-CNN	40-50
Fast R-CNN	2
Faster R-CNN	0.2

YOLO	0.02

When analysis was done based on the number of frames per second, YOLO performs far better than all the other algorithms as shown in fig.5, with 48 fps whereas, R-CNN processes 2 fps, fast R-CNN processes 5 fps and faster R-CNN processes 8 fps.

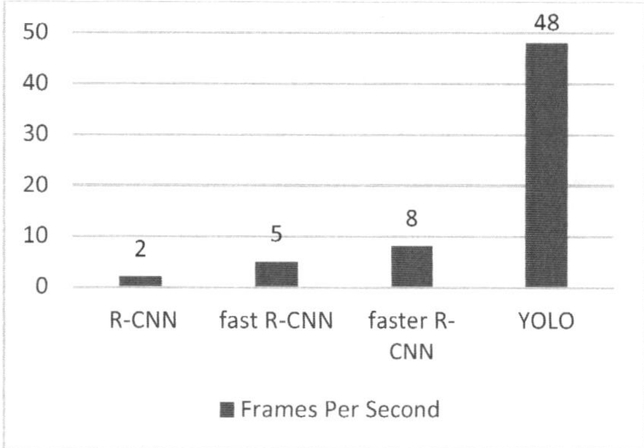

Fig.5 Performance analysis based on frames per second

When analysis was done based on the accuracy it is found that YOLO has lesser accuracy than the other three algorithms as shown in fig.6. So, it is not recommended to use YOLO for applications in which accuracy is the major concern.

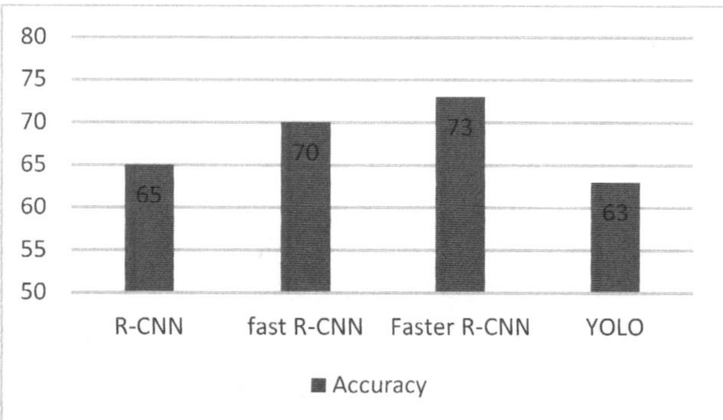

Fig. 6 Performance analysis based on accuracy

The model can be used in tracking objects for example tracking a ball during a football match, tracking movement of a cricket bat, tracking a person in a video, Video surveillance, Smart Class for students, Instructor for blind people to get details about unknown objects. It is also used in Pedestrian detection.

- **Face detection:** An example of object detection in daily life is that when we upload a new picture in Facebook or Instagram it detects our face using this method.
- **People Counting:** Object detection can be also used for people counting, it means that it is used for analyzing store performance or crowd statistics during festivals where the people spend a limited amount of time and other details .This type of analysis is little difficult as people move away from frame.
- **Vehicle detection:** When the object is a vehicle such as a bicycle or car or bus, object detection with tracking can prove effective in estimating the speed of the object. The type of ship entering a port can be determined by object detection based on the shape, size etc. This method of detecting ships has been developed in certain European Countries.

- **Manufacturing Industry:** Object detection is also used in industrial processes to identify products. If we want our machine to detect products which are only circular we can use Hough circle detection transform can be used for detection.
- **Online images:** Apart from these object detection can be used for classifying images found online. Obscene images are usually filtered out using object detection.
- **Security:** In the future we might be able to use object detection to identify anomalies in a scene such as bombs or explosives (by making use of a quadcopter).
- **Medical Diagnose:** Use of object detection and recognition in medical diagnose to detect the X-Ray report, brain tumors.

VI. Experimental Results

Since YOLO is a pretrained model it should have pre-trained YOLO V3 weights file, CFG file,text document containing the object classes in current program directory. This Script needs 4 Arguments.

- Input image – airplane.jpg
- YOLO config file - yolov3.cfg (contains details about layers or hidden layers used in Neural network)
- YOLO pretrained weights file – yolov3.weights
  (first few layers in Neural network have already learned some general factors applicable to all classes.
- Text file consists of object classes – yolov3.txt (contains object classes names)

This model is trained using COCO dataset so it has capability of detecting 80 objects.
Execute the script by this command,

Python yolo1.py --image airplane.jpg --config yolov3.cfg --weights yolov3.weights --classes yolov3.txt

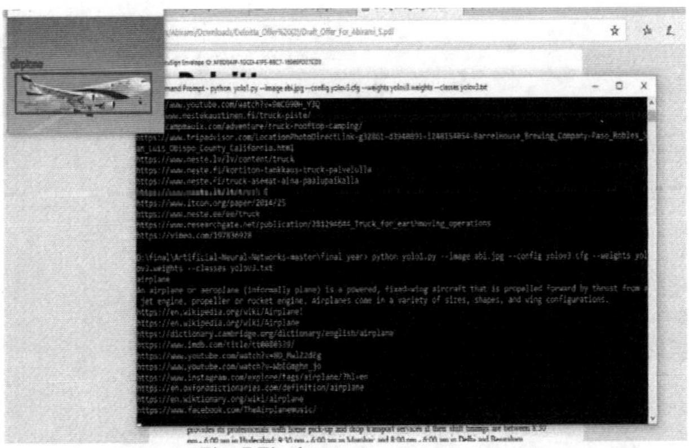

Fig. 7 Final output

Label (object class) will be printed and it will be converted to voice and then the content retrieval from Wikipedia gets executed as shown in fig.7 and similarly it will be converted to voice. Now the top Google links will be shown finally the image with the detected class label is seen with bounded box around the object. After closing this window the image with labelled object will be saved in the program directory.

Although the results are encouraging, the model has few limitations as follows:

- YOLO has strong spatial constraints and hence it cannot detect small objects which appear in groups.
- A small error in small grids can cause a greater impact in the result.
- The model struggles when object is in new aspect ratio or configuration.

Conclusion

An efficient model is developed in this research work that generates the audio of the descriptions for objects present in the given image. Identification of objects is achieved through YOLO and the descriptions for the detected objects are generated using Wikipedia package and the related URLs are retrieved using google search packages available in python library. The fetched descriptions are read out using the pyttsx3 package. Future avenue is to enhance this to an image based search engine.

REFERENCES

[1] Redmon, Joseph, Santosh Divvala, Ross Girshick, and Ali Farhadi. "You only look once: Unified, real-time object detection." In *Proceedings of the IEEE conference on computer vision and pattern recognition*, pp. 779-788. 2016.

[2] Impiombato, D., S. Giarrusso, T. Mineo, O. Catalano, C. Gargano, G. La Rosa, F. Russo et al. "You Only Look Once: Unified Real-Time Object Detection." *Nucl. Instruments Methods Phys. Res. Sect. A Accel. Spectrometers, Detect. Assoc. Equip.* 794 (2015): 185-192.

[3] Wang, Xiaolong, Abhinav Shrivastava, and Abhinav Gupta. "A-fast-rcnn: Hard positive generation via adversary for object detection." In *Proceedings of the IEEE Conference on Computer Vision and Pattern Recognition*, pp. 2606-2615. 2017.

[4] Girshick, Ross. "Fast r-cnn." In *Proceedings of the IEEE international conference on computer vision*, pp. 1440-1448. 2015.

[5] Ren, Shaoqing, Kaiming He, Ross Girshick, and Jian Sun. "Faster r-cnn: Towards real-time object detection with region proposal networks." In *Advances in neural information processing systems*, pp. 91-99. 2015.

[6] Vondrick, Carl, et al. "Hoggles: Visualizing object detection features." *Proceedings of the IEEE International Conference on Computer Vision*. 2013.

[7] Beena, M. V., MN Agnisarman Namboodiri, and P. G. Dean. "Automatic sign language finger spelling using convolution neural network: analysis." *International Journal of Pure and Applied Mathematics* 117.20 (2017): 9-15.

[8] Manaswi, Navin Kumar. "Speech to Text and Vice Versa." *Deep Learning with Applications Using Python*. Apress, Berkeley, CA, 2018. 127-144.

[9] Bharanipriya, V., and V. Kamakshi Prasad. "Web content mining tools: a comparative study." *International Journal of Information Technology and Knowledge Management* 4.1 (2011): 211-215.

Automatic Door Opening using Quantitative Gait Algorithm

[1]S.Mathumitha
PG Student ME Embedded System Technologies
Department of ECE
Sri Ramakrishna Engineering College
[1]mmswilbe@gmail.com

[2]Mr.A.Arunraja
Assistant Professor (Sr.G)
Department of ECE
Sri Ramakrishna Engineering College
[2]arunraja21@gmail.com

***Abstract*—**Quantitative Gait algorithm provides a system to identify a person from a distance. This project identifies three types of systems : Detection, Recognition and Control to open the Door.The initial process is extracting the forefront facial (human) images are converted into ordinary binary silhouette, it is captured from each frame in burst mode. The images are converted into values and stored in the database. In the process at real time, the facial recognition is done by comparing with the database. From the results it can also be observed that the person identified by raspberry pi camera and is done mainly for security purpose. This proposed system is more efficacious, consistently good in quality, and it utilize very less amount of data and power that is compared to the other existing systems. This project indicates that very better sensitivity , specificity and also the execution time of Quantitative Gait algorithms which is implemented in ARM processor therefore the results are obtained using MATLAB software.

Keywords— Raspberry Pi 3, Raspberry pi Camera, Motor driver L298, Quantitative Gait algorithm, MATLAB.

1. INTRODUCTION

The Automation technology is improving with time, internet and smart devices became more popular and reasonably priced, so the intension of this automation system is to develop an smart and secured way of automatic door access control system. People's expectancy about a smart security has modernized productively . Modern smart security system is a combination various prevalent computing devices and wireless sensor networks. The new complicated products in electronics world brought new model of security challenges to automation systems. The intensive progress of automation security system has enhanced with time period, wireless sensors and required actuators are integrated to detect and prevent from intrusions.

The facial image captured as an intruder image within a surrounding and accomplish by detecting the face with the help of Raspberry Pi camera and the further action will be executed. The Raspberry pi camera continuously captures the sequences of images instantly when the person appear near to door. The gain of this system is detection and recognition are performed by using Quantitative gait technique and the entire face recognition is concluded by comparing the values of face with the database which is already stored in the ARM Processor(Raspberry Pi 3).

2. LITERATURE SURVEY

The work of Arun Cyril Jose and Reza Malekian [2016] proposes the stages of process of verification in smart home technology using fingerprints and also the login credentials which is use to verify the other user for accessing the device through internet. The additional system is to consider a device's geographical location by using GPS and computing by fingerprint system.This system doesn't concentrated on gestures for smart security.

In their work SayidulMorsalin and A.M. Jahirul Islam [2016]developed a multi-layer home security system (HSS) is developed which maintains 5 to 6 levels of security and controls the performance against unwanted burglars and intruders. The security levels uses the Near Field Communication (NFC) tag, the other level uses a secured

IOP Conf. Series: Materials Science and Engineering **590** (2019) 012063 doi:10.1088/1757-899X/590/1/012063

password system and final level uses fingerprint system. This system makes secure and quality system gives strong work on fingerprint for security system.

Md. Raihaan Kamarudin et al [2013] developed a smart home automation system using voice recognition. The system uses home appliances from Graphical User Interface (GUI) using Microsoft Visual (MV) Basic software that use Microsoft Speech Recognition Engine (MSRE) as an input data and being controlled wirelessly. Atlast, the gesture based system analyzes to obtain better solution for the future enhancement.

Krishna Rathiet al [2017] proposed the Gesture Human-Machine Interface (GHMI) uses an accelerometer and flex sensor. GHMI mostly based on hand gesture recognition algorithm that uses flex sensor, Bluetooth model, raspberry pi and Arduino. The hand gestures are determined by the accelerometer & flex sensor. These signals are wirelessly transmitted to Bluetooth model HC-05.The Controller receives and process data send by the Bluetooth model. For the security reason unauthorized person photo is captured and send to owner via email. This paper succeeds in gesture based automation system.

Tejas Saraf and Ketan Shukla [2018] developed a system for door access system using a face recognition technique. This proposed system uses python language with cascade algorithm. And the another type of subsystem is used by Haar Cascade classifiers for better detection of object. The security device is developed by an Paul Viola and Michael Jones method. The disadvantage is that external battery supply is used for operating the system.

3. Quantitative Gait Algorithm

Quantitative Gait Algorithm is a technique used to identify user behavior to be authenticated. This technique is an combination of Detecting and Recognizing the facial parts of the person. The aid of using this technique is that the fastest method applicable for detection of face parts. The improvement of simple and adequate classifier put together for figuring the active features for the selection of better extracted feature. In the enhancement of this algorithm, the location and size of the real time face in the input database is recycled for specific case of class detection. The prevailing case of localizing the face is to examine the face recognition. The preeminent responsibility is to find the face location and size for detection and to strengthen the image to give as an input.

Figure : 1 shows flow of face recognitiondetected by quantitative gait algorithm in MATLAB Software.

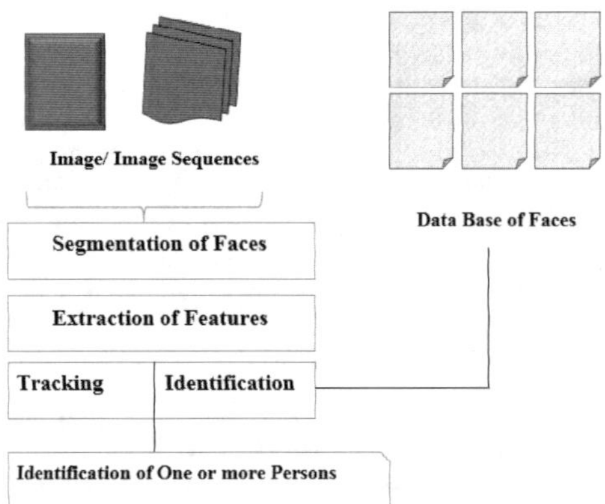

Fig.1.Flow of Quantitative Gait Algorithm

A. Face Detection

In this initial system process, the input facial images are grabbed from Raspberry Pi camera and are stored in database. Since accumulating the facial image it maintain the feature extraction and convert it into Feature-based Classifiers admit to the quantitative gait algorithm

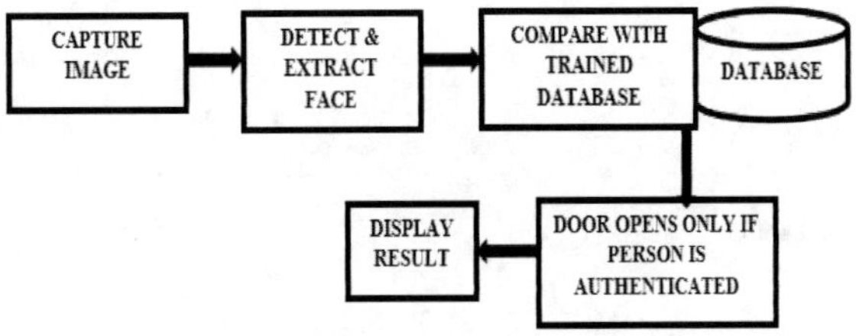

Fig.2 Flow of Face Detection

B. Face Recognition

In this segment of process, recognition of input facial images is done by comparing with database. This is the second process in identifying the person with database, where the image converted into binary values and Feature-based Classifiers matches with this feature extraction image to the database.

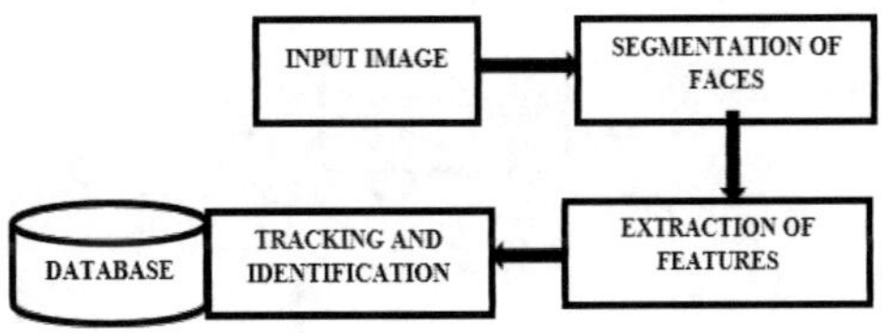

Fig.3 Flow of Face Recognition

C. Automatic Gait Algorithm

The proposed work for face recognition concept is to access the door opening system and it is implemented with Raspbian OS a popular computer vision library which can execute by Raspberry Pi 3 model B.The application of image processing is owing to important field called face recognition.

The system based on OpenCV is modeled for recognizing the facial images under best preference for overcoming the problems of biometrics in future life of the system. In the future consideration of the system is useful in identification, forensic sciences, authentication for banking system and security system also for authorized users like control for accessing the secured areas. Figure: 4 by face recognition system used here is from Quantitative gait algorithm (QGA) which is based on feature based classifiers in the project. The method used for face detection need a lot of negative and positive images to train the algorithm. This algorithm is like convolution method of matching the single value gained by subtracting the pixels under a required pixels of an image.

IOP Conf. Series: Materials Science and Engineering **590** (2019) 012063 doi:10.1088/1757-899X/590/1/012063

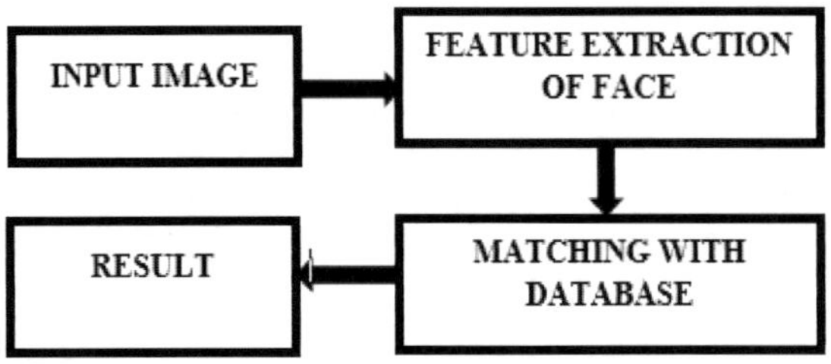

Fig.4 Flow of Door Access Control

4. HARDWARE DESCRIPTION

The proposed Embedded system shows the internal module of door accessing system. Which have the Raspberry Pi board with ARM Cortex Processor for Face recognition using Quantitative Gait Algorithm. , Motor driving circuit (L298 H – bridge), motor & Raspberry Pi camera of 8Mega Pixel.

The block diagram of Automatic door opening using Quantitative Gait algorithm using Raspberry Pi 3shown in the Figures: 5 is the controlling unit. The CD tray is driven using an servo motor which is used to open close. The Motor driver has two specific pins enabling or disabling which is connected to the GPIO pins of Raspberry pi . The authentication module which is used Hteik Htar Lwin, et al(2018) which will obtain the feature of face extraction using MATLAB.

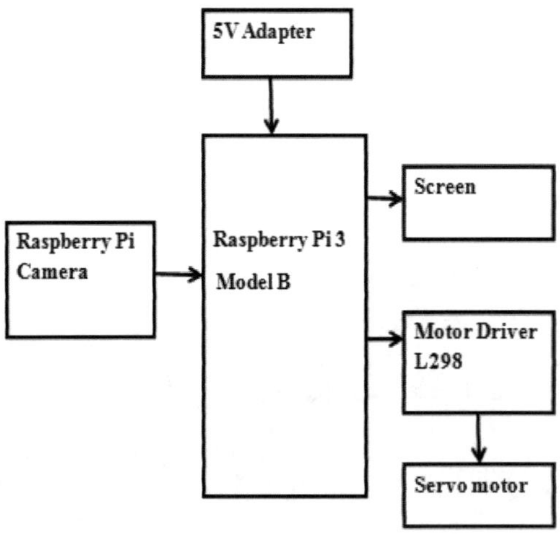

Fig.5 System Block Diagram

Fig.6 Experimental Setup

The Raspberry Pi Camera will capture images in the burst mode in real time. The captured images is compared with the respective database already saved in the OS memory card which is interfaced to Raspberry pi. The conversion takes place in the controller the images will be shown in the screen, After the completion of the comparison, the Raspberry pi allow the access to enter the person to enter. Only if the person is authenticated the door opens automatically or else it will not open.

5. EXPERIMENTAL RESULTS

A. Experimental Setup

Figure: 6 shows the Experimental setup of this project. Raspberry Pi 3 controller board with ARM Cortex processor is the main processing unit. Power supply to the Raspberry Pi 3 board is through the USB cable from PC. The board can also be powered from a DC adaptor. The algorithms are executed by the ARM Cortex processor and the images are processed and the performance measure values are displayed in the monitor. To design a Face identification from the databases by comparing where the frames are continuously sent at the time of recognition to the controlling unit for processing. The values of input image are then processed and compared with the trained database for the result.

The Raspberry Pi Camera is interfaced with the JTAG of the Raspberry Pi 3 where the camera captures the face and convert the frames and sent to the Raspberry Pi 3 as input. The output from the Raspberry Pi 3 is sent to the Servo motor interfaced with the door.

Power supply is for the Raspberry pi 3 to control the entire system. Raspberry pi camera V2 is an 8 megapixel camera capable of taking photographs of 3280 x 2464 pixels connected to J3 of Raspberry pi 3. Motor driver L298 is a high current and low voltage IC, the 2 specific pins enabling or disabling is connected to the controller.The motor is fixed in the CD tray to open close when the person is recognized.The output is displayed in the monitor.

B. MATLAB and Open CV results

A. Template Algorithm

The Template matching Algorithm compares the input images with stored patterns of faces or features in binary silhouette.

Appearance based method: A template matching method whose pattern database is learnt from a set of training images. This approach is simple to implement for face detection. It cannot achieve good results with variation in pose, scale and shape.

IOP Conf. Series: Materials Science and Engineering **590** (2019) 012063 doi:10.1088/1757-899X/590/1/012063

B. Quantitative Gait Algorithm

Quantitative Gait algorithm provides a way to automatic person identification at distance by monitor people without their cooperation. In this first step is extraction of foreground objects i.e. human's image is captured in the sequence and the binary silhouette of a walking person is detected from each frame. The sequences of images are segmented by face detector algorithm. Then extracting the gray scale images of the person and tracking the value in predefined data bases.

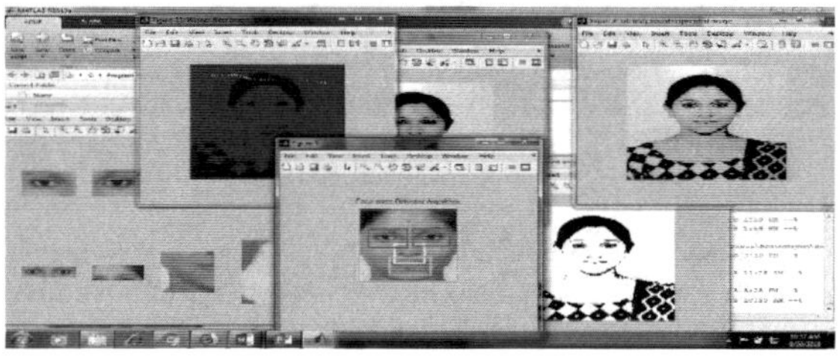

Fig.7 Extraction of individual parts and segmenting the feature

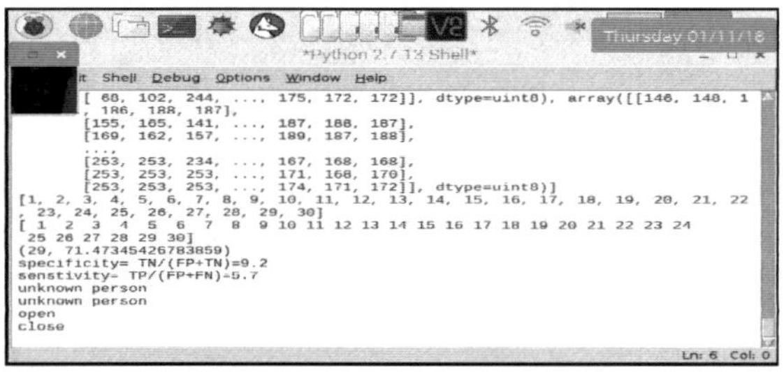

Figure 8 Output of Authentication by detecting and opening the door

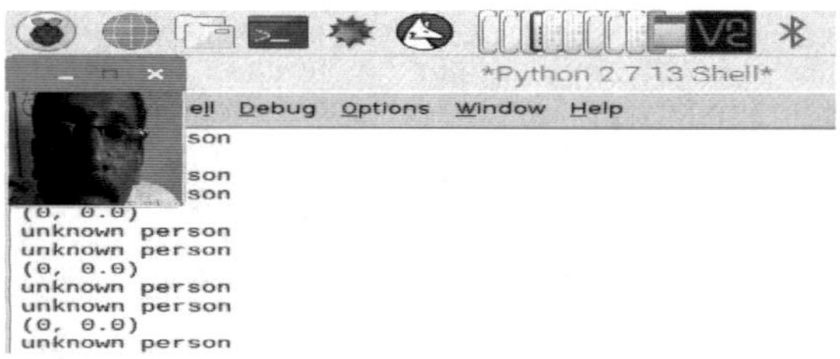

Figure 9 Output of Non-Authentication by detecting and opening the door

Figure: 7 shows the Authentication results in MATLAB using Quantitative gait algorithm .Where as figure: 8 shows the Output result of the authentication person by comparing with the databases (/home/pi/Desktop/mathu/db/student1.jpg) to (home/pi/Desktop/mathu/db/student1.jpg) which is already trained before and attained the values. In real time, the face is recognized through Pi camera and compared with the array values and opens the door automatically for the authenticated person.

IOP Conf. Series: Materials Science and Engineering **590** (2019) 012063 doi:10.1088/1757-899X/590/1/012063

6. Performance Measures

Around 20 datasets for different stages of cancer cells [1][6] have been processed in the proposed work. The algorithms were implemented both in MATLAB and ARM processor. Various parameters have been measured and tabulated for five types of images.

A.Accuracy: The measure of closeness of trained data to testing data is termed as accuracy. It is expressed in terms of percentage.

$$Accuracy = (TP + TN)/(P + N) \qquad (1)$$

*B.Sensitivity:*Sensitivity is a proportion of positive samples which are correctly classified.

$$Sensitivity = (TP)/(TP + FN) \qquad (2)$$

C. Specificity: Specificity is a measure of the proportion of negative samples that are correctly classified.

$$Specificity = (TN)/(TN + FP) \qquad (3)$$

Accuracy is used to find the exactness of facial images. Hence more accuracy means efficient in detecting the image at time. During the fourth image in Table :1 is a very good accuracy of 99% is obtained.Sensitivity finds the quality of datasets. Specificity finds the percentage of number of images captured in real time. So greater the sensitivity and specificity means there will be less error in distinguishing multiple facial images.

TABLE 1:Performance metrics in MATLAB

Images	**Accuracy**	**Specificity**	**Sensitivity**	**Execution Time**
Imge 1	88.8094	76.34	70.11	2.0019
Imge 2	91.0028	75.13	76.49	3.1013
Imge 3	90.2016	78.93	75.93	4.1911
Imge 4	91.5987	77.87	78.45	3.2387

TABLE 2:Performance metrics in ARM Processor

Images	**Accuracy**	**Specificity**	**Sensitivity**	**Execution Time**
Imge 1	97.3452	92.253	87.34	0.783
Imge 2	98.9342	87.248	84.85	0.638
Imge 3	97.8036	93.254	92.90	0.638
Imge 4	100.1423	96.389	86.13	0.438

Table 1 shows the Performance measure values of the input images processed using MATLAB. Table 2 shows the Performance measure values of the input images processed using MATLAB.From the above tabulations, that the proposed algorithm implemented in ARM processor gives much better results than MATLAB results by producing higher Accuracy value with less error. This algorithm is more robust to changes than color histogram. This is because, instead of validating several features of image pixels at a time, the algorithm is made on all the features for several stages using quantitative gait during detection stage. Therefore it is computationally simple and fast with higher detection rate. Also it is clear quantitative gait has shown the better result of performance for the images which also contain the complex background. The overall execution time is very much lesser when processed using ARM Processor than in MATLAB.

7. CONCLUSION

In this paper, design and implementation of quantitative gait algorithm based face recognition is identifying the role of multi view quantitative gait of facial images through normal Pi camera . The results shows that these features are more efficacious to identify people from distance.The result shows recognition rate of 97% with multiple databases trained in real time and recognized the results in real time manner.

This project also displays the low cost and the current used by the organization of multiple companies for security manner which can be monitored at any time. Quantitative gait algorithm can be combined with other biometrics system and this can be extended with multi modal system. So it can be used as one of the better way of authentication.

The future can be resulted in high computation and memory expandable system. The system can also be extend by fixing high quality web camera and by sending the information to cloud to convey about false intrusion and any of the changes in the environment.

REFERENCES

[1]. T .Indumathi and Dr.M.Pushparani, "Automatic Door Opening Using Gait Identification For Movement As Gesture", Journal of Engineering Technology (JET), Volume 4, Issue 1; 2016 pp. 132 - 140133, August 2016

[2]. Kunal Nayak, Siddhesh Jawale, Ramlalit Yadav " Smart Home Automation using Raspberry Pi, Motion Sensor and Android with Gesture based Controls " International Journal of Scientific & Engineering Research, Volume 8, Issue 2, ISSN 2229-5518, February-2017 .

[3]. Saad M. Darwish, "Design of adaptive biometric gait recognition algorithm with free walking directions 'The Institution of Engineering and Technology (IET) Biom., 2017, Vol. 6 Iss. 2, pp. 53-60, July 2016.

[4]. Krishna Rathi1, Dinesh Patil2, " Gesture Human-Machine Interface (GHMI) in Home Automation " International Research Journal of Engineering and Technology (IRJET) e-ISSN: 2395 -0056 Volume: 04 Issue: 06 | June -2017.

[5]. Emdad Hossain and Girija Chetty"Person Identity Verification Based on Multimodal Face-Gait Fusion" International Journal of Scientific & Engineering Research, Volume 8, Issue 2, February-2017 ISSN 2229-5518

[6]. Tejas Saraf, Ketan Shukla,"Automated Door Access Control System Using Face Recognition "International Research Journal of Engineering and Technology (IRJET) e-ISSN: 2395-0056 Volume: 05 Issue: 04 | Apr-2018

[7]. Santhoshkumar Sunderrajan and B. S. Manjunath, "Context-Aware Hypergraph Modeling for Re-identification and Summarization" IEEE Transactions On Multimedia, Vol. 18, No. 1, January 2016.

[8]. Prabhjot Singh and Anjana Sharma "Face Recognition Using Principal Component Analysis in MATLAB"International Journal of Scientific Research in Computer Science and Engineering, Vol-3(1), PP (1-5), E-ISSN: 2320-7639 Feb 2015.

IOP Conf. Series: Materials Science and Engineering **590** (2019) 012064 doi:10.1088/1757-899X/590/1/012064

Raspberry-Pi Based Secure Systems in Car

S.Sharmiladevi[1], M.Kasiselvanathan[2], Dr.S.P.Vimal[3]

[1,2]Assistant Professor, Sri Ramakrishna Engineering College, Coimbatore-22, Tamil Nadu, India.

[3]Associate Professor, Sri Ramakrishna Engineering College, Coimbatore-22, Tamil Nadu, India.

sharmiladevi.s@srec.ac.in, kasiselvanathan@gmail.com, vimal.sp@srec.ac.in

Abstract: Automotive accidents leads to not only mortality of people but also causes financial cost to both people and society. This special issue is focused on reviewing history and background of automotive system by presenting recent developments. This paper aims at giving an overview of implementing safety and security system in car for today and future development. The key enabling technology for active safety system includes drowsiness detection and Anti pinch window's effect. Other proposed concepts to be implemented are seat belt detection and alcohol detection. Driver's drowsiness is monitored using technique called image processing and it is processed through MATLAB but the main constraint in this technique is that processing speed on hardware. The idea proposed in this paper is that Open CV library is used for real time facial images analysis to warn the drowsiness of driver in order to prevent traffic accidents. Anti-pinch window system helps to avoid injuries because of power window. Seat belt alert system gives an audio output to wear seat belts for safety. Finally Alcohol detection detects a drunken driver and it will send a message to the nearby police station along with the details of the vehicle and its position. The proposed system does all the above features with the help of a Raspberry-Pi , camera and few sensors in its hard ware part along with coding computations through python which has the advantage of continuous monitoring at minimal cost thereby preventing accidental injuries to a great extent.

Keywords: open CV; Anti-Pinch; Drowsiness Detection; Power window, Raspberry-Pi.

1. Introduction

The main reason for majority of damage to society is because of road accident. As per recent survey, every year 1.34 lakh people die due to road accidents in which 70% of accidents are due to drunken drive. In USA drowsiness causes 20% accidents every year. Tens of thousands more are injured .Families are broken apart. The future of young people is clashed. Road accidents have become the leading cause of death. Drowsiness is also more or less similar to alcohol consumption where in both cases drivers concentration, alertness and attentiveness is reduced. Apart from traffic fatalities, injuries also occur inside the car. Enhancement of comfort features like power windows in car, also leads to hazardousness. Power windows in automobile have killed or injured thousands of children. In past 30 years more than 50 children died because to suffocation due to accidental usage of power windows. Power windows will exert from thirty to eighty pounds of force whereas only 22 pounds of force is enough to suffocate a children. Most drivers and passengers killed in crashes are unrestrained .Seat belts dramatically reduce risk of death and serious injury. It is proven that people who are not wearing their seat belts have high probability of being thrown out from the vehicle during crash. The above work can be implemented using CAN protocol also[10].But to overcome all these problems raspberry pi along with open CV and various sensors is used to reduce the injuries that cause because of power windows and carelessness of not wearing seatbelt. It also reduces the risk of drowsy and alcohol drunken drivers by monitoring and alerting them.

2. Related Works

A lot of research has been done about security systems in car. Nisi et al [1] have proposed a smart vehicle for prevention of accidents using eye blink sensing technology along with seat belt remainder system. This system takes Arduino Mega 2560 as a main controller and monitors with the help of various sensors and conveys message through GSM module. Roushan Kumar et al [2] aimed to design and developed the hardware for power window control mechanism using AVR ADMEGA16 microcontroller. WeslleyTorres et al [3] considered an approach for model based safety functionality for the embedded software of automobile power window system. Dian et al [4] have discussed about the features based on gradient orientation is employed to detect and describe the seat belt by using image pre-processing, the front window location and human face detection. Kiyami et al [5] have discussed the development of a new breath alcohol detector without mouthpiece to prevent alcohol impaired driving. Lea et al [6] proposed to develop a system that captures the iris image of the driver by detecting if the person is drunk and likewise to develop a reliable algorithm for iris recognition.

3. System Overview

The enhanced features helps to suppress the accidents that happen on road. The hardware setup consists of Raspberry pi ARM 11 processor, DC motor drive, LCD display, IR sensor, seat belt sensor,gas sensor and camera. Raspberry pi ARM 11 processor is a credit card sized, single board computer which acts as a heart of the system. The hardware parts like camera, IR sensor,seat belt sensor and gas sensorgives input to the Pi board .Camera is used to detect the drowsiness of the driver and IR sensor is used to find whether obstacle is present in the path of power windows. Gas sensor is used to identify thedrunken driver and Banana sensor is used for seat belt monitoring.The system arrangement is shown in figure 1.

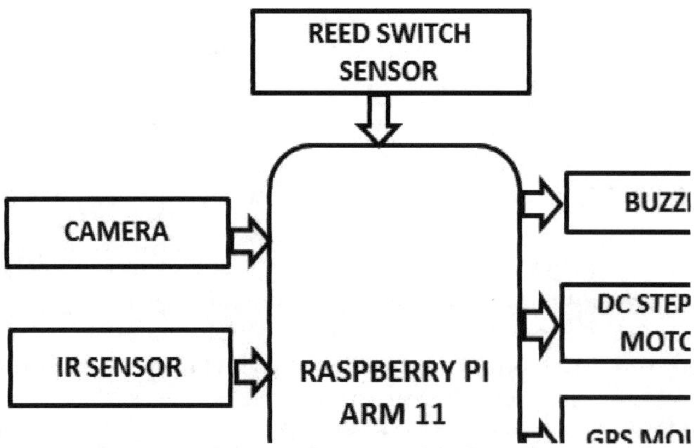

Figure .1. Block Diagram of Safety Systems

The system consists of various steps such as sensing parameters in peak time, comparing with threshold values and activating the corresponding output signal. The driver's drowsiness is detected by capturing his face and monitoring the eye movement continuously. Initially the threshold value

IOP Conf. Series: Materials Science and Engineering **590** (2019) 012064 doi:10.1088/1757-899X/590/1/012064

will be calculated by using horizontal and vertical eye points and then compared with the current real time value if it exceeds buzzer will alert. Seat belt reminder sensor detects occupants and rear seats. This information is used along with the banana sensor to trigger an audible output chime. Gas sensor which has a high sensitivity to alcohol with quick response speed is used to detect the drunken drivers and sends messages to the nearby police station through cloud.

4. Results and Discussion

4.1 Anti pinch window system

It is a safety system which is employed in power windows of modern cars. If the window finds any obstacle in its path, it prevents up further movement thus preventing any injuries. This system consists of motor which is used to operate the window. Usually, children watch out through windows of the car and they try will lend their hands or necks out through the window. In such a case, it will cause injuries or asphyxiate to children when someone operates the window accidentally. These things can also happen in such windows which are programmed without the necessity to press the switch. Toavoid such mishaps, engineers invented the anti-pinch.

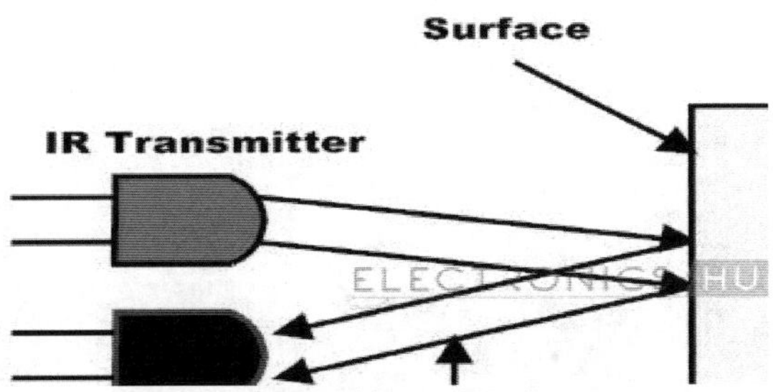

Figure.2. Working of IR Sensor

The window of the car is equipped with Infrared sensor which is connected to the stepper motor. If infrared sensor finds any obstacle it will make the motor to stop moving further and also to make it to rotate in reverse direction.The working of IR sensor is shown in Figure 2. Microcontroller is used as interface between sensor and stepper motor. Anti-Pinch window effect is shown in figure 3.

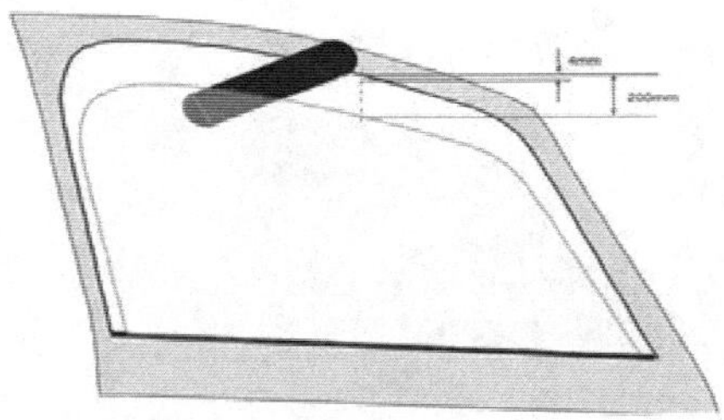

Figure.3. Anti- Pinch Technology

The controller will process the analog signal which is output of sensor to digital signal and this digital signal is fed to the motor which rotates according to that signal.Infrared technology has a wide range of applications in today's world. One of the important application area is sensing. Generally infrared region in electromagnetic spectrum is divided into near infrared region, mid infrared region and far infrared region.The sensor consists of transmitter and receiver. The transmitter will keep on transmitting infrared signal. The simulated output is shown in Figure If it finds any obstacle, the

IOP Conf. Series: Materials Science and Engineering **590** (2019) 012064 doi:10.1088/1757-899X/590/1/012064

infrared signal will get reflected after hitting the obstacle and it is received. The signal received is amplified by amplifier because the signal is very weak.It is a brushless DC motor which divides its full rotation into equal number of steps. Consist of a permanent magnet with rotating shaft, called the rotor.

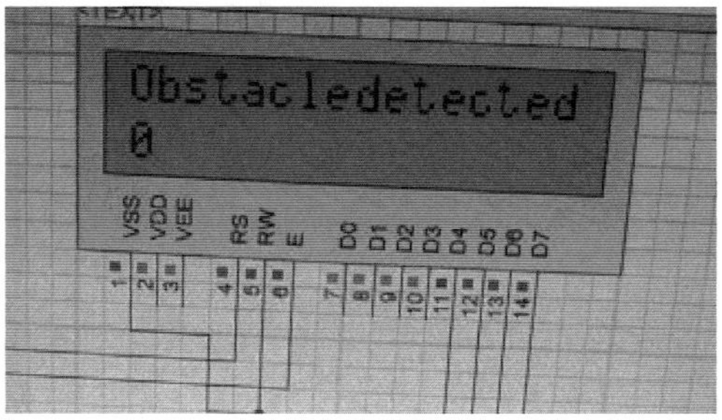

Figure.4. LCD Ouput For Anti-Pinch Window System

Depending upon the logic signal fed it rotates accordingly in clockwise or in anticlockwise direction, for each step it rotates 90 degree. For each stepper motor it has its own steps. The DC stepper motor along with PI interfacing is shown in Figure 5.

Figure.5. DC stepper motor interfacing with PI board

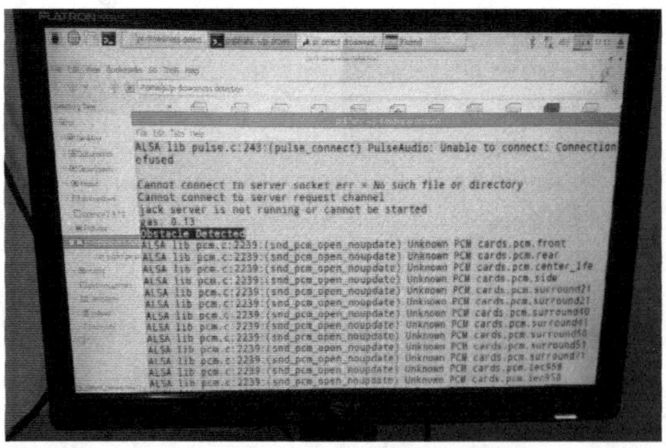

Figure.6. Simulated Anti-Pinch Window Output

4.2. Drowsiness Detection

IOP Conf. Series: Materials Science and Engineering **590** (2019) 012064 doi:10.1088/1757-899X/590/1/012064

Face detection is more difficult because of variability and dynamic nature of human face. There are many methods to monitor drowsiness, one of them is camera. It process radio stream in real time to compute the drowsiness of the driver. The calculation are made by monitoring the number of frames when eye is open and number of frames when eye is closed, The camera is placed on the vehicle dashboard for effective face capture. The system takes input image stream of a drivers eyes for analysis purpose. The frames of data stream where the driver's eyes are detected is used for calculation purpose. The output is sent to the alarm board, when it exceeds the index value, alarm gets activated. Tracking of face and eye depends on the face illumination and right intensity. Image processing is done on each frame to determine the state of driver's eye in a real time. Flow chart of drowsiness detection is shown in figure 7. The drowsiness index is calculated based on the frame.

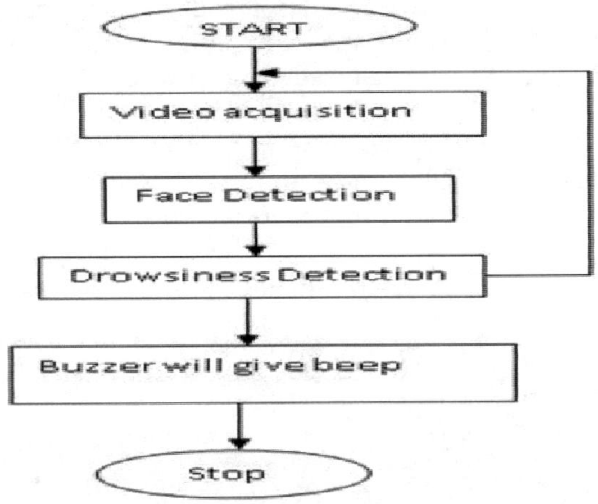

Figure.7. Flow Chart For Drowsiness Detection

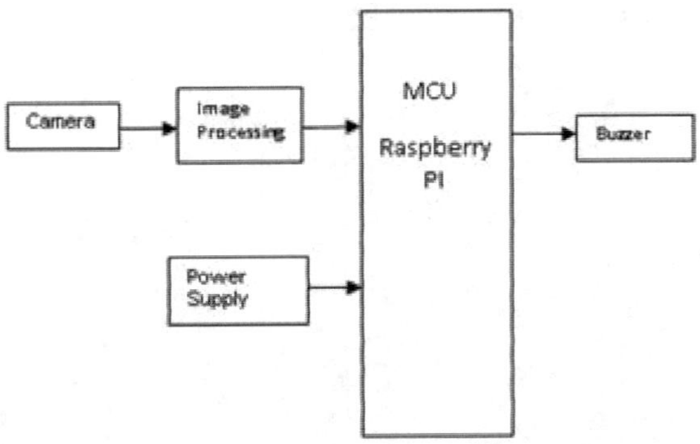

Figure.8. Block Diagram of Drowsiness Detection

The following technique is used for finding the drowsiness. Those are physiological characteristics, drive operation sensing, vehicle response sensing, and the response of driver. These can be implemented with the hardware support of raspberry pi and support of python and OS platform-Linux. The system block diagram of drowsiness detection is shown in figure 8. Due to secretion of adrenalin and acetylcholine, the area of opening eye changes. Basically, our eye movements are controlled by six muscles namely superior rectus, inferior rectus, lateral rectus, medial rectus, and superior oblique and inferior oblique. And this six muscles activity is controlled by the nerve. Due to the movements of these muscles, pupil part of eye is dilated or constricted and iris part of eye is also constricted or dilated. In order to monitor the eye movements along with the iris and pupil area, we have to use a camera for capturing the real time images of eye. For further processing on that image; we need to send the image to Raspberry-pi system. Our proposed system consists of open source 5 megapixel digital camera for capturing real time images of car driver. System is loaded with Raspbian OS and Python packages for Open CV (Computer Vision).

4.3. Alcohol Detection

Countless people's life has been cut short because of driving under alcohol influence. By drinking and driving you put not only your life at risk but also the life of passengers and life of people at road sides. So in this project we developed automatic engine locking system through alcohol detection. This system works on the principle, if a driver has drunk, the sensor will detect the alcohol consumption value in the driver's breathe and compares it with the reference value. If it goes beyond the reference value it will immediately send the location of the vehicle to the nearby police station. The sensor value is fed to the microcontroller which processes the value and compares with reference value and also operates the GPS module. Before and after detected threshold value is shown in Figure 9 and 10.Generally there are 24 satellites on the earth which gives us latitude longitude and other details of our current location. There are totally six orbits and each orbits consists of four satellites, so totally we use to get location details from four satellites with respect to our location.

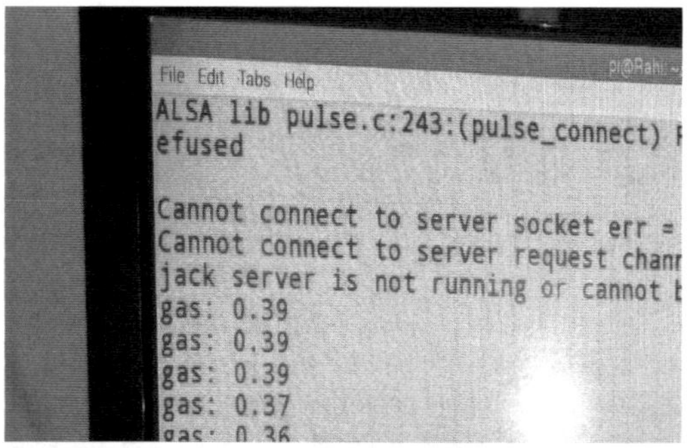

Figure.9. Threshold Value before Alcohol Detection

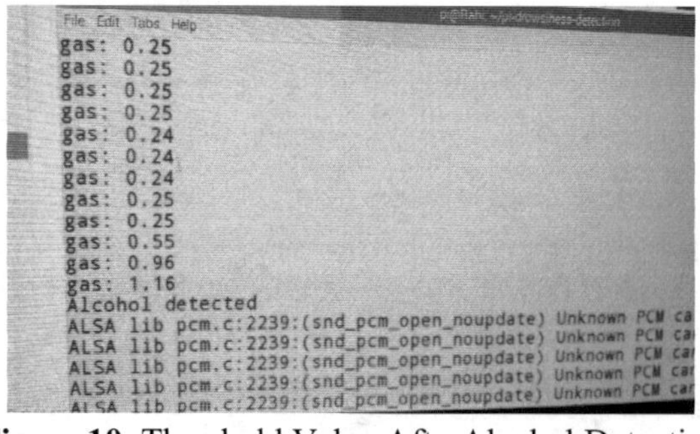

Figure.10. Threshold Value AfterAlcohol Detection

4.4. Seat Belt Sensor

As we all know, seat belts are now used by all airlines and in almost all automotive vehicles with the possibility that they may become mandatory for buses and trains as well. In recent years the Reed Switch sensor has been discovered by safety belt designers as being perhaps the best and most reliable way to detect when a seat belt has been engaged. Our seat belt remainder (SBR) sensor mat detects occupants in passenger and rear seats. This information is the vehicle to trigger a warning light and or an audible chime reminding unbuckled seat occupants to fasten their seat belts.LCD output for Seatbelt Detection is shown in Figure 13.

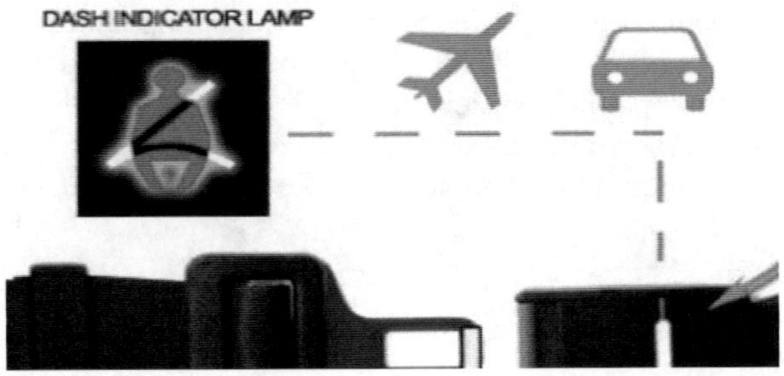

Figure.11. Reed Sensor Detection without Wearing Seatbelt

If you are getting traffic tickets for not wearing seatbelt and when you come to renewal your insurance, rates could skyrocket or even it can be canceled.Read sensor detection with and without wearing seat belt is shown in figure 11 and 12. According to recent survey still seat belts remain the number one vehicle safety device, despite new detection and accident avoiding technologies. Seat belts can also helps in keeping pregnant women alive, which is the most important thing a woman can do to protect her unborn child in a car. They are instructed to wear the lap belt low, so that it pulls downward on the pelvic bones.

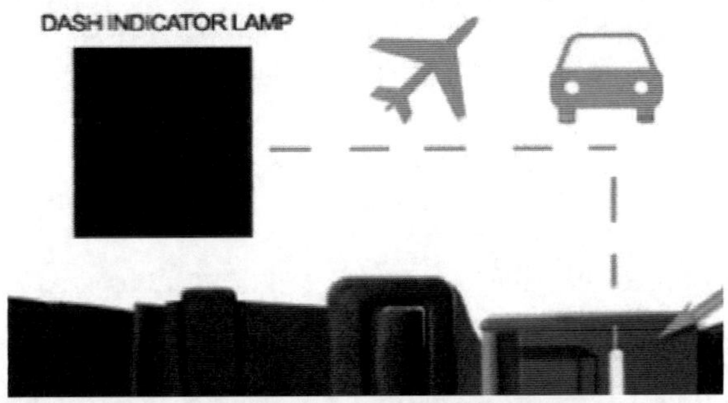

Figure.12. Reed Sensor Detection Wearing Seatbelt

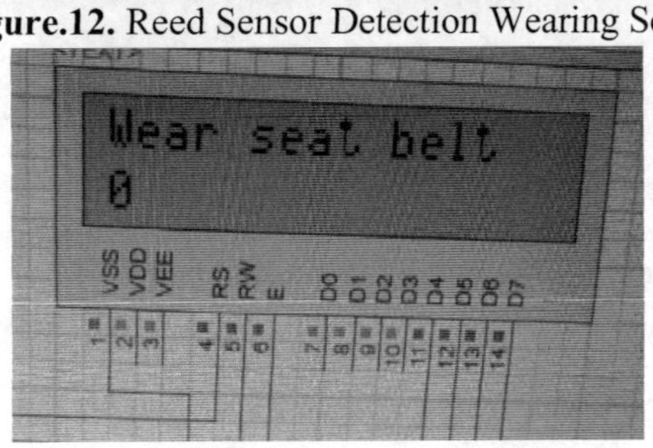

Figure.13. LCD Output for Seatbelt Detection

5. Conclusion

The proposed system is a demo version of car safety system using open CV software and other hardware. This detection is based on the high pixel camera therefore it will have more accuracy. Open CV is an open source software therefore the system is cost-effective compared to MATLAB based system. The use of ARM 11 processor also ensures the high speed. Later this system can be employed in most of the cars for enhanced safety. The system can be used for automobile applications by making required changes in the hardware setup according to the number of safety measures should perform simultaneously.The changes may include more no of cameras, high range IR sensor and advanced motors. This system provides detection of drowsiness, alcohol detection, Anti-pinch effect and seat belt monitoring system, all these can be modified according to the future need by reprogramming the requirements which includes interfacing of ignition system for enhanced safety.

References

[1] Dwipjoy Sarkar and Atanu Chowdhury, "A Real Time Embedded System Application for Driver Drowsiness and Alcoholic Intoxication Detection", International Journal of Engineering Trends and Technology (IJETT), Vol.10, No. 9, pp. 461-465, 2014.

[2] Varsha.E.dahiphale and Sathyanarayan.R, "Computer Vision System for Driver Fatigue Detection", International Journal of Advanced Research in Electronics and communication Engineering(IJARECE), Vol. 4, No.9, pp-2331-2334, 2015.

[3] Saeid F, and Parisa E, "Tracking Eye State for Fatigue Detection", International Conference on advanced in Computer and Electrical Engineering (ICACEE'2012), 2012.

[4] Lorraine Saju, Christeena Jestine, FarhanaYasmin, and Surekha Mariam Varghese, "Drowsiness Detection System for Drivers using Haartraining.and Template Matching", International Journal of Engineering Applied Sciences and Technology, Vol. 1, No. 6, pp. 106-110, 2016.

[5] Ma Weize, Zhang Shenke, Wang Hongjie, "The Design of Anti-pinch Car Windows Using Hall Sensor", Automotive Engineering, vol. 12, 2008.

[6] Hye-Jin Lee, Tae-Sung Yoon, Ra Won-Sang, Jin-Bae, "Park. Practical pinch detection algorithm for low-cost anti-pinch window control system", Proceedings of the IEEE International Conference on Industrial Technology, pp. 995-1000, 2005.

[7] Kan Fu, BO Liu, Li Ertao, "Research and Design of LIN Bus Based on Anti-pinched Electric Window in Vehicle", Journal of Hangzhou University, vol. 06, pp. 39-42, 2008.

[8] Guangmin Liu, Zhixue Wang, MA Jianhui, "Design of Sensor-less Anti-pinch Window Lifter Control Module", Modern Electronics Technique, vol. 23, pp. 198-200, 2007.

[9] Varoo, R, Prasanthi T.D, , Rajasekhar, K, Janardhanarao, T, and Satyanarayana, V, "Design of ARM based face Recognition system using Open CV library", International Journal of Advanced Research in Computer & Technology(IJARCET),vol.1,no.9, pp-233-240, 2012.

[10] Nale R.V., Srivastava S., Sinha A.K., Saravanakumar U. A CAN Protocol Based Driver Assistance System for Safety Manner. International Conference on Intelligent Data Communication Technologies and Internet of Things (ICICI), Lecture Notes on Data Engineering and Communications Technologies, vol. 26, 2018.

IOP Conf. Series: Materials Science and Engineering **590** (2019) 012065 doi:10.1088/1757-899X/590/1/012065

Early Detection of Chronicity Foot Lesion in Diabetic Neuropathy Patients

Mrs. S Saranya[1], Mrs. N Banupriya[2], Ms. S Sivaharni[3], Ms. E Suvalakshmi[4] and Ms. B Suganthi[5]

[1,2] Assistant Professor, Department of Electronics and Communication Engineering, Sri Ramakrishna Engineering College, Coimbatore - 641022, Tamil Nadu, India.

[3,4,5] Student, Department of Electronics and Communication Engineering, Sri Ramakrishna Engineering College, Coimbatore - 641022, Tamil Nadu, India.

Abstract This paper describes about the early detection of chronicity foot lesion in diabetic patients by monitoring their foot pressure. Patients with diabetic for long term and poor glycemic control are often leads to peripheral neurovascular disorder. This neurovascular disorder is caused due to Diabetes Triopathy (inadequate levels of growth factors & presence of biofilm). This neurovascular complication results in foot ulceration. 40% to 50% of diabetic patients are affected by foot ulceration. These ulceration leads to lower extremity exclusion even in the absence of critical branch ischemia. So, early detection and prevention of foot ulcer is needed. In this paper, hardware is setup to monitor the foot pressure in diabetic patients continuously with the help of flexi force sensor. For effective screening of foot ulceration, a graphical program environment LABVIEW is used. In this work, foot ulcer is detected earlier. So that it can be prevented earlier. Normalized PeakPressure (KPa)values are listed based on collected performa and had classified the diabetic patients and normal patients.

Keywords: Diabetic foot ulceration, neurovascular disorder, Ischemia, LabVIEW.

1.Introduction

The physical work like standing, walking and running are supported by our foot. Our foot is the foundation /basement of our whole body, it can adapt to uneven surface, it act as a shock absorber for our body. Loss of any one of these function will result in defect in the feet. It is often noticed in diabetic patients and it indicates that there is a risk of developing foot ulceration[5].Patients with diabetic for long term and poor glycemic control are often leads to peripheral neurovascular disorder. This neurovascular disorder is caused due to inadequate levels of growth factors and presence of biofilm[3].

In India, diabetics mellitus is the most common disease. In addition to diabetics mellitus there is another disease called diabetics insipidus. Diabetics insipidus occurs when our body can't properly balance the body's fluid levels where, Diabetics mellitus is a chronic disorder of carbohydrate, protein, and fat metabolism resulting from insulin deficiency or abnormality in the use of insulin[9]. The normal range of glucose level is 70 to 130 mg/dl.There are two main causes for diabetic mellitus: Our body fails to produce insulin; cells fail to use insulin. Diabetic mellitus patients should maintain their sugar level regularly or otherwise it leads to various problems like poor blood flow, poor sensation, and tissue damage and neuropathy or neurovascular disorder[4]. Diabetic mellitus with neurovascular disorder leads to foot ulceration. Foot ulceration is the major complication of diabetic mellitus. Diabetic foot ulcer (DFU) is mainly caused by neuropathy, ischemia, and infection lead to tissue breakdown or ulceration resulting in amputation. Foot ulceration often leads to amputation of the affected limb. If the numbness is present the foot pressure will be decreased.[2].

The initial steps for preventing diabetic foot ulcers are monitoring foot pressure daily using diabetic footwear and educating patients to take proper care of their feet. If there was any small variation in

IOP Conf. Series: Materials Science and Engineering **590** (2019) 012065 doi:10.1088/1757-899X/590/1/012065

their foot pressure it must be monitored and the patients should take their medications. It is the effective method for reducing the rate of foot ulceration in diabetic patients. With the help of current technology, these devices can be placed as an insole in our normal footwear. Since, the diabetic foot wear is cost effective the patients can use it and can prevent them from foot ulcer.

Datasets of 50 diabetic patients with and without foot ulceration were analyzed. The normalized foot pressure is measured between 400 to 750 kpa (kilo Pascal)[5]. The patients with pressure below 400 kpa are at the risk of developing diabetic foot and pressures below 280kpa are at higher risk of developing diabetic foot. Since, diabetics are one of the form of the genetic disease which comes in hereditary DNA. Overweight/Obesity is an another reason for diabetic foot ulceration.For those patients we haveto check Body mass index (BMI)[2]. If the patients BMI ranges from 25 to 29.9(kg/m2) can be considered as that they are in the early stage of foot ulcer.

1.1Classification and Management control of Diabetic foot

The stages and simple management plan [8] for each stages are briefly described:

Table 1: Stages and Management control of Diabetic Foot

Stages	**Lesion**	**Management control**
Stage 0	No open lesions	Wound control
Stage 1	Superficial ulcer	Microbiological control
Stage 2	Deep ulcer	Mechanical control
Stage 3	Deep ulcer with abscess	Vascular control
Stage 4	Local gangrene	Metabolic control
Stage 5	Gangrene of entire foot	Educational control

2.Diagonistic Process

2.1Glycoselated Hemoglobin (HbA1c)

HbA1c is a test to measure the amount of glycated hemoglobin in our red blood cells and it is formed by the combination of blood sugar and hemoglobin.The normal level ranges from 5.4-6.8mg/dl[6].

2.2Blood Pressure

This is one of the tests to diagnose foot ulcer. Blood pressure is measured using sphygmomanometer[6]. The variation in the systolic pressure of right and left leg indicates that the foot is at risk of developing foot ulcer.

2.3Doppler Test

This is another method to test foot ulcer. Doppler test is done to check the blood flow in the foot region[6]. In this test a gel is applied in foot region and a device is made to move on the foot area.Then the blood flow is recorded by sound.

IOP Conf. Series: Materials Science and Engineering **590** (2019) 012065 doi:10.1088/1757-899X/590/1/012065

3.Software used

3.1LabVIEW

National instruments is one to introduce Lab view which stands for Laboratory Virtual Instrument Engineering Workbench. LabVIEW is easy tool to create program since it uses icons instead of line of text[7].The main reason to use LabVIEW is for graphical representation of the output. The application of LabVIEW includes data acquisition and data analysis. In LabVIEW there is a tool called VISA(Virtual instrument software architecture), which was implemented by several Test and Measurements companies such as National Instruments.VISA is widely used as I/O API. Actually, VISA transmits/receives data from/to UART[10].

4.Hardware used

There are different types of sensors like strain gauge, piezoelectric force sensor, optical force sensor, flexi force sensor are available to measure pressure. Here, Flexi force sensor is used for foot pressure measurement. Force sensor consists of force gauge which is used to detect small force during a pull or push. Flexi force sensor has a 0.5" diameter, sensing area. The resistance of the Flexi force sensor will vary depending on pressure applied to the sensing area. This FSR can sense the applied force anywhere in the range of 100kg -10kg[1].PIC16F877A belongs to the family of microcontroller whichis manufactured microchip technology. There are 35 instructions, 35 input/output pins,8K ROM FLASH memory, 256 bytes EEPROM data memory, 368 bytes RAMmemory, 10 bit analog to digital converter are available in PIC16F877A[5]. The light modulating properties of liquid crystals are used by LCD(Liquid Crystal Display) also called as flat panel display which do not emit the light directly. Words, digits and 7- segment display can be displayed using LCD.LCD is mainly used for the applications such as signage, computer monitors,Televisions, aircraft cockpit displays.A computer hardware device used for asynchronous serial communication is named as UART(Universal Asynchronous Receiver and Transmitter). The transmission speed of UART is 921.6 kbits/second. UART receives/sends data to microprocessor/microcontroller through data bus. UART is used as an communication link between two devices.UART is mainly used for serial communication.

Figure 1: Hardware Setup

5.Proposed Methodology

Early detection of the foot ulceration in diabetic patients can be monitored by their foot pressure. Overweight/Obesity is the main risk factor for diabetic foot lesion[2]. The BMI (Body Mass Index) and foot pressure value directly dependent on each other. So we have to calculate the BMI value for the patients.Foot ulceration often leads to amputation of the affected part. so, early detection and prevention of foot ulcer is needed. A designed circuit is placed in foot wear of the patient which continuously measures foot pressure of the patient with the help of flexi force sensor. The output will be in terms of kilopascal(kpa). Pascal(pa) and Kilopascal(kpa) are the units of pressure. If the output is in terms of Kg/cm^2 then it should be converted to Kilopascal(kpa). The conversion is shown below. The effective screening of foot pressure, a graphical programmableenvironment is done in the LabVIEW. The hardware and software was interfaced using theUART (Universal Asynchronous Receiver and Transmitter).

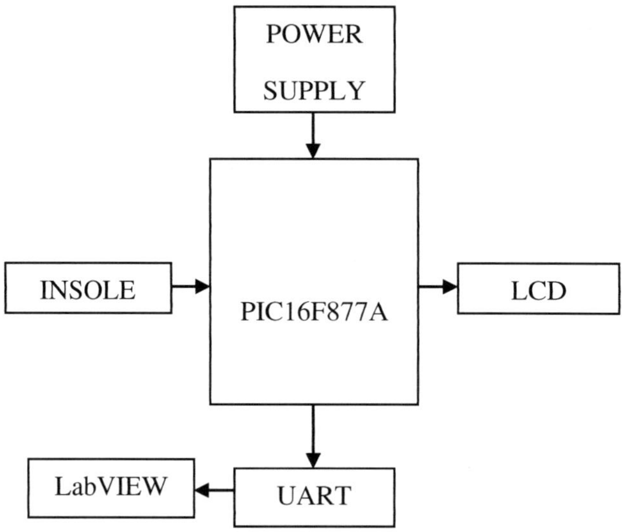

Figure 2: Block diagram

5V to 12V power supply is used to give supply to the circuit. Insole is a place where two sensors are placed. PIC16F877A is used as an analog to digital converter. LCD is used to display the digital output from the PIC16F877A.UART is used to interface PIC16F877A with LabVIEW. It transmits the dataserially to LabVIEW. LabVIEW receives the data from the UART and the graph was plotted.

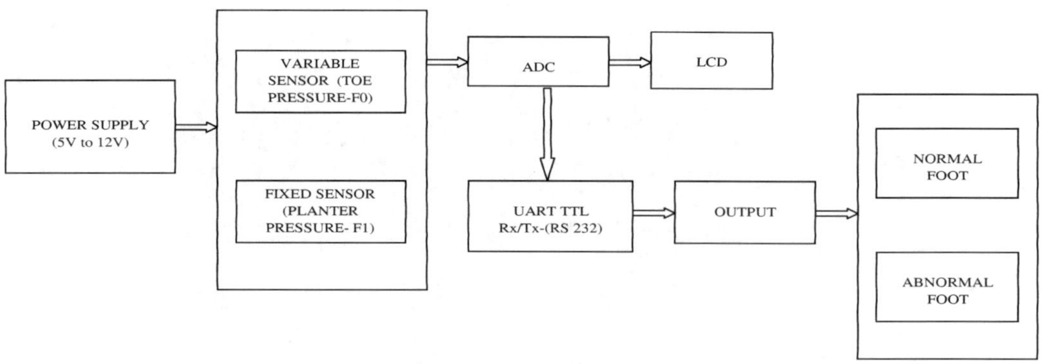

Figure3: Schematic diagram

5V to 12V supply is used to give supply to the circuit. Two force sensors used to measure the foot pressure. One sensor is placed in foot (F1) and another one is placed in toe (F0). The two sensors which is placed in insole senses the pressure given by the patients. The output of the sensor will be an

IOP Conf. Series: Materials Science and Engineering **590** (2019) 012065 doi:10.1088/1757-899X/590/1/012065

analog signal.PIC16F877A is used as an analog to digital converter. PIC16F877A has 5 ports(port A,B,C,D,E). Port A and E acts as an analog to digital converter. The two sensors are connected to the ADC pin RA0 and RA2. The analog signal from the sensor is fed to PIC16F877A. It converts the analog signal to digital signal. A 16X2 LCD is used to display the output values and it is connected with PIC16F877A. The digital output from the PIC16F877A will be displayed on the LCD display. UART is connected to both PIC16F877A and PC. UART receives data from the PIC16F877A and transmits the data to PC(LABVIEW) serially. The data from the PIC is fed to LabVIEW using UART. LabVIEW collects and Analyze the data and the graph was plotted. There is a control knob in front panel to indicate the doctors whether the patients is having normal or abnormal foot.

6.Results and Discussions

The age, weight, height, BMI, sugar level, foot pressure of diabetic patients are given below:

Table 2: Performa of the diabetic patients

S.no	**Sample**	**Age (yrs)**	**Weight (kg)**	**Height (cm)**	**BMI (kg/cm^2)**	**Sugar level(pre meals)** (mmol/L)	**Sugar level(post meals)** (mmol/L)	**Foot Pressure (kpa)**
1.	A	58	80	150	35.6	126	153	595
2.	B	59	73	156	30	80	176	634
3.	C	64	80	162	30.5	158	209	248
4.	D	47	78	157	31.6	98	185	627
5.	E	50	78	158	31.2	147	234	252
6.	F	53	76	168	26.9	152	195	465

From the above table we infer that the samples A, B, D, have normal blood sugar level and the foot pressure is also above 400kpa . The samples C, E, F have high blood sugar level 158 mmol/L ,147 mmol/L and 152 mmol/L respectively where C and E does not have a proper foot care which results in abnormal foot pressure value 242 kpa and 252 Kparesulting in foot numbness. If this is untreated it may result in foot ulcer. Even though the sample F have high blood sugar the foot pressure value is 465kpa which shows that the sample is very precautionary in foot care problems and follows proper medications regularly.

- **Formula to calculate BMI**:
  BMI = (Weight in kg / (Height in cm)^2)X10000 (1)
  Example:
  Weight in kg = 80
  Height in cm = 150
  BMI = 35.6(kg/cm^2)

- **Conversion of Kg/cm^2 to Kpa:**
  Kpa value = Kg/cm^2 value X 98.0665 (2)
  Example:
  If the flexi force sensor output is 60Kg/cm^2 then the pressure in Kpa is,
  Kpa value = 60 Kg/cm^2 X 98.0665 = 588Kpa.

IOP Conf. Series: Materials Science and Engineering **590** (2019) 012065 doi:10.1088/1757-899X/590/1/012065

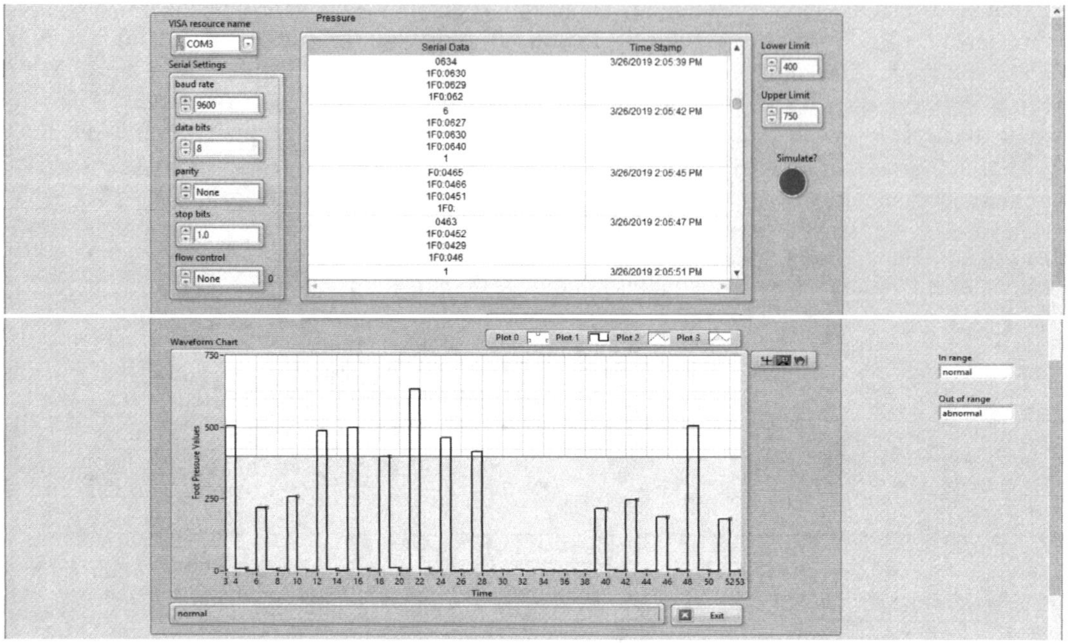

Figure 4a: Foot Pressure value of normal sample through LabVIEW Simulation

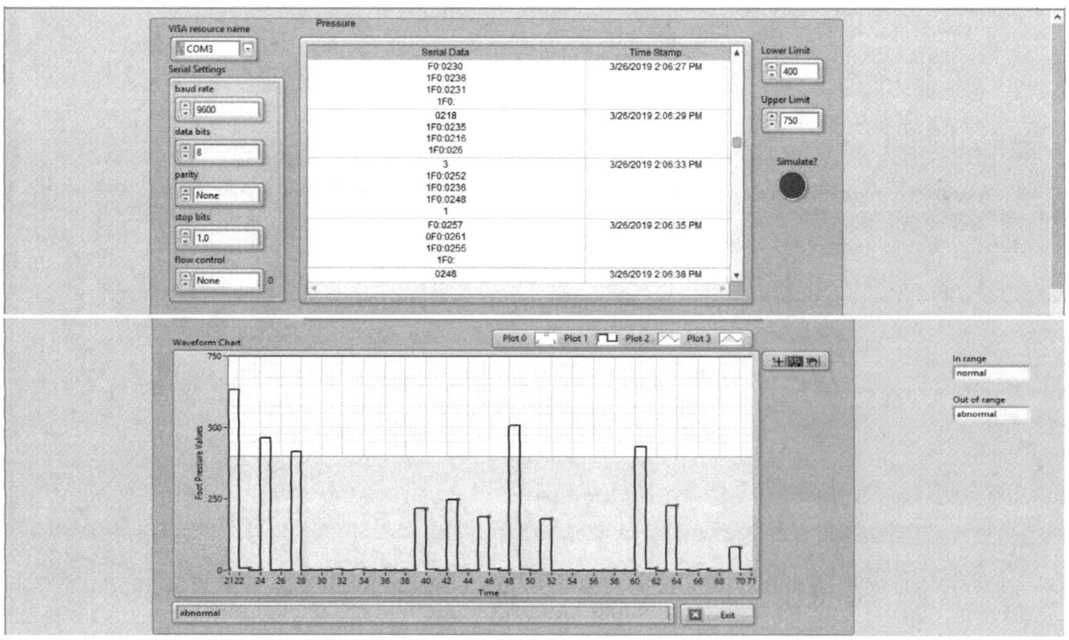

Figure 4b:Foot Pressure value of abnormal sample through LabVIEW Simulation

IOP Conf. Series: Materials Science and Engineering **590** (2019) 012065 doi:10.1088/1757-899X/590/1/012065

The above figure 4a and 4b represents the front panel of LABVIEW at two different scenarios. The graphical representation consists of time and Foot pressure in X and Y axis respectively. The figure 4a shows the pressure valuewhen a Diabetic Patient with Proper Foot care uses the hardwarewhere figure 4b showsthe pressure value when a Diabetic Patient with no Proper medication and meager foot care uses the hardware. Based on the above calculation and numbness relation with pressure, the upper and lower limit of pressure values are fixed as 750kpa and 400kpa respectively for normal patients. If the pressure given by the patient is below 400kpa then it will indicate as abnormal foot.

7.Conclusion

In this paper, foot pressure is measured to detect the foot ulcer earlier. The foot pressure for both normal and diabetic subjects were analyzed. The results indicates that the subjects with the foot pressure range between 400kpa to 750kpa are having normal foot and the subjects with the range below 400kpa are having abnormal foot. Monitoring and detecting of foot ulcer earlier reduces foot amputation problems.

In future, the footwear circuit can also be designed for normal person. For example, Now a days girls are used to wear heels. So, that many problems will arise like foot pain, unnatural position which inhibits blood circulation to the feet. So, they are also in need of monitoring their foot pressure.

References

[1]. Adan Saeed, Mehrdad Nourani, Sarah Ostadabbas, " Sensor architectural trade off for diabetic foot ulcer monitoring", 34th Annual International Conference of IEEE Engineering in Medicine and Biology Society, August 2012.

[2]. Antonio Abelha , Jose Machado, Jose Nevel, Paulo Novais, " Artificial Neural Networks in Diabetes Control", Science and Information Conference, Jul 28-30,2015.

[3]. Ashok Damir, " Why Diabetic foot ulcer do not heal?", JIMSA, Vol 24, Number 4, October to December 2011.

[4]. N.Banupriya, N.Sahanaa Sree, " A Cloud Based Risk Prediction of Coronary Heart Disease", International Journal of Applied Engineering Research ISSN 0973-4562,Vol 13, Number 5, Page number 2786-2790,2018.

[5]. U. M. Chaskar,Madhuri A. Thatte , S. L. Patil, ,"Development of Planter Foot Pressure Distribution System Using Flexi Force Sensors", Sensors & Transducers Journal, Vol. 108, Issue 9, September 2009.

[6]. S.Dhivya ,A.N.Nithya,R.Premkumar,M.Vennila, " A Real Time Foot Pressure Measurement For Early Detection Of Ulcer Formation In Diabetic Patient Using LabVIEW",International conference on design and manufacturing , Page number 1302-1309, 2013.

[7]. John Essick, *Hands on Introduction to LabVIEW for Scientists and Engineers*, Oxford University.

[8]. Michel E Edmond,Alethea VM Foster, Managing the Diabetic Foot,2nd Edition, Blackwell Publishing,London

[9]. Ning Wang, Guixia Kang, " A Monitoring System for Type 2 Diabetic Mellitus", 14th IEEE International conference on e- Health Networking, 2012.

[10]. S.Saranya, M. Shankar, N.Muthulingam, T.Sakthivarman, " Intelligent Automobile System For Accident Prevention and Detection", International Conference on Engineering Trends and Science & Humanities, ISSN: 2348-8549,,2015.

IOP Conf. Series: Materials Science and Engineering **590** (2019) 012066 doi:10.1088/1757-899X/590/1/012066

Design of phase coded spiral resonator based Chipless RFID tag for tracking applications

P Prabavathi[1], S Subha Rani[2], R Aiswarya Thangam [3]
[1]Assistant Professor, [2]Professor & Senior Member *IEEE*,
Department of Electronics and Communication Engineering, PSG College of Technology, Coimbatore, Tamil Nadu, India
[1]pprabavathi@gmail.com, [2]ssr.ece@psgtech.ac.in

Abstract. RFID (Radio identification) is a technology which deals with the wireless identification of people and products. This chipless RFID tag design is on research to replace chipped RFID tags, which employs high cost silicon wafer to store unique identification code. The main challenge in the design of chipless RFID tag is how to encode the data without a chip. This challenge is overcome by the use of electromagnetic properties of the resonators, filters etc to encode the data bits. In this work RFID tag is designed based on multiresonators with each resonator representing a data bit. The proposed tag consists of the multiresonating circuit to store information bits and transmitting and receiving UWB antenna for wireless communication with the RFID reader for identification. The multiresonating circuit designed consists of 4-spiral resonators electromagnetically coupled to the microstip coupled line. Each spiral resonates at unique resonating frequency representing the unique data bit. The data encoding is carried out with the help of phase ripples at the resonating frequency of the spirals. Three variations are observed and corresponding 3-logic states (0, 1, 2) are encoded. The designed tag is simulated in Advanced Design System (ADS) using RO-4003 substrate with the dielectric constant 3.38 and thickness of 20 mil. From the proposed methodology, a total of 3^4=81 unique tags can be designed that can be employed in cheap item tracking. The Monopole disc antenna is designed which operates for wide band covering all resonating frequency of the spiral resonator.

Keywords— chipless RFID tag, Multiresonators, microstip coupled line, Monopole disc antenna.

1. Introduction

RFID (Radio Frequency Identification) can be defined as the Automatic identification technology which used to identify objects carrying tags when they come close to a reader. It is recently being used in a wide range of applications such as Supply Chain Management (SCM), health care, traffic monitoring, retail, access control, etc. The idea behind the RFID tag is to store a unique identification number, same as that of a bar code or a magnetic strip. To retrieve information stored in the bar code or magnetic strip, the device must be scanned in a close proximity with its scanning device. But in RFID, the data transfer between RFID tag and RFID Reader is done wirelessly and hence enables remote identification. The paper [1] deals with the design of spiral resonator based RFID tag. Each spiral resonator resonates at unique frequency and three states in any resonant frequency (encoding in base-3) can be realized and more data bits can be encoded using the same bandwidth as base-2 encoding method[3].

Microstrip Spiral Resonator for the UWB chipless RFID Tag [2] deals with the simulation of the single spiral resonator coupled to the transmission line. Employment of circular and rectangular resonators in the multiresonating circuit of the chipless RFID is compared with its simulations results and the area occupied.

In [3], the tag encodes data based on spectral signature using a multiresonator. Both amplitude and phase components of the spectral signature are used for data encoding with base-2 coding methods. A novel chipless RFID tag using spiral resonator to achieve the pentamerous data encoding form [4] and deals with the new method of data encoding to increase the number of coded bits in multiresonator circuit. The phase of scattering parameters (S-parameters) at a certain frequency is considered for coding bits. The different phases of S parameters created by different coupling positions of the spiral resonator are considered for bit encoding.

This book [5] gives a complete review of the chipless RFID tag and the advantages of chipless tag over chipped tags and its future growth of replacing barcode. It also gives an complete overview of the employment of spiral resonators in chipless RFID design. In [6], a fully passive flexible chipless RFID system is presented in this paper. The designed chipless tag uses the magnitude and phase of the frequency signature of a multiresonator circuit for encoding of data bits. The tag comprises of a microstrip spiral multiresonator and cross-polarized transmitting and receiving microstrip UWB monopole antennas.

The chipless RFID transponder consists of three microstrip patch antennas, which are loaded with open circuited stubs. The antennas are resonant at nearby frequencies, and when excited with their respective resonant frequency signals, they re-radiate backscattered signals with distinct phase characteristics. This phase information is encoded as hexadecimal bits [7].

2. Proposed work

2.1. Multiresonating circuit design

The proposed circuit consists of transmission line with coupled spiral resonators along its sides. The line is matched to 50Ω at the both ends and the spiral resonators have different resonance frequencies. Every spiral resonators produces attenuation in magnitude and ripple in phase at its resonating frequency. The resonating frequency can be modified by changing the dimensions of the spiral resonators. Every resonating frequency carry a bit binary data bit, and by increasing the number of data bits more bits can be encoded. The following Table 1. shows the design specification of the proposed design.

2.2. Coupled line and Multiresonator design

The concept in the coupled transmission line is achieved using two couple line between the input and output ports with impedance matching of 50Ω at its end. The geometry is depicted as follows, Length of the coupled line=11mm, width of the coupled line=0.5 and spacing between coupled line=0.24mm.

Table 1. Design specification of the proposed tag

Design tool	ADS (Advanced design system)
No of spiral resonator	4
Bit capacity(Base-3 code)	$3^4 = 81$ unique ID's
Size of the tag	13mm X 13mm
Substrate	Ro-4003 substrate with the dielectric constant 3.38 and thickness of 20 mil
Resonating frequencies	2.95GHz, 3.19GHz, 3.37GHz, 3.60GHz

The 50Ω impedance matching is achieved by the following equations, for Z_0=50Ω, thickness of the substrate H=20 mil, Dielectric constant ε_r=3.38,

$$\frac{W}{H} = \frac{8e^{A}}{e^{2A}-2} \tag{1}$$

Where, w=width of the 50Ω line,

$$A = 2\pi\frac{Z_0}{Z_f}\sqrt{\frac{\epsilon_r+1}{2}} + \frac{\epsilon_r-1}{\epsilon_r-1}\left(0.23 + \frac{0.11}{\epsilon_r}\right) \tag{2}$$

The length of the 50Ω transmission line,

$$l = \frac{90^\circ \frac{\pi}{180}}{\sqrt{\epsilon_{eff}} K_0} \tag{3}$$

Where,

$$\epsilon_{eff} = \frac{\epsilon_r + 1}{2} + \frac{\epsilon_r - 1}{2}\left(1 + \frac{12H}{W}\right)^{\frac{-1}{2}} \tag{4}$$

From the above equations the length of the line=3mm and width of the line=1.15mm.

The 4 spiral resonators are designed which resonates for a unique frequencies, this could be achieved by using varying number of turns and length of the spiral resonators. The spiral resonators are designed to resonate at different frequencies. This unique frequency resonation is being achieved by the variation of the length of the spiral resonator. The designed spiral resonators specification is shown in the Table 2.

Table 2. Designed spiral resonator and its resonating frequencies

Spiral Resonators	No of turns	Length of the resonator (mm)	Width of the resonator (mm)	Space (mm)	Resonating Frequency (GHz)
Resonator 1	4	3.8	1.5	0.1	2.95
Resonator 2	4	3.6	1.5	0.1	3.19
Resonator 3	4	3.4	1.5	0.1	3.37
Resonator 4	4	3.2	1.5	0.1	3.60

2.3. *Antenna design*

The Monopole disc antenna designed covers all the resonating frequency of the spiral resonators. This Monopole UWB antenna is used for the reception of interrogating signals from the reader and transmission of the encoded information to the reader. The designed ultra wideband monopole antenna has the following specifications as shown in the Table 3.

Table 3. Designed antenna parameters

Design tool	ADS (Advanced design system)
Substrate	Ro-4003 substrate with the dielectric constant 3.38 and thickness of 20 mil
Radius of the disc	25mm
Length of the feed line	32mm
Width of the feed line	2mm
Length of the ground plane	34mm
Width of the ground plane	60mm

2.4. *Bit encoding technique*

The bit encoding method employed in this design is the presence and absence of resonating frequency band. In every literature discussed, absence or presence of the stop band in each frequency can be translated to logic states

- Presence of stop band -logic 1
- Absence of stop band -logic 0

In this design, for each resonating frequency 3-logic states are employed as the base-3 code (0, 1, 2).This is achieved by encoding the corresponding phase ripples of the each spiral resonator with respect to frequency. Hence coding capacity is improved with higher bandwidth efficiency.

3. Results and discussion

The designed coupled line with impedance matching is shown in the Figure1. Based on the operating frequency of the coupled transmission line, spiral resonators are designed. The Figure 2. which shows the S_{21} magnitude and phase variation for coupled transmission line of 7mm, the start frequency is from 1GHz up to the range of 12GHz.

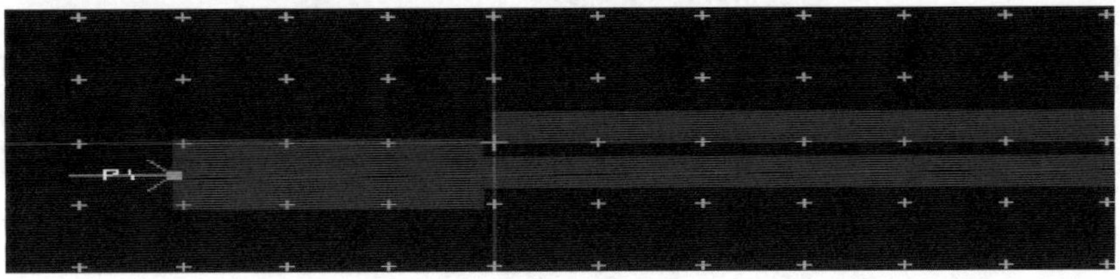

Figure 1. Layout of the proposed coupled line

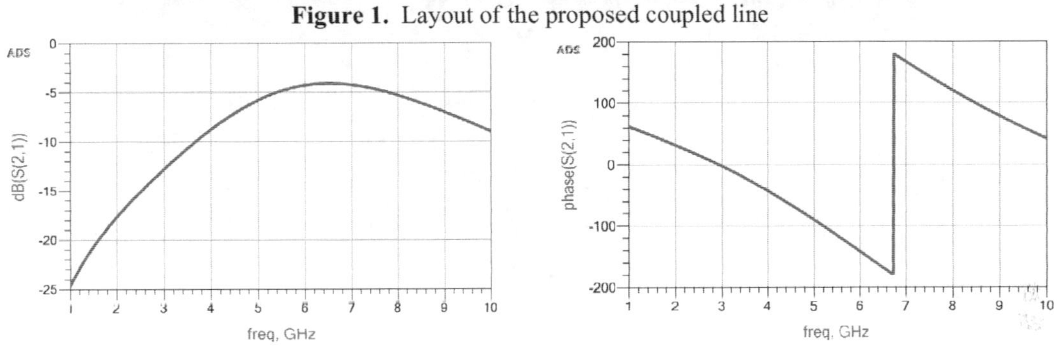

Figure 2. S_{21} magnitude and phase simulation for the designed coupled line

Each spiral resonator coupled to the transmission line with a distance of 0.1mm acts as the band stop filter, producing attenuation and phase ripples at resonating frequencies. This phase ripples corresponding to each spiral is taken for the corresponding bit encoding, when the spirals are coupled above the line, the current distribution in the coupled line and the spiral are in phase (current flow in same direction), hence maximum phase ripple occurs which corresponds to a single unique bit '1', similarly when the spirals are coupled below the line, the current distribution in the coupled line and the spiral are out of phase (current flow in opposite direction), hence minimum phase occurs which corresponds to a single unique bit '2', also absence of spiral produces no phase ripples which corresponds to bit '0' which are shown in below Figure .3, Figure .4, Figure .5 respectively.

Figure 3., Figure 4., Figure 5. shows the S_{12} magnitude and phase simulations for different specific codes of 1111, 2222 and 1102 for the corresponding layout designed in ADS. The magnitude simulation shows that the loss corresponding to the layout designed is optimum and corresponding phase simulation with the desired output of phase ripples which can be taken for encoding the base-3 code.

The steps involved in base-3 encoding is as follows,

- If minimum happened in S_{21} phase, this phase state is set to '2' in base-3 code
- If maximum happened in S_{21} phase, this phase state is set to '1' in base-3 code
- If no changes happened in S_{21} phase, this phase state is set to '0' in base-3 code

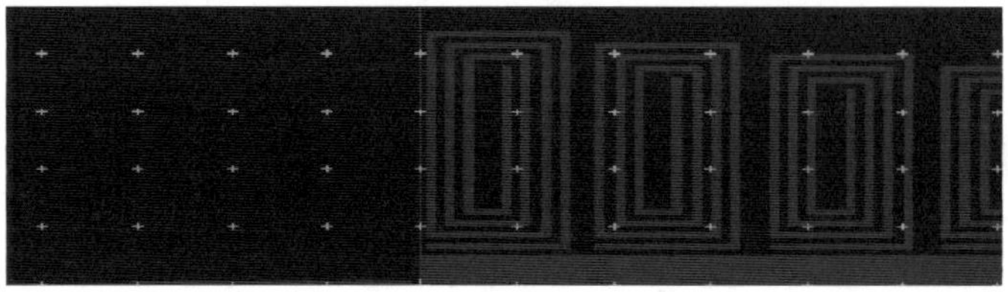

(a)

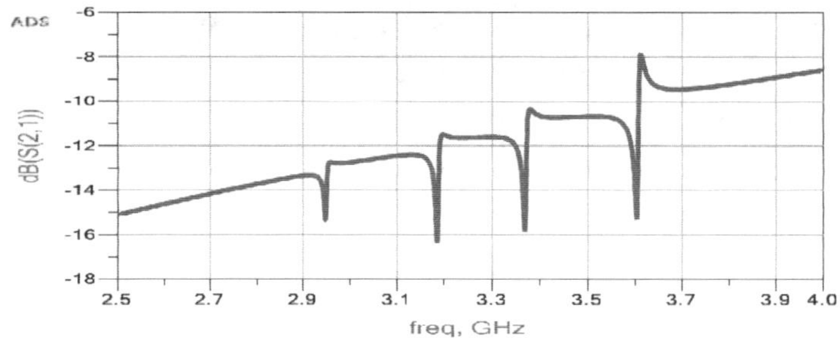

(b)

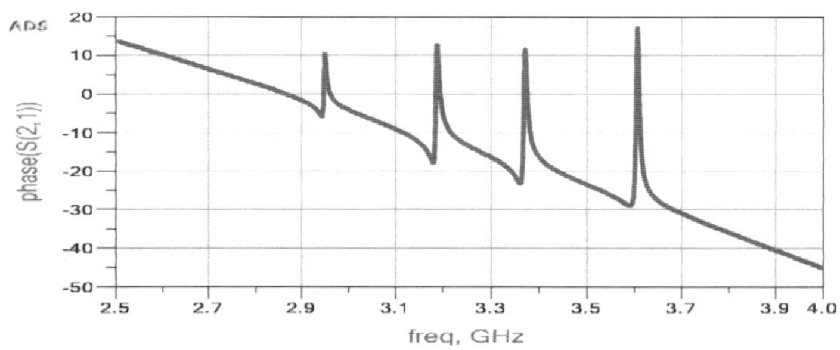

(c)

Figure 3. (a) Layout of the proposed tag with encoded bit (1111) (b) S_{21} Magnitude simulation for the corresponding layout (c) S_{21} phase simulation for the corresponding layout

The number of spiral resonators (M) according to the ID number (n) is calculated by the below equation 5,

$$M = log_3^n \quad (5)$$

In this design M=4, hence number of unique ID's generated are 81 Unique identification can be designed.

In the proposed design there is a reduction in the size of the tag, which is the major challenge in wireless identification. Also, there is increase in operating frequency of the spiral resonators and hence greater range of communication can be achieved. There is a efficient use of total frequency bandwidth.

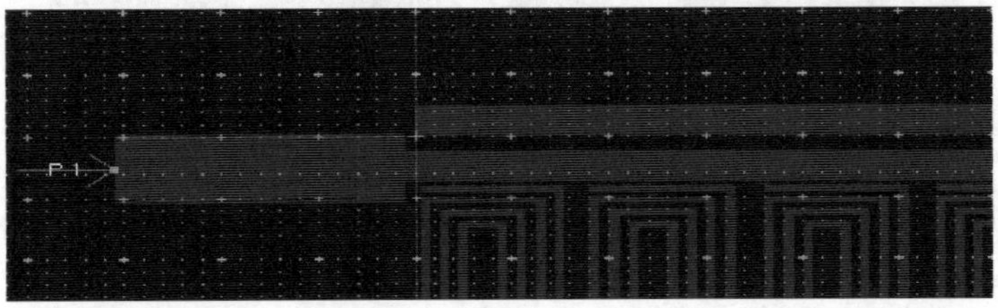

(a)

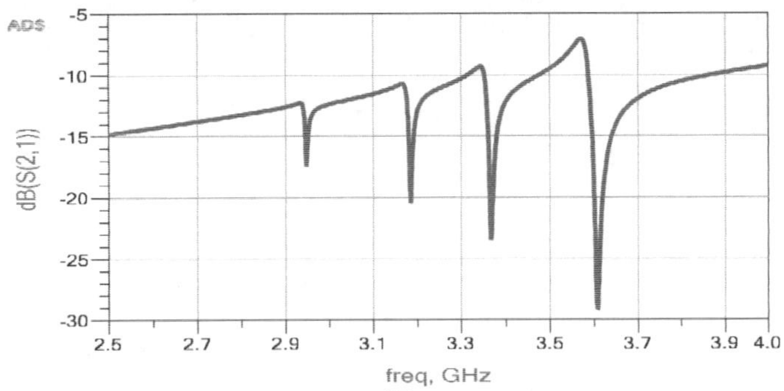

(b)

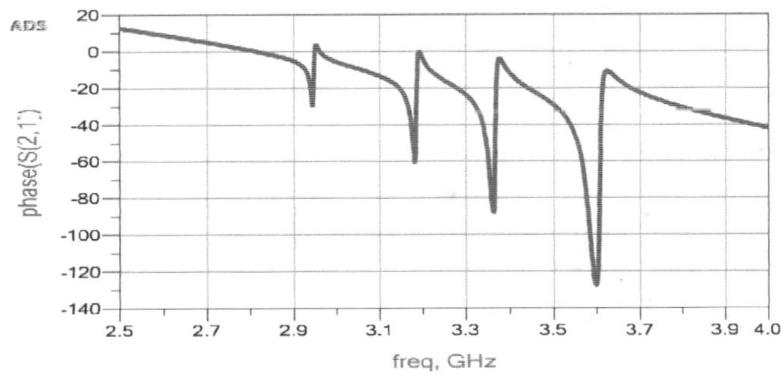

(c)

Figure 4. (a) Layout of the proposed tag with encoded bit (2222) (b) S_{21} Magnitude simulation for the corresponding layout (c) S_{21} phase simulation for the corresponding layout

The tag design also uses coupled transmission line which contributes to the expected current distribution which produces phase ripples required for encoding of the data in the chipless tag.

IOP Conf. Series: Materials Science and Engineering **590** (2019) 012066 doi:10.1088/1757-899X/590/1/012066

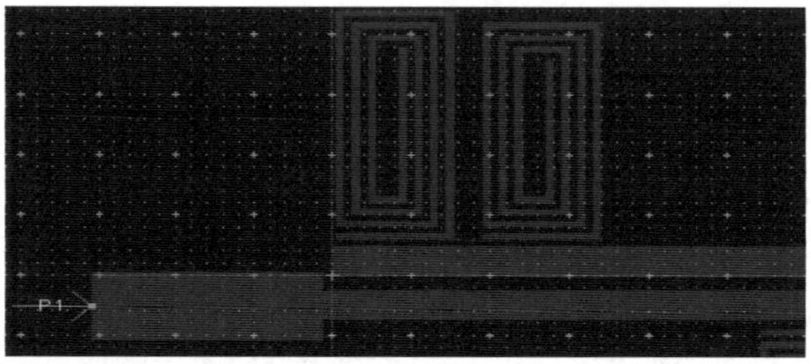

(a)

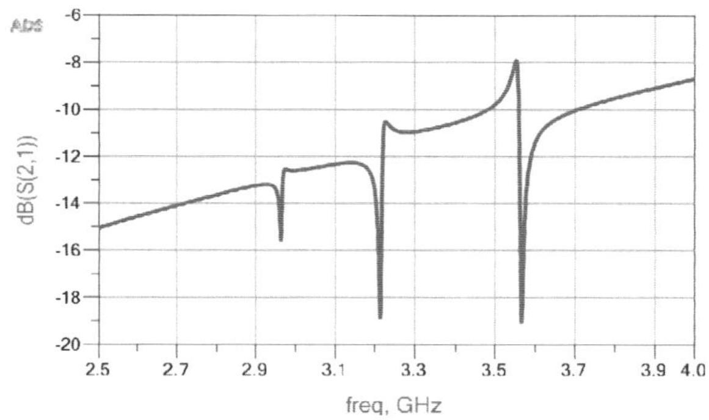

(b)

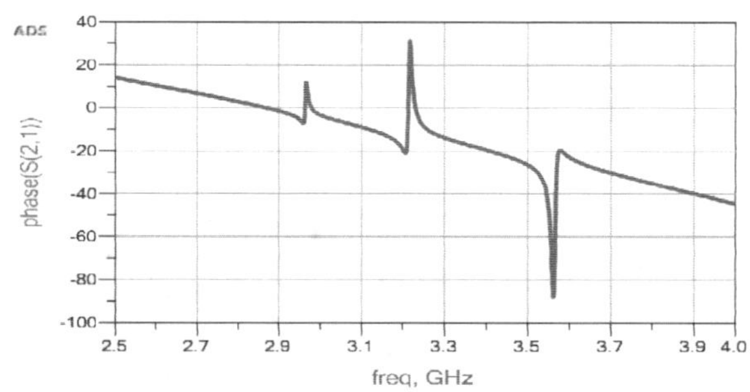

(c)

Figure 5. (a) Layout of the proposed tag with encoded bit (1102) (b) S_{21} Magnitude simulation for the corresponding layout (c) S_{21} phase simulation for the corresponding layout.

The layout of the designed monopole antenna is shown in the Figure 6. and the return loss (S11) is shown in the Figure 7., which covers all the operating frequency of the spiral resonators from 2.4 GHz to 3.8 GHz. The proposed tag can be used for variety of applications such as tracking applications like cheap and mass item tracking, Electronic surveillance to keep track of people, Toll bill collection and other wide variety of RFID applications.

IOP Conf. Series: Materials Science and Engineering **590** (2019) 012066 doi:10.1088/1757-899X/590/1/012066

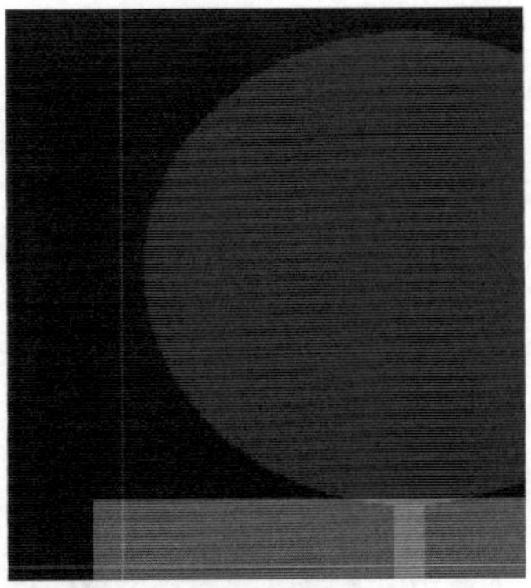

Figure 6. Layout of the designed UWB antenna.

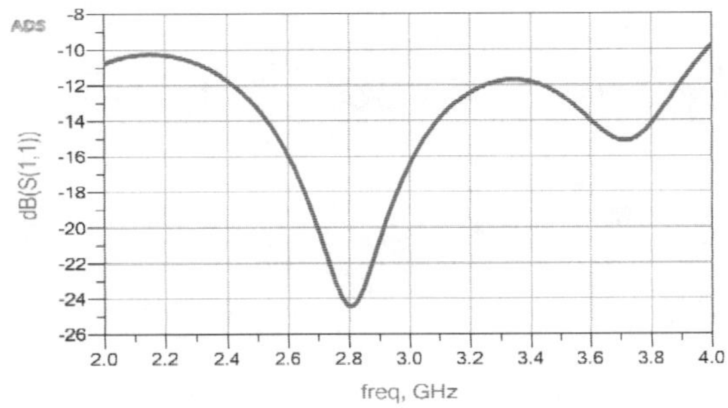

Figure 7. Simulated Return loss for the designed UWB antenna.

4. Conclusion

In the proposed work reduction of tag size has been successfully implemented with same encoding methods and substrate elements [1]. This method of chipless RFID tag design has many advantages in terms of printability/fabrication, wide range of applications, computation and cost compared to that of the chipped tag design. In future, the number of bits encoded can be increased by increasing the number of spiral resonators. The operating frequency of the spiral resonators can be made higher, which increases the range of communication. The resonating circuit part and the antenna part can be designed as a single prototype and wireless measurements can be made using signal analyzers. The alternative shape of the spiral which has the better performance can also be employed instead of the conventional rectangular spiral resonator design.

References

[1] Sohrab Majidifar , Arash Ahmadi , Omid Sadeghi-Fathabadi , Majid Ahmadi “A novel phase coding method in chipless RFID systems” , International Journal of Electronics and Communications, February 2015.

[2] Hossain A. K. M. Z, Motakabber S. M. A, and Ibrahimy M. I "Microstrip Spiral Resonator for the UWB Chipless RFID Tag", Progress in Systems Engineering, Springer International Publishing Switzerland 2015.

[3] O. Necibi,S.Beldiand ,A. Gharsallah “Design of a Chipless RFID TAG Using Cascaded and Parallel Spiral Resonators at 30 GHz”, IEEE journal, September 2014.

[4] Karimi G, Majidifar S ”A novel chipless RFID tag using spiral resonator to achieve the pentamerous data encoding form”. Journal of Electromagnetic Wave, 2014;28:13–27.

[5] Preradovic.S and Karmakar.N.C.,2012, "Multiresonator-Based Chipless RFID: Barcode of the Future", Springer Science & Business Media,pp.25-51.

[6] Preradovic S, Bablin I, Karmakar NC, Swiengers GF “Multiresonator based chip-less RFID system for low-cost item tracking”, IEEE Transactions on Microwave Theory Technology 2009;57:1411–9.

[7] Isaac Balbin, Student Member, IEEE, and Nemai Chandra Karmakar, Senior Member, IEEE "Phase-encoded chipless RFID transponder for large-scale low-cost applications", IEEE Microwave and Wireless Components Letters, Vol 19 Aug/2009.

Institute of Physics
Dirac House, Temple Back
Bristol BS1 6BE UK

ISSN: 1757-8981
ISBN 978-1-5108-9814-1

CURRAN ASSOCIATES INC.
proceedings.com